SOFT AND FRAGILE MATTER

NONEQUILIBRIUM DYNAMICS, METASTABILITY AND FLOW

SOFT AND FRAGILE MATTER

NONEQUILIBRIUM DYNAMICS, METASTABILITY AND FLOW

Proceedings of the Fifty Third Scottish Universities Summer School in Physics, St. Andrews, July 1999.

A NATO Advanced Study Institute

Edited by

M E Cates — University of Edinburgh
M R Evans — University of Edinburgh

Series Editor

P Osborne — University of Edinburgh

Copublished by
Scottish Universities Summer School in Physics &
Institute of Physics Publishing, Bristol and Philadelphia

British Library cataloguing-in-Publication Data:

*A catalogue record for this book is available
from the British Library*

ISBN 0-7503-0724-2

Library of Congress Cataloging-in-Publication Data are available.

Copublished by

SUSSP Publications
The Department of Physics, Edinburgh University,
The King's Buildings, Mayfield Road, Edinburgh EH9 3JZ, Scotland.

and

Institute of Physics Publishing, wholly owned by
The Institute of Physics, London.
Institute of Physics Publishing, Dirac House, Temple Back, Bristol BS1 6BE, UK.
US Editorial Office: Institute of Physics Publishing, The Public Ledger Building,
Suite 1035, 150 Independence Mall West, Philadelphia, PA 19106, USA.

Printed in Great Britain by J W Arrowsmith Ltd, Bristol.

SUSSP Proceedings

1	1960	Dispersion Relations
2	1961	Fluctuation, Relaxation and Resonance in Magnetic Systems
3	1962	Polarons and Excitons
4	1963	Strong Interactions and High Energy Physics
5	1964	Nuclear Structure and Electromagnetic Interactions
6	1965	Phonons in Perfect and Imperfect Lattices
7	1966	Particle Interactions at High Energy
8	1967	Methods in Solid State and Superfluid Theory
9	1968	Physics of Hot Plasmas
10	1969	Quantum Optics
11	1970	Hadronic Interactions of Photons and Electrons
12	1971	Atoms and Molecules in Astrophysics
13	1972	Properties of Amorphous Semiconductors
14	1973	Phenomenology of Particles at High Energy
15	1974	The Helium Liquids
16	1975	Non-linear Optics
17	1976	Fundamentals of Quark Models
18	1977	Nuclear Structure Physics
19	1978	Metal Non-metal Transitions in Disordered Solids
20	1979	Laser-Plasma Interactions: 1
21	1980	Gauge Theories and Experiments at High Energy
22	1981	Magnetism in Solids
23	1982	Lasers: Physics, Systems and Techniques
24	1982	Laser-Plasma Interactions: 2
25	1983	Quantitative Electron Microscopy
26	1983	Statistical and Particle Physics
27	1984	Fundamental Forces
28	1985	Superstrings and Supergravity
29	1985	Laser-Plasma Interactions: 3
30	1985	Synchrotron Radiation Sources and their Applications
31	1986	Localisation and Interaction
32	1987	Computational Physics
33	1987	Astrophysical and Laboratory Spectroscopy
34	1988	Optical Computing
35	1988	Laser-Plasma Interactions: 4

/continued

SUSSP Proceedings (continued)

36	1989	Physics of the Early Universe
37	1990	Pattern Recognition and Image Processing in Physics
38	1991	Physics of Nanostructures
39	1991	High Temperature Superconductivity
40	1992	Quantitative Microbeam Analysis
41	1992	Nonlinear Dynamics and Spatial Complexity in Optical Systems
42	1993	High Energy Phenomenology
43	1994	Determination of Geophysical Parameters from Space
44	1994	Quantum Dynamics of Simple Systems
45	1994	Laser-Plasma Interactions 5: Inertial Confinement Fusion
46	1995	General Relativity
47	1995	Laser Sources and Applications
48	1996	Generation and Application of High Power Microwaves
49	1997	Physical Processes in the Coastal Zone
50	1998	Semiconductor Quantum Optoelectronics
51	1998	Muon Science
52	1998	Advances in Lasers and Applications
53	1999	Soft and Fragile Matter
54	2000	Dynamics of the N-Body Gravitational problem

Lecturers

Jean-Philippe Bouchaud	CEA Saclay
Alan J Bray	University of Manchester
Michael E Cates	University of Edinburgh
Paul M Chaikin	Princeton University
Daan Frenkel	FOM Institute, Amsterdam
Alexei R Khokhlov	Moscow State University
Walter Kob	Johannes Gutenberg University, Mainz
Kurt Kremer	Max Planck Institute for Polymer Science, Mainz
Henk N W Lekkerkerker	University of Utrecht
Thomas C B McLeish	University of Leeds
David Mukamel	Weizmann Institute, Rehovot
Sidney R Nagel	University of Chicago
David J Pine	University of California, Santa Barbara
Wilson C K Poon	University of Edinburgh
Didier Roux	Centre de Recherche Paul Pascal, Pessac

Executive Committee

Prof Michael Cates	University of Edinburgh	*Director and Co-Editor*
Dr Stefan Egelhaaf	University of Edinburgh	*Treasurer*
Dr Martin Evans	University of Edinburgh	*Co-Editor*
Dr Wilson Poon	University of Edinburgh	*Co-Director*
Dr Nigel Wilding	University of Edinburgh	*Secretary*
Dr Peter Bladon	University of Edinburgh	

International Advisory Committee

Dr Kurt Kremer	MPI for Polymer Science, Mainz, Germany
Dr David Pine	University of California, Santa Barbara, USA
Dr Didier Roux	Centre de Recherche Paul Pascal, Pessac, France

Directors' Preface

'Soft and Fragile Matter' covers colloids, polymers, surfactant phases, emulsions, and granular media. Recent advances in all these areas have stemmed from enhanced experimental and simulation capabilities, and fundamental theoretical work on nonequilibrium systems. The aim of the 53rd Scottish Universities Summer School in Physics was to address experimental, simulation and theoretical studies of soft and fragile matter, focussing on unifying conceptual principles rather than specific materials or applications. In fact, several of these unifying principles are only just being recognised as such within the soft matter community. For example, 'jamming' in colloids under flow (and perhaps under gravity) is related to fundamental work on driven diffusive systems. Likewise 'aging', found experimentally in soft gels, dense emulsions *etc.*, relates to general concepts of glassy dynamics. Since these links are not yet fully worked out, several of the articles in this volume address relevant conceptual principles from a more general perspective.

The diversity of soft materials listed above was matched by that of participants at the School itself. Lecturers had been chosen, from among the leading international scientists in the field, with specific regard to their pedagogical skills. A careful attempt was then made to coordinate the content among the various courses. Most lecturers were asked to spend at least the first of their three lectures covering some particular area of the subject at an introductory level. The assignments were as follows: Pine, experimental methods; Khokhlov, polymers; McLeish, rheology; Frenkel, colloids; Kremer, simulation; Roux, surfactants; Bray, phase kinetics; Mukamel, driven systems; Kob, structural glasses; Bouchaud, slow dynamics; Nagel, granular matter. Collectively, the lecturers managed to carry their audience from the basic foundations of the subject to a representative sample of topics at the forefront of current research. Most participants felt that they had learned a great deal from the School.

SUSSP53 was held in the School of Physics and Astronomy and John Burnet Hall at the University of St Andrews, close to the ancient town's pubs, shops, beaches, historical monuments and golf courses. A busy social programme kept everybody occupied outside of the formal sessions, and featured a memorable ceilidh as well as a whisky-tasting evening. We are grateful for the help of many individuals (particularly Nigel Wilding and Stefan Egelhaaf) and organisations (particularly NATO, the EC, EPSRC and NSF) in contributing so much to the success of the School. The staff of John Burnet Hall provided a high quality and very friendly service. Secretarial assistance was ably provided by Leanne O'Donnell.

Michael Cates and Wilson Poon

Edinburgh, February 2000

Editors' Note

To achieve the same ease of communication in this Proceedings volume as occurred at the School itself is a challenging, if not impossible, goal! Nonetheless, the articles have been carefully edited with that aim in mind—for example by adding cross-references in many places where relevant introductory material is to be found in a different article. The sequence of the articles follows roughly that of the School's lectures, though of course the latter were interleaved in a way the articles cannot be. Thus, following an introductory survey (Poon) the volume can be informally subdivided into three sections: methodologies and phenomena of soft condensed matter (six chapters, Pine to Roux inclusive); modern concepts of nonequilibrium statistical physics (four chapters, Bray to Bouchaud); dynamics and metastability in colloidal and granular systems (four chapters, Lekkerkerker to Cates). The aim of this volume, like that of the School, is to lead the reader from basic principles to a selection of the most recent developments in this diverse and fascinating field. We hope this has been achieved, in many cases within the course of a single chapter.

Michael Cates and Martin Evans
Edinburgh, February 2000

Contents

A day in the life of a hard-sphere suspension

Wilson C K Poon

University of Edinburgh, UK

1 Introduction

This summer school has a very fashionable title: the terms 'soft matter' and 'fragile matter' are very recent additions to the physics vocabulary. P-G de Gennes was one of the first to use the term 'soft matter' to refer to the study of colloids, polymers and surfactants in his Nobel lecture [1] in 1991. 'Fragile matter', in at least one of the senses used in this School, is even more recent [2] (see Cates, this volume). The adjectives 'soft' and 'fragile' used to describe matter share another characteristic: they both refer to how materials respond to *mechanical* disturbances.

The school is devoted to understanding three particular aspects of soft and fragile matter, as detailed in the subtitle: *nonequilibrium dynamics, metastability* and *flow*. What I want to do in this introductory lecture is first to give some reasons why one might expect systems described as soft or fragile matter to be suitable for the investigation of these particular aspects of nonequilibrium physics. Then, in the main part of the lecture, I will illustrate all of these aspects with what is possibly the simplest model system: a colloidal suspension of hard spheres.

1.1 Nonequilibrium physics and soft matter

The understanding of systems in thermal equilibrium is one of the major achievements of twentieth-century physics. We have a recipe to do this starting from a knowledge of the microscopic pair interaction, $U(r)$. First calculate the partition function, given by $Z = \int dr_1 \ldots dr_N \exp[-U(r)/k_BT]$. Then take its logarithm to give the (Helmholtz) free energy: $F = -k_BT \ln Z$. The equilibrium behaviour of the system is obtained by minimising F. The calculation of Z is, of course, a hard mathematical problem, and a large amount of physical insight (and computer time!) is needed to make progress. Nevertheless, the general recipe is available.

The situation is very different when we come to the behaviour of systems away from equilibrium. The general question can be stated as follows. What happens when we apply a perturbation, transient or continuous, (change its temperature, shake it, *etc.*) to a system in thermal equilibrium? Here, relative to where we have got to with equilibrium statistical mechanics, we are still fumbling in the dark. Kubo's judgement [3] in this regard is an understatement: "The foundation of nonequilibrium statistical mechanics is perhaps far more difficult to establish than that of equilibrium statistical mechanics."

Apart from the lack of suitable theoretical foundation and mathematical tools, another reason for the slow progress in understanding nonequilibrium systems is the lack of experimental models. This is where 'soft matter' comes in. Colloids, polymers and surfactants, sometimes also known as 'complex fluids', have one characteristic in common: they involve a *mesoscopic* length scale between the atomic ($\sim$ 1nm) and the bulk ($\sim$ 1mm). On this intermediate length scale one finds structures such as suspended particles/droplets, macromolecular coils, and self-assembled structures such as micelles and bilayers. The presence of this intermediate length scale in complex fluids gives rise to three reasons why they make ideal candidates for the investigation of nonequilibrium physics.

Firstly, the upper end of the mesoscopic length scale, $R \sim 1\mu$m, is comparable to the wavelength of visible light, so that direct imaging using optical microscopy is fast becoming a standard tool in complex fluid investigations. Secondly, the relaxation time of complex fluids, τ_R, the time taken for an entity (*e.g.* a colloidal particle or a polymer coil) to diffuse over a length scale comparable to its size, scales according to $R^2 \sim D\tau$. We can estimate the diffusion coefficient D by using the Stokes–Einstein relation for a sphere of radius R suspended in a solvent of viscosity η: $D = k_BT/6\pi\eta R$. This gives

$$\tau_R \sim \frac{6\pi\eta R^3}{k_BT}\,, \tag{1}$$

with values in the region of 1ms to 1s. Corresponding relaxation times in atomic materials are in the picosecond range. These relaxation times are, of course, modified by interactions. Typical attractions in complex fluids are of the order 1 to $20k_BT$, so that the Boltzmann factor amplifying the elementary relaxation time τ is never much bigger than 10^8. The upshot is that the characteristic times over which nonequilibrium complex fluids evolve are likely to be in the range of 1ms to 1 year.

Thirdly, complex fluids are 'soft', an adjective to be discussed by McLeish, this volume. Here I note that their mechanical response is mainly governed by entropy, so that a typical modulus (of a colloid for example) is given by

$$G \sim k_BT/R^3\,, \tag{2}$$

which is of the order of 10^{-3}–1 Pa. We can also estimate the effect of a shear rate of $\dot{\gamma}$ by appealing to a dimensionless group known in the colloid community as the Peclet number: $\mathrm{Pe} \propto \tau_R\dot{\gamma}$. If $\mathrm{Pe} < 1$, Brownian relaxation dominates; if $\mathrm{Pe} > 1$, shear dominates. Using Equation 1, we get

$$\mathrm{Pe} \sim 6\pi\eta R^3\dot{\gamma}/k_BT\,. \tag{3}$$

In this expression, we recognise $\eta\dot{\gamma}$ to be a stress. For a stress equal to the mechanical modulus we have just estimated from Equation 2, we get $\mathrm{Pe} \sim 10$. Therefore relatively high shear rates are easily achieved, so that we can study highly nonequilibrium flow behaviour in complex fluids. In particular, we will later see that it is in thinking about the high-shear response of colloids that one definition of the concept of 'fragile matter' was first given.

For these reasons, as well as the stunning ingenuity of synthetic chemists in preparing well-characterised systems 'to order', there is now a growing realisation that complex fluids are ideal laboratories for nonequilibrium physics. In what follows, I want to introduce the themes of the Summer School by describing nonequilibrium dynamics, metastability and flow in perhaps the simplest possible complex fluid—a suspension of hard spheres.

Since my aim is simply to give examples of a range of phenomena, I will not attempt to provide a complete set of references; in each example, I will give representative references with a bias towards papers with extensive bibliographies for further reading.

2 Hard-sphere colloids

All of the experiments which I will describe are performed using a model system developed originally by ICI for paints, and first used for academic research by Ron Ottewill and his group in Bristol [4]. They are suspensions of polymethylmethacrylate (PMMA) spheres ($R \lesssim 1\mu$m) with chemically-grafted coatings of poly-12-hydroxystearic acid (PHSA) of thickness ~10nm. A large body of research over the last two decades has shown that the interaction between two such particles is almost perfectly hard-sphere like [5]: there is no interaction until the coated particle surfaces touch, whereupon over a very short spatial range (~ 10nm) a strong entropic repulsion develops. In particular, they show the equilibrium phase behaviour expected of hard spheres (to be reviewed in more detail by Frenkel, this volume). At low volume fractions (the fraction of the total volume V occupied N spheres of radius R is $\phi = 4\pi R^3 N/3V$) the equilibrium state is a colloidal fluid—particles adopt an amorphous arrangement and can (given time) diffuse throughout the sample volume. At high volume fractions, the equilibrium state is a colloidal crystal; this is easily detected because colloidal crystallites appear iridescent in white light due to the Bragg reflections from crystal planes. Within the interval $\phi_F = 0.494 < \phi < \phi_M = 0.545$, the fluid at ϕ_F and crystal at ϕ_M coexist.

Before moving on to describe nonequilibrium dynamics, metastability and flow in this model system, I just want to mention briefly two examples of on-going work on the equilibrium properties of hard spheres, if only to show that despite having a firm theoretical foundation, equilibrium statistical mechanics is far from a closed subject. First comes the structure of hard-sphere crystals. These are made of hexagonally-packed layers stacking on top of each other. Given the short-range nature of the interparticle potential, we expect very small free-energy differences between the infinitely many possible stacking sequences (the two most well-known ones being face-centred cubic $ABCABC$ and hexagonal close packed $ABAB$; random hexagonal stacking corresponds to a random sequence of A,B,C). Calculating these free energy differences is a big challenge for equilibrium simulations, which is only recently beginning to be met [6]. Secondly, real suspensions never have particles of uniform size, in which case they would be monodisperse. The effect of having a distribution of particle sizes, known as polydispersity, is to render the system an infinite-component one, giving rise to formidable challenges in attempting a theoretical description that are, again, only recently being attended to [7].

2.1 Metastability

Given the small mechanical moduli of colloidal crystals ($\sim k_BT/R^3$), they can easily be shear-melted to a metastable fluid state: the stresses involved in shaking a bottle of colloidal crystals are equivalent to putting a few hundred Mount Everests on top of a block of copper! Out of this metastable colloidal fluid are nucleated ordered domains of colloidal crystallites. Visually, in a test tube that has been shaken, one sees iridescent crystallites appearing throughout the bulk over times of minutes to hours. The emergence of crystalline order in a hard-sphere system may be the simplest symmetry-breaking transition open to study in the laboratory. Until recently, the decay of the metastable fluid towards equilibrium crystals has been studied exclusively by diffraction. This is a matter of necessity in atomic materials, and a matter of tradition in colloids [8]. A particular drawback of diffraction methods is that by the time Bragg peaks are visible, the initial symmetry-breaking nucleation step is already long over. What is observed is growth averaged over many crystal nuclei, with information on nucleation only available by more or less indirect inference and extrapolation. Recently direct microscopic observation has been used to study the nucleation of crystallites from metastable colloidal fluids: see Figure 1. For example, in the group in Edinburgh, Mark Elliot [9] has captured the genesis and evolution of an almost-critical nucleus in a PMMA colloid in real-time and with single-particle resolution.

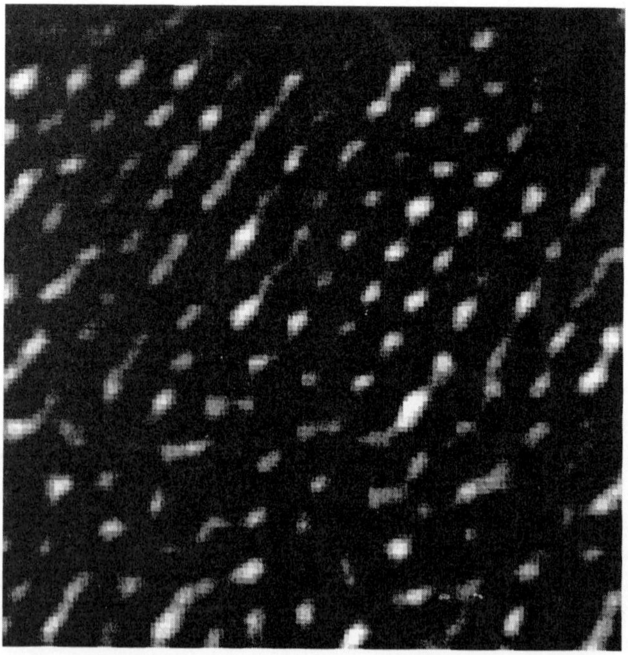

Figure 1. *Optical micrograph of a colloidal crystallite nucleating out of a surrounding disordered, metastable colloidal fluid. The particle diameter is $\sim 1\mu m$. This image was taken $48\mu m$ from the bottom of a suspension confined to a $100\mu m$-thick capillary. Taken from [9].*

Such microscopic observations have the potential of testing a number of intriguing theoretical predictions. For example, it has often been suggested that the structure of the initial crystal nucleus may well be different from that of the final bulk crystal [10]. Simulations have also cast doubt on the single-particle picture of nucleation implicit behind classical nucleation theory [11]. Neither of these results are particularly amenable to testing by diffraction experiments: direct observation of individual early-stage nuclei are necessary.

As formed, hard-sphere colloidal crystals are made of randomly-stacked hexagonal layers. In some experiments (see *e.g.* the preliminary report in [5]), these random-stacked crystallites were observed to 'ripen' towards a face-centred cubic structure over days and months. This is consistent with recent simulations reporting *fcc* stacking to be that with the lowest free energy (by $\sim 10^{-3}k_BT$ per particle or thereabouts) [6]. If this is so, then randomly-stacked colloidal crystals are long-lived metastable structures. The kinetics and mechanism of such 'ripening' is not well understood.

2.2 Nonequilibrium dynamics

The phase diagram of hard spheres has already been reviewed: fluid for $\phi < 0.494$, crystal for $\phi > 0.545$, and fluid-crystal coexistence between those two volume fractions. Thus, for all volume fractions above 0.545 all the way to the closest possible packing density ($\phi_{\text{max}} = \pi/3\sqrt{2} \approx 0.74$) the equilibrium thermodynamic state is crystalline. Experimentally, however, homogeneous nucleation of colloidal crystallites is *not* observed above $\phi \sim 0.58$ [12]. This has been interpreted as a glass transition. (Note that even above $\phi = 0.58$, *heterogeneous* crystallisation, *e.g.* at sample tube walls, is still observed.)

This 'glass transition' appears to be associated with a seizing up of dynamics at all but the shortest length scales, as revealed by dynamic light scattering (DLS). DLS measures the normalised intermediate scattering function

$$f(\mathbf{q},\tau) = \frac{F(\mathbf{q},\tau)}{F(\mathbf{q},0)}\,, \tag{4}$$

where

$$F(\mathbf{q},\tau) = \frac{1}{N}\sum_{j=1}^{N}\sum_{k=1}^{N}\langle \exp i\mathbf{q}\cdot[\mathbf{r}_j(t) - \mathbf{r}_k(t+\tau)]\rangle\,. \tag{5}$$

(This quantity is discussed in Section 2.2 of the article by Pine, this volume, where the notation $g_E(\tau)$ is used.) The static structure factor is $S(\mathbf{q}) \equiv F(\mathbf{q},0)$. N is the number of particles in the scattering volume, assumed to be large, and $\mathbf{r}_j(t)$ is the position of particle j at time t. The normalisation ensures that $f(\mathbf{q},\tau) = 1$ at short times. In a system that is ergodic over the experimental time window, *i.e.* one that explores all configurations many times over, $f(\mathbf{q},\tau) \to 0$ as $\tau \to \infty$. Essentially, the rate of this decay to zero gives information about the diffusive dynamics of density fluctuations at length scale $2\pi/q$. The DLS signature of the glass transition in a hard sphere suspension is that $f(\mathbf{q},\tau)$ fails to decay to zero at all scattering vectors. The most careful DLS studies of the hard-sphere glass transition to date have been performed by van Megen and co-workers using the PMMA system [13]. One interesting conclusion to emerge from these careful measurements is that many of the predictions of mode-coupling theory (MCT), a highly mathematical construction involving the structure factor $S(\mathbf{q})$ that implies a dynamical transition at high densities (see Kob, this volume), are substantially correct for this system. Of even more interest for this School (see Bouchaud, this volume) van Megen *et al.* [13] detected 'aging'—slower dynamics were observed at a longer 'waiting time', the time elapsed since the system was prepared before the commencement of the experimental (here DLS) measurements.

2.3 Flow and fragility

Concentrated suspensions, like other complex fluids, are non-Newtonian. In particular, the suspension viscosity is a function of shear rate. A convenient dimensionless shear rate, the Peclet number, has been introduced in Equation 3. Consider a hard-sphere suspension at $\phi \sim 0.5$. At Pe $\to 0$, the exists a well-defined low-shear limit viscosity. At Pe ~ 0.1, shear thinning starts to occur—the viscosity decreases rapidly with shear rate until Pe ~ 1, whereupon it remains more or less constant for many decades of Pe. At very high shear rate, a sudden and dramatic increase in viscosity (factor of 10 or more) is often observed [14]. This phenomenon is known as shear thickening.

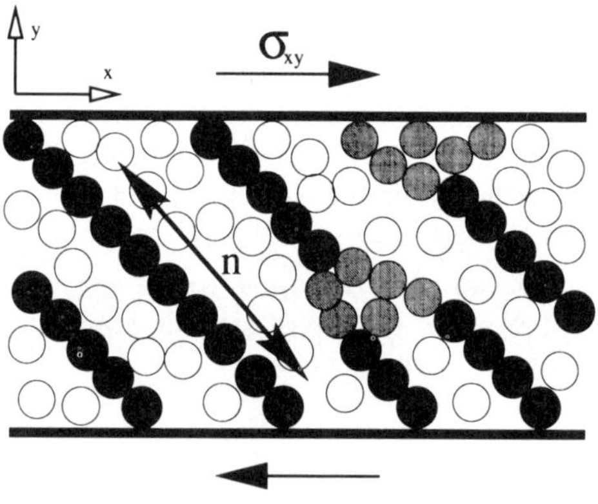

Figure 2. *A schematic representation of a suspension subject to shear σ_{xy}. Under strong shear, stress-bearing 'force chains' of particles form (dark and shaded circles), leading to 'jamming' of the suspension. Taken from [2].*

In a recent publication [2], Cates and co-workers interpreted shear thickening as due to the formation of 'force chains' in the system, leading to 'jamming'—see Figure 2. These stress-bearing force chains render the suspension solid-like with respect to the particular imposed shear stress, but *not* with respect to any other stress pattern; if the stress pattern is changed, the system will immediately flow and jam again, a characteristic which Cates *et al.* proposed to call *fragile*.

2.4 An example—DWS echo study of hard-sphere glasses

Recently, a UK-French team have carried out an experiment on hard-sphere suspensions that involves all three aspects of metastability, nonequilibrium dynamics and flow, as reviewed individually in the last three sections. Haw *et al.* [15] used the new technique of diffusing wave spectroscopy (DWS echo) to study the yielding and flow of a hard-sphere colloidal glass under oscillatory shear. DWS echo will be discussed in much more detail by its inventor, David Pine, in his lectures on experimental techniques (see Pine, section 2.2.3, this volume). Briefly, the technique studies the statistics of the speckle pattern formed when laser radiation (wavelength λ) from different multiple-scattering paths through a turbid medium (here a dense PMMA suspension at $\phi \sim$0.57–0.6) interfere with each other. If each of the N scatterers in the medium moves by a distance $\sim \lambda/N$, the new speckle pattern will be completely decorrelated with the original pattern—on average, we expect a bright speckle in the original pattern to become dark (and vice versa). If N is large, DWS echo then provides a means of detecting very small movements.

Now consider a scattering medium under oscillatory shear. The correlation function of the speckle pattern will decay from unity to zero once the shear has moved scatterers by a distance $\sim \lambda/N$, and remain at zero throughout the shear cycle. If the medium behaves elastically, so that at the beginning of the next cycle all scatterers are back at their positions at beginning of the previous cycle, then the correlation function will recover fully the value unity: the speckle pattern at $t = t_0 + T$ is exactly the same as that at $t = t_0$ (where T is the shear period). A plot of the correlation function against time will therefore give a series of peaks of unit height (echoes) with time period T. If, however, portions of the scattering medium deform plastically, so that scattering centres do not recover their positions after a shear cycle, the echo-peaks will have less than unit height.

Using this method, Haw *et al.* found that there is essentially no decrease in the height of the echoes in a hard-sphere colloidal glass until a volume-fraction-dependent critical shear amplitude is reached. At $\phi = 0.585$, for example, there is little decrease in the echoes until the peak-to-peak shear amplitude is ~ 0.3. Simultaneous static scattering experiments showed that at and above this amplitude, rapid crystallisation of the colloidal glass occurred. This behaviour is quite different from that of dense emulsions (see Pine, this volume).

3 Conclusion

The purpose of this brief survey of 'a day in the life of a hard-sphere suspension' is to show that even the simplest conceivable complex fluid shows fascinating physics in the areas of metastability, nonequilibrium dynamics and flow. Many of the ideas encountered in this survey will recur throughout the School.

Acknowledgements

Most of the work described in this overview has been done in the Soft Condensed Matter Group in Edinburgh. It is a pleasure to thank the past and present members of the group (too numerous to name individually) for their tireless effort. Valuable discussion partners beyond Edinburgh have included Paul Bartlett, Wilem Kegel, Henk Lekkerkerker and Patrick Warren.

References

[1] de Gennes P-G, *Rev. Mod. Phys.* **64**, 645 (1992)

[2] Cates M E, Wittmer J P, Bouchaud J-P and Claudin P, *Phys. Rev. Lett.* **81**, 1841 (1998)

[3] Kubo R, Toda M and Hashitsume N, *Statistical Physics* Volume 2, 2nd edition, Springer (1995)

[4] Antl L, Goodwin J W, Hill R D, Ottewill R H, Owen S M, Papworth S and Waters J A, *Colloids and Surfaces* **17**, 67 (1986)

[5] Poon W C K and Pusey P N, in *Observation, prediction and simulation of phase transitions in complex fluids*, ed. Baus M et al., Kleuwer (1995), p. 3

[6] Mau S C and Huse D A *Phys. Rev. E* **59**, 4396 (1999);
Pronk S and Frenkel D *J. Chem. Phys.* **110**, 4589 (1999);
Bruce A D, Jackson A N, Ackland G J and Wilding N B, *Phys. Rev. E* **61**, 906 (2000)

[7] Sollich P and Cates M E, *Phys. Rev. Lett.* **80**, 1365 (1998);
Warren P B, Evans R M L, Fairhurst D J and Poon W C K, *Phys. Rev. Lett.* **81**, 1326 (1998);
Bartlett P, *Mol. Phys.* **97**, 685 (1999);
Kofke D A and Bolhuis P G, *Phys. Rev. E* **59**, 618 (1999)

[8] Palberg T, J. Phys. Condens. Matter **11**, 323 (1999).

[9] Elliot M S, Ph.D. thesis, The University of Edinburgh (1999); Elliot M S and Poon W C K, (2000) *Adv. Colloid Interface Sci.* in the press.

[10] van Duijneveldt J S and Frenkel D, *J. Chem. Phys.* **96**, 4655 (1992);
Shen Y C and Oxtoby D W, *Phys. Rev. Lett.* **77**, 3583 (1996)

[11] Khanna R and Harrowell P *Phys. Rev. E* **56**, 1910 (1997)

[12] Pusey P N and van Megen W, *Nature* **320**, 340 (1986)

[13] van Megen W, Mortensen T C and Williams S R, *Phys. Rev. E* **58**, 6073 (1998)

[14] Frith W J, d'Haene P, Buscall R and Mewis J, *J. Rheology* **40**, 531 (1996)

[15] Haw M D, Poon W C K, Pusey P N, Hebraud P and Lequeux F, *Phys. Rev. E* **58**, 4673 (1998)

Light scattering and rheology of complex fluids driven far from equilibrium

David J Pine

University of California at Santa Barbara, USA

1 Introduction

In these lectures, we explore two examples of systems driven far from equilibrium by the application of shear. With these two examples, we investigate different experimental strategies which are designed to probe directly the connection between macroscopic non-linear rheology and the microscopic structure and dynamics of a broad range of soft materials. The study of these systems illustrates the importance of performing *simultaneous* measurements of the microscopic structure, flow, and rheological properties of soft materials when such systems are driven far from equilibrium by shear flows. There are several reasons for this. First, the flows are frequently inhomogeneous. Such inhomogeneities can arise from various mechanisms; the two most frequently observed and discussed are hydrodynamic instabilities and flow-induced phase transitions. Other difficulties can also arise for the case of virtually any flow that is not a pure shear flow. In pure extensional flows, for example, the nonlinear rheological properties of the fluid under study can modify the flow field in ways that are extremely difficult to predict. Thus, without a detailed knowledge of the flow field, it is virtually impossible to develop a meaningful theory. Second, systems do not always tend towards a steady state. Even when they do, the steady state is not necessarily characterised by any general principle of detailed balance to constrain the theory which one can construct. Furthermore, the structures that develop under shear often do not resemble the structures found in the same system in equilibrium. That is, the nonequilibrium structures frequently cannot be described as perturbations of the equilibrium structures. Therefore, as important as microscopic structural measurements are for understanding and developing theories for systems in equilibrium, they become even more important when systems are driven far from equilibrium.

The systems we study are solutions of worm-like micelles and oil-in-water emulsions. These two systems exhibit many of the generic properties that soft materials exhibit under shear flow including a shear-induced phase transition, inhomogeneous flows, plastic deformation, and yielding. We explore these phenomena in these two systems with a combination of optical and light scattering techniques, and with rheological measurements.

In Section 2, we review some important aspects of basic light scattering theory. We then discuss some general characteristics of light scattering when the system under study is subjected to a steady or oscillatory shear flow. We conclude our discussion of light scattering with an overview of diffusing-wave spectroscopy (DWS), that is, dynamic light scattering (DLS) in the multiple scattering limit.

Next (Section 3) we discuss the results of some recent experiments on shear thickening in dilute and semi-dilute solutions of wormlike micellar solutions. We also present a phenomenological theory for shear thickening in these systems which captures many of the salient features of our experiments. Our discussion of the experiments and theory is preceded by a brief overview of wormlike micellar solutions.

Finally, we present in Section 4 results from some recent experiments which examine microstructural changes in dense glassy emulsions when they are sheared beyond the limit of linear response.

2 Light and other scattering techniques

Scattering techniques are among the most powerful and widely used methods for probing the microscopic structure and dynamics of matter. In soft condensed matter, the most commonly used scattering techniques are X-ray, neutron, and light scattering. The choice of which scattering technique to use depends first and foremost on the length scale of the structures that one wishes to probe. The length scales directly probed by the various scattering techniques are set by the wavelength of the radiation. The smallest length scale that can be directly measured by scattering is $\lambda/2$ where λ is the wavelength. As discussed later, larger length scales are probed by varying the scattering angle. For X-ray and neutron scattering, where the wavelengths used are typically ~ 1Å, the upper limit is about 1000Å, which can be achieved by working at very small scattering angles. For light scattering, where the wavelength is $\sim 0.5\mu$m, the upper limit us usually several microns although length scales of up to $\sim 200\mu$m have been achieved recently.

Another consideration in choosing which scattering technique to use has to do with how the radiation interacts with matter. X-rays are scattered by fluctuations in the electron density. Therefore, substances containing heavier elements scatter more strongly than substances rich in the lighter elements. For example, substances containing a great deal of hydrogen and relatively low concentrations of heavier elements scatter X-rays weakly. By contrast, neutrons interact primarily through the nuclear interaction, which varies more or less randomly from one nuclear species to another. It so happens that neutrons are scattered by hydrogen much more strongly than are X-rays. In fact, hydrogen and its heavier isotope, deuterium, scatter in such a way that they partially cancel one another. Thus, by judiciously adjusting the concentration and location of hydrogen and deuterium within a molecule, one can adjust the overall scattering strength and even selectively scatter from hydrogen atoms at specific molecular locations. Such 'contrast matching' has proven to be a powerful tool for probing the structure of polymers and other complex fluids. Neutrons also have magnetic moments and therefore couple to the spin of nuclei.

Thus, systems with magnetic properties can be probed with neutrons. By contrast, light is scattered by fluctuations in the dielectric constant of a material. Light is an especially useful probe of soft materials because they frequently contain structures with length scales comparable to the wavelength of light. In fact, the 'softness' of many and perhaps even most soft materials is derived from the fact that they are made up of structures whose fundamental length scales are comparable to optical length scales.

One other characteristic of light scattering is that light is typically scattered much more strongly than are X-rays or neutrons. A simple quantitative measure of the scattering strengths of light and X-rays is the mean spatial fluctuation in the dielectric constant $\Delta\epsilon/\epsilon$. For a typical sample probed by light scattering $\Delta\epsilon/\epsilon \sim 0.1$; for X-ray scattering $\Delta\epsilon/\epsilon \sim 10^{-5}$. Thus, light is scattered much more strongly than X-rays. (Neutron scattering strengths are roughly comparable to X-rays.) One consequence of this is that multiple scattering is frequently an important consideration for light scattering experiments whereas it is almost never important for X-ray or neutron scattering. In fact, multiple scattering is such a common occurrence in light scattering experiments that techniques have been developed to cope with and in some cases even exploit multiple light scattering. Further on in these lectures (Section 2.2) we will discuss diffusing-wave spectroscopy (DWS), a technique which exploits multiple light scattering to probe very small particle movements. We now turn to a discussion of basic light scattering theory.

2.1 Static light scattering

The basic principles of light scattering can be understood by first considering scattering from two nearby particles as illustrated in Figure 1. Coherent light from a laser is incident

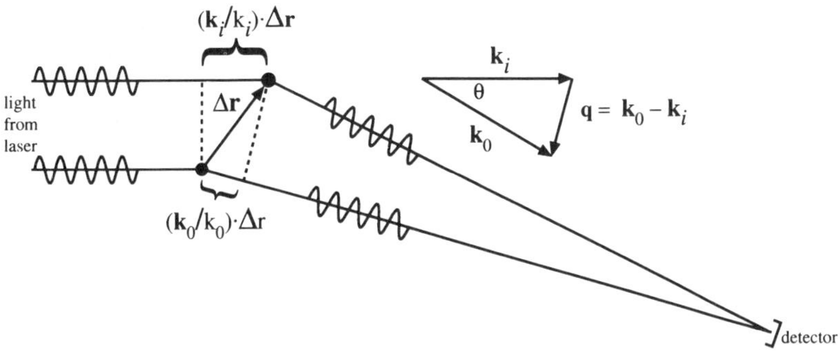

Figure 1. *Schematic for scattering of light from two particles.*

from the left onto the two particles. Light scattered through an arbitrary scattering angle θ is collected by a detector which is sensitive to the *intensity* of the light that falls on its surface. Light scattered from the top particle will in general have travelled a different distance from the laser to the detector than that from the bottom particle. Thus, the scattered electric fields from the two particles will not be in phase with each other. Since the wavelength of the scattered light is unchanged (*i.e.* the scattering is elastic), the difference in phase is given by the magnitude of the wavevector, $k \equiv 2\pi/\lambda$, times the difference in path lengths Δs, where λ is the wavelength of light in the sample. From

Figure 1, the difference in path lengths is $\Delta s = (\mathbf{k}_0/k_0) \cdot \Delta\mathbf{r} - (\mathbf{k}_i/k_i) \cdot \Delta\mathbf{r}$. In writing down this expression for Δs, we have made the approximation that the distance between particles is small compared to the distance between the particles and the detector. Thus, the paths from each of the two particles to the detector are essentially parallel. This is usually an excellent approximation. Noting that $k \equiv k_i = k_0$, the phase difference $\Delta\phi$ is

$$\Delta\phi = k\Delta s = (\mathbf{k}_0 - \mathbf{k}_i) \cdot \Delta\mathbf{r} = \mathbf{q} \cdot \Delta\mathbf{r}\,, \tag{1}$$

where the scattering vector is defined by $\mathbf{q} \equiv \mathbf{k}_0 - \mathbf{k}_i$. Clearly, if $\Delta\phi \sim \pi$, the light scattered from the different particles interferes destructively. If $\Delta\phi \sim 0$, the scattered light interferes constructively. Thus, the relative phase between the light scattered from different particles is sensitive to particle positions on the length scale of the wavelength of light. This is the essential physics which underlies the sensitivity of light scattering to the spatial structure of the scatterers. One additional note: as can be seen from the geometry of the scattering diagram in Figure 1, the magnitude of $\mathbf{q}$ is related to the scattering angle θ by

$$q = 2k \sin\frac{\theta}{2}\,. \tag{2}$$

To obtain a quantitative expression for the scattered intensity from N particles, we first add the contributions from all particles within the scattering volume to obtain the total electric field at the detector:

$$\mathbf{E}_d(\mathbf{q}) = \sum_{i=1}^{N} \mathbf{E}_i e^{i\mathbf{q}\cdot\mathbf{r}_i} = \mathbf{E}_s \sum_{i=1}^{N} e^{i\mathbf{q}\cdot\mathbf{r}_i}; \tag{3}$$

where the absolute phase for each path $\phi_i = \mathbf{q}\cdot\mathbf{r}_i$ is measured relative to an arbitrary fixed origin (as we will see below, the scattered intensity does not depend on the choice of the origin of the coordinate system). For simplicity, we have assumed that the amplitudes of the scattered fields $\mathbf{E}_i$ are all identical and equal to $\mathbf{E}_s$ as would be the case for identical spherical particles much smaller than the wavelength of light. The scattered *intensity* is proportional to the *square* modulus of the electric field:

$$I_d(\mathbf{q}) \propto |E_d|^2 = |E_s|^2 \sum_{i=1}^{N} e^{i\mathbf{q}\cdot\mathbf{r}_i} \sum_{j=1}^{N} e^{-i\mathbf{q}\cdot\mathbf{r}_j} = |E_s|^2 \sum_{i,j}^{N} e^{i\mathbf{q}\cdot(\mathbf{r}_i-\mathbf{r}_j)}\,. \tag{4}$$

Thus, it is apparent that the scattered intensity is dependent on the *relative* positions of the scatters and, as expected, is not sensitive to our choice of coordinate systems for calculating the phase of the scattered light. Static light scattering experiments measure the average of the scattered intensity. Therefore, it is useful to extract from the ensemble average of Equation 4 that part which contains the structural information in which we are interested. To this end, we define the *static structure factor*,

$$S(\mathbf{q}) \equiv \frac{1}{N} \sum_{i,j}^{N} \left\langle e^{i\mathbf{q}\cdot(\mathbf{r}_i-\mathbf{r}_j)} \right\rangle\,, \tag{5}$$

and note that $I(\mathbf{q}) \propto S(\mathbf{q})$. The static structure factor $S(\mathbf{q})$ can be calculated without recourse to scattering theory as it contains only information about the average relative positions of particles. Thus, $S(\mathbf{q})$ is the quantity that connects static light scattering measurements with theory.

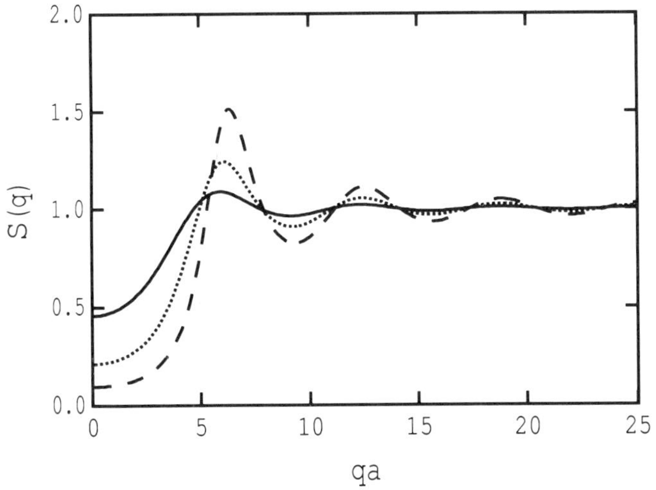

Figure 2. *Static liquid structure factor for 0different volume fractions: solid line, $\phi = 0.1$; dotted line, $\phi = 0.2$; dashed line, $\phi = 0.3$ (calculated for hard spheres using the Percus-Yevick approximation).*

2.1.1 Liquid structure factors

In order to develop some intuition about the results of scattering experiments, it is useful to consider a few examples. First, we consider scattering from a disordered liquid [1, 2]. In Figure 2, we show $S(q)$ for a liquid of hard spheres at three different volume fractions. Because a liquid is isotropic, the structure factor can only depend on the magnitude of q. We can better understand the origin of the oscillations in $S(q)$ by considering its relationship to the radial distribution function $g(r)$,

$$S(q) = 1 + n \int_V d\mathbf{r}\; g(r)\; e^{i\mathbf{q}\cdot\mathbf{r}}\,, \tag{6}$$

where $n = N/V$ is the average particle density. Physically, $n\,g(r)$ can be thought of as the average density of particles a distance r from the centre of an arbitrary particle. Thus, if we consider the spatial structure of a liquid as illustrated in Figure 3, we see that $g(r)$ must be zero near the origin since no other particle can occupy the same space as our reference particle out to some finite distance, roughly comparable to the particle diameter. At a radius corresponding to the first coordination shell, there is a higher than average probability of finding another particle so $g(r)$ must exceed unity. Between the first and second coordination shells, the density will again fall below the average density in a dense liquid because of packing constraints. At the second coordination shell, $g(r)$ will once again rise above unity but not as high as the first peak. This diminishing of the height of the peaks (and depth of the valleys), as the distance from the centre of the labelled particle increases, arises because the correlations in the particle positions die off due to the accumulation of space in which they can fluctuate relative to a central labelled particle. At large distances, all correlations die off in a liquid and $g(r)$ tends towards unity.

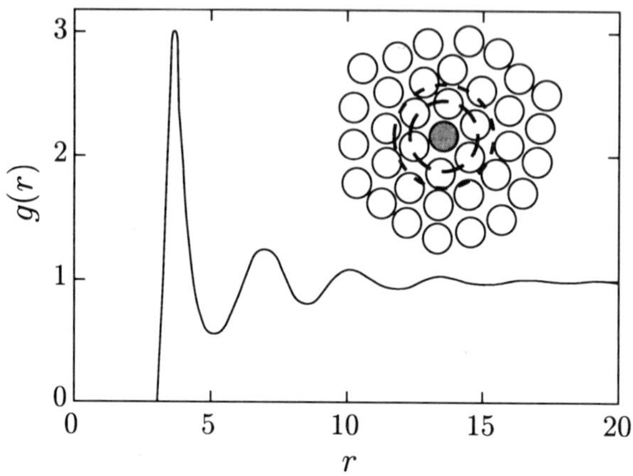

Figure 3. *Radial distribution function $g(r)$ and real space distribution of particles (inset). Dashed circles indicate the location of the first peak (long dashes) and first minimum (short dashes).*

The static structure factor $S(q)$ is, according to Equation 6, the Fourier transform of $g(r)$. Therefore, we can understand the origin of the oscillations in $S(q)$ at finite q as merely reflecting the short-range correlations between particles due primarily to packing constraints arising from the repulsive core of the potential. Thus, the first peak in $S(q)$ occurs in the vicinity of $2\pi/d$ where d is the interparticle spacing (for the special case of hard spheres, the peak is near $2\pi/a$ where a is the particle radius since particles have no reason to prefer the mean interparticle over any other spacing greater than the particle diameter). The structure factor is most interesting for diatomic and more complex molecules, as well as for mixtures of particles, since they show non-trivial correlations for wavevectors exceeding $\sim 2\pi/a$. For spherical particles, such correlations reflect the details of the interparticle potential and are otherwise not particularly interesting. While it may not be apparent from Figure 2, the interesting part of $S(q)$ occurs for values of q much less than $2\pi/d$ since these smaller values of q reflect the long range interparticle correlations.

2.1.2 Scattering from fractal clusters

A useful and intuitive way of characterising the structure of many disordered materials is to specify their fractal dimension. A structure's fractal dimension is defined according to how the mass of the object scales with its radius. Trivial examples are given by: (1) a line for which $m \sim r^1$, (2) a flat sheet of paper for which $m \sim r^2$, and (3) a dense solid object for which $m \sim r^3$. In these examples of simple one, two, and three dimensional objects, $m \sim r^{d_f}$, where d_f is the dimensionality of the object. This concept can be generalised to include many structures found in nature for which $m \sim r^{d_f}$, where d_f is not an integer. A compelling example, studied extensively by light scattering, is clusters of colloidal spheres formed by irreversible aggregation when they collide while undergoing Brownian motion in a solvent. Their structure is illustrated schematically in Figure 4. From experiment and extensive computer simulation it is found that when the potential barrier to the formation of aggregates is small, such that particles almost always stick irreversibly the first time

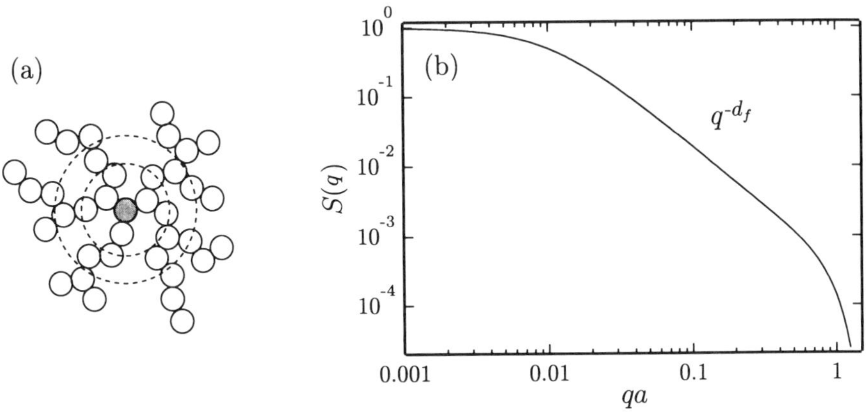

Figure 4. *(a) Fractal cluster. The amount of mass m enclosed within increasingly larger spheres of radius r scales as $m \sim r^{d_f}$ where d_f is the fractal dimension. (b) Static structure factor $S(q)$ for a fractal cluster.*

they come in contact, fractal clusters with a fractal dimension of $d_f \approx 1.7$ form [3]. When the barrier to the formation of aggregates is large, such particles stick irreversibly only after many close encounters and fractal clusters with a fractal dimension of $d_f \approx 2.1$ form.

To determine the scaling properties of the radial distribution function of a fractal object, recall that $n\,g(r)$ is the average density of particles a distance r from a given particle. Thus, taking $m(r)$ to be the total mass within a sphere of radius r, we can write

$$\begin{aligned} n\,g(r) &= \frac{\text{mass in a spherical shell of radius } r}{\text{volume of a spherical shell of radius } r} \\ &= \frac{m(r+dr)-m(r)}{4\pi r^2\,dr} = \frac{1}{4\pi r^2}\frac{dm}{dr} \propto \frac{1}{r^2} r^{d_f-1}\,. \end{aligned}$$

Thus, we see that $g(r)$ scales with radius according to

$$g(r) \sim \frac{1}{r^{3-d_f}}\,. \tag{7}$$

When this result is substituted into Equation 6, we find that

$$S(q) \sim q^{-d_f}\,. \tag{8}$$

This result applies to a wide variety of structures. For example, it is well known that a random walk has a fractal dimension of $d_f = 2$. Thus, for an isolated polymer chain whose conformation is well described by a random walk, it is found experimentally that $S(q) \sim q^{-2}$ over a wide range of q [4, 5]. Such conformations only occur at a specific temperature T_Θ, called the theta temperature, where the net effective interactions between monomers in the chain vanishes (*i.e.* the second virial coefficient $B_2(T_\Theta)$ is zero; see Khokhlov, this volume). As the temperature is increased, the polymer coil generally expands, because of an increased favourable interaction between the monomers and the solvent. In this range, experiments show that $S(q) \sim q^{-5/3}$ indicating a smaller fractal

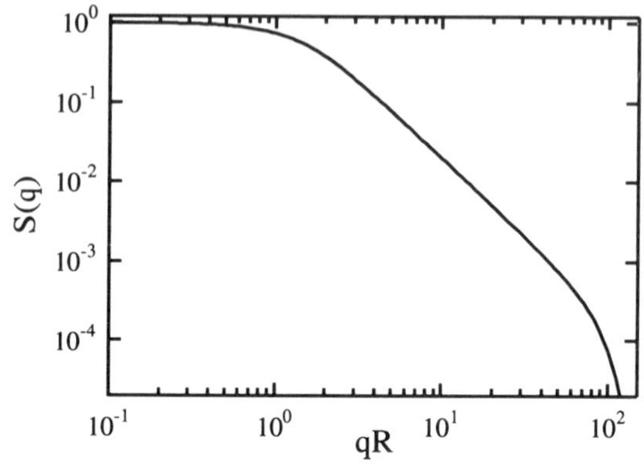

Figure 5. *Structure factor for a random polymer coil.*

dimension consistent with an expanded conformation, *i.e.* closer to a straight line [4, 5]. A schematic representation of $S(q)$ for an isolated polymer coil is shown in Figure 5. Note that $S(q)$ exhibits the q^{-d_f} scaling only over a finite range of q. At large q, when q is comparable to an inverse monomer diameter, the scaling behaviour ceases and $S(q)$ reflects the microscopic correlations between monomers in a chain. At small q, when q is comparable to the inverse radius of gyration R_g^{-1} of the polymer chain, $S(q)$ flattens out, reflecting the featureless random correlations of isolated polymer chains. Thus, on length scales greater than R_g, the isolated chains behave like an ideal gas. Although we have used the example of an isolated polymer chain, the concepts discussed here are applicable to many other systems. For example, the structure factor for a fractal aggregate exhibits similar cutoffs at small and large values of q because of the finite size of the cluster and the structure of the individual particles, respectively. Such cutoffs are observed in all physical realisations of fractal structures.

2.1.3 Scattering from density fluctuations

Up until now, we have considered the scattering of light only by particles. More generally, light is scattering by spatial fluctuations in the dielectric constant. From this point of view, the scattering of light by particles arises because the particles cause fluctuations in the dielectric constant. Indeed, if particles are suspended in a solvent with the same dielectric constant as the particles, there will be no scattering of light by the particles. In most systems, spatial fluctuations in the dielectric constant are, to within a very good approximation, equivalent to fluctuations in the particle concentration or fluctuations in the density. More importantly, useful insights into light scattering can be gained by viewing the scattering as originating from specific Fourier components of the spatial fluctuations in the density (of particles, molecules, *etc.*). In fact, it can be shown that scattering at a particular value of $\mathbf{q}$ corresponds to scattering from sinusoidal density fluctuations of the form $\exp(i\mathbf{q}\cdot\mathbf{r})$.

Consider scattering from a particular Fourier component $\exp(i\mathbf{q}\cdot\mathbf{r})$ as illustrated in Figure 6. Light is incident from the left and is scattered by sinusoidal fluctuations with

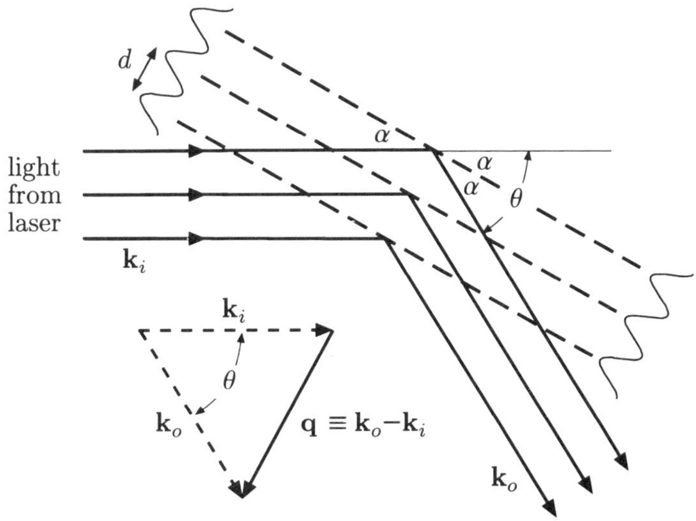

Figure 6. *Scattering of light by sinusoidal density fluctuations. Light scattered at a scattering wavevector of* $\mathbf{q} = \mathbf{k}_0 - \mathbf{k}_i$ *is scattered by sinusoidal density fluctuations* $\Delta n \sim \exp(i\mathbf{q} \cdot \mathbf{r})$ *with wavelength* $d = 2\pi/q$.

wavelength $d = 2\pi/q$ in the dielectric constant. The orientation of the fluctuations is determined by the direction of the wavevector $\mathbf{q}$. The planes of constant phase are oriented at an angle $\alpha = \theta/2$ with respect the direction of the incident light. Thus, one can view the light as being reflected from the fluctuations in the dielectric constant with the angle of incidence α equal to the angle of reflection α. The scattering of light can be viewed as Bragg scattering from these sinusoidal fluctuations. In this case the Bragg condition can be expressed as

$$2d \sin\alpha \;=\; m\lambda\,. \tag{9}$$

In this equation we take $m = 1$ since higher order Fourier components are absent in a *sinusoidal* fluctuation. Thus substituting $\alpha = \theta/2$, $d = 2\pi/q$, and $\lambda = 2\pi/k$ gives the equation $q = 2k\sin\theta/2$ which is Equation 2. This illustrates how light scattering from the sinusoidal fluctuations in the dielectric constant is consistent with the idea that such fluctuations are effectively at the Bragg condition for scattering. Note how this also illustrates that scattering of light at a particular wavevector $\mathbf{q}$ specifies not only the wavelength of the fluctuation that is probed by light scattering but also its spatial orientation. We are now in a position to consider how changes in the microstructure of a complex fluid caused by shear flow can be probed by light scattering.

2.1.4 The effects of shear flow on fluid structure

As a simple example of how shear flow can affect the structure of a complex fluid, we consider a droplet of oil suspended in water. In the absence of flow, the droplet will assume a spherical shape in order to minimise the interfacial (or surface tension) energy between the droplet and the water. Upon the application of a planar shear flow, $v_x = \dot{\gamma}y$,

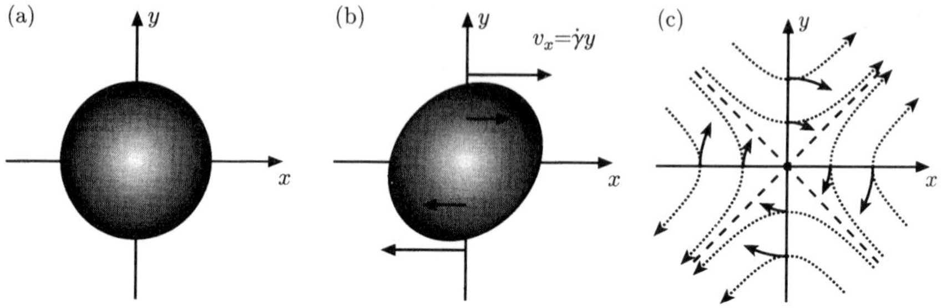

Figure 7. *Effect of shear flow on the shape of a droplet. (a) A spherical droplet (b) is deformed by shear flow. (c) Planar shear flow can be decomposed into a linear superposition of pure extensional (dotted lines) and pure rotational flows (solid lines).*

the droplet will distort. To understand how the droplet distorts, it is useful to write the shear flow as a linear superposition of pure extensional flow and pure rotation (see McLeish, this volume) As illustrated in Figure 7, planar shear flow can be decomposed into a linear superposition of pure extensional and pure rotational flows such that fluid elements are transformed according to $\dot{\mathbf{r}} = \underline{\underline{\mathbf{K}}} \cdot \mathbf{r} = (\underline{\underline{\mathbf{S}}} + \underline{\underline{\mathbf{\Omega}}}) \cdot \mathbf{r}$ where:

$$\underline{\underline{\mathbf{K}}} = \begin{pmatrix} 0 & \dot{\gamma} & 0 \\ 0 & 0 & 0 \\ 0 & 0 & 0 \end{pmatrix} \tag{10}$$

and

$$\underline{\underline{\mathbf{S}}} = \begin{pmatrix} 0 & \dot{\gamma}/2 & 0 \\ \dot{\gamma}/2 & 0 & 0 \\ 0 & 0 & 0 \end{pmatrix}, \quad \underline{\underline{\mathbf{\Omega}}} = \begin{pmatrix} 0 & \omega & 0 \\ -\omega & 0 & 0 \\ 0 & 0 & 0 \end{pmatrix}, \tag{11}$$

where the rotation frequency ω is half the strain rate $\dot{\gamma} = \partial v_x/\partial y$. The effect of the extensional flow $\underline{\underline{\mathbf{S}}}$ is to distort the droplet along a line oriented 45° to the x-axis while the effect of the rotational flow $\underline{\underline{\mathbf{\Omega}}}$ is merely to rotate the droplet.

To understand the effect of flow on a concentration fluctuation, consider the following thought experiment. Imagine that a spherical fluctuation instantaneously comes into existence in a shear flow at time $t = 0$. The initial effect of the shear flow will be to stretch the droplet along a line oriented 45° to the x-axis and then to rotate it slightly towards the x-axis. How far the droplet is ultimately stretched and rotated depends on the relaxation rate Γ or lifetime $\tau \equiv 1/\Gamma$ of the fluctuation compared to the shear rate $\dot{\gamma}$. If $\Gamma \gg \dot{\gamma}$, then the fluctuation will be stretched only slightly and hardly rotated at all away from 45° before it disappears. If $\Gamma \ll \dot{\gamma}$, then the fluctuation can be stretched much more and can be rotated until it is essentially aligned with the x-axis. Of course, the degree to which the droplet is stretched also depends on the surface tension of the droplet, its radius, and on the relative viscosities of the fluid inside and outside the fluctuation. If the fluctuation in concentration is not very large, as is typically the case, then the droplet can be expected to deform affinely (which means that it follows locally the macroscopic applied shear flow field). In any case, the degree to which the droplet is rotated depends primarily on the whether its lifetime τ is short or long compared to the time it takes for the droplet to be distorted and rotated towards the x-axis. The two limiting cases, in which $\Gamma \gg \dot{\gamma}$ and $\Gamma \ll \dot{\gamma}$ are illustrated in Figure 8(a) and (b).

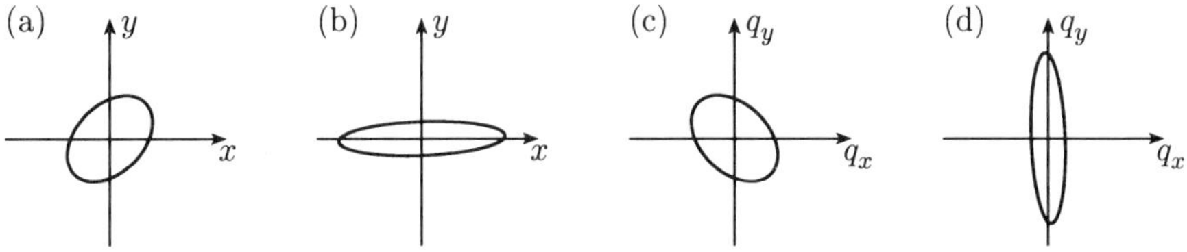

Figure 8. *Effect of shear flow on the shape of fluctuations and the resulting scattering patterns. (a) A fluctuation where $\Gamma \gg \dot{\gamma}$. (b) A fluctuation where $\Gamma \ll \dot{\gamma}$. (c) Scattering pattern corresponding to (a). (d) Scattering pattern corresponding to (b).*

It is also useful to consider how light is scattered from such fluctuations and what the resulting scattering patterns would be. Thus, we imagine that we perform a light scattering experiment on the fluctuations illustrated in Figure 8. We arrange the experiment so that the scattering wavevector $\mathbf{q}$ always lies in the x-y plane. Following our discussion in Section 2.1.1, we expect that the scattering patterns will be related to the Fourier transforms of the real-space distribution of matter. In Figure 8(c) and (d), we illustrate schematically the basic symmetries of the scattering patterns that would result from scattering from the fluctuations shown in Figure 8(a) and (b). One can view the scattering as being qualitatively similar to what one would obtain from diffraction from a slit oriented in the same fashion as the concentration fluctuation. Thus, the narrow parts of the fluctuations result in scattering over a broad range of angles and the wide parts of the fluctuations result in scattering over a narrow range of angles (or equivalently, a broad or narrow range of $\mathbf{q}$ vector—recall Equation 2).

2.2 Dynamic light scattering

Dynamic light scattering (DLS), as its name suggests, probes the temporal evolution of the concentration fluctuations measured in static light scattering. To understand the basic ideas behind dynamic light scattering we once again consider scattering from two particles as illustrated in Figure 9. As in the case of static scattering, the relative phase at

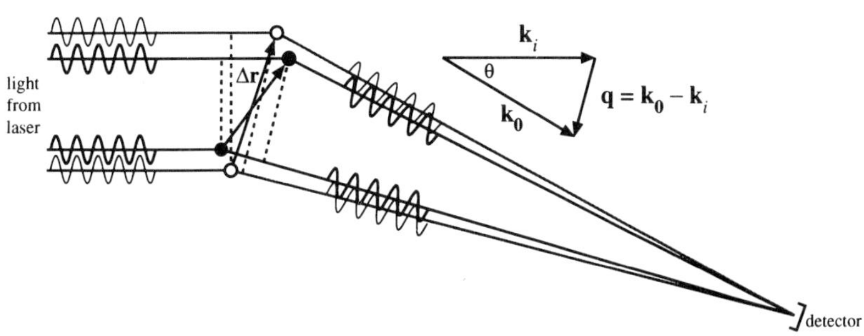

Figure 9. *Schematic for dynamic light scattering of two light paths from two particles. The filled and open circles indicate the positions of the two particles at times t and $t + \tau$, respectively.*

the detector of the light scattered from the two particles determines the degree to which there is constructive or destructive interference. As the particles move, the differences in the path lengths $\Delta\mathbf{r}$ between the pair of particles changes, causing their relative phases at the detector to change. Thus, as the particles move, the intensity of light at the detector fluctuates in time. The typical time scale for the duration of a fluctuation is determined by the time it takes the relative phase difference between the two paths to change by approximately unity. This means that $\Delta\mathbf{r}(t+\tau)-\Delta\mathbf{r}(t) \simeq \lambda/\sin(\theta/2)$. If we assume that each particle moves randomly and independently of every other particle, then to within a factor of order unity this condition can be expressed more simply in terms of the motion of a single particle as $\mathbf{r}(t+\tau)-\mathbf{r}(t) \simeq \lambda/\sin(\theta/2)$. Thus the lifetime of a fluctuation is determined by the time it takes particles to move approximately the wavelength of light, or somewhat farther depending on the scattering angle θ.

We can generalise this analysis to a collection of N scatterers. In that case the electric field at the detector becomes

$$\mathbf{E}_d(t) = \mathbf{E}_s \sum_{i=1}^{N} e^{i\mathbf{q}\cdot\mathbf{r}_i(t)} , \tag{12}$$

where for simplicity we take the scattering amplitude to be the same for all scatterers as would be the case for a collection of identical spherical particles. The intensity of the scattered light is proportional to the square modulus of the electric field at the detector:

$$I(t) \propto |\mathbf{E}_d|^2 = |\mathbf{E}_s|^2 \sum_{i,j}^{N} e^{i\mathbf{q}\cdot[\mathbf{r}_i(t)-\mathbf{r}_j(t)]} . \tag{13}$$

We see that for N scatterers the scattered intensity is determined by the differences in phases between pairs of light paths, just as for the case of a pair of particles discussed above. Since the scattering volume (*i.e.* the volume of sample from which scattered light is collected) is typically much larger than spatial extent of fluctuations, the sum in Equation 12 represents a sum over many independent fluctuations. Thus, the electric field $\mathbf{E}_d(t)$ in Equation 12 is the sum of many independent random variables, and, by the central limit theorem [6], is a random Gaussian variable. Since $I(t) \propto |\mathbf{E}_d|^2$, this means that the intensity of scattered light is distributed according to $P(I) = \exp(-I/\langle I\rangle)/\langle I\rangle$. In Figure 10, we plot the intensity of the scattered light as a function of time obtained from Equation 13 for 2000 randomly diffusing particles. It is interesting and important to note that the fluctuations do not diminish as the number of particles increases; in fact, the amplitude of the intensity fluctuations actually increases. It is this feature of scattered light that makes dynamic light scattering feasible, since there are on the order of 10^{12} or more scatterers in a typical scattering experiment. As stated previously, the duration of a typical fluctuation is given by the time it takes for the phase of the light scattered from a particle to change by order unity, *i.e.* $\mathbf{q}\cdot[\mathbf{r}(t+\tau)-\mathbf{r}(t)] \equiv \mathbf{q}\cdot\Delta\mathbf{r}(\tau) \sim 1$.

The temporal evolution of the intensity fluctuations of the scattering light reflects the stochastic motion of the scatterers. For example, if we heat the sample so that the scatterers move more rapidly, the intensity of the scattered light will fluctuate more rapidly. To extract this information, we need some quantitative means for characterising the statistics of the temporal fluctuations of the scattered light. This is most frequently done by calculating the temporal autocorrelation function $g_I(t,\tau)$ of the scattered light:

$$g_I(t,\tau) \equiv \frac{\langle I(t+\tau)I(t)\rangle}{\langle I(t)\rangle^2} , \tag{14}$$

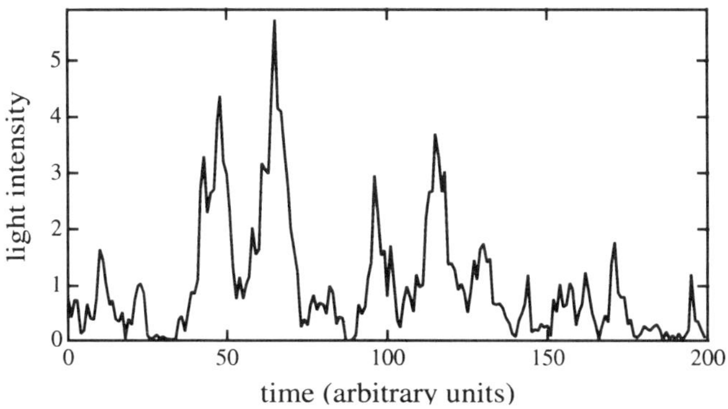

Figure 10. *Intensity vs. time for light scattering from a suspension of 2000 diffusion particles. The average intensity of this plot is unity and the characteristic time of the fluctuations in approximately 15 time units.*

where the brackets indicate a temporal average taken over the duration of the experiment. Alternatively, it is often convenient to introduce the temporal autocorrelation function of the scattered electric field $g_E(t,\tau)$ defined as

$$g_E(t,\tau) \equiv \frac{\langle E(t+\tau)E(t)\rangle}{\langle I(t)\rangle} . \tag{15}$$

For scattered fields with Gaussian statistics, these two correlation functions are related by the Siegert relation [6]:

$$g_I(t,\tau) = 1 + |g_E(t,\tau)|^2 . \tag{16}$$

To obtain a statistically meaningful sampling of the temporal fluctuations, an experiment should ideally acquire data over a time scale which is long compared to the time scale of the longest relaxation time of the system. If the system is stationary, that is, if its dynamics do not change with the passage of time, then $g_I(t,\tau)$ will be independent of t and will depend only on τ. In this case, we can write $g_I(t,\tau) = g_I(\tau)$. For $\tau \to 0$, $g_I(\tau) \to \langle I^2(t)\rangle / \langle I(t)\rangle^2 = 2$, where the last equality follows for the typical case where the scattered electric field obeys Gaussian statistics (as discussed above). For τ much greater than the duration of the longest lived fluctuation of the system τ_M, the scattered intensity at time $t+\tau$ becomes independent of the scattered intensity at time t, and $\langle I(t+\tau)I(t)\rangle$ factorises into $\langle I(t+\tau)\rangle\langle I(t)\rangle = \langle I(t)\rangle^2$. Thus, for $\tau \to \infty$, $g_I(\tau) \to 1$.

Therefore, we expect that the correlation function $g_I(t,\tau)$ will in general decay from a value of two for $\tau = 0$ to unity for $\tau \gg \tau_M$ or, equivalently, that $g_E(t,\tau)$ will decay from unity for $\tau = 0$ to zero for $\tau \gg \tau_M$. The time over which these correlation functions decay and the functional form of the decay will depend on the dynamics of the system. As an example, we consider a system whose dynamics are governed by simple diffusion. In this case, we imagine that fluctuations in the concentration of particles (scatterers) given by $\delta c(\mathbf{r},t) \equiv c(\mathbf{r},t) - \langle c\rangle$ is governed by the diffusion equation

$$\frac{\partial}{\partial t}\delta c(\mathbf{r},t) = D\nabla^2 \delta c(\mathbf{r},t) , \tag{17}$$

where D is the diffusion coefficient of the concentration fluctuations. Because light is scattered by sinusoidal fluctuations in the concentration of scatterers, it is useful to consider the spatial Fourier transform of Equation 17:

$$\frac{\partial}{\partial t}\delta c(\mathbf{q},t) = -Dq^2\delta c(\mathbf{q},t)\,, \tag{18}$$

where

$$\delta c(\mathbf{q},t) = \int_V \delta c(\mathbf{r},t)\, e^{i\mathbf{q}\cdot\mathbf{r}}\, d\mathbf{r}\,. \tag{19}$$

Solving Equation 18 for $\delta c(\mathbf{q},t)$, we find

$$\delta c(\mathbf{q},t) = \delta c(\mathbf{r},0)\, e^{-Dq^2 t}\,. \tag{20}$$

Thus, we find that spatial fluctuations in the concentration with a wavelength of $2\pi/q$ relax with a time constant of $1/Dq^2$. The temporal autocorrelation functions for scattered light reflect this dynamics.

Starting from Equations 12 and 15 and writing the phase of the scattered light as $\phi(t) \equiv \mathbf{q}\cdot\mathbf{r}_i(t)$, we can calculate $g_E(\tau)$ and, via Equation 16, $g_I(\tau)$:

$$g_E(\tau) = \frac{1}{N}\left\langle \sum_{i,j}^{N} e^{i\mathbf{q}\cdot[\mathbf{r}_i(t+\tau)-\mathbf{r}_j(t)]} \right\rangle \tag{21}$$

$$= \frac{1}{N}\left\langle \sum_{i}^{N} e^{i\mathbf{q}\cdot[\mathbf{r}_i(t+\tau)-\mathbf{r}_i(t)]} \right\rangle \tag{22}$$

$$= \frac{1}{N}\left\langle \sum_{i}^{N} e^{i\Delta\phi_i(\tau)} \right\rangle\,, \tag{23}$$

where we define the change in phase for a scattering from a single particle as $\Delta\phi_i(\tau) \equiv \mathbf{q}\cdot[\mathbf{r}_i(t+\tau) - \mathbf{r}_i(t)]$. In passing from Equation 21 to Equation 22, we have assumed that interactions between particles are not important so we can ignore terms in the sum where $i \neq j$. The sum in Equation 23 is a sum over light paths through the sample where each path involves exactly one scattering event from a particle and at a wavevector $\mathbf{q}$. Thus, if all the particles are identical, then all the paths are statistically equivalent and we can simplify Equation 22 by writing it as the ensemble average over a single path:

$$g_E(\tau) = \left\langle e^{i\Delta\phi(\tau)} \right\rangle \tag{24}$$

$$= e^{-\frac{1}{2}\langle\Delta\phi^2(\tau)\rangle}\,. \tag{25}$$

Equation 25 follows from Equation 24 if, as is usually the case, the distribution of phase differences $\Delta\phi$ is Gaussian. Since $\Delta\phi(\tau) = \mathbf{q}\cdot\mathbf{r}$

$$\langle\Delta\phi^2(\tau)\rangle = \frac{1}{3}q^2\langle\Delta r^2(\tau)\rangle\,, \tag{26}$$

where the factor of $1/3$ comes from performing the 3-d angular average $\langle\cos^2\theta\rangle_{\theta,\phi}$ over the unit sphere. For simple particle diffusion, the mean square displacement is given by $\langle\Delta r^2(\tau)\rangle = 6D\tau$ which leads to the result

$$g_E(\mathbf{q},\tau) = e^{-\frac{1}{6}q^2\langle\Delta r^2(\tau)\rangle} = e^{-Dq^2\tau}\,, \tag{27}$$

or equivalently

$$g_I(\mathbf{q}, \tau) = 1 + e^{-2Dq^2\tau} . \tag{28}$$

Note the similarity of these results to Equation 20. For interacting particles, similar results are often obtained with the additional modification that the diffusion coefficient D becomes q-dependent. Thus, fluctuations of different wavelengths relax with a q-dependent relaxation rate of $q^2D(q)$.

2.2.1 Dynamic light scattering in steady shear flow

In these lectures, we are particularly interested in examining the dynamics of systems driven away from equilibrium by the application of shear flow. Thus, it is natural to ask what happens to $g_I(t, \tau)$ for a system subjected to shear. To simplify the discussion, we first consider the case where the scatterers are randomly distributed throughout the sample and move only in response to an imposed shear flow (*e.g.* there is no Brownian motion). The detailed analysis of dynamic scattering from a system undergoing shear is complex. But the basic ideas can be understood by noting that dynamic light scattering is sensitive to *differences* in particle velocities. This is easily appreciated by studying Figure 9 from which it is evident that the relative phases of light scattered from different particles do not change if all the particles move with the same uniform motion. In a homogeneous shear flow, however, particles move with different velocities depending upon their relative positions within the scattering volume. The particles the farthest apart along the velocity gradient and within the scattering volume have the greatest velocity difference $\Delta\mathbf{v}$. This can be written as $\Delta\mathbf{v} = \dot{\gamma}\mathbf{d}$, where $\mathbf{d}$ is a vector, directed along the velocity, whose magnitude is the distance across the scattering volume in the velocity gradient direction. Thus, to within numerical factors of order unity, the characteristic decay time τ_S of $g_E(\tau)$ for a sheared system is $1/\mathbf{q} \cdot \Delta\mathbf{v} = (\dot{\gamma}\mathbf{q} \cdot \mathbf{d})^{-1}$. To within the same level of approximation, the decay of the correlation function is given by $g_E(\tau) \sim \exp[-(\mathbf{q} \cdot \mathbf{d}\dot{\gamma}\tau)^2]$ Note that for homogeneous shear flow, the correlation function decays with a Gaussian time dependence rather than the simple exponential time dependence found for diffusion. This result simply reflects the fact that in a shear flow, the separation between pairs of particles grows linearly in time whereas for diffusion, the separation between particles grows as the square root of time.

We now consider what happens when there is Brownian motion. For shear rates much smaller than the slowest relaxation rate τ_M^{-1}, that is for $\dot{\gamma}\tau_M \ll 1$, the internal dynamics of the system relaxes on time scales much faster than the rate at which shear can alter the structure of the system. In this case, the Brownian motion is unaffected by the presence of the shear flow. The more interesting situation is the case where the shear rate is sufficiently high to alter the structure of the system before it can relax by its usual equilibrium dynamics, that is, when $\dot{\gamma}\tau_M \gg 1$. In this case, there will be two contributions to the decay of $g_E(\tau)$: the decay resulting from the shear flow itself (discussed in the previous paragraph) and the decay resulting from the Brownian motion (possibly modified by the presence of the shear flow). The ratio of the characteristic decay times for these processes is given by

$$\frac{\tau_M}{\tau_S} = \frac{\tau_M}{(\dot{\gamma}\mathbf{q} \cdot \mathbf{d})^{-1}} = \dot{\gamma}\tau_M\mathbf{q} \cdot \mathbf{d} . \tag{29}$$

Scattering volumes for sheared systems are typically no smaller than 50μm. If we take

$d \sim 50\mu\text{m}$ and $q \sim 10^5\text{cm}^{-1}$, then $\mathbf{q} \cdot \mathbf{d} \sim 50$. Thus, in the physically interesting case when $\dot{\gamma}\tau_M \sim 1$, the time scale for the decay of $g_E(\tau)$ from the shear motion τ_S can be expected to be approximately 50 times faster than the time scale τ_M of decay due to the intrinsic dynamics of the system. Thus, the physically interesting change in the system dynamics due to shear flow is masked by the effect of the shear flow itself on the decay of $g_E(\tau)$. In principle, this problem can be circumvented by making $\mathbf{q}$ perpendicular to $\mathbf{d}$. In practice, however, this is very difficult to achieve. Moreover, it limits measurement of the change in the system dynamics only to those fluctuations which are in the direction perpendicular to the velocity. Therefore, one must seek other methods for measuring the change in the system dynamics arising from the application of shear flow. Such methods have been developed, and are discussed in the next section.

2.2.2 Dynamic light scattering in oscillatory shear flow

To develop a method to measure the dynamics of a system under shear flow, we exploit the fact that shear flow is deterministic and reversible, while Brownian motion is not. First, we consider light scattering from a system undergoing oscillatory shear flow where, as in the previous section, there is no Brownian motion. Suppose our system consists of randomly distributed non-Brownian spheres suspended in a liquid undergoing sinusoidal planar shear flow with a period $T \ll \tau_S$. For decay times such that $\tau_S < \tau < T$ we can ignore the fact that the shear flow is oscillatory. In this case, $g_E(\tau)$ decays on a time scale $\tau_S \sim (\dot{\gamma}\mathbf{q} \cdot \mathbf{d})^{-1}$ in the same way as it would for steady shear flow (here, $\dot{\gamma}$ should be taken to be some characteristic shear such as the RMS shear rate).

Upon the reversal of the flow field, however, the situation becomes qualitatively different from that of steady shear flow; the suspended particles retrace their trajectories so that they return to their same exact positions every period T. Therefore, the scattered light will always be perfectly correlated with itself an integral number of periods T ago. Thus, the temporal autocorrelation function will consist of a series of echoes spaced one period apart. The height of the echoes will be unity as long as the particles return to precisely the same position they were at one period ago. The width of the echoes will be twice the width of the initial decay arising from the shearing motion. A correlation function corresponding to this situation is illustrated in Figure 11 by the dotted lines.

With the addition of Brownian motion, there is irreversible movement of the scatterers between echoes and the echoes do not return to their full height. The attenuation of the echos is a quantitative measure of the irreversible motion and is determined by the nonequilibrium dynamics of the system, as illustrated in Figure 11. In the simplest case, when $\dot{\gamma}\tau_S < \dot{\gamma}\tau_M \ll 1$, the envelope which determines the attenuation of the echoes is exponential and simply given by the equilibrium decay $g_I(\tau) = 1 + \exp(-2Dq^2\tau)$. The physically interesting case occurs when $\dot{\gamma}\tau_M$ becomes comparable to or exceeds unity. In this case, one generally expects the dynamics of the system to depart from their equilibrium behaviour; normally, relaxation rates increase as the shear provides more efficient paths for fluctuations to relax as compared to the equilibrium case. We will provide examples of how such measurements can be used to probe glassy emulsions in Section 4.

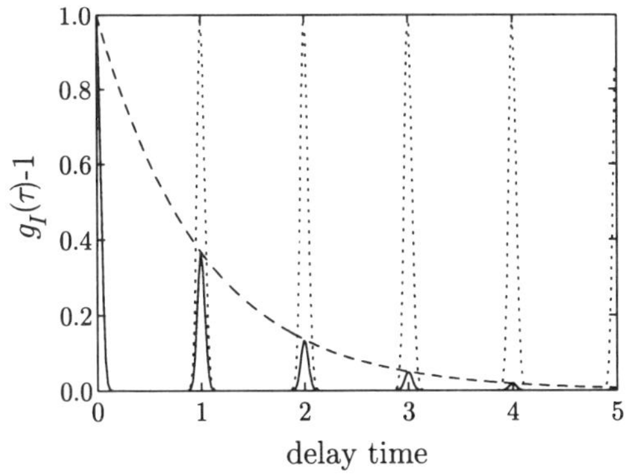

Figure 11. *Appearance of echoes in the intensity autocorrelation function for oscillatory shear flow. In the absence of irreversible motion of the scatterers, the echoes are unattenuated (dotted line). When there is irreversible motion, the echoes are attenuated (solid line). The envelope of the echo heights (dashed line) is determined by the amount of irreversible motion.*

2.2.3 Diffusing-wave spectroscopy

Many complex fluids and fragile materials consist of structures with characteristic length scales comparable to or larger than the wavelength of light. Examples of such systems include colloidal suspensions, emulsions, and foams which are typically made up of micron-sized solid, liquid, or gas particles, respectively, in a liquid matrix. The presence of such structures leads to spatial fluctuations in the refractive index (or dielectric constant) on a length scale comparable to light. As a result, there is multiple scattering of light such that all but the thinnest samples of such materials are opaque. If such a material does not absorb light significantly, these materials appear white (this is typically the case). Thus normal light scattering, which requires that the light be scattered no more than once, is not feasible.

Presently, there are two methods that have proven useful for extracting quantitative information from light scattering on systems which multiply scatter light. The first is a dynamic light scattering technique which uses two lasers with different wavelengths. The lasers and optics are aligned in very clever way so as to discriminate against multiply scattered light. In this scheme, only singly scattered light contributes significantly to the dynamical signal even though there may be significant multiple scattering. The technique is difficult to set up and costly to instrument. Nevertheless, it is a powerful tool provided there is sufficient singly scattered light to provide a measurable signal.

The second approach, called diffusing-wave spectroscopy (DWS) [7], is completely different. Instead of discriminating against multiply scattered light, one exploits it. The primary task is to describe the transport of light sufficiently well that quantitative information can be extracted from a measurement. Fortunately, and perhaps surprisingly, such an enterprise is not difficult. In fact, the experiments are exceedingly simple to instrument—even simpler and less costly than conventional single scattering experiments.

The determination of the electromagnetic fields inside a sample which exhibits a high

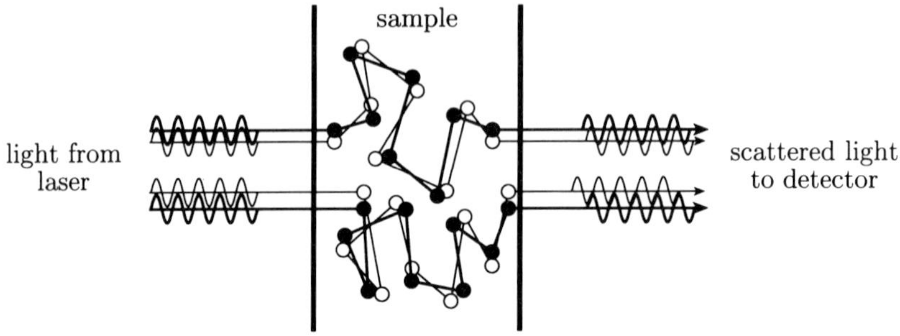

Figure 12. *Schematic for multiple dynamic light scattering of two light paths from many particles. The filled and open circles indicate the positions of the particles for the two paths at times t and $t+\tau$, respectively. Particles not involved in the scattering of the two represented light paths are present but not shown for clarity.*

degree of multiple scattering is a complex task. Fortunately, it is not necessary to specify the electromagnetic field everywhere within the sample in order to extract useful information about the sample from the multiply scattered light. Instead, it is sufficient to consider a single pair of light paths through the sample, in much the same way as we did for singly scattered light in Figure 9. Consider two light paths consisting of light scattered many times by different particles while passing through a sample as illustrated in Figure 12. The time of flight for the light through the sample is essentially instantaneous since it occurs on a time scale of 100ps or less, which is much less than any time scale we will be interested in for the motion of the particles. Nevertheless, light emerging from the sample after having scattered many times will have a phase that depends on the precise optical path length through the sample. The intensity of the scattered light at the detector will depend on the exact relationship between all these phases of the light coming from different paths through the sample. As the particles within the sample move, the path lengths for the light through the sample change. This, in turn, changes the phase relationships between the different pairs of light paths incident on the detector and causes the light intensity to fluctuate, just as in the case of single scattering DLS.

As for DLS, we seek to characterise the fluctuations in the scattered light arising from the motion of particles by calculating the correlation function $g_E(\tau)$ (recall that we can obtain $g_I(\tau)$ using the Siegert relation given by Equation 16 The calculation proceeds similarly to our calculation in the single scattering case. In fact, we write $g_E(\tau)$ just as we did before, starting with the sum over scattering paths represented in Equation 23:

$$g_E(\tau) = \frac{1}{N_p}\left\langle \sum_i^{N_p} e^{i\Delta\phi_i(\tau)} \right\rangle , \tag{30}$$

where we emphasise that here, as in the single scattering case, the sum is over the number of light scattering *paths* N_p through the sample. In contrast to the situation for single scattering, however, here each path consists of many scattering events, as depicted in Figure 12. Moreover, different scattering events within a path occur at different wavevectors whereas for the case of single scattering, each path involved scattering from a single particle and at a single wavevector which was the same for all paths.

Let us denote the number of scattering events in the i-th path as n_i. Then, the phase difference $\Delta\phi_i(\tau)$ for a given path in Equation 30 involves a sum over all the n_i scattering events for that path. That is,

$$\Delta\phi_i(\tau) = \sum_{j=1}^{n_i} \mathbf{q}_j \cdot \mathbf{r}_j(t+\tau) - \sum_{j=1}^{n_i} \mathbf{q}_j \cdot \mathbf{r}_j(t) = \sum_{j=1}^{n_i} \mathbf{q}_j \cdot \Delta\mathbf{r}(\tau) \tag{31}$$

where $\Delta\mathbf{r} = \mathbf{r}_j(t+\tau) - \mathbf{r}_j(t)$. In the case of multiple scattering, paths may have any number of scattering events but only those paths with the same number of scattering events can be regarded as statistically equivalent. Thus, the analysis of multiple light scattering is somewhat more complicated than for the case of single scattering.

The first step in the analysis of Equation 30 is to consider only paths with a given number of scattering events $n = n_i$ (dropping the now superfluous subscript). Next, we note that the statistical distribution of phases $\Delta\phi_n$ for paths of a given length n is Gaussian. In the case of multiple scattering, this is an even better approximation than for the case of single scattering because here the phase is the sum of many random variables; by the central limit theorem, such a sum should obey Gaussian statistics (in the limit of large n). The contribution to the total correlation function $g_E(\tau)$ for all the paths having a given number of scattering events n is:

$$g_E^n(\tau) = \frac{N_n}{N_p}\left\langle e^{i\Delta\phi_n(\tau)}\right\rangle = \frac{N_n}{N_p} e^{-\frac{1}{2}\langle\Delta\phi_n^2(\tau)\rangle}, \tag{32}$$

where N_n/N_p is the fraction of paths with exactly n scattering events. The mean square phase difference for paths with n steps is obtained by squaring and averaging over Equation 31:

$$\langle\Delta\phi_n^2(\tau)\rangle = \sum_{j=1}^{n}\left\langle[\mathbf{q}_j \cdot \Delta\mathbf{r}_j(\tau)]^2\right\rangle \tag{33}$$

$$= n\frac{1}{3}\langle q^2\rangle\langle\Delta r^2(\tau)\rangle, \tag{34}$$

where we have made the assumption that only the diagonal terms in the squared sum are non-zero, consistent with our assumption that the position and motion of different particles are independent. The averages over q^2 and $\Delta r(\tau)^2$ factorise because the scattering wavevectors are independent of the particle motion.

To obtain the full correlation function for light paths of all orders, we sum Equation 32 over all path lengths:

$$g_E(\tau) = \sum_n^{\infty} \frac{N_n}{N_p} e^{-\frac{1}{2}\langle\Delta\phi_n^2(\tau)\rangle} \tag{35}$$

$$= \sum_n^{\infty} \frac{N_n}{N_p} e^{-n\langle q^2\rangle\langle\Delta r(\tau)^2\rangle/6}. \tag{36}$$

Note that in passing from Equation 30 to Equation 36 we have changed the sum from a sum over all paths to a sum over all *path lengths* with each path length weighted by the fraction of paths N_n/N_p with a given number of scattering events n.

The sum in Equation 36 cannot in general be performed analytically. Therefore, we approximate the sum over the number of scattering events n in a path by passing to the continuum limit and performing an integral over the length of a path $s = nl$, where l is the *mean free path* between scattering events. The fraction N_n/N_p of paths consisting of n scattering events becomes the fraction (or probability) $P(s)$ of paths of length s. The mean square phase difference undergoes the following transformation:

$$\frac{1}{2}\langle\Delta\phi_n^2(\tau)\rangle = \frac{1}{6}n\langle q^2\rangle\langle\Delta r^2(\tau)\rangle \tag{37}$$

$$= \frac{s}{6l}2k^2\langle 1-\cos\theta\rangle\,\langle\Delta r^2(\tau)\rangle \tag{38}$$

$$= \frac{s}{3l^\star}k^2\langle\Delta r^2(\tau)\rangle\,, \tag{39}$$

where we have used the fact that $\langle q^2\rangle = 2k^2\langle 1-\cos\theta\rangle$ and that the *transport* mean free path is given by $l^\star \equiv l/\langle 1-\cos\theta\rangle$. Here θ is the scattering angle and $k \equiv 2\pi/\lambda$. From the definition of $l^\star$ it is clear that $l^\star \geq l$. Physically, $l^\star$ is the length scale over which the direction of scattered light is randomised, *i.e.* the characteristic length scale over which scattered light loses memory of its initial direction of propagation. Since scatterers comparable to or larger than the wavelength of light tend to scatter preferentially in the forward direction, several scattering events may be required to randomise the direction of light propagation. In this case, $l^\star > l$. For small particles where the scattering is essentially isotropic, $l^\star \simeq l$.

Using these results, we can convert the sum in Equation 36 to an integral over path lengths:

$$g_E(\tau) = \int_0^\infty P(s)\,e^{-k^2\langle\Delta r^2(\tau)\rangle s/(3l^\star)}\,ds\,. \tag{40}$$

For diffusing particles where $\langle\Delta r^2(\tau)\rangle = 6D\tau$, Equation 40 can be rewritten as

$$g_E(\tau) = \int_0^\infty P(s)\,e^{-2Dk^2\tau s/l^\star}\,ds\,. \tag{41}$$

To evaluate Equation 41, one must determine the distribution of path lengths $P(s)$. For samples which exhibit a high degree of multiple scattering, the path the light takes in traversing the sample can be described as a random walk. Typically, the transport occurs over a length scale much greater than the mean free path $l^\star$ (typically $l^\star \sim 10^2\mu$m and sample dimensions $\sim 10^3\mu$m). In this limit, where the characteristic dimension traversed by random walk is much larger than the basic step length, the random walk can be described by diffusion. Using these ideas, Equations 40 and 41 can be solved for a variety of situations.

Consider, for example, a sample confined between two parallel glass plates a distance L apart with light from a laser incident on one side. If one detects scattered light emerging from the opposite side of the samples, then one obtains [7]

$$g_E(\tau) \simeq \frac{x}{\sinh x} \tag{42}$$

where

$$x = \frac{L}{l^\star}\sqrt{k^2\langle\Delta r^2(\tau)\rangle}\,. \tag{43}$$

For the case of particles (scatterers) which diffuse with a diffusion coefficient D,

$$x = \frac{L}{l^\star}\sqrt{6Dk^2\tau}\,. \tag{44}$$

In this case, the decay of the correlation function is approximately exponential with characteristic decay time of $(l^{\star}/L)^2/Dk^2$. For single scattering, the characteristic decay time is $1/Dq^2 \sim 1/Dk^2$. Thus, the decay of the correlation function for multiple scattering is faster than the decay for single scattering by a factor of approximately $(L/l^{\star})^2$. Physically, this acceleration of the decay is easy to understand. For both single and multiple scattering, the correlation function decays in the time that it takes the phase $\Delta\phi(\tau)$ of the scattered light to change by approximately 1. For the case of single scattering this means that a particle must move by a distance $\sim 1/q \sim \lambda$ or roughly the wavelength of light. For the case of multiple scattering, each particle in a given path must move only λ/n, where n is the number of scattering events in a typical light path, in order for the entire path length to change by approximately the wavelength of light. This is reflected in Equations 26 and 34 for the mean square phase change for single and multiple light scattering, respectively; Equation 34 has a factor of n which is not present in Equation 26. Since the end-to-end distance for a random walk scales as the square root of the number of steps, the decay of $g_E(\tau)$ is a factor of $(L/l^{\star})^2$ faster for multiple scattering than for single scattering.

Thus, perhaps the single most important difference between single and multiple dynamic light scattering is the fact that multiple dynamic light scattering, or DWS, is much more sensitive to very small particle motions. For a typical DWS transmission experiment where $(L/l^{\star}) \sim 10$, the characteristic distance a typical particle moves in a decay time decay is $\lambda/n \sim \lambda/(L/l^{\star})^2 \sim \lambda/100$ or about 50Å. With some effort and care, RMS particle motions on much smaller scales can be resolved, with the current record being somewhat less than 1Å. We will return to our discussion of DWS in Section 4 where we illustrate the use and sensitivity of DWS in a study of the response of disordered emulsions to oscillatory shearing motion. In the next section, we present the results of some experiments on shear thickening which demonstrate, among other things, how single light scattering can be used to study complex fluids and fragile materials.

3 Shear thickening in wormlike micellar solutions

Solutions of wormlike micelles exhibit a fascinating range of rheological behaviour. Above the overlap concentration, their linear rheological behaviour is deceptively simple. A fairly complete theoretical description based on a modified reptation picture is available and has been remarkably successful in describing a wide variety of experiments [8]. In this section, however, we are concerned with another class of wormlike micellar systems, micellar solutions near and below the overlap concentration which exhibit nonlinear shear thickening. The shear-thickening behaviour is quite dramatic and continues to puzzle researchers after more than 15 years of intensive research [9, 10]. Before reviewing the behaviour of these systems, we briefly review some basic ideas concerning wormlike micellar solutions.

3.1 Basic properties of shear-thickening micellar solutions

Surfactants are molecules with a dual personality: one part of the molecule is hydrophilic or water-loving and the other part is hydrophobic or water-hating. They are useful in a variety of contexts, most notably perhaps in promoting the mixing of chemically incompatible liquids by reducing the interfacial tension between the two liquids. Our interests, however, lie elsewhere. The surfactants we are interested in consist of molecules with fairly compact hydrophilic ionic polar head groups, and hydrophobic hydrocarbon tails which typically have about 16 carbon molecules. Below a certain concentration called the *critical micelle concentration* or CMC, the surfactants exist as single molecules in aqueous solution; the CMC is typically on the order of 1mM but can be significantly lower or higher. Above the CMC, these molecules form small aggregates, typically spherical just above the CMC, but often taking on other shapes as the concentration is increased.

Aggregates of surfactants form in order to hide their hydrophobic tails from the surrounding water. They do this by forming a sphere, for example, with all the tails on the inside of the sphere and all the polar heads at the surface of the sphere where they are in contact with the water. They can accomplish the same thing by forming other shapes as well, including cylinders, lamellae, and other more complex structures. Which structure forms depends on the surfactant concentration, the size of the head group relative to the tail, and the surfactant and solvent interactions; see Roux, this volume, for a fuller discussion. A cylindrical micelle is illustrated schematically in Figure 13.

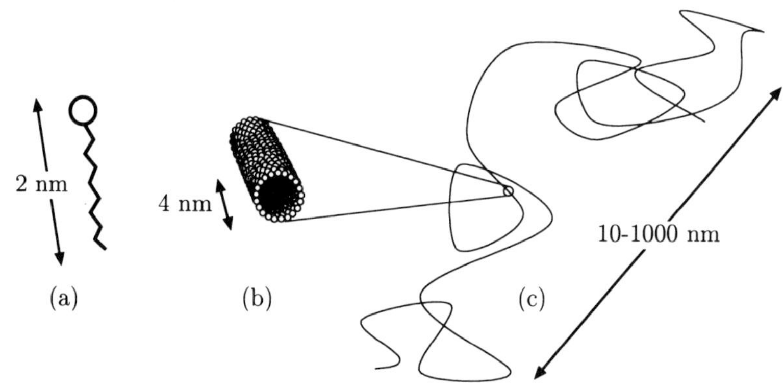

Figure 13. *Surfactants and micelles at different length scales. (a) Surfactant molecule with hydrophilic head group and polar tail. (b) Cross section of a cylindrical micelle with the hydrocarbon tails shielded from the water by the polar head groups at the surface. (c) Random coil formed by a long cylindrical wormlike micelle.*

The question of whether aggregates form or not involves a competition between energy and entropy. When aggregates form, the overall energy is reduced because the surfactant tails are shielded from the water. However, the formation of aggregates reduces the number of possible configurations and decreases the entropy. At low concentrations, entropy almost always wins and there are no micelles. As the concentration in increased, however, energetic considerations become increasingly important such that micelles begin to form above the CMC.

We are interested in cylindrical wormlike micellar solutions. These are systems where the shape of the surfactant head group, size of the tail, and interactions favour the formation of *cylindrical* aggregates. These cylinders can grow very long and flexible such that they resemble a long linear polymer chain. The basic differences between wormlike micelles and polymers are: (1) micelles typically have a diameter of about 4nm, or about ten times greater than a typical polymer; (2) micelles are dynamical entities whose length is determined by an equilibrium process—by contrast, the length of a polymer is fixed at the time of synthesis. The dynamical nature of wormlike micelles has several important ramifications. First, the distribution of the length L of wormlike micelles is thought to be broad, typically exponential $[P(L) \sim \exp(-L/\langle L\rangle)]$. Moreover, $\langle L\rangle$ in general increases with surfactant concentration. Thus, as surfactant concentration increases, the mean size of micelles increase leading to a situation where different micelles begin to overlap. As for conventional polymers, at concentrations above the overlap concentration $c^\star$, there is a dramatic increase in viscosity and in the concentration dependence of the viscosity. A second important consequence of the dynamical nature of micelles is that they spontaneously break and reform in equilibrium. The rate at which this process occurs depends on the scission energy and the temperature; external disturbances such as shear flow can be expected to alter this process.

The specific systems we are concerned with here are ionic wormlike micelles formed from CTAB (cetyltrimethylammonium bromide) or closely related surfactants, and NaSal (sodium salicylate), typically at or near a 1:1 molar ratio. NMR measurements reveal that when the CTAB forms cylindrical micelles in the presence of NaSal, the NaSal is incorporated into the micelle at nearly a 1:1 molar ratio. This means that the micelle has both positively and negatively charged ions, resulting from the dissociation of Br^- from the CTAB and Na^+ from the NaSal. This leaves a highly ionic solution where Coulomb interactions are likely to be important.

The basic shear-thickening rheology which interests us is illustrated in the two plots in Figure 14. In Figure 14(a) we show the response of a wormlike micellar solution to

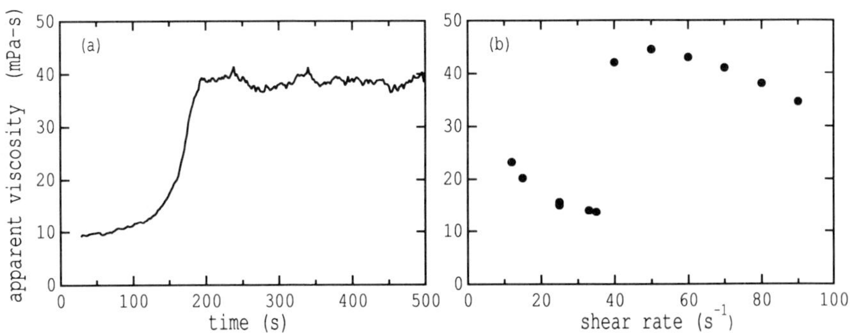

Figure 14. *Basic rheology illustrating shear-thickening in solutions of wormlike micelles. (a) Slow increase in the viscosity measured after the application of a steady shear rate of approximately* $80 s^{-1}$. *(b) Long-time steady-state measured viscosity exhibiting sharp increase above a critical shear rate of approximately* $37 s^{-1}$.

the sudden application of a steady shear rate [11, 12]. At first, nothing unusual occurs, but after tens of seconds the measured viscosity begins to rise until, after approximately 200s, the system reaches a steady state plateau with a viscosity which is about 3 times larger than the viscosity of the original solution. The long-time steady-state viscosity obtained by repeating this experiment for different shear rates yields the data plotted in

Figure 14(b). The most striking feature of these data is the existence of a critical shear rate $\dot{\gamma}$ above which shear thickening is observed and below which nothing extraordinary happens. This shear-thickening is observed for concentrations well below the overlap concentration $c^\star$ up to concentrations which are 2–3 times $c^\star$.

3.2 Light scattering microscopy and rheology

Shear-thickening systems such as those discussed above have proven notoriously difficult to understand. One problem that was not well appreciated until recently is that such systems frequently become spatially inhomogeneous on length scales comparable to the sample dimensions when they undergo a shear-thickening transition. When this occurs, the system can develop large-scale zones or 'bands' with different rheological properties. Thus, conventional macroscopic rheological measurements alone are not sufficient to understand the mechanical behaviour of the system; one must be able to probe the spatial structure as well. Moreover, since such transitions often exhibit slowly evolving rheological changes accompanied by simultaneous structural changes, it becomes paramount to have a means for making *simultaneous* rheological and structural measurements.

Because of these concerns, we developed a transparent Couette cell rheometer and a light scattering technique for following rheological and structural changes in shear-thickening systems as they occur. Our apparatus is illustrated schematically in Figure 15 [11, 12]. The sample is contained between two concentric quartz cylinders having

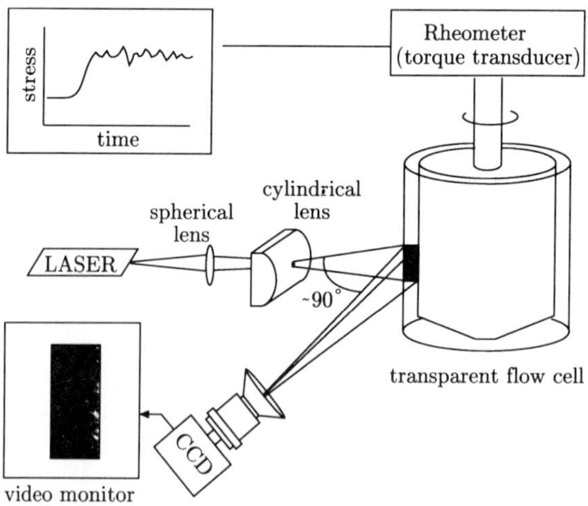

Figure 15. *Experimental setup for light scattering microscopy.*

diameters on the order of 25mm and a gap between them of approximately 1mm. A portion of the sample is illuminated by a sheet of laser light formed by passing the laser beam through a spherical and then through a cylindrical lens as illustrated in Figure 15. The laser beam is directed radially inward towards the cell such that the illuminated volume has a width determined by the gap, is several millimetres high, and is about 50μm thick. Light scattered through approximately 90° from this sheet of light is collected by

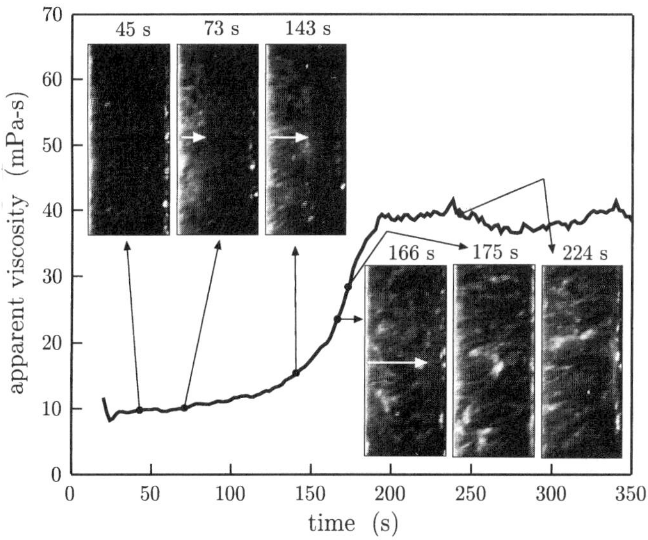

Figure 16. *Effective viscosity vs. time after the commencement of shear flow. Data obtained at constant average shear rate. Images obtained at a succession of times show the growth of another phase as a region of increased scattering (brighter) which proceeds from the inner cylinder (left side of images) and grows towards the outer cylinder (right side of images). White arrows indicate the progression of the interface of the bright phase.*

a CCD camera such that an image of the scattering volume is formed on the CCD array. Regions of the sample which scatter light more strongly than other regions will appear brighter on the image. The resolution of the image is about 30μm. At this resolution, an equilibrium sample is expected to be spatially homogeneous and therefore should exhibit uniform brightness across the illuminated volume.

In Figure 16, we show measurements of the transient effective viscosity of a micellar solution along with a succession of images obtained from our light scattering microscope. The left sides of the images correspond to the inner cylinder and the right sides to the outer cylinder of the Couette cell. After about 73s, a bright region appears on the left side of the image just as the apparent viscosity begins to increase. As time proceeds, the measured viscosity increases and the bright region moves to the right towards the outer cylinder such that it fills an ever increasing fraction of the gap. Eventually, the bright phase appears to fill the entire gap and all the shear is confined to a very thin layer. It also appears that the bright phase is more viscous than the dark phase.

If one of the two phases which appears in Figure 16 is more viscous than the other, then the velocity profile will not be constant across the gap. To investigate this possibility, we measure the velocity profile across the gap of the Couette cell using our light scattering microscopy apparatus. To effect these measurements, we rotate the plane of the sheet of laser light incident on the cell by 90° about the axis along the direction of propagation of the laser beam—see Figure 15. Thus, the sheet of light is oriented horizontally across the gap rather than vertically as in the previous measurements. We then seed the sample with a very small concentration of polystyrene microspheres and follow their motion in time as

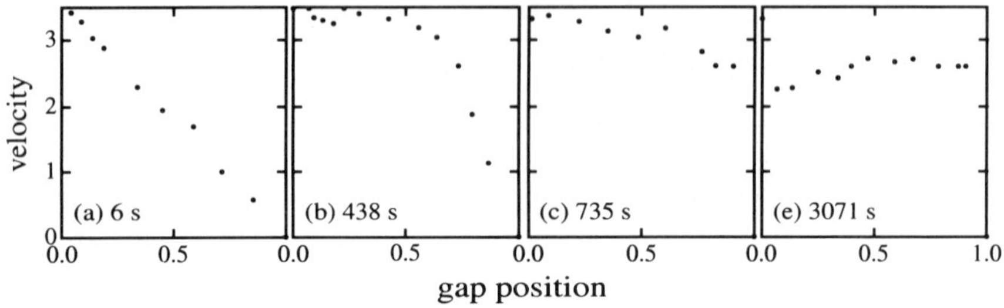

Figure 17. *Velocity profiles in a Couette cell at 4 times after the commencement of shear flow. The inner cylinder is on the left (0.0) and the outer cylinder is on the right (1.0) in each of the plots. The numbers at the lower left corners indicate the time in seconds that have passed after the commencement of shear flow.*

they traverse the illuminated sheet. By measuring the distance that each particle moves in successive video frames (30 frames/s), we are able to map out the velocity profile across the gap of the Couette cell. Figure 17 shows the results of such a measurement for an experiment in which a steady shear flow is applied to an equilibrium sample. The velocity profile is shown at 4 different times. Immediately after the commencement of shear flow, the velocity profile is linear across the gap as can be seen in Figure 17(a); thus, the velocity gradient is constant. The subsequent velocity profiles shown in Figures 17(b)–(d) reveal that the velocity gradient in the bright phase is much smaller than in the dark phase near the outer portion of the cylinder; that is, the bright phase is more viscous than the dark phase. In fact, to within the experimental precision, the velocity gradient in the bright phase is zero. Thus, we refer to the bright phase as a 'gel.' We emphasise, however, that our referring to it as a gel is simply a matter of convenience as the experiments strictly reveal only that the velocity gradient within the bright phase is very small. After about 10 minutes, the gel appears to fill the gap and all the flow is confined to a narrow slip layer next to the outer cylinder as can be seen in Figure 17. Later on, a slip layer appears at the inner cylinder as well although we do not wish to focus on such features here.

With the picture provided by our measurements of a gel growing from the inner cylinder and a few simple assumptions, we can already begin to understand some of the qualitative features of the discontinuous jump in the apparent viscosity. First, we see that the gel only begins to grow when a certain critical shear rate $\dot{\gamma}_c$ is exceeded. Consider then what happens when a shear rate $\dot{\gamma}_0 > \dot{\gamma}_c$ is applied to a sample which has been at equilibrium. We note, first of all, the stress in a Couette cell is greatest at the inner cylinder (we shall explain this in greater detail a little further on). Thus, it is not surprising that the gel begins to form at the inner cylinder. Since there is essentially no flow in the gel, and the average shear rate across the gap of the Couette cell is held constant, the shear rate $\dot{\gamma}_1$ in the remaining liquid must increase such that $\dot{\gamma}_1 > \dot{\gamma}_0 > \dot{\gamma}_c$. Thus, even more gel is formed. We might expect this process to continue until the surfactant available to form gel from the fluid phase is exhausted. In this way the gel phase can grow across the cell and cause the observed increase in the apparent solution viscosity.

Suppose that instead of keeping the shear rate constant that we keep the stress constant

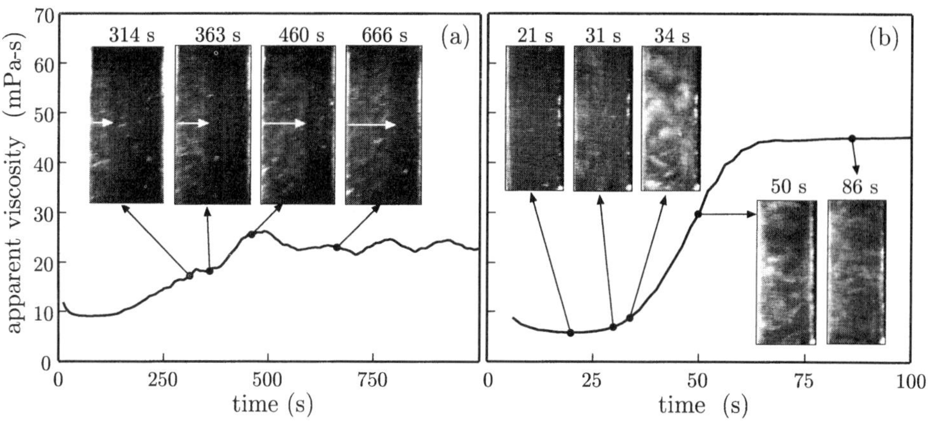

Figure 18. *Effective viscosity vs. time after the commencement of shear flow. Data obtained at constant shear stress. Images obtained at a succession of times. (a) Stress = 0.6Pa: White arrows indicate the progression of the interface of the bright phase as it grows from the inner cylinder. (b) Stress = 1.6Pa: White phase appears homogeneously throughout the gap without a visible interface.*

what happens to the rheology and structure of the sample. The results of one experiment where we hold the stress fixed at 0.6Pa is shown in Figure 18. Initially after startup, the rheology and light scattering images look similar to those obtained under constant shear rate; the apparent viscosity increases as the viscous phase grows from the inner cylinder. An important difference, however, is that the gel phase stops short of the outer cylinder and remains in approximately the same position indefinitely. Thus, the gel and liquid phases can coexist at steady state for a constant applied stress.

If the experiment is repeated at a higher stress but still less than 1.2Pa, the data are qualitatively similar to those obtained at 0.6Pa and displayed in Figure 18 with the apparent viscosity increasing in proportion to the growth of the gel phase across the gap. At a stress of 1.2Pa, the gel phase appears to fill the gap, just as in the case for constant shear rate. Once again, there is a very thin slip layer near the outer cylinder.

The situation changes somewhat when the experiment is repeated at higher stresses. Consider the data in Figure 18 obtained for a startup experiment performed at a constant stress of 1.6Pa. In this case, the bright phase begins to appear after approximately 30s but without any readily visible interface just as the apparent viscosity begins to rise. Indeed, by the time 34s have passed, the bright phase is clearly visible throughout the gap but with only about a 20% rise in the apparent viscosity. As time passes, the apparent viscosity increases dramatically. Meanwhile the bright phase continues to fill the gap but its brightness appears to decrease somewhat in intensity. Thus, it appears that in contrast to data obtained at lower shear rates, at these higher shear rates the bright viscous phase nucleates homogeneously throughout the gap in small isolated droplets. As time progresses, these droplets connect and percolate throughout the gap causing the dramatic observed increase in the apparent viscosity. This interpretation of the data is also consistent with the observed brightness of the gel phase which is brightest around 34s before the apparent viscosity has risen significantly. At this stage, the isolated droplets

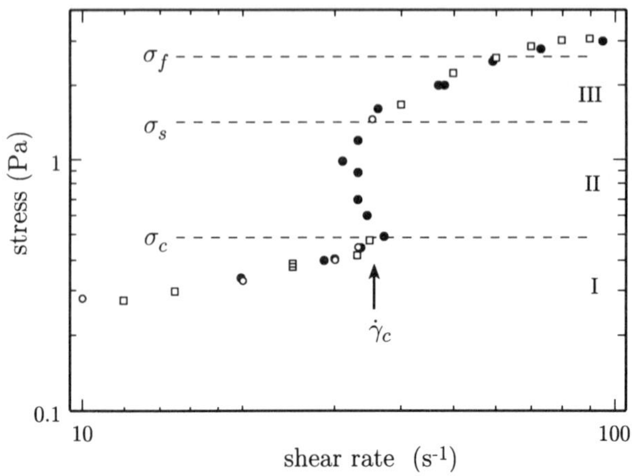

Figure 19. *Steady state stress vs. average shear rate for data obtained under constant shear rate (open symbols) and constant stress (closed symbols).*

create large *fluctuations* in the concentration of surfactant. When the droplets of gel phase begin to interconnect, the fluctuations in concentration are reduced thus reducing the strength of the light scattering.

Figure 19 summarises the shear thickening behaviour discussed thus far. Note that there is a range of stresses between the critical stress σ_c and a higher stress σ_s where there exist steady states under controlled stress but not under controlled shear rate. These data correspond to the situation represented in Figure 18(a) where the gel and fluid phases coexist and both occupy a finite fraction of the gap. It is clear from the data in Figures 19 and 18 that the shear thickening seems to be better characterised by a critical shear *stress* rather than a critical shear *rate*. As can been seen in Figure 19, the average shear rate actually *decreases* when the system begins to shear thicken. Moreover, inside the gel, the shear rate is essentially zero. If it were the shear rate that controls the formation of gel, then the gel would immediately dissolve upon formation because of the low shear rate inside the gel. Nevertheless, the gel remains. Note, however, that under all circumstances, the gel phase is only observed when the shear stress is greater than a critical shear stress σ_c as shown in Figure 19. Thus, the data strongly suggest that it is the shear stress and not the shear rate which controls determines the overall steady state of the system. The reader should be cautioned that this does not mean that the shear rate does not play any role in the kinetics of the nucleation and growth of the gel. It would seem, however, that the shear rate cannot act as a state variable for the long time steady behaviour of the system.

The determinative role of stress is also evident in the cases where homogeneous rather than heterogeneous nucleation occurs. The main point here is simply that stress decreases as r^{-2} in a Couette cell. This follows from the fact that the torque must be balanced at every point in the gap. (The torque Γ is given by $\Gamma = rF$, where r is the radial distance from the axis of the Couette cell of a point in the gap and F is the force at that point. Since the force is related to the stress by $\sigma = F/(2\pi rh)$, where h is the height of sample in the gap, $\sigma = \Gamma/(2\pi r^2 h)$ or $\sigma \propto r^{-2}$.) Thus, when the applied stress is near the critical

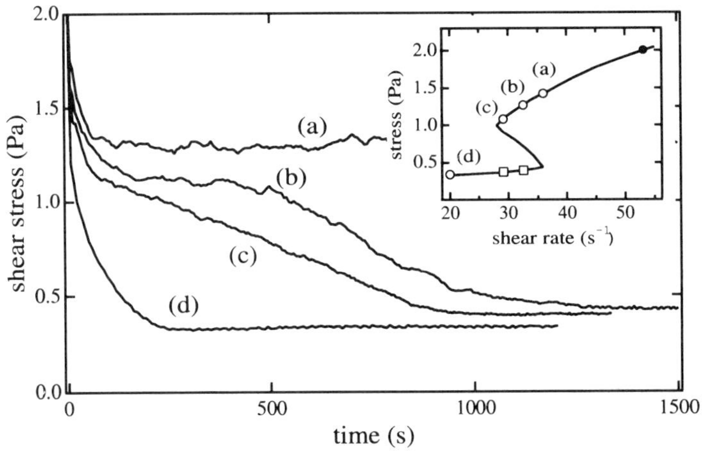

Figure 20. *Quench experiments showing the response of the stress after a sudden decrease in shear rate. The system is prepared at a steady state shear rate of* $\dot{\gamma}_0 = 44s^{-1}$ *as indicated by the solid circle in the inset; the shear rate is then suddenly reduced to one of four lower shear rates indicated by the open circles and squares (see text).*

in the gap, $\sigma = \Gamma/(2\pi r^2 h)$ or $\sigma \propto r^{-2}$.) Thus, when the applied stress is near the critical stress, gel can only be nucleated where the stress is greatest—near the inner cylinder. When the stress in increased so that the stress is well above the critical stress throughout the gap, gel nucleates everywhere. While this is an appealing argument and probably at least partially correct, it must also be pointed out that the shear rate also decreases with radius, though with the weaker dependence of $\dot{\gamma} \sim r^{-1}$. For this and reasons previously discussed, one should not ignore the role of the shear rate in nucleation.

Thus far we have examined only the formation of the gel phase upon the commencement of shear flow. We can also ask what happens if, after a steady state has been reached, the shear rate is suddenly decreased to a value below the critical shear rate. In Figure 20 we show the results of several experiments in which the shear is suddenly decreased after the system has achieved a steady state at a shear rate well above the critical shear rate. In each experiment, the stress falls immediately by an amount which is linearly proportional to the decrease in the applied shear rate. This essentially instantaneous response is followed by a slower decrease in stress which can last several minutes or longer. In the first experiment, labelled (a) in Figure 20, the shear rate is suddenly decreased to a shear rate just a little greater than the critical shear rate. In this case, the gel remains intact and reaches a new steady state in about 100s. When the shear rate is decreased to a value just below the critical stress, as shown in trace (b) in Figure 20, the stress appears to come to a new steady state at a point along the upper branch of the steady state curve shown in the inset (open circle). The system remains there for a few minutes before relaxing to the lowest stress on the steady state curve consistent with the applied shear rate (open square). In this state, there is no gel. The data shown in trace (c) are similar to trace (b) though there is no obvious plateau as the stress relaxes to the low-stress state where the gel has disappeared. Finally, when the shear rate is reduced to a value well below the relaxes fairly rapidly (within a few minutes) to its fluid gel-free state.

At the present time, the microscopic origin of shear thickening in these micellar systems is unknown. Various mechanisms have been suggested, the most developed of which is based on the idea that alignment of the micelles by the imposed flow causes the length of the micelles to grow and to thereby increase the viscosity of the solution. Here, the basic idea has to do with a competition between entropy, which favours many short micelles over fewer longer micelles, and end cap energy, which seeks to minimise the number of micelles ends by forming fewer long micelles. The basic problem is that these theories produce only a mild shear thickening which is much smaller than that observed in experiments. These theories do not take into account the Coulomb interactions which are thought to be important in these systems because of the high degree of ionised species which are typically present in micellar solutions which shear thicken. Other suggestions have been offered as well, including a very interesting proposal involving the existence of closed micellar loops which are broken by shear flow. Presently, there is insufficient experimental evidence which strongly points to any particular theory.

Recently, there has been some progress in developing phenomenological theories which seem capable of reproducing much of the observed rheological behaviour, including co-existing phases of a very viscous and less viscous phase under controlled stress. No microscopic mechanism for shear thickening is proposed in these models. Instead, these models posit the existence of a shear-induced phase transition along with an equation of motion for the interface between the two phases. Then, by employing simple constitutive equations for the rheological behaviour and enforcing conservation of mass, one can reproduce much of the observed phase and rheological behaviour of these systems, including re-entrant rheological curves similar to Figure 19 and discontinuities in the steady-state stress observed under constant shear rate. These models are significant because they can significantly limit the kinds of phase behaviour possible, and thus can serve as a guide to the development of more sophisticated microscopic models. In this way, they resemble descriptions of equilibrium systems based on classical thermodynamics rather than statistical mechanics.

In concluding this section, we summarise the most significant results of these experiments. First is the observation that shear thickening occurs by the nucleation of a new viscous phase. At shear rates and stresses just above $\dot{\gamma}_c$ and σ_c, the nucleation of the new phase is inhomogeneous and the system divides itself into macroscopically distinct regions of high and low viscosity. Second, shear stress rather than shear rate seems to be the more useful variable for describing the steady state of the system. Third, the shear-thickening transition seems to have the character of a first order phase transition. One should exercise caution, however, in applying equilibrium concepts such as phase transitions to systems and processes which are manifestly nonequilibrium. Nevertheless, for these systems the concept of a nonequilibrium phase transition seems to have some value. (See also Roux, this volume.)

Finally, it is useful to emphasise the utility of augmenting rheological measurements with structural measurements in systems which shear thicken. Indeed, simply by learning that these systems divide themselves into two distinct phases when they shear thicken, we have made significant progress towards our ultimate understanding of them.

4 Yielding and rearrangements in glassy emulsions

Fragile materials are characterised by a high degree of sensitivity to external forces and typically yield under very mild shear stresses. In most cases, the response of the material is elastic up to some yield stress beyond which the response is more complex, depending upon the system under consideration. In this section, we consider the response of one realisation of such a material, namely emulsions, and study the structural rearrangements which occur when such a system is strained beyond its elastic limit. Before discussing the specific experiments, we provide a brief review of emulsions and their properties.

4.1 Emulsions

Emulsions are multicomponent systems which in their most basic form consist of three components: oil, water, and a surfactant. In fact, the two liquid components need not be oil and water, but can be any two liquids which are mutually insoluble (or which at most exhibit only very limited mutual solubility). Nevertheless, the vast majority of emulsions consist of water and some insoluble oil.

Emulsions are usually formed by mechanical mixing which creates a dispersion of oil droplets in a continuous background of water or a dispersion of water in a background of oil. The latter system is often called an 'inverse emulsion'. The smallest droplet size that can be achieved by mechanical mixing is typically about 0.1μm. Such a mixture is not in equilibrium, however, and will demix unless measures are taken to suppress the subsequent coalescence of droplets. Suppressing coalescence is the role of the surfactant which, when mixed with the oil and water, goes to the interface between the droplets and the surrounding fluid. In some cases, the surfactant is electrically charged which results in a repulsive interaction between droplets. In other cases, the surfactant provides a steric or entropic barrier between droplets. In either case, the surfactant provides an effective repulsive interaction which acts as a barrier to coalescence by keeping the droplets from coming into contact. Thus, emulsions are kinetically stabilised against coalescence and do not represent the lowest energy state of the system. Normally, the lower energy state consists of a system which is completely phase-separated into macroscopic regions of oil and water with the surfactant dissolved in one of the phases, and perhaps existing as micellar solution. Alternatively, the system may form a thermodynamically stable microemulsion, in which case it can be difficult to maintain the systems as a nonequilibrium emulsion. Thermodynamically stable microemulsions differ from emulsions in that they consist of much smaller droplets (typically about 10nm), are thermodynamically stable, and can usually form spontaneously without the addition of mechanical energy. Nevertheless, our interest here is focused on emulsions which are metastable and which, with proper selection of materials, can remain stable indefinitely.

When prepared by mechanical mixing, emulsions usually have a wide distribution of droplet sizes. Numerous methods have been developed for producing emulsions with a high degree of monodispersity. For example, various fractionation schemes have been developed by which a polydisperse emulsion can be successively divided into fractions consisting of particles all within a fairly narrow range of diameters. With some effort, samples with 10% polydispersity can be achieve in this manner. Other schemes, mostly mechanical, also exist whereby emulsions can be produced with polydispersities in the

10–30% range. These schemes require more specialised equipment but are capable of producing much greater quantities of monodisperse material. In the experiments we discuss in this section, the emulsions have a polydispersity of about 10%. This level of polydispersity prevents the emulsion droplets from forming an ordered crystal. Thus, the emulsions discussed here are amorphous at all concentrations.

When the volume fraction ϕ of droplets in an emulsion is not too high, emulsions behave very similarly to colloidal dispersions of solid particles. They are subject to the same thermal forces, for example, and exhibit Brownian motion just as do solid colloidal particles. The situation changes, however, when the volume fraction of droplets approaches and exceeds random close packing. For nearly monodisperse spheres, random close packing occurs at volume fraction $\phi_{\rm rcp}$ of about 0.63. For $\phi \ll \phi_{\rm rcp}$, emulsion droplets exist as isolated spheres. But as ϕ approaches $\phi_{\rm rcp}$, increasingly less space is available to each particle. For $\phi > \phi_{\rm rcp}$, no more room is available and particle motion is arrested. It is still possible to mechanically deform the system, however, because the droplets themselves are deformable. We now review the mechanical behaviour of random close packed emulsions.

4.2 Mechanical properties of random close packed emulsions

For sufficiently small strains, one expects a random close packed emulsion to exhibit linear viscoelastic behaviour. That is, one expects the system to respond elastically, but not without some viscous dissipation arising from shearing of the liquid in the emulsion. Such behaviour can be characterised by a complex frequency dependent elastic shear modulus $G(\omega)$ (see McLeish, this volume). To understand the physical meaning of $G(\omega)$ we consider the following simple experiment. Suppose an emulsion is confined between two parallel plates which are spaced a distance apart which is much greater than the droplet diameter (a spacing of $\sim$ 1mm is typical). The top plate is moved back and forth sinusoidally producing a time-dependent shear strain across the sample which is given by $\gamma(\omega,t) = \mathrm{Re}[\gamma_0 \exp(i\omega t)]$ where ω is the frequency and γ_0 is the strain amplitude. One then measures the time-dependent stress $\sigma(\omega,t)$ on the bottom plate which for a linear viscoelastic material can be written as $\sigma(\omega,t) = G(\omega)\gamma(\omega,t)$. Because the system exhibits both viscous dissipation and elastic response, the stress is in general not completely in phase with the applied strain. Thus, $G(\omega)$ is complex and is written as $G(\omega) = G'(\omega) + G''(\omega)$, where $G'(\omega)$ characterises the in-phase elastic response of the system and $G''(\omega)$ characterises the out-of-phase dissipative or viscous response of the system.

At small strains, close packed emulsions deform elastically like any elastic solid with an elastic modulus $G'(\omega) = G'_0$ which is independent of frequency. By contrast, the dissipative response which is characterised by the loss modulus $G''(\omega)$ becomes smaller as the frequency is reduced reflecting the fact that viscous dissipation depends on the *velocity* gradient rather than the *displacement*. In the limit low frequencies, $G''(\omega) = \eta\omega$ where η is the zero-frequency (or zero-shear-rate) viscosity of the emulsion.

At low strains, measurements of $G'(\omega)$ and $G''(\omega)$ are independent of the strain amplitude γ_0 as expected for a linear viscoelastic material. As the strain is increased, however, the response becomes nonlinear and amplitude dependent, signalling the onset of yielding and plastic flow. Mason *et al.* [13] have studied the linear and nonlinear rheology of concentrated disordered emulsions as well as yielding and flow. As expected they find normal linear viscoelastic behaviour at low strain amplitudes consistent with an elastic solid as described above. Above a concentration-dependent strain amplitude, they find that the emulsions do yield. They also find that there is a dramatic increase in the dissipation associated with the onset of nonlinear behaviour and yielding. One expects that this increased dissipation is associated with irreversible rearrangements of droplets. Unfortunately, the rheological measurements do not provide any direct measurement of such droplet motions. For this, we turn to light scattering.

4.3 Light scattering in emulsions in an oscillatory shear flow

The basic phenomenon we wish to investigate is the irreversible movement of emulsion droplets subjected to an oscillatory shear flow; the basic idea is to use dynamic light scattering. As discussed previously in Section 2.2.2, a light scattering measurement performed on a sample undergoing oscillatory shear flow leads to a series of echoes in the temporal correlation function of the scattered light. Although the experiments described in Section 2.2.2 were discussed in the context of single light scattering, all the concepts apply equally well to multiple light scattering, that is, to DWS. The only pertinent difference is that multiple light scattering is much more sensitive to particle motion and can therefore detect much smaller irreversible particle movements. Therefore, we expect to obtain data qualitatively similar to that displayed in Figure 11.

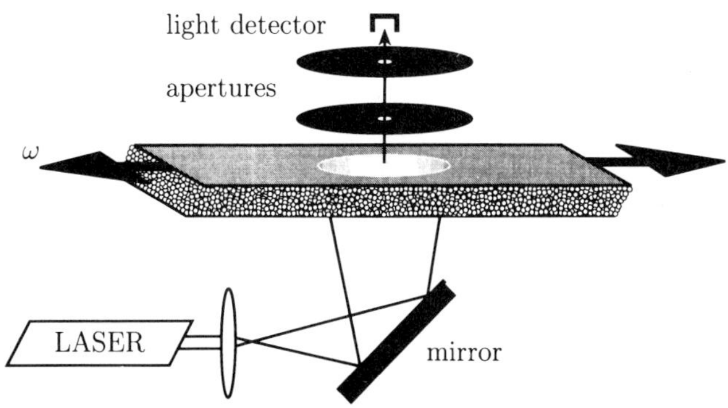

Figure 21. *Schematic of DWS transmission measurement of sheared emulsion. Coherent light from a laser is expanded and directed towards the bottom glass plate on which the emulsion is placed. The upper glass plate is moved back and forth using a precision piezo-electric device. Apertures assure that light from on the order of one speckle is collected by the detector.*

In Figure 21, we show a schematic of the experimental setup for a DWS transmission experiment. Light from a laser is multiply scattered by the emulsion contained between two glass slides. For the case shown, multiply scattered light which is transmitted through the sample is detected and sent to an electronic correlator. The glass slides are roughened to ensure that the emulsion does not slip when the upper slide is oscillated back and forth. Backscattering DWS experiments are carried out using the same cell but, in that case, multiply scattered light is collected from the same side of the sample as on which the light is incident.

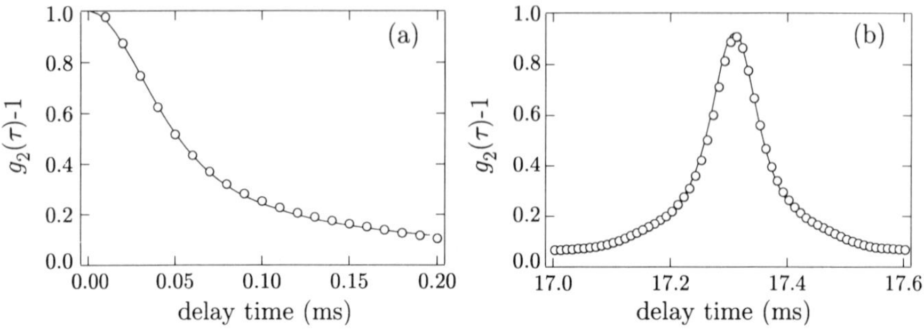

Figure 22. *Temporal correlation function obtained using diffusing-wave spectroscopy on an emulsion undergoing oscillatory flow. (a) Initial decay of the correlation function arising from the shearing motion. (b) First echo in correlation function centred at a delay time of one period of the imposed oscillating shear flow.*

In Figure 22, we show data obtained from an emulsion subjected to an oscillatory shear flow with a strain amplitude of $\gamma_0 = 0.010$ and frequency of 57.8Hz ($\phi_{oil} = 0.85$). The figure shows the initial decay of the correlation function and the first echo. Between these two features, the correlation function is essentially zero. Note that the widths of the initial decay and the echo are much narrower than the delay time between them. The width of the peaks is set by the characteristic shear rate $\gamma_0\omega$ and the thickness of the cell [14]. Because of the narrow widths of the peaks, it is essential that the clocks running the correlator and the shear flow be synchronised. This can be accomplished, for example, by using the clock for the correlator as the master clock to which the shear flow is synchronised using a phase-locked loop. Alternatively, two separate clocks may be used if they are both sufficiently stable over the duration of the experiment.

To extract useful information from the correlation functions we measure, we need an expression for the correlation function. Recall that in Section 2.2.3 we derived expressions for $g_E(\tau)$ as an integral over light paths through the sample which had the form:

$$g_E(\tau) = \int_0^\infty P(s)\, e^{-x(\tau)s/l^\star}\, ds\,, \tag{45}$$

where x depended on the type of motion that the scatterers execute. For example, in Equation 40, $x = (k^2/3)\langle\Delta r^2(\tau)\rangle$ where $\langle\Delta r^2(\tau)\rangle$ is the mean square displacement of the scatterers. For oscillating shear flow, it can be shown that this reduces to,

$$x(\tau, \tau_0) = \frac{1}{15}\left[kl^\star\,|\gamma(\tau+\tau_0) - \gamma(\tau_0)|\right]^2\,, \tag{46}$$

where $\gamma(\tau_0)$ is the initial value of the strain [14, 15]. For oscillatory flow, the particle motion is not stationary but depends on which part of the strain cycle the systems is at. A typical electronic correlator, such as the one used in these experiments, calculates $g_I(\tau)$ from the data stream in a manner which essentially assumes the process producing the data is stationary. Thus, it continually updates the time τ_0 at which it starts calculating the correlation function. Therefore, to account for the fact that an oscillating shear flow is not stationary, we must integrate the theoretical correlation function over all initial values of the strain. This will allow comparison with data taken from the correlator. Thus, we substitute Equation 46 into Equation 45, and integrate the intensity correlation function over all initial strains (one oscillation period):

$$g_I(\tau) - 1 = \frac{\omega}{2\pi} \int_0^{2\pi/\omega} |g_E[x(\tau, \tau_0)]|^2 \, d\tau_0 \,, \tag{47}$$

The integral in Equation 47 can be performed numerically to obtain the shapes of the correlation functions we measure. We determine the transport mean free path $l^\star$ by measuring the total transmitted intensity [7, 16]. As stated previously, $g_I(\tau)$ is insensitive to $l^\star$ in the transmission geometry and the initial decay of the correlation function essentially depends only on the strain amplitude γ_0 and frequency ω. Thus, we can compare our data to the theoretical expression given in Equation 47 without any adjustable parameters. The result of this comparison is shown in Figure 22(a) where the circles represent the data from the experiment and the solid line the theoretical result obtained from Equation 47. The agreement between theory and experiment is remarkable and confirms our theoretical description of the decay of the correlation function due purely to shear flow.

In writing Equations 46 and 47, we have assumed that there is no motion other than the affine displacement of scatterers with the applied strain. If this is the case, then the scatterers should all return to their exact same positions when the shear is reversed thereby causing an echo in the correlation function at a delay time of the period of oscillation. Furthermore, the shape of the echo should be governed by the same process that governs the initial decay discussed above. Thus, the shape of the echo should be described by Equations 46 and 47. In Figure 22(b) we show a fit of the data to Equations 46 and 47 where the only adjustable parameter is the echo height about which we have no *a priori* knowledge. Once again, the theory fits the data very well.

We now turn to the decay of the echo heights under the application of the oscillating shear. In Figure 23(a), we show a correlation function for a sinusoidal shear flow with a strain amplitude of $\gamma_0 = 0.05$. Because of technical limitations in the instruments

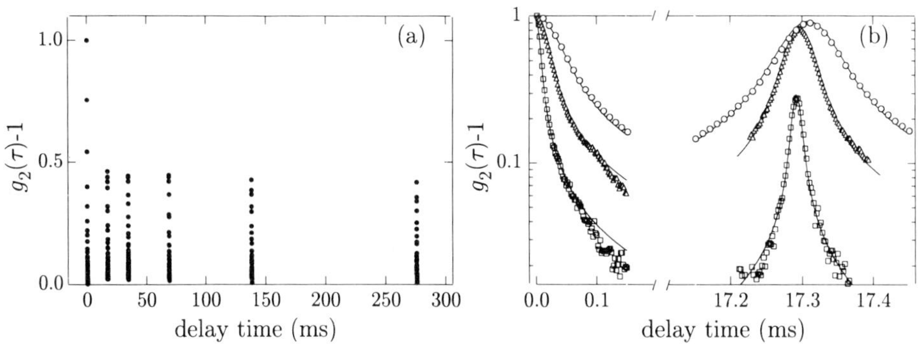

Figure 23. *Temporal correlation function obtained using diffusing-wave spectroscopy on an emulsion undergoing oscillatory flow. (a) Correlation function showing initial decay and multiple echoes. The strain amplitude is* $\gamma_0 = 0.05$. *(b) Initial decay and first echo in correlation function on an expanded time scale. The strain amplitudes are* $\gamma_0 = 0.01$ *(circles), 0.02 (triangles), and 0.06 (squares). The use of slightly different strain frequencies leads to the slightly shifted peak positions of the echoes.*

used to calculate the correlation function from the experimental data train, only the first, second, fourth, eighth, and sixteenth echoes were determined in these experiments. The other echoes exist, as confirmed by other experiments, but were not determined in the measurement displayed. The first echo is less than unity, as expected, because of irreversible motion of at least some of the droplets. There is one quite unexpected feature of these data, however, and that is that *all the echos have the same height.* This is in stark contrast to the behaviour illustrated in Figure 11 where the echo heights decayed exponentially consistent with particle diffusion. Indeed, light scattering experiments on colloidal suspensions under an oscillating shear exhibit the expected exponential decay [17].

Although all the echoes have the same height for a given strain amplitude γ_0, the height of the echoes decrease with increasing γ_0, as shown in Figure 23(b). For all strain amplitudes, however, the echo height is constant for as large of delay times as we can measure. We also note that this same behaviour is observed for backscattering DWS measurements as well.

The fact that the echoes do not decay after the first echo means that there is a finite fraction of the emulsion which undergoes reversible periodic motion. If this were not the case, there would not be any echoes. It also means that there is another fraction of the emulsion for which undergoes irreversible motion. That is, the trajectories of some fraction of the emulsion droplets are chaotic. This is why the echo heights are less than unity. Finally, the fact that the echo heights do not change in time means that these two fractions of emulsion droplets are disjoint sets. If a droplet undergoes a reversible trajectory after one shear cycle, then it does so indefinitely. Similarly, if a droplet undergoes an irreversible trajectory after one shear cycle, then it does so indefinitely. Thus, under oscillatory shear *the system partitions itself into fragile regions which are fluid-like and elastic regions which are solid-like.* Furthermore, these regions maintain their identity and integrity over time.

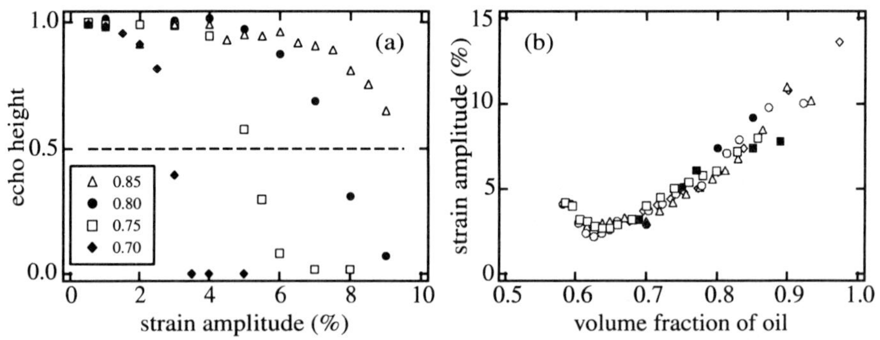

Figure 24. *(a) Echo height from DWS backscattering experiments vs. strain amplitude for different volume fractions. (b) Comparison of the critical strain amplitude vs. volume fraction obtained from DWS measurements (solid symbols) to yield strains obtained from rheological measurements by Mason et al. [13] (open symbols).*

The dependence of the echo heights on strain amplitude is shown in Figure 24. This plot shows that the echo height decays monotonically as a function of increasing strain amplitude. Thus the volume fraction of droplets that rearrange under strain increases as the amplitude of the strain increases.

Mason *et al.* [13] found using rheological measurements that the emulsion *yields* above some critical value of the strain amplitude. To make contact with their measurements, we plot the data they obtained for the yield strain vs. volume fraction of droplets in Figure 24(b). In the same Figure, we also plot the volume fraction at which the echo heights drop to one half their initial value. While this criterion is somewhat arbitrary, the good agreement between the two data sets support the idea that yielding is associated with particle rearrangements in the emulsion.

To obtain a quantitative measure of the fraction of emulsion that undergoes rearrangements we must analyse the shape of the correlation function and the echo heights in more detail. Basically, there are three contributions to the shape of the correlation function: (1) the decay arising from the oscillatory straining motion which we have already discussed, (2) the spontaneous decay of the correlation function caused by Brownian motion, and (3) droplet rearrangements caused by the oscillating strain. The spontaneous decay arising from Brownian motion is observable only at the lower volume fractions, where there remains some room for the particles to move. This contribution to $x(t)$ is given by $(k^2/3)\langle\Delta r^2(t)\rangle$ where $\langle\Delta r^2(t)\rangle$ is the mean square displacement caused by Brownian motion and can be determined from the decay of $g_2(\tau)$ in the absence of shear.

The contribution to the decay of the correlation function arising from rearrangements can be determined in the following way [18]. In the absence of a rearrangement event, we assume that there is no decay of the height of the echo. That is, the length of a light path through the sample is exactly the same as it was one oscillation period ago unless the path intersects a region in the sample that undergoes a rearrangement. In the case that it does intersect a region that has undergone a rearrangement, the phase of that path is completely randomised. This assumption is justified by the fact that the droplet sizes are comparable to the wavelength of light and any rearrangement of even a small group of droplets is virtually certain to change the path length through the sample by at least one wavelength. Thus, the height of the echo is simply determined by the fraction of paths that have not been randomised after a delay time τ, where τ is an integral number of oscillating periods. If the number of rearrangements of a given size a occur randomly at a rate R per unit volume, then the rate at which rearrangements occur within a light path α is proportional to the product of R and the volume swept out by the light path $sl^{\star 2}$, since $l^\star$ is the shortest length scale over which one can describe the transport of light within the diffusion approximation. In addition, α will scale as $a^3/l^{\star 3}$ since larger rearrangement events will randomise more paths. Thus, $\alpha \approx R(sl^{\star 2})(a^3/l^{\star 3}) = (a^3R)(s/l^\star)$. This means that the contribution to the decay of the correlation function of paths of a given length s is $\exp(-\alpha\tau) = \exp[-(a^3R)(s/l^\star)\tau]$. This can be written in simpler form by noting that at any given delay time, the volume fraction Φ of the sample that has undergone an irreversible rearrangement is $\Phi = aR^3\tau$. Thus, in Equation 45, to within factors of order unity, $x(\tau) \approx \Phi(\tau)$.

Putting together the three contributions to the decay of the correlation, we obtain

$$x(t) = \frac{1}{3}k^2\langle\Delta r^2(t)\rangle + \frac{1}{15}\left[kl^\star\left|\gamma(\tau+\tau_0) - \gamma(\tau_0)\right|\right]^2 + \Phi(\tau) \tag{48}$$

Substituting this into Equation 45, we can use this result to extract from the echo height the volume fraction which has undergone irreversible rearrangements $\Phi(\tau)$. Note that in

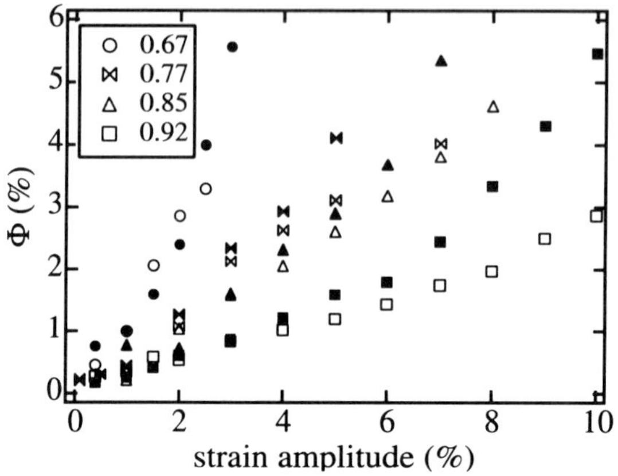

Figure 25. *Volume fraction Φ of emulsion sample undergoing irreversible rearrangements vs. strain amplitude. The solid symbols were obtained from backscattering data and the open symbols from transmission data. The volume fraction of droplets for each sample is indicated in the legend.*

general, this quantity can have a time dependence, and is written explicitly as having one in Equation 48. The fact that the echo heights do not change in time simply reflects the fact that the volume fraction which undergoes irreversible rearrangements remains constant in time.

In Figure 25, we plot the volume fraction of emulsion Φ that has rearranged as a function of strain amplitude γ_0. Data obtained in both transmission and backscattering are plotted. At large strain amplitudes, there is some discrepancy between the transmission and backscattering measurements suggesting that the volume fraction of emulsion which undergoes irreversible rearrangements increases near the walls of the sample. As expected, the emulsions which have a lower volume fraction of droplets undergo more rearrangements at a given strain amplitude than do emulsions with a higher volume fraction of droplets. By comparing our measurements of $\Phi(\gamma_0)$ shown in Figure 25 to the measurements of yield strain by Mason *et al.* [13] we estimate that yielding occurs in these emulsions when approximately 4% of the emulsion droplets have undergone irreversible rearrangements.

5 Acknowledgments

The work described in this article represents collaborations with many people over a number of years. In particular, the sections on light scattering represent the collaborations with P.M Chaikin, D.J. Durian, D.A Weitz, X.-L. Wu, and A.J. Yodh, as well as productive interactions with many others too numerous to mention here. The shear thickening work reflects collaborations with P. Boltenhagen, E. Matthys, and Y.T. Hu. I have also benefited greatly from discussions with J. Goveas and P. Olmsted, and David Lu on this topic. Finally, the section on sheared emulsions represents a collaboration with P. Hébraud, F. Lequeux, and J.P. Munch.

References

[1] Hansen J P and McDonald I R, 1986. *Theory of Simple Liquids* (London ; Orlando: Academic Press) second edn.

[2] Sirota E B, Ou-Yang H D, Sinha S K, Chaikin P M, Axe J D and Fujii Y, 1989. Complete phase diagram of a charged colloidal system: a synchrotron X-ray scattering study *Phys Rev Lett* **62** 1524–7

[3] Lin M Y, Lindsay H M, Weitz D A, Ball R C, Klein R and Meakin P, 1989. Universality of fractal aggregates as probed by light scattering *Proc R Soc Lond A, Math Phys Sci* **423** 71–87

[4] de Gennes P G, 1979. *Scaling Concepts in Polymer Physics* (Ithaca, N.Y.: Cornell University Press)

[5] Strobl G R, 1997. *The Physics of Polymers : Concepts for Understanding their Structures and Behavior* (Berlin ; New York: Springer) second edn.

[6] Berne B J and Pecora R, 1976. *Dynamic Light Scattering : with Applications to Chemistry, Biology, and Physics* (New York: Wiley)

[7] Weitz D and Pine D, 1993. Diffusing-Wave Spectroscopy in *Dynamic Light Scattering: The Method and Some Applications*, ed. W Brown vol. 49 of *Monographs on the Physics and Chemistry of Materials* 652–720 (Oxford: Oxford University Press)

[8] Cates M E and Candau S J, 1990. Statics and dynamics of worm-like surfactant micelles *J Physics: Cond Mat* **2** 6869–6892

[9] Wunderlich I, Hoffmann H and Rehage H, 1987. Flow birefringence and rheological measurements on shear induced micellar structures *Rheol Acta* **26** 532–542

[10] Rehage H and Hoffmann H, 1982. Shear induced phase transitions in highly dilute aqueous detergent solutions *Rheol Acta* **21** 561–563

[11] Hu Y, Boltenhagen P and Pine D, 1998. Shear-thickening in low-concentration solutions of worm-like micelles I: Direct visualisation of transient behavior and phase transitions *J Rheol* **42** 1185–1208

[12] Hu Y, Boltenhagen P, Matthys E and Pine D, 1998. Shear-thickening in low-concentration solutions of worm-like micelles II: Slip, fracture, and stability of the shear-induced phase *J Rheol* **42** 1209–1226

[13] Mason T G, Bibette J and Weitz D A, 1996. Yielding and Flow of Monodisperse Emulsions *J Colloid Interface Sci* **179** 439–448

[14] Wu X L, Pine D, Chaikin P, Huang J and Weitz D, 1990. Diffusing-Wave Spectroscopy in a Shear Flow *J Opt Soc B* **7** 15–20

[15] Hébraud P, Lequeux F, Munch J and Pine D, 1997. Yielding and rearrangements in disordered emulsions *Phys Rev Lett* **78** 4657–4660

[16] Durian D J, 1995. Penetration depth for diffusing-wave spectroscopy *Appl Opt* **34** 7100–7105

[17] Qiu X, Ou-Yang H, Pine D and Chaikin P, 1988. Self-Diffusion of Interacting Colloids Far from Equilibrium *Phys Rev Lett* **61** 2554–2557

[18] Durian D, Weitz D and Pine D, 1991. Multiple Light-Scattering Probes of Foam Structure and Dynamics *Science* **252** 686–688

Polymer physics: from basic concepts to modern developments

Alexei Khokhlov

Moscow State University, Russia

1 Basic concepts of polymer physics

1.1 Fundamentals of the physical viewpoint in polymer science

Polymer chains of different chemical structure have, of course, different properties. However, there are many common properties characteristic of large classes of polymer systems. For example, all rubbers (cross-linked polymer networks, see below) exhibit the property of *high elasticity*, all polymer melts are *viscoelastic*, all polyelectrolyte gels absorb a large amount of water, *etc.* Such properties can be described on a molecular level by taking into account only the general polymeric nature of constituent molecules, rather than the details of their chemical structure. It is these properties that are studied using polymer physics. For a more complete introduction, and many further references, see Grosberg and Khokhiov 1994; Grosberg and Khokhlov 1997.

What are the main factors governing the general physical behaviour of polymer systems? Three of them should be mentioned in the first place.

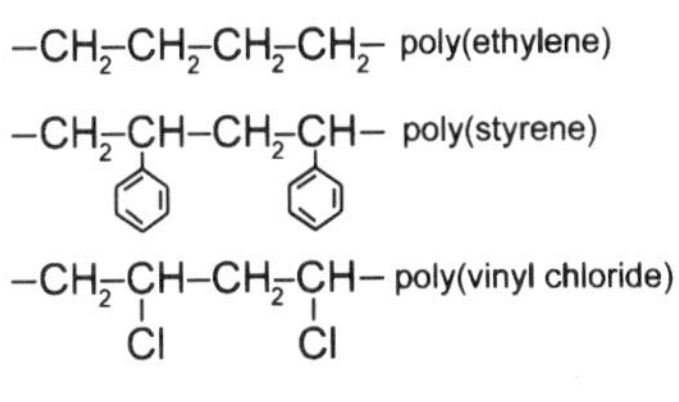

Figure 1. *Common polymer chains.*

First of all, polymers are long molecular chains. In Figure 1 three of the most common polymer chains with carbon backbones are shown. One can see that small atomic groups (monomer units) are connected in linear chains by covalent chemical bonds. The *chain structure* of constituent molecules is the first fundamental feature of polymer systems.

In particular, this means that monomer units do not have the freedom of independent translational motion, and therefore polymers do not possess the entropy associated with this motion (the so-called translational entropy). This is sometimes expressed as follows: polymer systems are *poor in entropy.*

Second, the number of monomer units in the chain N, is large: $N \gg 1$ (otherwise we have an 'oligomer', not a polymer). For macromolecules synthesised in the chemical laboratory, normally $N = 10^2$–10^4. For biological macromolecules the values of N can be much larger, for example, the longest polymer chains are those of DNA molecules: $N \simeq 10^9$–10^{10}. Such large objects can be seen by a normal optical microscope (if DNA is labelled with fluorescence dyes), since the linear size of DNA coil turn out to be larger than the wavelength of light.

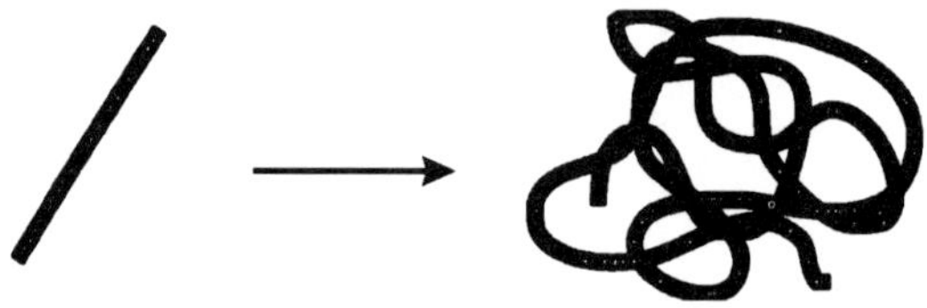

Figure 2. *Polymer chains are generally flexible, they normally take the configuration of the coil (right), not of the rigid rod (left).*

Third, polymer chains are generally *flexible* (see Figure 2), they normally take the conformation of a random coil, rather than that of a rigid rod. We will discuss in detail the notion of polymer chain flexibility in Section 1.2.

In summary, their chain structure, the large number of monomer units in each chain, and chain flexibility are the three main factors responsible for the special properties of polymer systems.

1.2 Flexibility mechanisms of a polymer chain

1.2.1 Rotational-isomeric flexibility mechanism

Let us consider the simplest polyethylene chain (Figure 1) and let us ask ourselves for which conformation do we have the absolute energetic minimum? Such a conformation corresponds to a straight line and is shown schematically in Figure 3. For this conformation all the monomer units are in the so-called *trans* position. This would be the equilibrium conformation at $T = 0$.

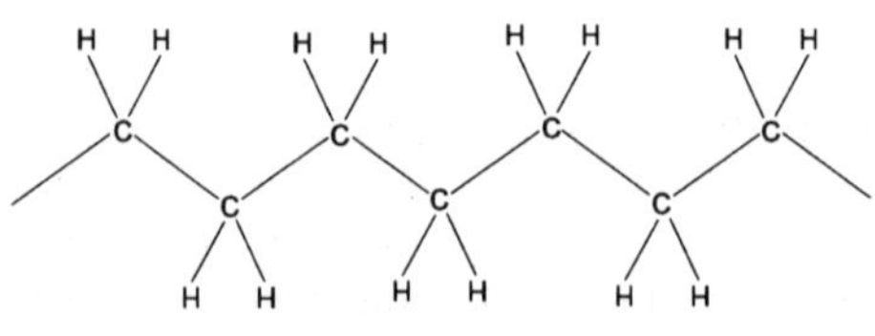

Figure 3. *The rectilinear (all trans) conformation of a polyethylene chain.*

At $T \neq 0$, due to the thermal motion, deviations from the minimum-energy conformation are possible. According to the Boltzmann law, the probability of realisation of a conformation with the excess energy U over the minimum-energy conformation is

$$P(U) \sim \exp\left(-\frac{U}{k_B T}\right) . \tag{1}$$

What are the possible conformational deviations from the structure shown in Figure 3? For a carbon backbone the *valency angle* γ (see Figure 4) should be normally considered as fixed (for different chains $50° < \gamma < 80°$). However, rotation with fixed γ by changing the angle of internal rotation φ (see Figure 4), is possible. Any value $\varphi \neq 0$ gives rise to deviations from the rectilinear conformation, *i.e.* to chain flexibility, though usually there are only two or three preferred values of φ corresponding to different rotational isomers. This kind of flexibility is called the *rotational-isomeric flexibility mechanism.*

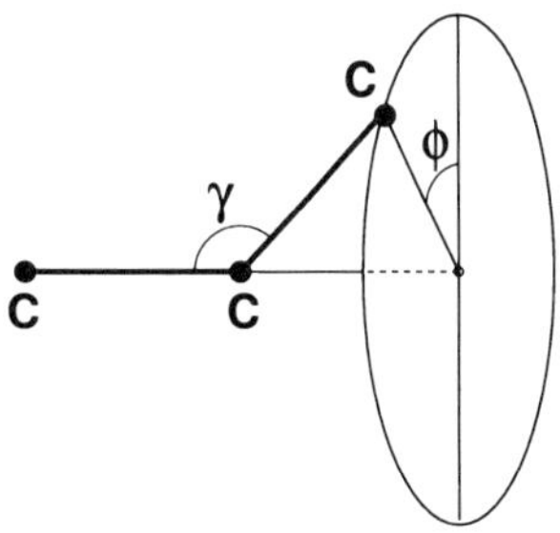

Figure 4. *The valency angle γ and angle of internal rotation φ for a carbon backbone.*

1.2.2 Persistent flexibility mechanism

Another flexibility mechanism can be realised when rotational isomers are not allowed, e.g. in α-helical polypeptides or DNA double helix. The conformations of these chains are stabilised by hydrogen bonds and internal rotation is impossible. In this case *small thermal vibrations* around the equilibrium conformation play the most important role. Via their accumulation over large distances along the chain, these vibrations give rise to the deviations from the rectilinear conformation, *i.e.* to the chain flexibility. This is a *persistent flexibility mechanism*; it is analogous to the flexibility of a homogeneous elastic filament.

1.2.3 Freely-jointed flexibility mechanism

Another mechanism of flexibility is realised in the so-called *freely-jointed model* of a polymer chain. In this model the flexibility is located in freely-rotating junction points: φ in Figure 4 takes any value. This mechanism is not very characteristic of real chains, but it is frequently used for model theoretical calculations.

Figure 5. *Typical conformation of a polymer coil of freely-jointed segments.*

1.3 Portrait of a polymer coil

In Figure 5 a typical conformation of a polymeric coil is presented. It was constructed on the computer for the freely-jointed model by allowing each subsequent segment to be oriented in an arbitrary direction with respect to the previous one. From this picture one can draw the following conclusions:

- The volume fraction occupied by the monomer units inside the coil is very small. There are many 'holes' inside.
- By the manner of our construction of Figure 5 it is easy to realise that chain trajectory is analogous to the trajectory of a Brownian particle.

A single coil conformations of the type shown in Figure 5 can be realised in real experiment in *dilute polymer solutions* when polymer coils do not overlap.

1.4 Size of an ideal freely-jointed chain: the random coil

By definition, in the *ideal polymer chain* we take into account only the interactions of close neighbours along the chain. The interactions of monomer units which are far from each other along the chain are neglected. Polymer chains behave as ideal ones in so-called Θ-conditions (see below).

Consider an ideal N-segment freely-jointed chain with each segment of length l (see Figure 6). The size of such a chain can be characterised by its end-to-end vector $\mathbf{R}$. However, this vector will change rapidly in the course of the thermal motion. An important characteristic is the average size R of a polymer coil. This average cannot be defined as $\langle \mathbf{R} \rangle$ since all the segments orientations are equiprobable, therefore $\langle \mathbf{R} \rangle = 0$. That is why the size R of coil is usually characterised by the root-mean-square end-to-end distance: $R \simeq \sqrt{\langle |\mathbf{R}|^2 \rangle}$. Let us calculate this value for our model. The end-to-end vector is the sum of the segments vectors (see Figure 6):

$$\mathbf{R} = \sum_{i=1}^{N} \mathbf{u}_i \,. \tag{2}$$

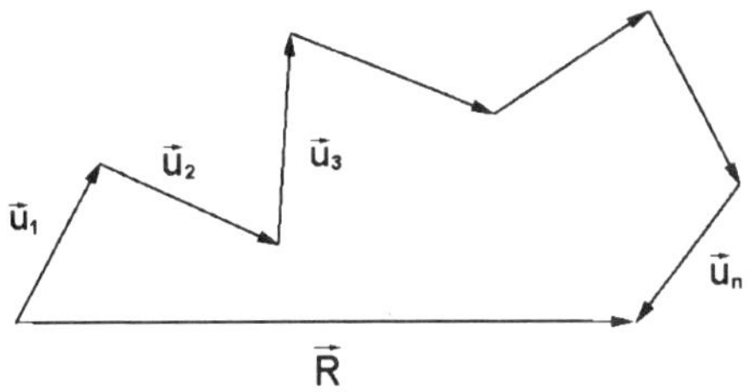

Figure 6. *The model of freely-jointed chain.*

Thus the square of the end-to-end distance is:

$$|\mathbf{R}|^2 = \left(\sum_{i=1}^{N} \mathbf{u}_i\right) \cdot \left(\sum_{j=1}^{N} \mathbf{u}_j\right) = \sum_{i=1}^{N}\sum_{j=1}^{N} \mathbf{u}_i \cdot \mathbf{u}_j \,, \tag{3}$$

and the average of this value

$$\langle R^2 \rangle = \sum_{i=1}^{N}\sum_{j=1}^{N} \langle \mathbf{u}_i \cdot \mathbf{u}_j \rangle = \sum_{i=1}^{N} \langle |\mathbf{u}_i|^2 \rangle + \sum_{i=1}^{N} \sum_{j=1, j\neq i}^{N} \langle \mathbf{u}_i \cdot \mathbf{u}_j \rangle \,. \tag{4}$$

In the last equality in Equation 4 we have separated the terms with $i = j$ from all the other terms. Taking into account that $\langle |\mathbf{u}_i|^2 \rangle = l^2$ and $\langle \mathbf{u}_i \cdot \mathbf{u}_j \rangle_{i \neq j} = 0$ (because the orientations of different chain segments in the freely-jointed chain model are not correlated), the final result is

$$R \simeq \sqrt{\langle R^2 \rangle} = N^{1/2} l \,. \tag{5}$$

Note that the mean square end-to-end distance is much less than the contour length L of the chain: $R \ll L = Nl$. Thus, the conformation is far from the rectilinear one. Because **R** in Equation 2 is the sum of many random variables, its three components are each gaussian distributed; an ideal chain forms a random walk or *gaussian coil* (Figure 2). Note that Equation 5 can be also rewritten as $\langle R^2 \rangle = Ll$ since $L = Nl$.

1.5 Size of an ideal chain with fixed valency angle

The conclusion $R \sim N^{1/2}$ is actually valid for an ideal chain with *any* flexibility mechanism (not only for a freely-jointed chain model). For example, let us consider the model with fixed valency angle γ between the segments of length b and free internal rotation (see Figure 4). As follows from Section 1.2, this model is close to a real chain with the rotational-isomeric flexibility mechanism.

It can be shown that for this model

$$R \simeq \sqrt{\langle R^2 \rangle} = N^{1/2} b \sqrt{\frac{1 - \cos\gamma}{1 + \cos\gamma}} \tag{6}$$

where b is the bond length. We see that the average size of the chain is still proportional to $N^{1/2}$, and in this model the chain is also in the *gaussian coil* conformation. This is a general property of ideal polymer chains, independent of the model for chain flexibility. At $\gamma < 90°$ the value of R is larger than for a freely-jointed chain, while at $\gamma > 90°$ the reverse is true.

1.6 Kuhn segment length of a polymer chain

We have seen above that for any ideal chain $\langle R^2 \rangle \sim N \sim L$ (at large values of the contour length L). Therefore, the ratio $\langle R^2 \rangle / L$ should be independent of L and should give a measure of chain flexibility. By definition, the *Kuhn segment length* of a polymer chain is introduced as

$$l = \frac{\langle R^2 \rangle}{L} \quad \text{(at large } L\text{)} \tag{7}$$

so that the equality $\langle R^2 \rangle = Ll$ is exact by definition.

The physical meaning of l follows from comparison of this equality with Equation 5, valid for a freely-jointed chain. The comparison shows that if we try to choose a *freely-jointed equivalent* to a given chain with the same values of $\langle R^2 \rangle$ and L, the segment length for this equivalent chain should be equal to l. Thus l, the length of an *equivalent segment*, represents an approximately straight subunit of the chain. It is a quantitative characteristic of chain flexibility.

1.7 Persistence length of a polymer chain

It can be shown that for the model of Figure 4, as well as for most of other polymer models, the orientational correlations of the chain as a function of the contour distance obey

$$\langle \cos \vartheta \rangle \sim \exp(-s/\tilde{l}) , \tag{8}$$

where ϑ is the angle between unit vectors $\mathbf{u}(0)$ and $\mathbf{u}(s)$ (see Figure 7). We see that these correlations decay exponentially along the chain; the characteristic length of the decay, $\tilde{l}$, is called the *persistence length* of the chain. The physical meaning of this characteristic

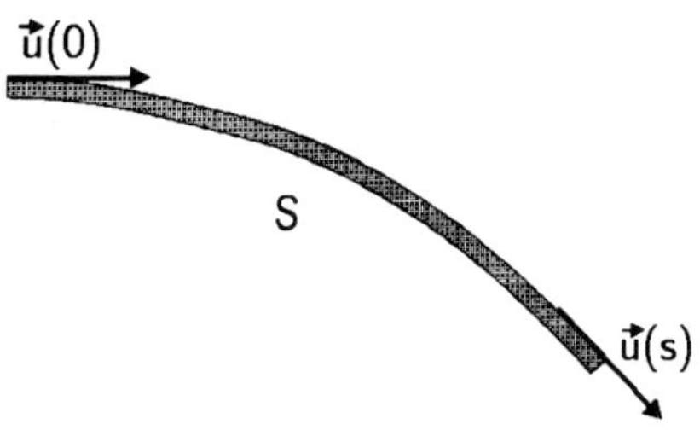

Figure 7. *Illustration for the definition of persistence length of a polymer chain.*

follows from Equation 8. At $s \ll \tilde{l}$ we have $\langle \cos \vartheta \rangle \approx 1$, so the chain is approximately rectilinear, while at $s \gg \tilde{l}$ we obtain $\langle \cos \vartheta \rangle \approx 0$, so that the memory of chain orientation is lost.

It is worthwhile to note the advantages and disadvantages of using l and $\tilde{l}$ as quantitative characteristics of chain flexibility. The advantage of l is that it can be directly experimentally measured (the values of $\langle R^2 \rangle$ and L can be determined from light scattering (Grosberg and Khokhlov 1994)). The advantage of $\tilde{l}$ is that it has a direct microscopic meaning (see Figure 7). Depending on what is more important in the specific problem, one may choose $\tilde{l}$ or l. One can show that always $\tilde{l} \simeq l$; in most cases, $\tilde{l}$ is approximately $l/2$.

1.8 Stiff and flexible chains

Let us choose the Kuhn segment length l to characterise stiffness. The value of l is normally larger than the *contour length per monomer unit* l_0. The ratios l/l_0 for some common polymers are shown below.

poly(ethylene oxide)	2.5
poly(propylene)	3
poly(methyl methacrylate)	4
poly(vinyl chloride)	4
poly(styrene)	5
poly(acrylamide)	6.5
cellulose diacetate	26
poly(para-benzamide)	200
DNA (in double helix)	300
poly(benzyl glutamate) (in α-helix)	500

From a macroscopic viewpoint, a polymer chain can be always represented locally as some filament which is characterised by two microscopic lengths: the *Kuhn segment length* l and the filament's *characteristic diameter* d. (This describes the thickness of the filament.) Depending on the ratio between these two lengths, we can now introduce the notion of stiff and flexible chains. *Stiff* chains are those for which $l \gg d$, while for *flexible* chains $l \simeq d$. Some examples of stiff chains are DNA, helical polypeptides, aromatic polyamides *etc.* Examples of flexible chains are polyethylene, polystyrene, *etc.* —in fact, most polymers having a single-chain carbon backbone.

2 Swelling and collapse of single chains and of gels

2.1 Basic physical effects

If polymer chains are not ideal, interactions of non-neighbouring monomer units (the so-called *volume interactions*) should be taken into account. If these interactions are repulsive, the coil *swells* with respect to its ideal dimensions. If monomer units attract each other, *contraction* of the macromolecule is observed. In many cases this contraction leads to the 'condensation' of polymer chain upon itself, with the formation of a 'dense droplet' conformation, which is called a *polymer globule*. The transition to this collapsed conformation (Figure 8) is called the *coil-globule transition.*

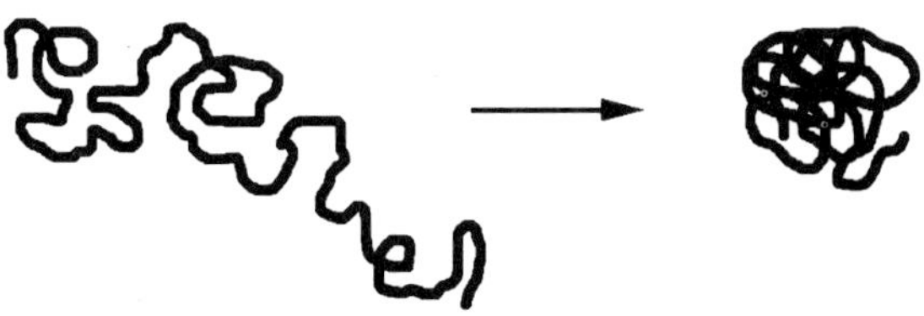

Figure 8. *Coil-globule transition.*

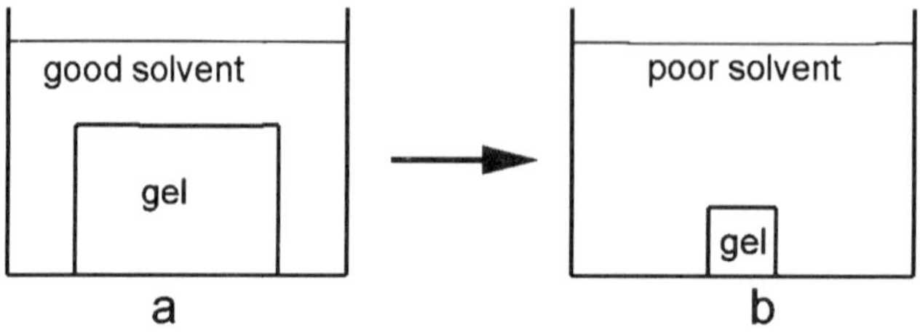

Figure 9. *Polymer gel in a good (a) and poor (b) solvents.*

The same type of phenomenon can be observed in *polymer gels*, which are cross-linked polymer networks swollen in a solvent. Gels of linear size of 1mm to 1cm are normally used in experiments, although it is possible to prepare 'microgels' of up to 100nm size. The gel as a whole is actually one giant *three-dimensional molecule.* Suppose that such a gel is in contact with a large volume of solvent (Figure 9). If the monomer units of the gel repel each other (in other words, polymer chains prefer to be in contact with solvent molecules, rather than with each other), the gel should swell to minimise polymer-polymer contacts (Figure 9a). This is the so-called situation of a *good solvent.* If, on the contrary, monomer units are mutually attracting (*poor solvent* case), the gel should collapse (Figure 9b). In this way the number of unfavourable polymer-solvent contacts is minimised.

One can see that the phenomena shown in Figures 8 and 9 are similar in nature. There is some advantage in studying conformational transitions in the gels. Since all the gel chains are connected in one integrated spatial framework, one can observe the corresponding molecular processes just by direct visual recording of the macroscopic dimensions of the gel sample. The main disadvantage here is connected with the extremely slow equilibration in macroscopic gels: for samples with a size of about 1cm the equilibration time τ may be several days. This can be diminished by using smaller gels: one can show that $\tau \sim L^2$, where L is the linear dimension of the gel sample.

2.2 Concept of the Θ-temperature

Let us consider the simplest model of polymer chain shown in Figure 10. In this model, the chain is represented as a chain of N beads each of volume v on an immaterial filament. This model is not very realistic, although it would become so if only a small fraction of monomers on a real chain were subject to volume interactions, so that each was separated by several persistence lengths from its neighbours along the chain. This allows one to assume a short stretch of gaussian chain between successive beads, whose mean-square

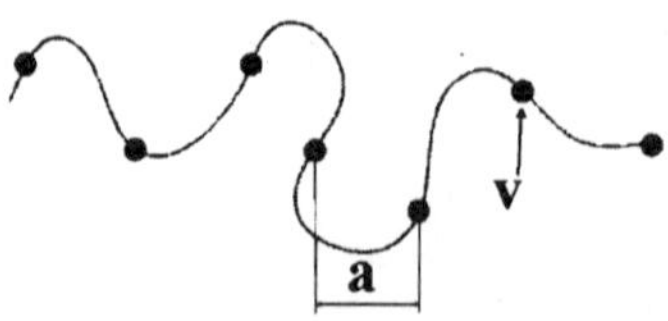

Figure 10. *Model of beads on a Gaussian filament.*

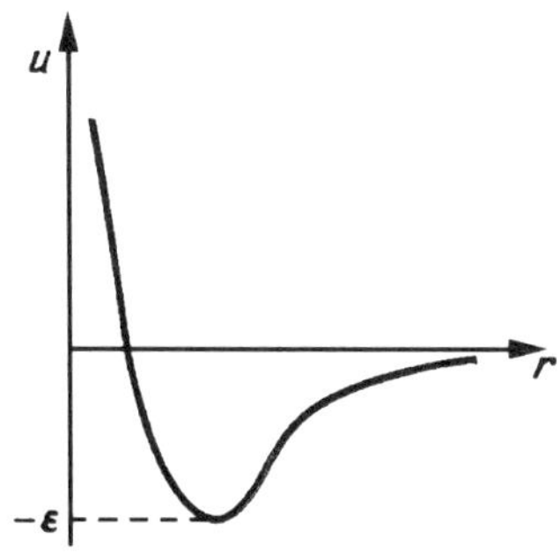

Figure 11. *Typical interaction potential for the model of beads.*

end-to-end distance is denoted a^2.

The beads interact with a pair potential $U(r)$. A typical 'Lennard-Jones type' potential is shown in Figure 11. The dependence of $U(r)$ at small values of r corresponds to the repulsion due to the hard-core volume of the beads (excluded volume) and the tail at large r describes the Van-der-Waals attraction between the beads. The characteristic energy scale of this potential can be estimated as its value at the minimum ε. It is easy to see that:

1. At high values of T, $\varepsilon/k_BT \ll 1$ and only repulsion matters. The coil should swell with respect to the ideal dimensions; this phenomenon is called the *excluded volume effect*. In this case the so-called *swelling coefficient* of the coil, α, is larger than unity:
$$\alpha^2 = \frac{\langle R^2 \rangle}{\langle R^2 \rangle_0} > 1\,. \tag{9}$$
In the history of polymer science the dependence of $\langle R^2 \rangle$ (or α^2) on the number of monomer units, for chains with excluded volume, has been studied in detail by experiment, theory and simulation. Computer experiments (e.g. for lattice models) give in three dimensions $\langle R^2 \rangle \sim N^{6/5}$, *i.e.* $\alpha^2 \sim N^{1/5}$. Therefore, the excluded volume effect is very significant, it even changes the character of the dependence of $\langle R^2 \rangle$ on N from that of ideal chains ($\langle R^2 \rangle \sim N$).

2. At low values of T, such that $\varepsilon/k_BT \gg 1$, attraction dominates. The coil should shrink and form a condensed globule (the coil-globule transition).

3. At intermediate values of T, the effect of repulsion and attraction should *compensate* each other and the coil should adopt ideal-chain (unperturbed) size. This happens at the so-called Θ-*temperature.*

Let us consider the concept of Θ-temperature in more detail. The free energy of a coil is the sum of energetic and entropic contributions:

$$F = E - TS\,. \tag{10}$$

The entropy of an ideal coil can be written in the well-known Flory form (see McLeish, this volume):

$$S = -\frac{3k_B R^2}{2a^2 N} \,. \tag{11}$$

As to the energy E, contrary to the case of ideal coil, for the present case it is not equal to zero (or to a constant).

To write down the expression for E we should note that the concentration of monomer units inside the polymer coil $n \simeq N/R^3$ is very small for long chains (proportional to $N^{-1/2}$ if they are nearly ideal). Therefore, we can write the expression for E as an expansion in powers of n:

$$E = Nk_B T(Bn + Cn^2 + \ldots) \,. \tag{12}$$

This is the *virial expansion*, and the coefficients B, C, ... are the *virial coefficients.* From the statistical physics of nearly-ideal gases, we know that coefficients B, C, ... describe binary, ternary, *etc.* interactions of monomer units, and that they can be expressed in terms of the interaction potential $U(r)$. For example, the second virial coefficient $B(T)$ is given by

$$B(T) = \frac{1}{2}\int [1 - \exp(-U(\mathbf{r})/k_B T)]\, d^3 r \,. \tag{13}$$

The typical behaviour of $B(T)$ is shown in Figure 12. At high temperatures $B \sim v > 0$, where v is the volume of a monomer unit. When the temperature is lowered the value of B decreases due to the contribution of the second term in the integrand of Equation 13 and

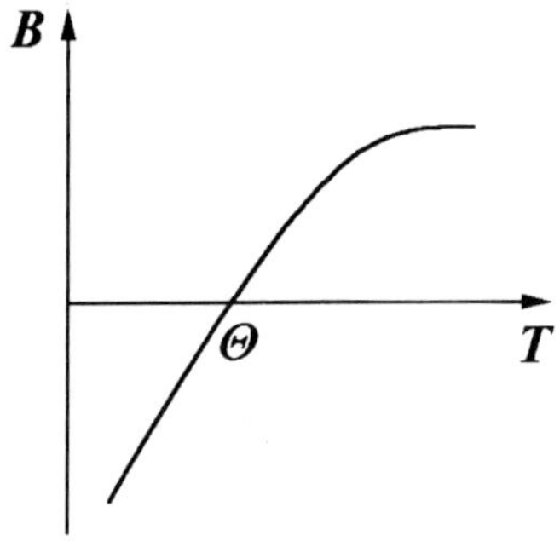

Figure 12. *The characteristic dependence of second virial coefficient on temperature.*

finally at some $T = \Theta$ we have $B(\Theta) = 0$. (In gases, this is called the Boyle temperature.) Since $B(T)$ is approximately linear through the Θ-temperature, we may approximate in this region

$$B \simeq v\tau, \qquad \tau = \frac{T - \Theta}{T} \,. \tag{14}$$

Since for large values of N the concentration n inside the coil is very small, we can with very good accuracy retain only the first term of the virial expansion (Equation 12):

$$E = Nk_B TBn \,. \tag{15}$$

But at the temperature $T = \Theta$ the value of B is equal to zero, therefore $E = 0$, thus $F = -TS$, and the chain adopts the conformation of an ideal coil. At $T > \Theta$ the energy of interaction, Equation 15, is positive, repulsion dominates and the coil should swell. According to the terminology introduced above, this is the regime of *good solvent.* On the other hand, at $T < \Theta$ the free energy of interactions is negative, which means that the attraction prevails and therefore a coil shrinks to a globule. Thus, $T < \Theta$ is the regime of *poor solvent.* In this case it is necessary to go beyond Equation 15 and include higher terms.

Note that the complete compensation of interactions at the Θ-point is a *specific polymer property* (not valid for gases) connected with low polymer concentration in the coil. Only for polymer coils (in at least three dimensions) can the third and higher virial coefficients can be neglected at the Θ-point itself.

2.3 The excluded volume problem

Let us consider the polymer coil far above the Θ-point in the good solvent region, and let us calculate the swelling of the coil by excluded volume. The free energy for this system can be approximated in the form

$$F = E - TS = Nk_BTBn - TS = Nk_BTB\frac{N}{\left(\frac{4}{3}\pi R^3\right)} + \frac{3k_BTR^2}{2Na^2} + \text{const}\,. \qquad (16)$$

Where for the second term the Flory expression for the entropy of an expanded coil was used (see Equation 11). The excluded volume repulsion (first term) induces the coil swelling, while entropic elasticity (second term) opposes it. Minimisation of F with respect to R gives the equilibrium coil size. Setting $\partial F/\partial R = 0$ gives (omitting all numerical coefficients)

$$-\frac{k_BTBN^2}{R^4} + \frac{k_BTR}{Na^2} = 0\,. \qquad (17)$$

Therefore,

$$R \simeq (Ba^2)^{1/5} N^{3/5} \simeq (va^2)^{1/5} N^{3/5}\,. \qquad (18)$$

The last equation may be written in the second form because at high temperatures $B \simeq v$. It follows that

$$\alpha \simeq R/N^{1/2}a \simeq (v/a^3)^{1/5} N^{1/10} \gg 1\,. \qquad (19)$$

Equations 18 and 19 are in agreement with the results of computer experiments for the lattice models mentioned earlier, and also with real experimental data on dilute swollen chains. (Much more sophisticated theories are possible, but these barely change the $R \sim N^{3/5}$ dependence: see, *e.g.*, Grosberg and Khokhlov 1994.) Therefore, we conclude that the polymeric coil swells due to the excluded volume, in spite of the extremely low polymer concentration in the coil. This illustrates the *high susceptibility* of long polymer chains to any perturbation, in particular to the excluded volume interactions.

2.4 Coil-globule transition

Now let us consider the whole range of temperatures below the good solvent region of the previous section. When the temperature is lowered below the Θ-point, the coil-globule transition (or polymer collapse) should take place as is depicted in Figure 8. The interest in the globular form of macromolecules was initially aroused by molecular biophysics, since most protein-enzymes are polymeric globules. Denaturation of globular proteins was sometimes considered to be analogous to the transition from globule to coil.

The theory of the coil-globule transition developed by Lifshitz et al. (1978) gives the following results:

1. The coil-globule transition takes place in the vicinity of the Θ-point, in fact at $-\tau \simeq a^3/(vN^{1/2}) \ll 1$. This is only very slightly lower than the Θ-temperature. It is enough to have a *very weak attraction* to induce the transition into a globule, in contrast to the case of condensation in gases. The reason is that, due to the chain connectivity, independent motion of monomer units is impossible: a polymer coil is *poor in entropy* compared to a gas (or a solution) of its monomers.

2. For a rigid chain the collapse transition is discrete, while for flexible chains it is continuous.

3. For a negative second virial coefficient, $B < 0$, the globular state is stable and

   (a) The size of the globule $R \sim N^{1/3}$ (cf. with $R \sim N^{1/2}$ for ideal coil and $R \sim N^{3/5}$ for the coil with excluded volume). So the monomer density within a globule is independent of N.

   (b) In the globule far from the Θ-point ($|\tau| \simeq 1$) the volume fraction of monomer units is generally not small. This is a *dense liquid droplet.*

   (c) The globule *swells* as the Θ-point (and the coil-globule transition point) is approached, so that a description in terms of only B and C (the second and third virial coefficients) becomes valid in the vicinity of the transition point.

4. Experimentally, the coil-globule transition was observed for many polymer-solvent systems. A very convenient system is polystyrene in cyclohexane, since the Θ-temperature for this case corresponds to $\Theta = 35°C$.

   The main difficulty for the experimental observation of the coil-globule transition of individual chains is the possibility of intermolecular aggregation and formation of a precipitate. To avoid this, the concentration of polymer in the solution should be very small (*e.g.* for polystyrene-cyclohexane system it should be less than 10^{-4}g/l).

3 Statistical physics of polyelectrolyte systems

3.1 Introduction

Polyelectrolytes are polymer molecules having charged units. A monomer link can be charged following dissociation which results in a charged unit and a low molecular weight *counterion.* The number of counterions is then equal to the number of charged units. As a rule, dissociation occurs when molecules are dissolved in highly polar solvents of which water is the most important one (dielectric constant $\varepsilon \approx 81$).

Polyelectrolytes are classified into strongly and weakly charged ones. In strongly charged polyelectrolytes, every link (or a considerable fraction of the links) carries a charge. Therefore, Coulomb interactions among monomers predominates over the non-electrostatic (e.g., Van der Waals) interactions. Charged macromolecules with a small fraction of charged units are called weakly charged polyelectrolytes. For weakly charged polyelectrolytes a nontrivial competition between Coulomb and non-electrostatic interactions takes place.

It is normally assumed that the potential arising from Coulomb interactions involving two elementary charges e separated by a distance r is given by the *Debye-Hückel potential*

$$U(r) = \frac{e^2}{\varepsilon r} \exp(-r/r_D)\,, \tag{20}$$

where ε is the dielectric constant of the solvent and $r_D = (\varepsilon k_B T/4\pi n e^2)^{1/2}$ is the so-called *Debye-Hückel radius*, or screening length. Here n is the total concentration of low molecular weight ions in the solution, including not only dissociated counterions, but also the counterions and coions of any low molecular weight salt that has been added. The screened Coulomb potential (20) is the main result of the Debye-Hückel theory (Landau and Lifshitz 1980). However, this theory is only applicable if the electrostatic effects are sufficiently weak (see below); specifically, the expression for r_D assumes that the Coulomb potential experienced by the counterions and coions is small compared to $k_B T$.

3.2 Main complications in the theory of polyelectrolytes

Compared with the theory of neutral polymers, polyelectrolytes pose complications:

1. There are additional parameters (linear charge density of the chain, salt concentration, pH *etc.*) which essentially influence the polyelectrolyte behaviour.

2. Coulomb interactions are generally not weak (so that Debye-Hückel approximation may be not valid). This is usually a problem for strongly charged polyelectrolytes. The most important new effect emerging as a result of this fact is the phenomenon of counterion condensation (see below).

3. In addition to screening of Coulomb interactions due to point-like ions there is also screening by extended polymer chains themselves. This complicates the character of electrostatic interactions in polyelectrolyte systems.

4. Interplay of Coulomb and Van-der-Waals interactions for weakly charged polyelectrolytes can lead to the formation of regular nanostructures with different morphology (spherical micelles, cylinders, lamellae, *etc.*) controlled by slight modulation of external parameters (salt concentration, pH, temperature, *etc.*).

These complications can be addressed using several basic concepts. Among these, the most important are: counterion condensation, the electrostatic persistence length, and translational entropy of counterions. These concepts will be considered in the following sections.

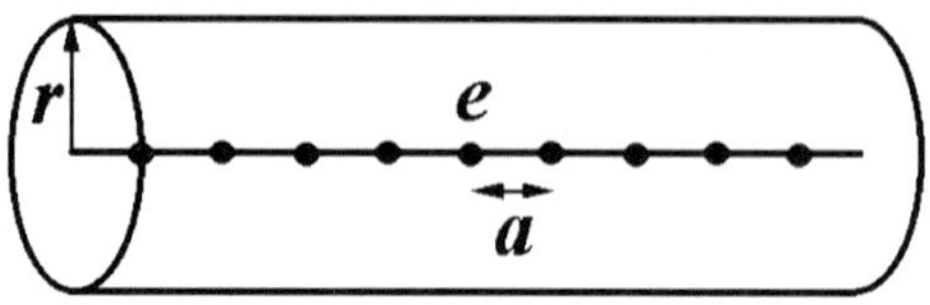

Figure 13. *Illustration of the phenomenon of counterion condensation.*

3.3 Counterion condensation: Manning mechanism

The condensation phenomenon can be explained as follows. Let us suppose that we have a charged line (polymer chain) of charges e separated by a distance a along the chain (Figure 13). The linear charge density of such a chain is $\rho = e/a$. Let us assume that some counterion experiences the field of this charged line. The question is, whether it is thermodynamically favourable for such a counterion to approach the charged line (minimising its electrostatic energy) or whether the opposite option is preferable (maximising the entropy of translation).

Let us suppose that in some initial state the counterion is confined to a cylinder of radius r_1 around the charged line (see Figure 13), while in a proposed 'final' state it is confined within a cylinder of radius $r_2 > r_1$. The gain in the free energy of translational motion is then

$$\Delta F_1 \simeq k_B T \ln \frac{V_2}{V_1} \simeq k_B T \ln \frac{r_2}{r_1}, \qquad (21)$$

where V_1 and V_2 are the volumes of the initial and final cylinders. On the other hand, the increase of the radius of the cylinder leads to a decrease of the average energy of attraction of counterions to the charged line. According to the formulae of electrostatics this decrease is

$$\Delta F_2 \simeq -e\frac{\rho}{\varepsilon} \ln \frac{r_2}{r_1} \simeq -\frac{e^2}{\varepsilon a} \ln \frac{r_2}{r_1}. \qquad (22)$$

One can see that both ΔF_1 and ΔF_2 are proportional to $\ln(r_2/r_1)$. Therefore the net result depends on the coefficients before the logarithm. If

$$u \equiv \frac{e^2}{\varepsilon a k_B T} < 1, \qquad \text{then} \qquad \Delta F_1 > |\Delta F_2| \qquad (23)$$

and this means that the gain in translational entropy is more important; the *counterion goes to infinity.* On the other hand, if

$$u > 1, \qquad \text{then} \qquad |\Delta F_2| > \Delta F_1 \qquad (24)$$

and the counterion should approach the charged line and *condense* on it.

Now we take the second, third and further counterions, and repeat for them the above considerations. As long as the linear charge density ρ_{eff} of the original charges plus its condensed counterions satisfies the inequality $(\rho_{\text{eff}} e/\varepsilon k_B T) > 1$ (compare Equation 24) the counterions will condense on the charged line, decreasing ρ_{eff}. When the number of condensed counterions neutralises the charge of the line to such an extent that

$$u_{\text{eff}} = \frac{\rho_{\text{eff}} e}{\varepsilon k_B T} = 1 \qquad (25)$$

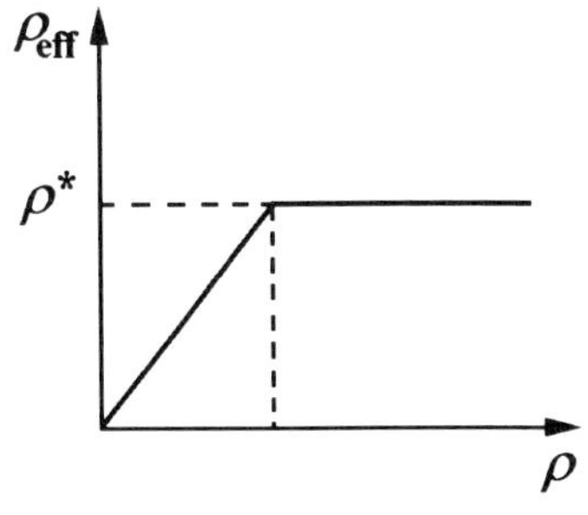

Figure 14. *The dependence of the effective charge on the line as a function of its initial charge.*

the condensation of counterions stops. All the remaining counterions are floating freely in the solution. The plot of the effective linear charge of the line as a function of initial linear charge is shown in Figure 14.

One can see that in the presence of counterions there is a threshold

$$\rho^* = \frac{\varepsilon k_B T}{e} \tag{26}$$

such that it is impossible to have a charged line with an effective linear charge density above this threshold.

The effect of condensation of counterions should be always taken into account in the consideration of polyelectrolytes in solution. Especially for strongly charged polyelectrolytes, some fraction of the counterions is always condensed and neutralises partially the strong electric field of the charged chain. Only the remaining fraction is free to participate in the Debye screening of electrostatic interactions according to the formula (20).

3.4 Electrostatic persistence length

The Coulomb interaction stiffens the chain of a strongly charged polyelectrolyte, that is, it leads to an increase in the persistence length l of the chain from l_0 to a new value $l_0 + l_e$. The contribution l_e is called the *electrostatic persistence length.*

That the Coulomb interaction leads to an effective renormalisation of the persistence length can be illustrating using Figure 15, in which a chain is shown for the case $a \ll r_D \ll l$. (This corresponds to a strongly charged polyelectrolyte, with a moderate concentration of a low-molecular-weight salt in the solution.) In this case two types of Coulomb interaction are possible:

1. Interaction between the charges separated by a distance $< r_D$ along the chain. (This is a short range repulsion, tending to increase the persistence length.)

2. Interaction between the charges separated by a distance $> l$ along the chain. (Such charges approach one another closer than the distance r_D as a result of random bending of the chain; their interaction should naturally be classified with the volume interaction.)

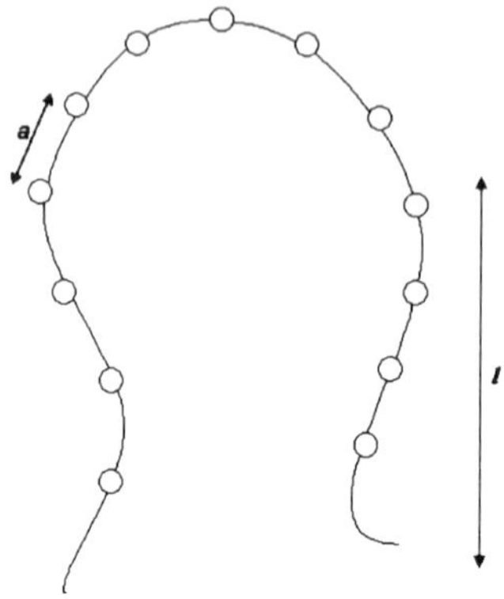

Figure 15. *Persistence length renormalisation in a polyelectrolyte chain.*

This subdivision into short-range and volume interactions is quite unambiguous for $r_D \ll l$, because links separated by a distance exceeding r_D but less than l can neither interact directly (because of the Debye screening) nor draw together as a result of chain bending. Thus, for $a \ll r_D \ll l$, it is the short-range part of the Coulomb interaction that brings about an increase in the persistence length.

To determine the quantity l_e, let us first derive the persistence length for an uncharged persistent chain from the energy of slight bending of a rod-like section of the chain. The bending energy of a section having length L described by a tangent $\mathbf{u}(s)$ (Figure 7) can be written as

$$\Delta F_0 = \frac{b}{2}\int_0^L |\bar{\boldsymbol{\rho}}(s)|^2 ds, \qquad \bar{\boldsymbol{\rho}}(s) = \frac{\partial \mathbf{u}}{\partial s}, \tag{27}$$

where b is an effective bending modulus for a unit length of the chain, and $\bar{\boldsymbol{\rho}}(s)$ is the curvature of the chain fragment as a function of contour length s (with $0 < s < L$). Because the filament can be assumed to bend with an approximately constant radius of curvature, $|\bar{\boldsymbol{\rho}}| \approx \varphi/L$ (with φ the total angular deflection) we can find from Equation 27 that $\Delta F_0(L, \varphi) \approx b\varphi^2/2L$. The elastic persistence length l_0 can be defined according to

$$\Delta F_0(L \simeq l_0, \varphi \simeq 1) \simeq k_B T\,, \tag{28}$$

i.e. $l_0 = b/k_BT$; it is proportional to the bending modulus and grows as the temperature decreases.

In the charged chain, an additional repulsion of the links (20) occurs so that the bending energy cost increases:

$$\Delta F = \Delta F_0 + \Delta F_e \simeq \frac{b'}{2}\frac{\varphi^2}{L}, \qquad l = \frac{b'}{k_BT} = l_0 + l_e \tag{29}$$

and as a consequence the persistence length is increased, too. The expression for the electrostatic persistence length l_e has the form (Odijk 1977)

$$l_e = \begin{cases} ur_D^2/(4a), & \text{at} \quad u < 1 \quad \text{(no condensation)} \\ r_D^2/(4ua), & \text{at} \quad u > 1 \quad \text{(condensation takes place).} \end{cases} \tag{30}$$

Taking into account that for typical cases $u \sim 1$ and $r_D \gg a$ (if the salt concentration is not very high), we reach the conclusion that $l_e \gg r_D$, so that the stiffening of the polymer chain because of electrostatic interactions occurs on length scales much larger than the Debye radius r_D (despite the fact that this interaction only acts over a radius r_D). In many cases, provided that the corresponding uncharged chains are not too stiff, we obtain $l_e \gg l_0$ so that the electrostatic contribution to the persistence length prevails.

Note that in the regime of counterion condensation, the quantity l_e is independent of the linear charge density $\rho = e/a$ of the polymer chain, as it should be, because the charge density in excess of e/a is compensated by the counterions condensing onto the chain.

3.5 Ionic gels: role of translational entropy of counterions

A schematic representation of a polyelectrolyte gel is shown in Figure 16. It consists of polyelectrolyte molecules cross-linked by covalent chemical bonds into an integrated spatial network swelling in a solvent (usually in water).

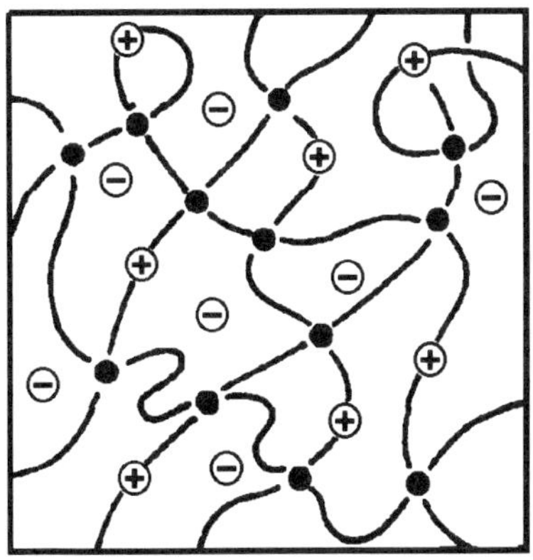

Figure 16. *Schematic picture of a polyelectrolyte gel.*

Figure 16 shows that a polyelectrolyte gel contains counterions together with charged monomer units that together ensure that a macroscopic gel sample as a whole is electrically neutral. When a gel swells in a large volume of water, it appears advantageous for the counterions to abandon the network and go out of the gel into the surrounding solution, because it would lead to substantial translational entropy gain. However, this is not the case since the *principle of electro-neutrality* of macroscopic gel samples would be violated. So, the counterions have to remain inside the network where they exert an *osmotic pressure.* This counterion pressure is responsible for two of the most important physical effects taking place in polyelectrolyte gels swelling in water.

Firstly, the osmotic pressure results in very considerable gel swelling: one gram of a dry polymer may consume hundred of grams of water absorbed by the gel. This allows polyelectrolyte gels to be used as so-called *super-absorbents* of water. Uses include diapers, retention of soil moisture, dust immobilisation, *etc.*

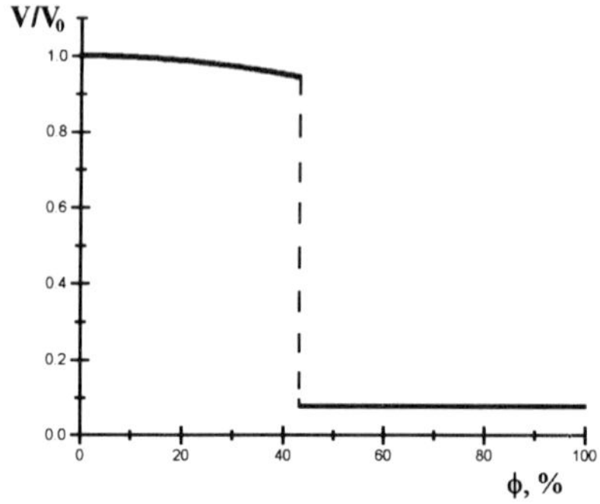

Figure 17. *The dependence of the volume of a polyelectrolyte gel V on the volume fraction ϕ of poor solvent added to water. V_0 is the volume corresponding to the gel swollen in pure water.*

Secondly, the excessive swelling of polyelectrolyte networks in water explains their sharp contraction upon deterioration of solvent quality and leading to a several-hundred-fold jumpwise decrease in the gel volume (see Figure 17).

This phenomenon is called *gel collapse.* It is connected with the coil-globule transition of the polymer chains in a gel, translated into collapse of a gel as a whole. The stronger the charge of the gel, the sharper is the collapse. This is easy to understand since the collapsed phase is stabilised by attractive forces between uncharged units. Therefore, the collapsed gel volume displays only weak charge-dependence whereas the volume of swollen gel grows considerably with increasing charge due to osmotic pressure of counterions. Thus the amplitude of the jump in volume between these two states should increase with the increase of the degree of charging of the gel chains.

3.6 Possible states of counterions in a polyelectrolyte network

The properties of weakly charged ion-containing polymer networks differ essentially depending on the state of the counterions.

First, depending on the surrounding medium (*e.g.* its dielectric constant) counterions can either be free, or stay bound to the chains forming ion pairs with the corresponding ions on the polymer chain, Figure 18. The first regime is called the polyelectrolyte regime and was considered in the previous section. In this case the counterions can move freely (independently of the polymer chains) and the properties of the gel are essentially determined by the translational entropy of counterions.

The second regime, when counterions stay closely bound to the chain, is called the *ionomer* regime. This corresponds to counterion condensation, except that we allow for the presence of discrete charges on the chain backbone instead of treating it as simply a charged line. (A counterion binding to an opposite charge on the backbone forms an *ion pair.*) This regime is realised by decreasing the dielectric constant ε (increasing u), which can be achieved by increasing of the volume fraction of polymer in the system: usually the dielectric constant is a decreasing function of polymer concentration in water solutions.

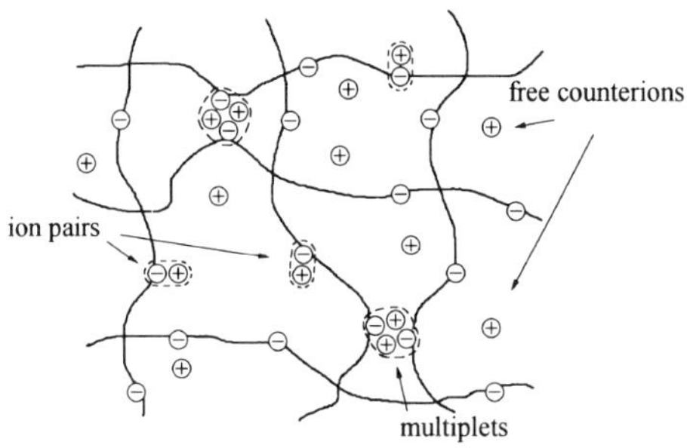

Figure 18. *Possible states of counterions in a polyelectrolyte network.*

The formation of ion pairs in polyelectrolyte gels leads to two simultaneous effects:

1. Due to the decrease of the concentration of mobile counterions inside the gel the corresponding osmotic pressure also decreases; this fact favours the formation of collapsed phase.
2. Ion pairs attract each other due to the dipole-dipole interactions, and form so-called *multiplets* which act as additional physical cross-links.

Ion pairing with subsequent aggregation of ion pairs into multiplet structure leads to the appearance of a new super-collapsed state of the gel. In contrast to the ordinary collapsed state of gels in poor solvents, the super-collapsed state corresponds to the state of a practically dry gel (Khokhlov and Kramarenko 1996).

3.7 Truly free versus trapped counterions in solutions and gels

The inhomogeneous spatial distribution of the immobilised charges on polymer chains creates potential wells for the counterions. As a result some of the counterions becomes trapped, and only a subset of them remain 'truly free'. The simplest example of such inhomogeneous systems is the dilute polyelectrolyte solution, Figure 19.

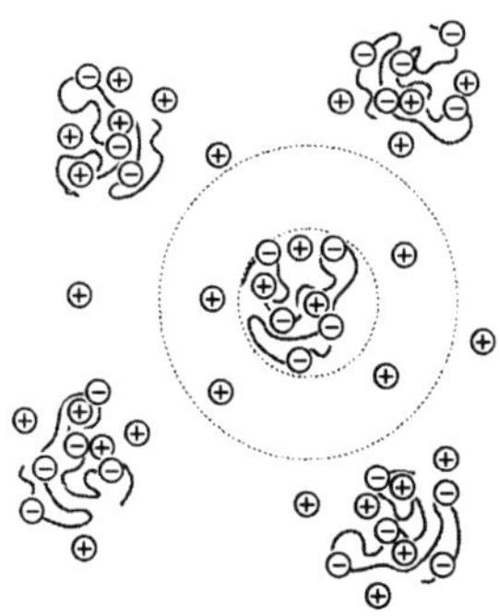

Figure 19. *Dilute polyelectrolyte solution as an example of an inhomogeneous polyelectrolyte system.*

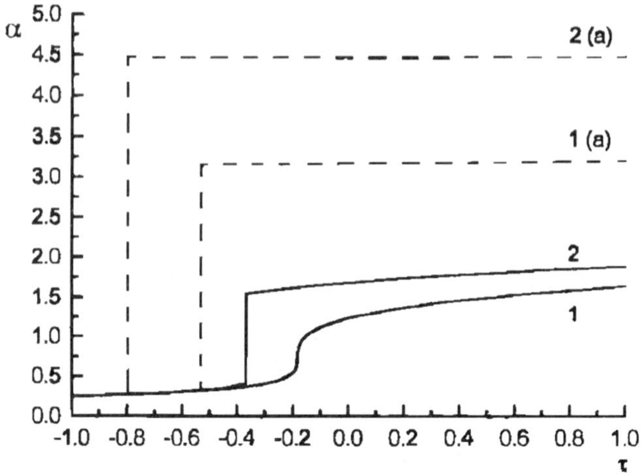

Figure 20. *Gel collapse curves for individual polyelectrolyte macromolecules (solid lines) and for macromolecules in the gel (dotted lines).*

The conformational behaviour of each individual molecule in the solution is determined mainly by the balance between the following physical factors. The presence of some fraction of counterions within the molecules, as well as non-screened charges on the chain, exerts osmotic pressure leading to the swelling. On the other hand, the elasticity of the chain and the osmotic pressure of truly free counterions (those moving in the solution exterior to the chain) oppose the swelling. The interaction of neutral monomeric links can change the balance between swelling and collapse of the chain, depending on the solvent quality. The swelling ratio α as a function of relative temperature deviation from the Θ-point, τ, is presented in Figure 20 (Kramarenko *et al.* 1997). In the good solvent regime, $\tau > 0$, the fraction β of counterions retained within the coil is close to zero; the condition of electro-neutrality of a single chain is totally violated, and the conformation of the chain is determined by electrostatic repulsion between charges along the chain. For comparison, the curves for $\beta = 1$ (chain swelling in gel) are shown by dotted lines. In this case the main reason for the increase of α is the osmotic pressure of the counterions. On the other hand, in a poor solvent, $\tau < 0$, the collapse of the chain induces localisation of some of the counterions within the chain to compensate the increasing charge density of the globule. However, the swelling ratio is determined mainly by non-Coulomb interactions of uncharged monomer links, and α does not depend strongly on the value of β in this regime.

Another polyelectrolyte system where inhomogeneities play an important role is a charged gel with the electric charge unevenly distributed in space. This inhomogeneity is an inherent property of most charged gels, connected with the statistical nature of distribution of subchain lengths and crosslink points (and therefore of the local polymer density). Experiments show that osmotic pressure and the swelling ratio of an inhomogeneous gel is less than that of a homogeneous gel, and this difference increases with increasing inhomogeneity (this can be studied, for example, by decreasing monomer concentration during gel synthesis).

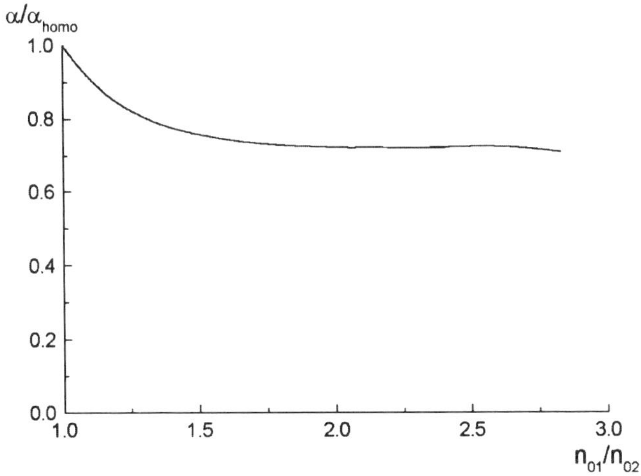

Figure 21. *Comparison of the swelling ratios of an inhomogeneous gel* α *and homogeneous gel* α_{homo} *as function of the degree of gel inhomogeneity.*

The explanation of this effect is the fact, already mentioned, that an inhomogeneous profile of polymer concentration creates also an inhomogeneous profile of immobilised charge. In this way, electrostatic potential wells are formed, and some of the counterions are trapped in these potential wells and do not contribute to the osmotic pressure. Such counterions can be called *osmotically passive*, since only the truly free counterions create the osmotic pressure. The fraction of osmotically passive ions can be rather high, and consequently the osmotic pressure and the swelling ratio of an inhomogeneous gel can be significantly smaller than those of a homogeneous gel.

One should note that there are several mechanisms leading to the existence of osmotically passive ions. The traditional mechanism is the Manning condensation of counterions on charged network subchains, described in Section 3.3 above, when the condensed ions are confined to the molecular vicinity of the polymer chains, and obviously are osmotically passive. Another mechanism is counterion trapping in large-scale inhomogeneities, as just described. In the first approximation, one can assume that the Manning condensation simply renormalises the charge of the polymer chains (or the fraction of charged monomer units), and then the remaining mobile ions can be partitioned between osmotically active and osmotically passive ones.

A simple theory of collapse of inhomogeneous gels taking into account the existence of osmotically passive ions has been developed by Zeldovich and Khokhlov (1999) within the framework of a two-phase model similar to those used for single polyelectrolyte chains and microgels. Figure 21 compares the swelling ratio of an inhomogeneous gel α to the swelling ratio of a similar homogeneous gel α_{homo} as a function of the degree of gel inhomogeneity. The degree of inhomogeneity is defined here as the ratio of a maximal $n_{0,1}$ and minimal $n_{0,2}$ local density of gel. We see that the higher is the inhomogeneity, the less the inhomogeneous gel is swollen with respect to a homogeneous gel. This is a manifestation of counterion trapping: the higher is the inhomogeneity, the deeper are the potential wells, the more counterions they trap, and the less the gel swells.

4 Conformation-dependent sequence design (engineering) of AB-copolymers

4.1 Introduction

The history of the polymer industry in the twentieth century shows a shift in main emphasis from the development of polymers as construction materials, in the beginning and in the middle of the century (1920-1980), to the development of polymers as functional materials. The best known applications of such an approach are super-absorbents, systems for controlled drug release, polymer electrolytes for lithium batteries, polymer membranes with enhanced selectivity, polymer adhesives *etc.* In the last ten years the so-called 'smart' polymeric materials have gained increasing attention (responsive gels as soft manipulators, polymer sensor systems, field-responsive polymer materials *etc.*). For these functional polymer systems the functions are becoming more complex and diverse; many involve *copolymers* in which different monomers are of different chemical species. If now we ask ourselves which of the known polymers perform the most complex and diverse functions, the answer is clear: these are biopolymers (*e.g.* DNA and globular proteins) which are responsible for extremely complicated functions in living systems. Therefore, if we want to move in that direction for synthetic polymer systems, we must look at the ideas implemented already in nature. Here we discuss one of the ideas of this kind.

We will consider several particular examples of a novel approach to the design of specific primary sequences for copolymer chains where the basic idea came from comparison with real proteins. This approach is based on the concept of *'colouring'* with two 'colours' (A and B) the monomeric units of a homopolymer, taken initially in some well-defined conformation (globular conformation, conformation of an adsorbed chain, *etc.*). The choice of colour depends on the spatial position of the unit in this *'parent'* conformation. Our computer simulations show that copolymers with AB-sequences generated in this way acquire a number of special functional properties, which distinguish them from the AB-copolymers with random or block primary structures. ('Primary' structure refers to the permanent chemical structure along the chain backbone; a 'block' is a stretch of pure A or pure B.) In a sense, we can say that some functional features of the parent conformation are *'memorised'* (or 'inherited') and then manifested in other conditions. This special conformation-dependent AB-sequence design (engineering of AB-copolymers) can in principle be achieved not only in computer simulations, but in the chemical laboratory as well. Further studies in this direction may have an important impact both in the problem of obtaining of AB-copolymers with special functional properties and in the problem of understanding of basic principles of biomolecular evolution at its early stages.

4.2 Protein-like AB-copolymers

The general idea just outlined above will be first illustrated taking as an example a globular conformation of a polymer chain.

The primary structure in real globular proteins is known to be highly specific. Globular proteins/enzymes functioning in living systems are the products of molecular evolution. Their primary structures involve 20 possible types of monomeric units (20 amino-acid residues), therefore globular proteins are much more complicated objects than typical AB-copolymers. However, the most essential distinction between different monomeric unit of proteins is that some of these units are hydrophobic, while others are hydrophilic or charged (Dickerson *et al.* 1969, Grosberg 1997). Thus in a very rough approximation it is possible to represent a globular protein as a kind of AB-copolymer. The spatial (ternary) structure of such a copolymer in the native state would then normally correspond to the structure in which hydrophilic units (A-type) cover the globular surface and prevent different globules from aggregation, while hydrophobic units (B-type) constitute the dense globular core.

Because the hydrophobic links should be in the core and hydrophilic ones should lie on the surface, there should be some long-range correlations within the primary structure, since these correlations are related to the spatial conformation of the globule. It is therefore interesting to ask whether it is possible to have a synthetic AB-copolymer with long-range correlations in primary structure analogous to the correlations in evolved proteins.

So, we are looking for preparation of such an AB-sequence that in the most dense globular conformation, all the B-units form a dense core while all the A-units are on the surface, *i.e.* such a sequence mimics one of the properties of a real protein what allows us to call the desired copolymer chain a *protein-like* AB-copolymer. Such copolymers should have interesting physical properties, for example, they should not precipitate when the dense globular conformation is formed. This is not the case for ordinary macromolecules (Lifshitz 1978, Grosberg *et al.* 1994; see also Section 2.4).

4.2.1 Preparation scheme for protein-like AB-copolymers

It is very easy to prepare such protein-like sequences in computer simulation (Khokhlov *et al.* 1998a,1998b,1998c,1999); the main steps of the real laboratory experiments should probably be the same. A computer realisation of the procedure is illustrated in Figure 22.

Figure 22. *Colouring procedure for preparation of protein-like copolymer.*

We take some particular conformation of a usual homopolymer coil with excluded volume, and switch on a strong attraction of monomeric links to let the chain collapse into the conformation of a homopolymer globule (Figure 22, left part). Next we take an 'instant snapshot' of the globule and assign the colour index A to those units that are on the surface of the globule and call these units hydrophilic, and assign the index B to the units in the core of the globule and call these units hydrophobic. Then we fix this primary structure (Figure 22, middle part; the structure shown in this figure will be further referred to as the parent globule). Finally, the last step is to remove the uniform strong attraction of monomeric units, and to add different interaction potentials for A- and B-units (Figure 22, right part).

4.2.2 Coil-globule transition for AB-copolymers with different structures

Let us consider the coil-globule transition for the protein-like copolymer thus formed, which occurs when one introduces attraction between hydrophobic B-units (hydrophilic A-units remaining repulsive to each other and to hydrophobic units). The Monte Carlo computer simulation was performed using the bond fluctuation algorithm (Carmesin 1988; see also Kremer, this volume). The coil-globule transition was realised upon the increase in attraction interaction between B-links (which is equivalent to a decrease of temperature). We studied primary sequences with an A/B composition ratio of 50/50.

We compared the coil-globule transition for three types of sequence: (i) our protein-like sequences; (ii) random AB-copolymers of the same A/B ratio but an entirely random sequence; (iii) random-block AB-copolymers, with the same A/B ratio and also the same degree of blockiness, *i.e.* with the same mean length of the blocks of A- and B-links as for protein-like copolymers. This comparison was done to distinguish the effects connected with the block lengths (visible in Figure 22) from the effects coming from the presence of long-range correlations between blocks, within the primary sequence. (Such correlations are, by construction, absent in case (iii).)

The data for temperature dependences of the mean energy per monomeric unit and of the specific heat are presented in Figure 23. We have found the transition for protein-like copolymers to occur at higher temperatures and to be more abrupt than that for either random or random-block copolymers (the peak of specific heat is narrower and higher for the protein-like copolymer). The kinetics of the coil-globule transition is also faster for protein-like copolymers (*i.e.* protein-like copolymers form the dense globule faster than random and random-block ones, under similar conditions).

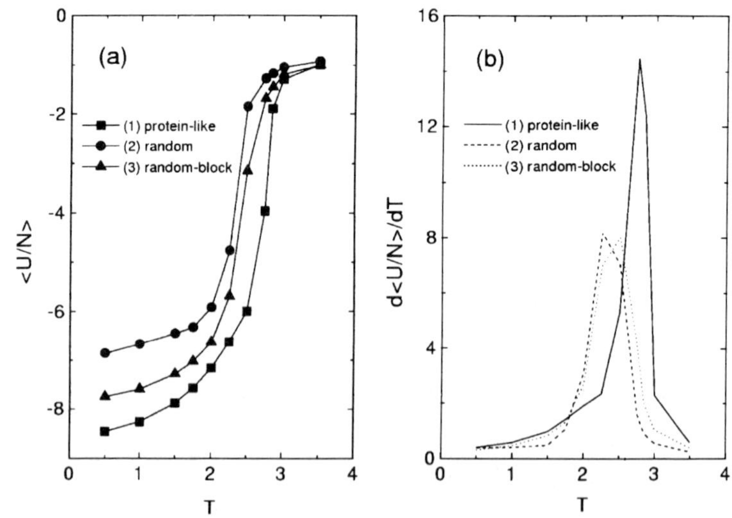

Figure 23. *The temperature dependences of the mean energy per monomeric unit (a) and of the specific heat (b) for copolymers with chain length $N = 512$ and primary sequences as indicated.*

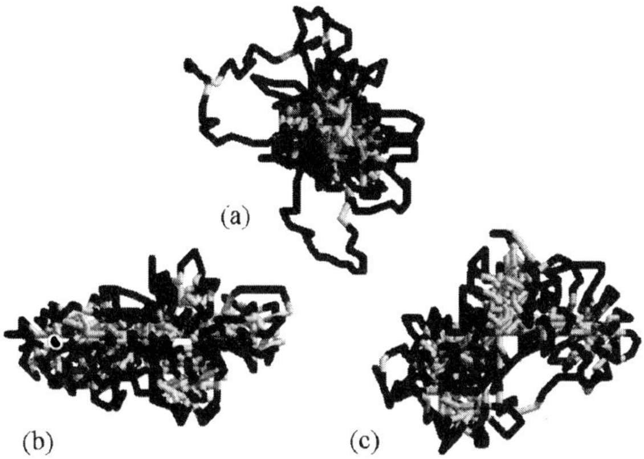

Figure 24. *Typical snapshots of the globular structures for (a) protein-like, (b) random, and (c) random-block copolymers.*

What is the reason for such effects? To answer this question let us look at the morphology of different globules. In Figure 24, typical snapshots are shown for globular conformations obtained for the three types of copolymers at equilibrium at low temperature. (The hydrophilic A-links are shown in a darker colour than the hydrophobic ones.) It can be seen that for protein-like AB-copolymers, practically all B-units are concentrated in the dense core of the globule which is stabilised by long dangling loops of hydrophilic A-links. On the other hand, the core of the globules formed by random and random-block AB-copolymers is much looser, and approximately 30% larger in average size than that for our designed AB-copolymers. A fraction of the hydrophilic A-links are now inside the core, and those of them which belong to the surface form very short dangling loops which apparently are not sufficient to prevent the aggregation of such globules in the solution.

It is reasonable to assume that the formation of the dense core shown in Figure 24 for protein-like copolymer globules is facilitated by the fact that the dense globule pre-existed in the parent conditions shown in Figure 22 (middle part). Since all the B-links in this parent core are fitting next to each other, there is no connectivity obstacle to *reassembly* of most of this core when the effective attraction between the B-units is switched on. In other words, we can say that the protein-like copolymer *inherited* some important features of the parent globule which were then reproduced in the other conditions.

4.3 Membrane-protein-like copolymers

As a second criterion for colouring monomeric units inside a dense homopolymer globule, we have introduced a model for AB-copolymers which mimic some properties of membrane proteins. It is well known that real membrane proteins are located inside the cell membrane in such a way that some fraction of the amino-acid units (mainly the hydrophobic and uncharged ones) are located inside the bilipid layer of the membrane, while the other amino-acid units are located in water environments inside and outside the cell.

In our simple model, we assigned colour index B to monomeric units which lie within

Figure 25. *The parent conformation of the membrane-protein-like copolymer (left part) and the conformation obtained after equilibration in computer simulation (right part).*

the intersection of a parent globule and a narrow flat slab. So, the B-part of the parent conformation takes the form of a narrow disk. We have taken 30% of all links to be of B-type. In the left part of Figure 25 we present a snapshot of an original parent conformation of our AB-copolymer globule. We marked both hemispheres of outer A-links (70% of the whole amount) of the original globule into two different colours (black and grey) to see whether the parent micro-segregated structure can be reestablished after the equilibration procedure.

In our Monte Carlo computer simulation (performed for chains of length of $N = 256$ monomeric units) we have indeed found that such a chain shows the effect of stability of the parent micro-segregated structure. A typical conformation obtained (after the procedure of decollapse, and recollapse under the influence of selective interactions) of the same chain is shown in the right part of Figure 25. A spherical B-core is formed instead of original disk-like B-core which is, of course, natural due to isotropy of the selective interaction potential. But one can see definitely, that the grey units have many more contacts with each other than with the black units and vice versa, *i.e.* the grey and black units are segregated from each other, even though both are of species A. In other words, we can again say that the copolymer chain with a specially designed primary sequence has inherited or memorised some important structural features of the parent globule, which were then reproduced under other conditions.

4.4 ABC-copolymers: proteins with an active enzymatic centre

As the third criterion for preparation of the primary structure of a copolymer chain, we have studied ABC-copolymers prepared by a 'triple colouring' of some particular homopolymer globule in the following way: we assigned colour index A to the surface monomeric units, index B to the inner monomeric units (as was previously described for protein-like copolymers), and index C to those inner monomeric units which lie inside a small sphere whose centre does not coincide, however, with the centre of mass of the parent homopolymer globule (see Figure 26).

Our idea was to prove whether such a parent conformation can be reassembled in the course of an equilibration procedure, for suitable sets of interaction parameters, so as to restore the originally given distance between the centres of the B-core and C-core. This would show whether the position of the C-links within the primary sequence, along with a specially chosen interaction potential, can lead to a stable reconstruction of the spatial conformation of the whole chain.

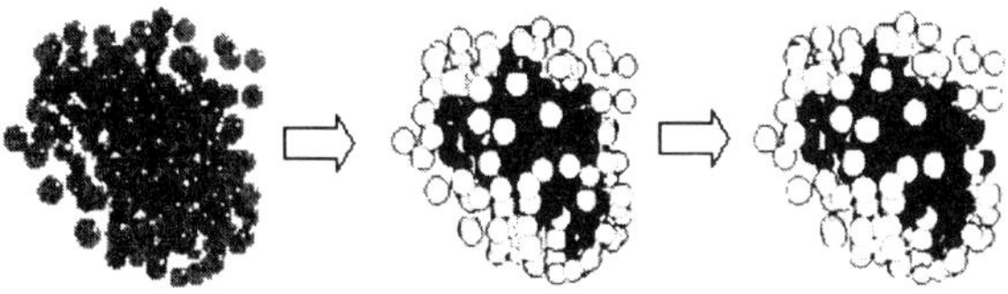

Figure 26. *The colouring procedure for ABC-copolymer modelling proteins with active enzymatic centre.*

We performed computer simulations for a chain of N=256 monomeric units using attractive interaction potentials for B and C-links (the attraction for C-link was taken stronger that that for B-links). We have found in our computer experiment that such ABC- copolymers normally restore their original structure with B- and C-cores both present, although we have not succeeded up to now in finding an interaction potential which would allow us to get the centre of C-core at the same distance from the centre of B-core as in the original conformation. Nevertheless, we have definitely found effective restoration of the 'active centre' (C-core) after the following procedure: we switch off the attraction between C-links and let them dissolve inside the dense B-core. If the attraction between C-links is restored we observe the reassembly of the C-core once again.

4.5 Adsorption-tuned AB-copolymers

Let us now generalise the above idea. The primary structure of protein-like copolymer was generated by a colouring procedure for a homopolymer chain in the globular state. However, special primary sequences can be obtained not only from globulars conformation; *any specific polymer chain conformation can play the role of a parent.*

The simplest example of this kind is connected with the conformation of a homopolymer chain adsorbed onto a plane surface. Let us colour the links in direct contact with the surface in some typical instant snapshot conformation (see Figure 27). This corresponds to the assumption that the surface catalyses some chemical transformation of the adsorbed links. Then we will end up with AB-copolymer for which the sequence design was performed in the parent adsorbed state. After desorption such AB-copolymer will have special functional properties: it will be *tuned to adsorption.*

Indeed, we have performed Monte-Carlo computer experiments along the lines of the sequence design scheme outlined above (Zheligovskaya *et al.* 1999) for chains of 32

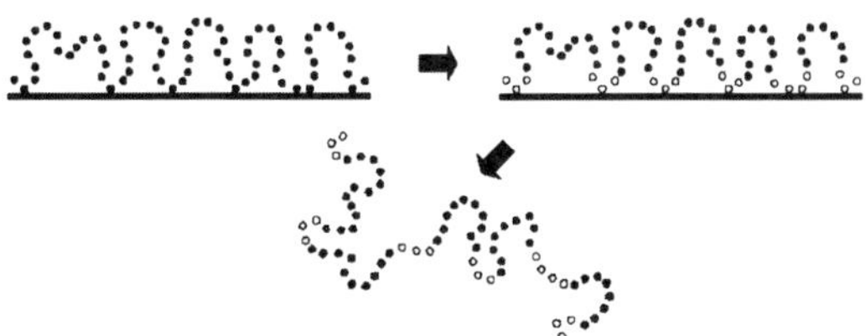

Figure 27. *Preparation of adsorption-tuned primary sequence.*

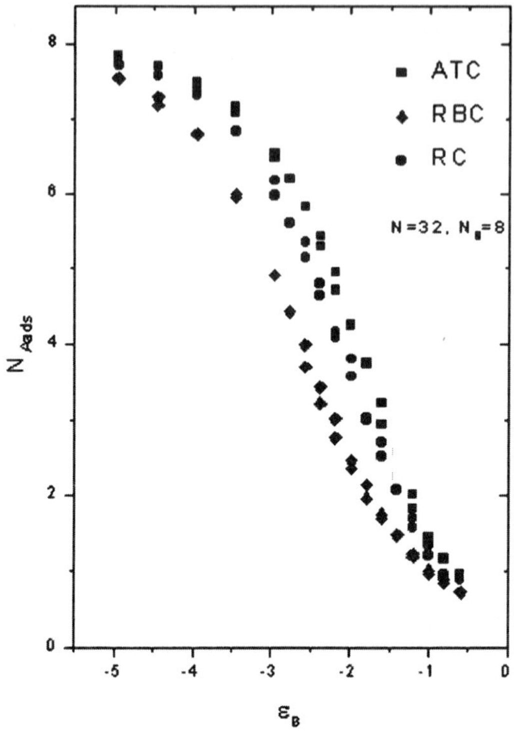

Figure 28. *The average number of adsorbed type-B segments vs. the attraction energy to the surface, ε_B for the adsorption-tuned (ATC), random-block (RBC), and random (RC) copolymer chains of length $N = 32$ with the number of type-B segments $N_B = 8$.*

monomer units. In the conformations of the adsorbed homopolymer chain 8 units which are closest to the surface were identified and denoted as B-units, the others were designated as A-units. Then we studied the adsorption behaviour of the AB-copolymer chain obtained in this way on a plane surface with a specific attraction for B-units, and compared it with the behaviour of the corresponding random and random-block copolymers (for details, see Zheligovskaya *et al.* 1999).

In Figure 28 we plot the average number of adsorbed B-units versus the energy of their attraction to the surface, ε_B. It can be seen that the number of adsorbed segments (at a given value of ε_B) is always highest for the designed AB-copolymers. In other words, due to the memorising of some functional features of the parent conformation, we have indeed obtained an AB-copolymer 'tuned to adsorption' on a plane surface.

4.6 Some generalisations and conclusions

One can imagine the analogous colouring procedure for a chain adsorbed on a small spherical colloidal particle. In this case a copolymer chain with the primary sequence tuned to the absorption of a small droplet of organic solvent, or a colloidal particle of given size, could be obtained. Such a copolymer could be called molecular dispenser. Indeed, when in equilibrium contact with an organic fluid, such a copolymer will absorb a small droplet of this fluid, with a volume approximately equal to the volume of the parent colloidal particle, because for such size of the droplet the maximal number of hydrophobic links will be in contact with the fluid, leading to maximal gain in interaction energy per monomer. If exposed instead to contact with a solution of colloidal particles of different sizes such a molecular dispenser will select the particles of size equal to the parent particle.

In conclusion, we have presented several evidences for the fact that an AB-copolymer chain, with a primary sequence prepared on the basis of a particular conformation of a homopolymer chain by some colouring procedure, preserves a memory of its parent spatial conformation. These memorised features are then manifested under other conditions. Such an interrelation can be regarded as one of the possible mechanisms of molecular evolution: a biopolymer acquires some special primary sequence in the parent conditions and then (in other conditions) uses the fact that primary structure is tuned to perform certain functions.

Acknowledgements

I am grateful to V.A. Ivanov, A.E. Likhtman, I.I. Potemkin, K.B. Zeldovich, A.V. Chertovich, A.A. Lazutin, R.E. Limberger for the help in preparation of this chapter.

References

Carmesin I and Kremer K, 1988, *Macromolecules* **21** 2819.

Dickerson R E and Geis I, 1969, *The Structure and Action of Proteins* (Harper & Row, NY).

Grosberg A Yu and Khokhlov A R, 1994, *Statistical Physics of Macromolecules* (American Institute of Physics, New York).

Grosberg A Yu and Khokhlov A R, 1997, *Giant Molecules: Here, There, and Everywhere...* (Academic Press).

Landau LD and Lifshitz EM, 1980, *Statistical Physics (part 1), 3rd Edn., (Pergamon, Oxford).*

Grosberg A Yu, 1997, *Physics-Uspekhi* **40** 109.

Khokhlov A R and Kramarenko E Yu, 1996, *Macromolecules* **29** 681.

Khokhlov A R and Khalatur P G, 1998a, *Physica A* **249** 253.

Khokhlov A R, Ivanov V A, Shusharina N P, and Khalatur P G, 1998b, Engineering of Synthetic Copolymers: Protein-Like Copolymers, in *The Physics of Complex Liquids*, editors F.Yonezawa *et al.* 155-171 (World Scientific, Singapore).

Khokhlov A R, Ivanov V A, Shusharina N P, and Khalatur P G, 1998c, *Russian Chemical Bulletin*, **47** 855.

Khokhlov A R and Khalatur P G, 1999, *Physical Review Letters* **82** 3456.

Kramarenko E Yu, Khokhlov A R, and Yoshikawa K, 1997, *Macromolecules* **30** 3383.

Lifshitz I M, Grosberg A Yu, and Khokhlov A R, 1978, *Rev. Mod. Phys.* **50** 683.

Odijk T, 1977, *J. Polym. Sci., Polym. Phys. Ed.* **15** 477.

Zeldovich K B and Khokhlov A R, 1999, *Macromolecules* **32** 3488.

Zheligovskaya E A, Khalatur P G, and Khokhlov A R, 1999, *Physical Review E* **59** 3071.

Rheology of linear and branched polymers

Tom McLeish

University of Leeds, UK

1 Introduction to rheology

From our earliest days we explore the physical properties of the world around us, by exciting mechanical deformations in it and observing the response. Dropping ceramic basins on a hard floor, pouring water from one bottle into another, pouring shampoo from a soon-to-be-empty bottle onto anything, turning free-running sand into a sludge by adding water—all these are experiments in the 3-year old's rheology laboratory. When we get older we add some numbers, a few equations, some sophisticated experimental methods and some impressive terminology like 'spectroscopy'. If you still enjoy playing around in sandpits then you might also enjoy those aspects of molecular theory which form the main topic of these lectures.

In the most general terms, rheology is the measurement and study of the relationship between the deformation of a material (measured by its *strain*) and its mechanical response (measured by *stress*). The relation between the strain (or strain history) of the material and its present state of stress is called the *constitutive equation.* A central goal of molecular rheology is to derive such equations from models of the underlying mesoscopic or microscopic physics, and perhaps also predict the results of direct structural measurements on systems under flow. An alternative, pragmatic approach, sometimes suitable for engineering applications, is to look for phenomenological constitutive equations that approximate to the behaviour of a given material. For a more thorough introduction to both approaches, see [1].

1.1 Why is rheology a good probe of soft matter?

Condensed matter falls broadly into two classes: *hard matter* and *soft matter*. Hard matter comprises most metals, ceramics, minerals, and materials below their glass transition temperature. It has the following attributes:

- All bulk deformations, including volume-preserving ones, couple directly to the strain of covalent (or metallic) bonds.
- There is no observable 'viscoelastic' regime of time scales with behaviour intermediate between solid and liquid.
- The material breaks catastrophically at strains that are still very small: typically less than one percent.

On the other hand, soft matter (polymers, gels, colloids, liquid crystals, foams, crystals with defects) reacts to strain in a qualitatively different way:

- Bulk deformations that are volume-preserving, such as simple shear, do not necessarily deform covalent bonds. (Such deformations are locally 'non-affine': bonds can rotate, rather than stretch or bend.) They are offered far less resistance by the material than volume-changing strains.
- Soft materials can maintain large bulk strains (tens or hundreds of percent) without failure, so exhibiting interesting nonlinear response.
- After a deformation, structural equilibrium is recovered by a series of processes spanning a wide range of time scales (*e.g.* entangled polymers).
- Flow itself may induce structural transitions that are non-catastrophic, but continuously or discontinuously change the material properties (*e.g.* flow-aligning block copolymer phases, shear-thinning in polymers, flow-induced nematic transitions in liquid crystals).

In summary, a continuous response or evolution of the structure in soft matter exists over far wider ranges of time scales and strains than in hard matter. This remark applies in particular to volume-preserving strains, which we now examine in more detail.

1.2 Volume preserving and non-preserving strains

In Figure 1 we consider examples of two types of strain that give rise to qualitatively different stress response in soft matter. Weak response arises in volume-preserving deformations (that do not couple to stretch of covalent or metallic bonds), and strong response in volume-changing ones (that do). For our purposes, the weak responses are the interesting ones. In terms of the displaced lengths l and the original side lengths L of the cube of material, both shear strain (left) and bulk compressive strain (right) are given by the dimensionless ratio l/L. The shear stress (force in strain direction per unit area of displaced side) is σ in the shear case, and the bulk stress (normal force per unit area on the strained side) is Δp in the compression case. These are given in terms of the shear modulus (G) and bulk modulus (B) by respectively:

$$\sigma = \frac{Gl}{L}\,, \qquad \Delta p = \frac{Bl}{L}\,. \tag{1}$$

In 'hard' condensed matter, both these moduli arise from distortion of covalent or metallic bonds; estimating the density and stiffness of these we find $G \approx B \approx 10^{11}$Pa.

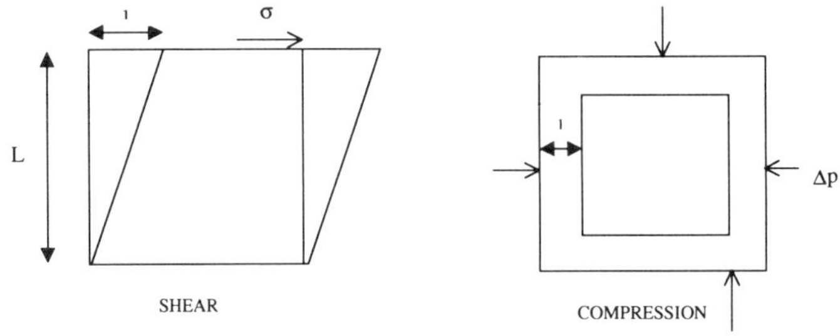

Figure 1. *Shear and compressional strains.*

On the other hand, in 'soft' condensed matter this is only true of the bulk modulus, so that $G \ll B \approx 10^{11}$Pa. Often the shear response is dominated by structures with a characteristic length scale much larger than a covalent bond, whose free energy is dominated by entropy (or else by an even balance of entropy and enthalpy). A good first estimate of G in such cases is $G \simeq k_B T/V_s$, where V_s is the 'structural volume' (or inverse number density) of structures whose principal degree of freedom couples to the shear strain and controls the free energy cost of deformation.

Example: A weak polymer gel has approximately one cross-link per $(10\text{nm})^3$, implying that $G \approx k_B T/10^{-24}\text{m}^3 \approx 500$Pa.

1.3 Strain and strain rate

We now refine and quantify our description of deformation in soft materials. Both strain and stress are 2nd-rank tensor quantities as each relates two vectors. In the case of strain, these two vectors can be chosen to be an embedded spatial vector and its displacement by the deformation. In the deformation of a continuous body, any embedded vector $\mathbf{X}(\mathbf{r})$ is transformed to a new vector $\mathbf{X}'(\mathbf{r})$ (Figure 2).

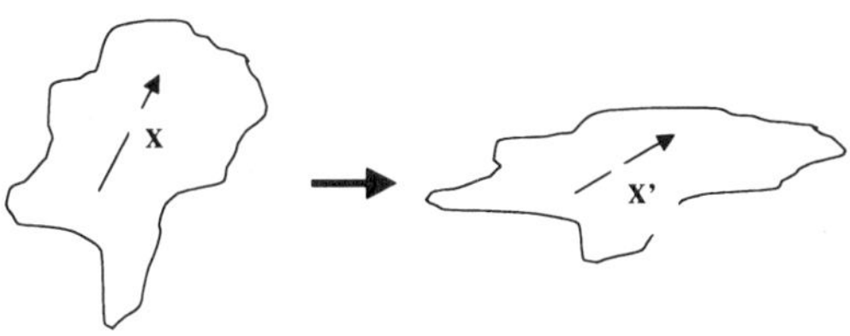

Figure 2. *Illustrating the deformation of a continuous body.*

We define the strain tensor $\underline{\underline{\mathbf{E}}}$ by this transformation, as follows:

$$\mathbf{X}' = \underline{\underline{\mathbf{E}}} \cdot \mathbf{X} \,. \tag{2}$$

For small enough deformations we may write $\underline{\underline{\mathbf{E}}} = \underline{\underline{\mathbf{I}}} + \underline{\underline{\mathbf{e}}}$ so that the field of embedded *displacement* vectors $\mathbf{u}$ obeys $\mathbf{u} \equiv \mathbf{X}' - \mathbf{X} = \underline{\underline{\mathbf{e}}} \cdot \mathbf{X}$. For a uniform deformation we may write this as $\underline{\underline{\mathbf{e}}} = \nabla \mathbf{u}$ or $e_{ij} = \nabla_i u_j$. Volume-preserving deformations have $\det(\underline{\underline{\mathbf{E}}}) = 1$, which becomes, for small displacements, the condition that $\text{Tr}(\underline{\underline{\mathbf{e}}}) = 0$.

If the strain is time dependent, there is a velocity field $\mathbf{v}(\mathbf{r})$ that generates the local deformation *rate* through its spatial gradient $\underline{\underline{\mathbf{K}}} = \nabla\mathbf{v}$. Embedded vectors now change with time, $\mathbf{X}' = \mathbf{X}'(t)$, so that $\mathbf{v}(\mathbf{X}') = \underline{\underline{\dot{\mathbf{e}}}} \cdot \mathbf{X}'$ and

$$\mathbf{X}'(t+\delta t) = (\underline{\mathbf{I}} + \underline{\underline{\dot{\mathbf{e}}}}\,\delta t)\cdot\mathbf{X}'(t) = (\underline{\mathbf{I}} + \underline{\underline{\dot{\mathbf{e}}}}\,\delta t)\cdot\underline{\underline{\mathbf{E}}}(t)\cdot\mathbf{X}\,. \tag{3}$$

But we also have $\mathbf{X}'(t+\delta t) = \underline{\underline{\mathbf{E}}}(t+\delta t)\cdot\mathbf{X}$ by definition of $\underline{\underline{\mathbf{E}}}$. Comparing this with Equation 3 and defining the time derivative of $\underline{\underline{\mathbf{E}}}(t)$ by the usual limit, gives

$$\frac{\partial\underline{\underline{\mathbf{E}}}}{\partial t} = \underline{\underline{\dot{\mathbf{e}}}}\cdot\underline{\underline{\mathbf{E}}}\,. \tag{4}$$

Note that the tensor $\underline{\underline{\dot{\mathbf{e}}}}$ is by definition $\partial\nabla\mathbf{u}/\partial t = \nabla\mathbf{v}$ where $\mathbf{v}(\mathbf{r})$ is the velocity field, defined above. Therefore we can identify $\underline{\underline{\dot{\mathbf{e}}}}$ with $\underline{\underline{\mathbf{K}}}$.

The differential Equation 4 is just a tensorial version of the familiar first-order linear equation $\partial f/\partial t = Kf$, and, in the case where $\underline{\underline{\mathbf{K}}}$ is constant (steady flow), has the solution

$$\underline{\underline{\mathbf{E}}}(t) = \exp(\underline{\underline{\mathbf{K}}}t)\,, \tag{5}$$

where we use the initial condition that $\underline{\underline{\mathbf{E}}}(0) = \underline{\mathbf{I}}$, and where the exponential of a tensor is defined by its series expansion

$$\exp(\underline{\underline{\mathbf{M}}}) \equiv \underline{\mathbf{I}} + \underline{\underline{\mathbf{M}}} + \frac{1}{2!}\underline{\underline{\mathbf{M}}}^2 + \frac{1}{3!}\underline{\underline{\mathbf{M}}}^3 + \ldots.\,.$$

1.3.1 Examples

There are two very important examples of volume-preserving deformations in soft matter, shear and extension. Shear occurs in sliding, or lubricating flows; extension in stretching flows such as the forming of fibres and films.

Shear

A shear flow with velocity along x and gradient along y, of shear rate $\dot\gamma = \partial v_x/\partial y$, has a deformation rate tensor (in Cartesians)

$$\underline{\underline{\mathbf{K}}} = \begin{pmatrix} 0 & \dot\gamma & 0 \\ 0 & 0 & 0 \\ 0 & 0 & 0 \end{pmatrix} \qquad \Longrightarrow \qquad \underline{\underline{\mathbf{E}}}(t) = \exp\left(\underline{\underline{\mathbf{K}}}t\right) = \begin{pmatrix} 1 & \gamma t & 0 \\ 0 & 1 & 0 \\ 0 & 0 & 1 \end{pmatrix}.$$

The shear flow does not generate exponential separation of embedded points (as Equation 5 might suggest) because the displacements of embedded vectors are always orthogonal to the vectors themselves.

Uniaxial extension

In contrast, a uniaxial extensional deformation (as occurs when pulling out a thread, during fibre-spinning for example), with extension along the x-axis and extension rate $\dot\varepsilon = \partial v_x/\partial x$, gives exponential separation of points and has the representation

$$\underline{\underline{\mathbf{K}}} = \begin{pmatrix} \dot\varepsilon & 0 & 0 \\ 0 & -\dot\varepsilon/2 & 0 \\ 0 & 0 & -\dot\varepsilon/2 \end{pmatrix} \qquad \Longrightarrow \qquad \underline{\underline{\mathbf{E}}}(t) = \begin{pmatrix} e^{\dot\varepsilon t} & 0 & 0 \\ 0 & e^{-\dot\varepsilon t/2} & 0 \\ 0 & 0 & e^{-\dot\varepsilon t/2} \end{pmatrix}.$$

Planar extension

There is a second important extensional flow, termed planar extension (as occurs in some film-forming processes). This has a neutral direction in which there is no deformation (in common with shear flow), yet exponential separation of embedded points (in common with extensional flow):

$$\underline{\underline{\mathbf{K}}} = \begin{pmatrix} \dot{\varepsilon} & 0 & 0 \\ 0 & -\dot{\varepsilon} & 0 \\ 0 & 0 & 0 \end{pmatrix} \quad \Longrightarrow \quad \underline{\underline{\mathbf{E}}}(t) = \begin{pmatrix} e^{\dot{\varepsilon}t} & 0 & 0 \\ 0 & e^{-\dot{\varepsilon}t} & 0 \\ 0 & 0 & 1 \end{pmatrix} .$$

In such a deformation (with $\dot{\varepsilon}$ positive) fluid is pulled inward along $\pm y$ and stretched outward along $\pm x$, with z neutral.

1.4 Stress

In deformed matter, forces are transmitted across any surface embedded in the material. The stress, like strain, is tensorial because both the locally transmitted force per unit area, and the local surface element (characterised by its normal) are vectors. We therefore define the stress tensor $\underline{\underline{\boldsymbol{\sigma}}}$ so that the force $d\mathbf{F}$ acting across a small area element dA of unit normal $\mathbf{n}$ is given by (Figure 3)

$$d\mathbf{F} = \underline{\underline{\boldsymbol{\sigma}}} \cdot \mathbf{n}\, dA\,. \tag{6}$$

Alternatively, σ_{ij} gives the i-th cartesian component of the force per unit area across the j-th face of a small cube embedded locally in the material.

Like many physical rank-2 tensors, $\underline{\underline{\boldsymbol{\sigma}}}$ is symmetric. Indeed, the torque (in the z-direction) on a small cube of side l is $l^3(\sigma_{xy} - \sigma_{yx})$. These two stress components must cancel, because the moment of inertia of such a cube scales as $Ml^2 \sim l^5$, which would otherwise lead to a divergent angular acceleration as $l \to 0$.

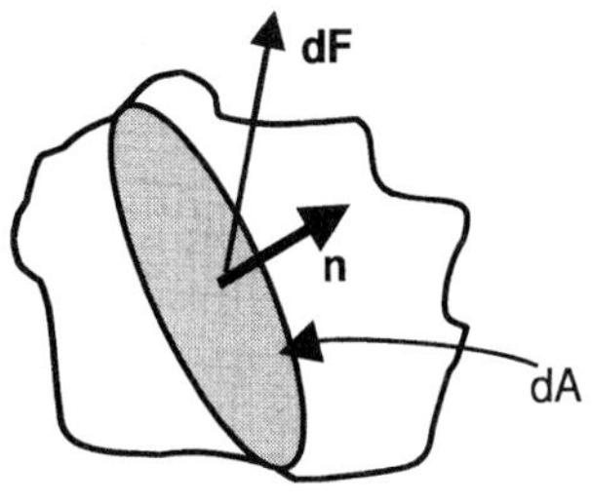

Figure 3. *Illustrating the definition of the stress tensor.*

1.4.1 Examples

Hydrostatic pressure

For a fluid at static equilibrium, one has $\underline{\underline{\boldsymbol{\sigma}}} = -p\underline{\underline{\mathbf{I}}}$, so that $p = -(1/3)\,\mathrm{Tr}(\underline{\underline{\boldsymbol{\sigma}}})$. The pressure field is not normally interesting in soft matter, as it acts as a Lagrange multiplier for the conservation of volume.

Newtonian fluid

The simplest constitutive equation for a fluid is that suggested by the tensorial symmetry of the strain rate and stress tensors, and the requirement of symmetry in $\underline{\underline{\boldsymbol{\sigma}}}$. These, combined with an assumption of linear response, and of incompressible flow, give in general $\underline{\underline{\boldsymbol{\sigma}}} = \eta\left(\underline{\underline{\mathbf{K}}} + \underline{\underline{\mathbf{K}}}^T\right)$. This characterises the fluid by a single number only, η, the viscosity. So in the case of simple shear

$$\sigma = \begin{pmatrix} -p & \eta\dot{\gamma} & 0 \\ \eta\dot{\gamma} & -p & 0 \\ 0 & 0 & -p \end{pmatrix}$$

Note that the pressure term just adds to the stress arising from the shear. Often we speak of the 'deviatoric stress' $\underline{\underline{\boldsymbol{\sigma}}}' = \underline{\underline{\boldsymbol{\sigma}}} - (1/3)\,\mathrm{Tr}(\underline{\underline{\boldsymbol{\sigma}}})\underline{\underline{\mathbf{I}}}$, which captures the stress arising from the material structure as it responds to a volume preserving deformation. We will often drop the prime in what follows.

Rubbery solid

A similar argument can be applied to an isotropic solid, allowing for the fact that the stress is now linearly proportional to the strain itself, rather than the strain rate. At small strains, this gives $\underline{\underline{\boldsymbol{\sigma}}} = G(\underline{\underline{\mathbf{e}}} + \underline{\underline{\mathbf{e}}}^T)$ for the deviatoric stress; G is the elastic modulus. An extension of this model to large strains, useful for rubbery solids, is to write $\underline{\underline{\boldsymbol{\sigma}}} = G(\underline{\underline{\mathbf{E}}} \cdot \underline{\underline{\mathbf{E}}}^T)$. Expanding for small $\underline{\underline{\mathbf{e}}} = \underline{\underline{\mathbf{E}}} - \underline{\underline{\mathbf{I}}}$ recovers the previous result to linear order, plus a contribution to the isotropic pressure.

A Maxwell model

The 'rubbery solid' constitutive equation just found may be generalised to a continuously-deformed material with a single viscoelastic relaxation time τ, by writing the following:

$$\frac{d}{dt}\underline{\underline{\boldsymbol{\sigma}}} = \underline{\underline{\mathbf{K}}} \cdot \underline{\underline{\boldsymbol{\sigma}}} + \underline{\underline{\boldsymbol{\sigma}}} \cdot \underline{\underline{\mathbf{K}}}^T - \frac{1}{\tau}(\underline{\underline{\boldsymbol{\sigma}}} - G\underline{\underline{\mathbf{I}}})\,. \tag{7}$$

Thus, if $\underline{\underline{\mathbf{K}}}$ vanishes (for example after a step-strain measurement: see below), the deviatoric stress decays to zero like $\exp[-t/\tau]$. In steady shear (with $\dot{\gamma} = \partial v_x/\partial y$) the stress tensor becomes:

$$\underline{\underline{\boldsymbol{\sigma}}} = \begin{pmatrix} G(1 + 2(\tau\dot{\gamma})^2) & G\tau\dot{\gamma} & 0 \\ G\tau\dot{\gamma} & G & 0 \\ 0 & 0 & G \end{pmatrix}$$

Exercise: Check this last result, and think about the physics of the proposed constitutive equation and the consequences of the predicted 'first normal stress difference': $\sigma_{xx} - \sigma_{yy} \neq 0$.

1.5 Rheometry

Rheometers are designed to impose on a material either shear flow (easy) or extensional flow (more difficult). A rotational device that generates a spatially uniform shear flow is the 'cone-and-plate' rheometer, Figure 4.

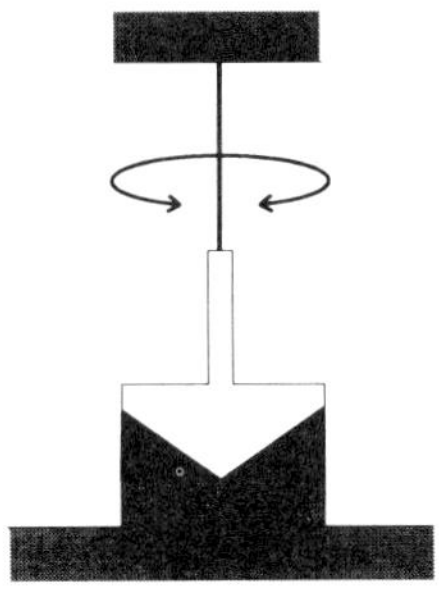

Figure 4. *Schematic of a cone-and-plate shear rheometer. The sample (black) lies between cone (white) and plate (grey).*

The cone is rotated at instantaneous angular velocity ω. The material in the gap at distance r from the axis has a velocity in the tangential direction that is zero at the bottom plate and ωr at the top plate. (We assume non-slip boundary conditions.) The local separation of the plates obeys $h(r) = \alpha r$ where α is the angle between plate and cone, which must be small. If so, the local shear rate is $\dot{\gamma} = \partial v_\theta / \partial z = \omega r / \alpha r = \omega / \alpha$: a uniform shear field. Maintaining such uniformity is especially important in non-linear deformation, where the material response may differ for different strains and strain rates. The shear force is measured from the torque on the rotor, and normal stress differences can also, in principle, be monitored (*e.g.* from the upthrust on the cone).

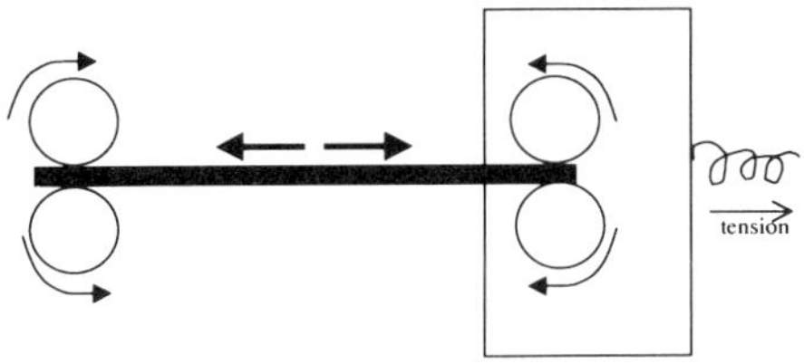

Figure 5. *Schematic of a moving-belt extensional rheometer (sample in black).*

Extensional rheometers (Figure 5) have been much harder to develop to the point where reproducible data is obtainable. This is due to the necessity of free surfaces over most of the sample in an extensional flow. However, extensional rheometry gives an important measure of the non-linear flow of many materials, that is independent of the shear response. For example, branched entangled polymers (see Section 3 below) may be strain-hardening in extension (the effective 'viscosity', which is the ratio of stress to strain rate, increases with strain), but strain-softening in simple shear. An illustration is in Figure 6. Here the extensional stress difference $\sigma_{xx} - \sigma_{yy}$, divided by the extensional strain rate $\dot{\varepsilon}$, is plotted against time, for two experiments at constant strain rate (started suddenly at time zero). On the same graph are the corresponding transient shear experiments, showing $\sigma_{xy}/\dot{\gamma}$. This way of representing data ensures that the curves within each set superimpose at early times, when the deformations are purely linear. The upper curves show strong extensional 'hardening' at the higher of two extension rates, but no hardening at the lower rate. The lower curves (for shear) all show softening. In this

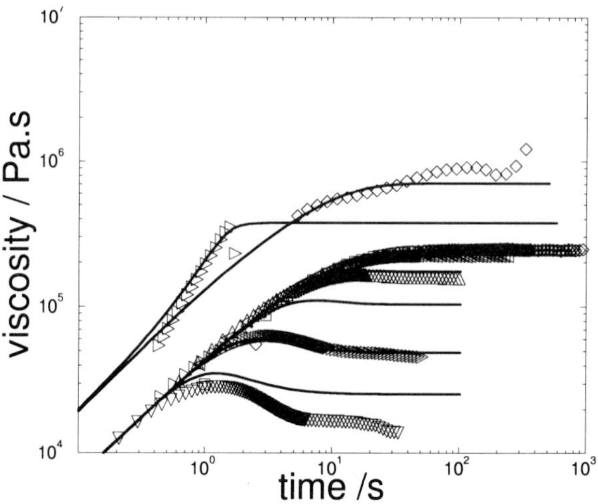

Figure 6. *Time dependent shear (lower curves) and extensional (upper curves) stresses normalised by deformation rates over a range of rates. Lines are from a non-linear generalisation of a model of branched polymers discussed in Section 3.*

case the polymer melt is composed of monodisperse molecules of identical ('H-shaped') branched structure.

1.6 Time dependence

The stress-growth curves of Figure 6 indicate that viscoelastic materials do not achieve a steady state of stress in a steady flow (or vice versa) until a certain relaxation time has elapsed. This relaxation arises from dynamic processes intrinsic to the materials themselves and can be a very sensitive (if indirect) probe of structural dynamics in soft matter. The most common experiments measure the time dependence of materials in linear deformation.

1.6.1 Step-strain response and relaxation modulus

In a step strain measurement, at time $t = 0$ a small strain γ (usually shear) is suddenly imposed and sustained. The resulting (shear) stress component $\sigma(t)$ decays with time, and is measured. If the material is in a true linear response regime, one has $\sigma(t) = G(t)\gamma$. The function $G(t)$ is the time dependent *relaxation modulus*, and is monotonically decreasing with time.

We will normally restrict ourself to isotropic materials, in which $G(t)$ is a scalar function of time. Lamellar, nematic and other ordered phases of surfactants and block copolymers will have special directions in which measurements of $G(t)$ may give very different results. However, polycrystalline samples of these materials recover isotropic rheology. Very few materials exhibit a single-exponential relaxation modulus $G(t) \propto \exp[-t/\tau]$, which is the linear response result of the Maxwell model (Equation 7). Many more can be described in terms of a sum of relaxation modes:

$$G(t) = \sum_{i=1}^{N} g_i e^{-t/\tau_i} . \tag{8}$$

The relaxation modulus $G(t)$ may be measured directly, but this suffers from two major drawbacks: (i) the initial step-strain is never quite instantaneous, degrading measurements of short relaxation times; (ii) the signal to noise ratio at long times is very weak, degrading measurements of long relaxation times.

The same information may instead be extracted from other flow histories, so long as the material properties are time-independent. That is, to each incremental strain $d\gamma(t')$ applied prior to time t there is a corresponding incremental stress $d\sigma(t) = G(t-t')d\gamma(t')$. We say that the material then has Time Translation Invariance (TTI—see the lectures of Bouchaud, this volume). Exceptions to this class are materials that are not in equilibrium (even in the absence of a flow), but which 'age' towards it on time scales longer than the length of the experiment. Using TTI we may write (suppressing tensor indices)

$$\sigma(t) = \lim_{\delta\gamma\to 0} \sum_{\delta\gamma} G(t-t')\delta\gamma(t') = \int_{-\infty}^{t} G(t-t')\frac{d\gamma}{dt'}dt' , \tag{9}$$

which is the linearised constitutive equation between shear strain $\gamma(t)$ and stress $\sigma(t)$.

1.6.2 Frequency-dependent modulus

The most common strain history used to extract the equivalent of $G(t)$ is the harmonic oscillation $\gamma(t) = \mathrm{Re}\,(\gamma_0 e^{i\omega t})$. Then using Equation 9 we write

$$\sigma_{xy}(t) = \mathrm{Re}\left(\int_{-\infty}^{t} G(t-t')\gamma_0 i\omega e^{i\omega t'} dt'\right) = \mathrm{Re}\left(\gamma_0 G^*(\omega)e^{i\omega t}\right) , \tag{10}$$

with the 'complex modulus' $G^*(\omega)$ defined by

$$G^*(\omega) = i\omega \int_0^{\infty} G(t)e^{-i\omega t}dt . \tag{11}$$

The form of Equation 10 means that the stress will be simple harmonic at frequency ω, but not in phase with the strain. If we write $G^*(\omega) = G'(\omega) + iG''(\omega)$, then we can identify the real part G' as the in-phase (elastic) part of the modulus and the imaginary part G'' as the out-of-phase (dissipative) part. In general both will be frequency-dependent, crossing over from viscous (dissipative) behaviour at low frequencies to elastic behaviour at high frequencies. Before giving examples, let us summarise these two ideal limits:

Ideal newtonian fluid (viscosity η)

$$\sigma_{xy} = \eta\frac{\partial\gamma}{\partial t} = \eta i\omega\gamma_0 e^{i\omega t} \qquad \Longrightarrow \qquad \begin{cases} G'(\omega) = 0 \\ G''(\omega) = \omega\eta \end{cases}$$

Ideal elastic solid (modulus G_0)

$$\sigma_{xy} = G_0\gamma = G_0\gamma_0 e^{i\omega t} \qquad \Longrightarrow \qquad \begin{cases} G'(\omega) = G_0 \\ G''(\omega) = 0 \end{cases}$$

1.6.3 Examples

The Maxwell model

Now we can interpret what the frequency-dependent experiment will give us for the simplest model of a viscoelastic fluid, with a single relaxation time $G(t) = G_0 e^{-t/\tau}$. The integral over t is readily done to yield

$$G'(\omega) = G_0 \frac{\omega^2\tau^2}{1+\omega^2\tau^2} \quad ; \quad G''(\omega) = G_0 \frac{\omega\tau}{1+\omega^2\tau^2} . \tag{12}$$

Note that the correct elastic and viscous behaviour are recovered at high and low frequency respectively. The characteristic time emerges as the inverse of the frequency at which the curves for G', G'' cross (or where G'' is maximum, in this case). The result for the steady state viscosity is $\eta = G\tau$. More generally, Equation 9 gives the exact integral for the ratio of stress to strain rate in steady state as $\eta = \int_0^\infty G(t)dt$, so it is always true that $\eta \simeq G\tau$ where G is an effective modulus and τ a characteristic relaxation time.

Polymeric matter

We finish this survey with a few examples of the elastic and loss modulus for polymeric materials. It is possible in many such cases to extract effective information on relaxations covering many decades of frequency, because of *time-temperature superposition.* For most polymers above both their melting point and glass temperature T_g, the time scales of *all* viscoelastic relaxations shift with temperature by the same factor $a_T = \exp[A/(T - T_0)]$, with material-dependent values of A and T_0. (This is the Volgel-Fulcher, or WLF form; see *e.g.* the lectures by Kob, this volume). Up to 12 decades in frequency are then accessible for polymers with very low T_g, by superposing data of different T.

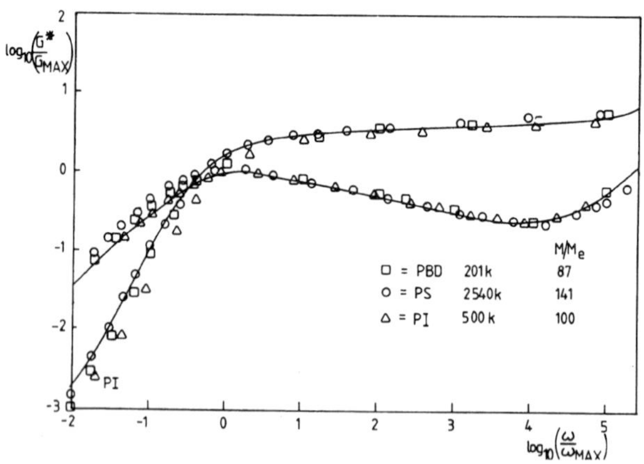

Figure 7. *Linear viscoelastic moduli G′ and G″ as functions of oscillation frequency ω, of monodisperse melts of polystyrene, polyisoprene and polybutadiene of similar degree of entanglement (M/M$_e$).*

In Figure 7 we show results for three chemistries of near-monodisperse linear polymer melts. Note that the data are, as usual for such experiments, plotted on log-log axes in which the Maxwell model would have G'' with slopes of 1 and -1 each side of the

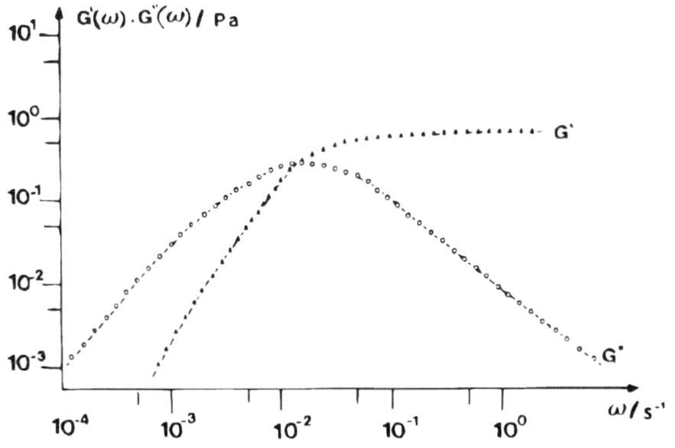

Figure 8. *Near-Maxwell behaviour of a wormlike surfactant solution.*

maximum (see Equation 12). The slope in the data is much shallower on the right, indicating the presence of some shorter relaxation times (Equation 8) but there is still clearly a dominant time at the crossover from viscous to elastic behaviour.

There is one family of polymer-like systems with a near-Maxwell behaviour: the self-assembled wormlike surfactant micelles. These entangled polymers support an additional dynamics of breaking and reforming, that narrows the viscoelastic spectrum towards a single exponential (Figure 8); see [1].

Finally we examine the effect of a change of molecular topology on the linear rheology. Figure 9 compares $G^*(\omega)$ for a linear and three-arm star architecture of polyisoprene melt. As before, the linear polymer has a strong dominant relaxation time, but the branched variety is quite different: the maximum in $G''(\omega)$ is no longer anywhere near the crossover point, indicating a much broader superposition of relaxation modes. The terminal time is also much longer in the case of the star polymer. We will examine the reasons for this critical effect of branching in Section 3.

2 Rheology of linear polymer chains

For a fuller account of the material in this section see [7].

2.1 Entropic elasticity

First we recap briefly the statistical physics of a polymer chain (as covered in the lectures by Khokhlov, this volume). Each chain is a *random walk* in space modelled by some local rule for spatial links; an example is the freely jointed chain. The step length of the chain corresponds to the Kuhn length of the polymer which we denote b. (This is the shortest independently orientable segment length, usually 4 or 5 monomers long.) Suppose the whole walk has N links and end-to-end vector $\mathbf{R}(N)$. From the theory of random walks, $\langle R(N)^2 \rangle = Nb^2$ where $R = |\mathbf{R}|$. Also $P(\mathbf{R})$ must have a Gaussian form (since $\mathbf{R}$ is a sum

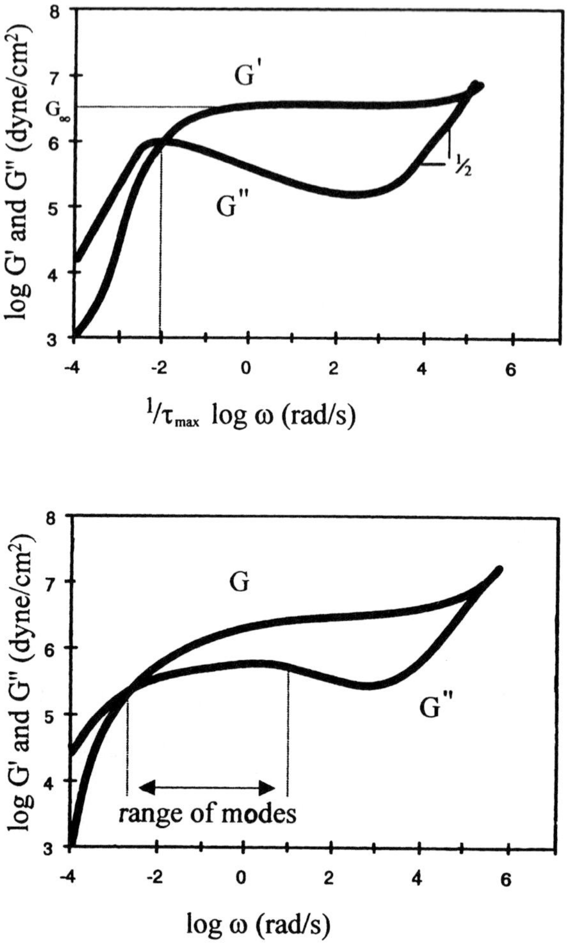

Figure 9. *Comparison of G′ and G″ for a linear (top) and star polymer melt of similar molecular weight polyisoprene [3]. Note the much broader range of relaxation times for the star polymer.*

of many independent random vectors). So

$$P(\mathbf{R})d^3\mathbf{R} = \left(\frac{3}{2\pi N b^2}\right)^{3/2} \exp\left(\frac{-3R^2}{2Nb^2}\right) d^3\mathbf{R}\,. \tag{13}$$

Now define a macrostate by the end-to-end vector $\mathbf{R}$. The microstates are the different random walks of given $\mathbf{R}$. In a freely flexible chain, each has the same energy, so the number of microstates obeys $\Omega(\mathbf{R}) = \Omega_{\text{total}} P(\mathbf{R})$. Since the entropy of the walk is given by $S = k_B \ln \Omega$ we have $S(R) = S(0) - 3k_B R^2/2Nb^2$. The free energy of the chain is then $F(R) = U - TS$ where U, the internal energy, is a constant: therefore we have $F(\mathbf{R}) = \text{const.} + 3k_B T R^2/2Nb^2$. Finally the entropic force ('Brownian tension') on the chain is

$$\mathbf{f} = -\frac{\partial F(\mathbf{R})}{\partial \mathbf{R}} = -\frac{3k_BT}{Nb^2}\mathbf{R} \equiv -\kappa\mathbf{R}\,. \tag{14}$$

Thus a random walk polymer, or 'Gaussian coil' is like a Hookean spring with stiffness κ proportional to T/N. In dilute solutions, polymer chains are not gaussian (Khokhlov, this volume), except at the Theta temperature. However, it transpires [7] that in concentrated solutions and polymer melts, the excluded volume interaction responsible for chain swelling is *screened*. (Although such chains remain selfavoiding at short distances, the driving force for chain swelling, is to decrease the probability of contacts within the chain. This is removed at high density: most collisions are with other chains, and swelling does not reduce the probability of these.) Chains in concentrated solutions and melts are gaussian at large enough distances, and Equation 14 applies to them.

2.1.1 Stress tensor

Equation 14 will enable us to calculate the *stress tensor* in any polymeric fluid provided the following conditions are met: (i) we know the instantaneous configuration of the chains at scales above some characteristic number $\widetilde{N}$ of links; (ii) the configurations have achieved a *local* equilibrium for chain segments at smaller scales than this; (iii) we may average over many subchains (of $\widetilde{N}$ links) within a local volume large enough to define a macroscopic stress, but small enough to define uniform physical conditions for the polymer chains within it.

Recall that component σ_{ij} of the stress tensor $\underline{\underline{\sigma}}$ is the i-th component of total force per unit area transmitted across a plane whose normal lies in the j-th direction. Now consider a small cubic volume in a polymeric fluid of side L (Figure 10). It contains $\mathbb{C}/N$

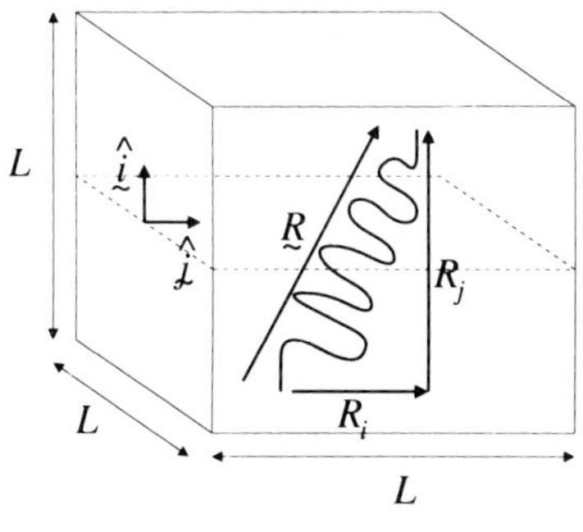

Figure 10. *Contribution of a single subchain to the stess tensor.*

subchains of length N, where $\mathbb{C}$ is the monomer concentration (we drop the tilde on $\tilde{N}$). The probability that one subchain of end-to-end vector $\mathbf{R}$ cuts a given j-plane across the volume is just R_j/L (the fraction of the sample length L in the j direction spanned by its end-to-end vector). The i-th component of force transmitted by this chain across the j-plane is, from Equation 14, κR_i. So its contribution to the mean local stress σ_{ij} is $\kappa R_i R_j/L^3$. The sum over all subchains may be replaced by the average $\langle ... \rangle$ over the ensemble multiplied by the number of subchains, $\mathbb{C}L^3/N$:

$$\sigma_{ij} = \frac{3k_BT\mathbb{C}}{N^2b^2}\langle R_iR_j\rangle\,. \tag{15}$$

We will find it convenient to work with a continuous representation of the chains $\mathbf{R}(n)$ that maps the arclength position of the n-th monomer onto its spatial position $\mathbf{R}$. Then we may identify $\mathbf{R}/N$ for a (small) subchain with $\partial\mathbf{R}/\partial n$. The formula for the stress tensor becomes pleasingly simple:

$$\sigma_{ij} = \frac{3k_BT}{b^2}\mathbb{C}\left\langle \frac{\partial R_i}{\partial n}\frac{\partial R_j}{\partial n} \right\rangle . \tag{16}$$

The second moment average $\langle ... \rangle$ that governs the stress now needs to be calculated under various different assumptions for the dynamics. For example, it is sometimes possible to identify subchains containing N monomers that have end-to-end distributions $P(\mathbf{R})$ fixed by external constraints (as in a network, when N is the number of monomers between cross links) or by dynamics at a particular time scale (as in an entangled melt, when N is the number of monomers between entanglements), but which are equilibrated at all smaller length scales. In this case the natural unit of arc length is the coarse-grained step length of the segments $\sqrt{N}b$. Writing $nb = s'\sqrt{N}$ the stress may be calculated from any known distribution of (coarse-grained) chain tangent vectors as:

$$\sigma_{ij} = 3k_BT\frac{\mathbb{C}}{N}\left\langle \frac{\partial R_i}{\partial s'}\frac{\partial R_j}{\partial s'} \right\rangle . \tag{17}$$

Each sub-chain thus contributes k_BT of stress, distributed tensorially via the second moment of its orientation distribution.

2.2 Dynamics

In polymer solutions and melts, the stress formula (Equation 16 or 17 above) is always appropriate given the validity of the three criteria listed at the start of Section 2.1.1, and the applicability of the Gaussian chain approximation. But there are important physical regimes in which the dynamics themselves differ qualitatively.

(i) *Unentangled Chains.* In the first regime, topological interactions between chains are not important because the chains are not sufficiently overlapped. Note that entanglement is only achieved at remarkable degrees of (spatial) overlap: even in the melt, chains must be several hundred monomers long, in order to see entanglement. The unentangled regime divides into two classes depending on whether long-range hydrodynamic interactions are important for the drag on the chains. If not, there is just local dissipation due to frictional forces as the chains slide past one another. Rouse [8] proposed this simplest case as a model for dilute solution, but it actually finds its realisation in low molecular weight melts and concentrated solutions. In dilute solution the more complex issue of hydrodynamic interaction dominates. We will not deal with this subject here, but the relevant model was devised by Zimm [9].

(ii) *Entangled Chains.* In this case the dissipation is local on the scale of the entanglement spacing (whether in melt or concentrated solution) but the chains' motion is severely restricted by the topological constraints of their surroundings—two chains may not cross each other. Rouse's formulation of the local drag needs to be supplemented by a model of these topological restrictions. The most powerful approach has proved to be the *tube model* of Doi, Edwards and de Gennes (see below and [7]). This entangled regime also divides into two classes, but now depending on the topological structure of the chains themselves, that is, whether they are linear or branched. The branched case offers a nice example of *hierarchical dynamics* in soft condensed matter (see Section 3 and [2]).

Starting below for unentangled chains, we will outline the calculation of two aspects of the polymer dynamics, one microscopic and one macroscopic. The first is the mean-square monomer displacement as a function of time $\phi(t) = \langle|\mathbf{R}(n,t) - \mathbf{R}(n,0)|^2\rangle$ averaged over all chains and monomer positions. This may be measured directly via NMR in some circumstances, and by scattering experiments indirectly [7]. The second aspect is the linear rheological response $G(t)$ and its frequency-dependent representations, $G'(\omega)$ and $G''(\omega)$. In each case, we first use a formal approach in which the Brownian motion of chains is handled using a random thermal force on the monomers (a 'Langevin' equation). Then we discuss the result using simple physical arguments.

2.3 The Rouse model

In this simplest fundamental model of polymer dynamics we assume:

- Gaussian chains, in which the force on a monomer or subchain n is the net entropic force from its neighbours. In the continuum language, this is equivalent to a thermodynamic force at each point on the chain $(\partial/\partial n)\kappa\partial\mathbf{R}/\partial n = \kappa\partial^2\mathbf{R}/\partial n^2$.
- Local drag: the drag force on a Kuhn segment comes from frictional drag against the background; this force is $\zeta\partial\mathbf{R}/\partial t$, with ζ a drag coefficient.
- Brownian motion: a random force $\mathbf{f}$ acts on each monomer or subchain, with correlation time much faster than any polymer dynamics.

2.3.1 A toy calculation: the Rouse-dumbell model

Suppose for a moment that the drag acts only on two points, at the extremities of a (sub)chain of N segments, $\mathbf{R}_1$ and $\mathbf{R}_2$. This simplified model will help us solve the full Rouse model below. The force balance for the two drag centres is:

$$\zeta\frac{\partial\mathbf{R}_1}{\partial t} = \kappa(\mathbf{R}_2 - \mathbf{R}_1) + \mathbf{f}_1\,, \tag{18}$$

$$\zeta\frac{\partial\mathbf{R}_2}{\partial t} = \kappa(\mathbf{R}_1 - \mathbf{R}_2) + \mathbf{f}_2\,. \tag{19}$$

The random forces have correlations in time that are just delta-functions on the polymer time scale: $\langle\mathbf{f}_1(t)\mathbf{f}_1(t')\rangle = \eta_1\underline{\mathbf{I}}\,\delta(t-t')$ with η_1 a constant (likewise for $\mathbf{f}_2$). This coupled system of equations is easily diagonalised with the following co-ordinates:

$$\mathbf{R}_{CM} = \frac{1}{2}(\mathbf{R}_1 + \mathbf{R}_2)\,, \tag{20}$$

$$\mathbf{r} = (\mathbf{R}_1 - \mathbf{R}_2)\,. \tag{21}$$

These represent the centre-of-mass motion of the molecule and the spatial separation of the drag points, respectively. In these variables the equations read:

$$\zeta_{CM}\frac{\partial \mathbf{R}_{CM}(t)}{\partial t} = \mathbf{f}_{CM}(t)\,, \tag{22}$$

$$\zeta\frac{\partial \mathbf{r}(t)}{\partial t} = -2\kappa\mathbf{r}(t) + \mathbf{f}_r(t)\,, \tag{23}$$

with new random forces defined appropriately, and $\zeta_{CM} = 2\zeta$.

The first (centre-of-mass) co-ordinate is subjected to a history of random forces that generate random displacements. The final value of $\mathbf{R}_{CM}(t)$ after such a history is clearly a sum of a large number of random variables; it will therefore have a Gaussian distribution. We recognise the physics of simple diffusion. The mean square displacement can be calculated by direct integration of the dynamical Equation 22:

$$\langle R_{CM}^2(t)\rangle = \frac{1}{\zeta_{CM}^2}\int_0^t dt' \int_0^t dt'' \,\langle \mathbf{f}_{CM}\cdot(t')\mathbf{f}_{CM}(t'')\rangle \tag{24}$$

$$= \frac{1}{\zeta_{CM}^2}\int_0^t \eta_{CM}\,\mathrm{Tr}(\underline{\underline{\mathbf{I}}})dt' = 6D_{CM}t\,, \tag{25}$$

where the diffusion constant is given in terms of the noise by $D_{CM} = \eta_{CM}/2\zeta_{CM}^2$. By the Einstein relation $D = k_BT/\zeta$, this sets the variance of the noise as $\eta_{CM} = 2k_BT\zeta_{CM}$.

The second co-ordinate is the relative separation of the chain ends, and describes an overdamped Hookean spring with a Brownian force. This time the solution is via the Green function for the 1st order ODE, Equation 23:

$$\mathbf{r}(t) = \mathbf{r}(0)e^{-t/\tau} + \int_0^t G(t,t')\mathbf{f}_r(t')\,dt' = \mathbf{r}(0)e^{-t/\tau} + \frac{1}{\zeta}\int_0^t e^{-(t-t')/\tau}\mathbf{f}_r(t')dt'\,, \tag{26}$$

with $\tau = \zeta/2\kappa$ a relaxation time. The second moment of $\mathbf{r}$ is found via a double integral (compare Equation 24):

$$\langle r(t)^2\rangle = r^2(0)e^{-2t/\tau} + \frac{1}{\zeta^2}\int_0^t\int_0^t dt'dt''e^{-(2t-t'-t'')/\tau}\,\langle \mathbf{f}_r(t')\cdot\mathbf{f}_r(t'')\rangle \tag{27}$$

$$= r^2(0)e^{-2t/\tau} + \frac{1}{\zeta^2}\int_0^t e^{-2(t-t')/\tau}\eta_r\,\mathrm{Tr}(\underline{\underline{\mathbf{I}}})dt' \tag{28}$$

$$= r^2(0)e^{-2t/\tau} + \frac{3\eta_r}{4\zeta\kappa}(1-e^{-2t/\tau}) \tag{29}$$

(where $r = |\mathbf{r}|$). Equation 29 says that the initial separation is 'forgotten' in a characteristic time given by $\tau = \zeta Nb^2/6k_BT$. As $t \to \infty$, we must recover the equilibrium value of the chain end separation, as given by the equipartition theorem:

$$\frac{\kappa}{2}\langle r(\infty)^2\rangle = \frac{\kappa}{2}\frac{3\eta_r}{4\zeta\kappa} = \frac{3k_BT}{2}\,.$$

So we find that the noise variance must be $\eta_r = 4k_BT\zeta$.

Exercise: Calculate the (tensorial) correlation function $\langle \mathbf{r}(t)\mathbf{r}(t')\rangle$ for the 'dumbell' molecule. [Answer: $(k_BT/\kappa)\underline{\underline{\mathbf{I}}}\,e^{-|t-t'|/\tau}$.]

2.3.2 The Rouse model and its normal modes

Now we have the tools we need to attack the Rouse model proper, in which frictional drag is uniformly distributed along the chain. The balance of entropic, drag and random forces on a chain of N segments is the Rouse equation:

$$\zeta_0 \frac{\partial \mathbf{R}}{\partial t} = \kappa \frac{\partial^2 \mathbf{R}}{\partial n^2} + \mathbf{f}(n,t)\,. \tag{30}$$

As before, the noise force on each monomer is related to its frictional drag by a (generalised) Einstein relation:

$$\langle \mathbf{f}(n,t)\mathbf{f}(m,t')\rangle = 2\zeta_0 k_B T \underline{\underline{\mathbf{I}}}\,\delta(n-m)\delta(t-t')\,. \tag{31}$$

The Rouse dynamical Equation 30 is diagonalised by the transformation:

$$\begin{aligned} \mathbf{R}(n,t) &= \mathbf{X}_0(t) + 2\sum_{p=1}^{\infty} \mathbf{X}_p(t)\cos\left(\frac{p\pi n}{N}\right), \\ \mathbf{X}_p(t) &= \frac{1}{N}\int_0^N \mathbf{R}(n,t)\cos\left(\frac{p\pi n}{N}\right)dn\,. \end{aligned} \tag{32}$$

The $\mathbf{X}_p(t)$ are the time-dependent amplitudes of the 'Rouse modes' of the polymer chain. These are just the (vector-amplitude) Fourier components of the chain path $\mathbf{R}(n,t)$ with respect to the arclength co-ordinate. We may re-write the dynamics by substituting Equation 32 for the Rouse modes into the Rouse Equation 30. The essential point is that the operator $\partial^2/\partial n^2$ becomes just $(p\pi/N)^2$ in the new modes. Then we operate from the left with the integral operator $(2/N)\int_0^N dm\,\cos(p\pi m/N)$ and use the orthogonality result for Fourier modes, $(2/N)\int_0^N \cos(p\pi m/N)\cos(p'\pi m/N)\,dm = \delta_{pp'}(1+\delta_{p0})$.

Each mode amplitude is then found to obey a decoupled Langevin equation, which reads (for $p \geq 1$) :

$$\zeta_p \frac{\partial \mathbf{X}_p}{\partial t} = -k_p \mathbf{X}_p + \mathbf{f}_p(t) \quad \text{with} \quad k_p = \frac{6k_B T p^2 \pi^2}{N b^2} \quad \text{and} \quad \zeta_p = 2N\zeta_0\,, \tag{33}$$

whereas the decoupled centre-of-mass mode ($p = 0$) satisfies

$$\zeta_{CM} \frac{\partial \mathbf{X}_0}{\partial t} = \mathbf{f}_0(t) \qquad ; \qquad \zeta_{CM} = N\zeta_0\,, \tag{34}$$

and undergoes simple diffusion. Each of the internal modes behaves exactly like the one internal mode of the dumbell molecule, with a noise term which can be calculated either by Fourier-transforming the spatial noise terms, $\mathbf{f}(n,t)$, or by observing that their strength must be sufficient to maintain an energy equipartition of $k_BT/2$ (for each Cartesian component) per mode. Either calculation gives

$$\langle \mathbf{f}_p\,\mathbf{f}_q\rangle = 2\zeta_p k_B T \underline{\underline{\mathbf{I}}}\,\delta_{pq}\delta(t-t')\,. \tag{35}$$

A key result is the time correlation function of the mode amplitudes (see the exercise on the Rouse dumbell above), which is:

$$\langle \mathbf{X}_p(t)\,\mathbf{X}_q(t')\rangle = \underline{\underline{\mathbf{I}}}\left(\frac{k_B T}{k_p}\right)\delta_{pq} e^{-|t-t'|/\tau_p}\,. \tag{36}$$

Each mode has its own relaxation time $\tau_p = \zeta/k_p$ that decrease rapidly (as $1/p^2$) with p. The longest of these relaxation times, $\tau_1 = \zeta N^2 b^2 / 3\pi^2 k_B T$, has special significance. It is known as the *Rouse time,* and often denoted τ_R. It is the time for relaxation of the overall shape of the molecule, and is also the time for a Rouse chain to diffuse a distance of order its own size.

2.3.3 Monomer motion in the Rouse model

What does the local motion of this model chain look like? We expect for short intervals that the chain contour may have adjusted locally, but retain a very similar global configuration (see Figure 11 for an illustration from video microscopy of giant DNA molecules).

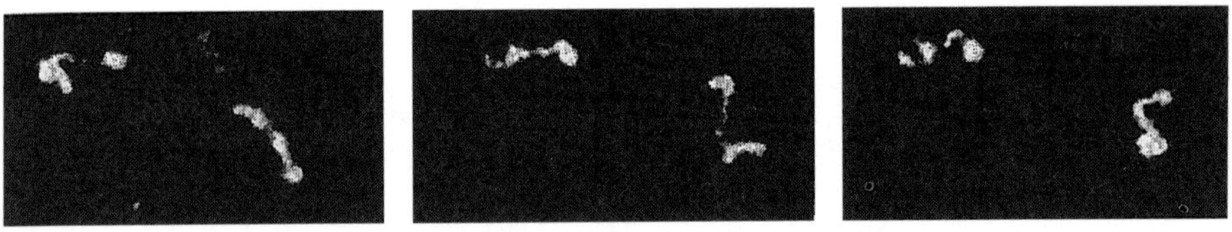

Figure 11. *Two fluorescence-labelled DNA chains in solution undergoing Rouse motion. The time lapse between frames is about 1 second.*

We need to calculate the correlation function $\phi(t) \equiv \langle |\mathbf{R}(n,t) - \mathbf{R}(n,0)|^2 \rangle$, so write it in terms of the Rouse modes whose dynamics we already know:

$$\begin{aligned}\phi(t) &= \left\langle \left| \mathbf{X}_0(t) + 2\sum_{p=1}^{\infty} \mathbf{X}_p(t)\cos\left(\frac{p\pi n}{N}\right) - \mathbf{X}_0(0) - 2\sum_{q=1}^{\infty} \mathbf{X}_q(0)\cos\left(\frac{q\pi n}{N}\right) \right|^2 \right\rangle \qquad (37)\\ &= \langle |\mathbf{X}_0(t) - \mathbf{X}_0(0)|^2 \rangle \\ &\quad + \left\langle 4\sum_{pq} \left\{ \mathbf{X}_p(t)\cdot\mathbf{X}_q(t) + \mathbf{X}_p(0)\cdot\mathbf{X}_q(0) - 2\mathbf{X}_p(t)\cdot\mathbf{X}_q(0) \right\} \cos\left(\frac{p\pi n}{N}\right)\cos\left(\frac{q\pi n}{N}\right) \right\rangle\end{aligned}$$

where the first term in the second expression is centre-of-mass diffusion.

From the last section, we know all the correlations (and only those with $p = q$ are non-zero), so direct substitution gives

$$\phi(t) = 6D_{CM}t + \frac{4k_BT}{k_1}\sum_{p=1}^{\infty}\frac{1}{p^2}\cos^2\left(\frac{p\pi n}{N}\right)\left(1 - e^{-p^2 t/\tau_R}\right). \qquad (38)$$

The first term is just the centre-of-mass diffusion of the entire chain, the second the contribution from the internal modes. Now, for times $t \ll \tau_R$, the amplitudes of the modes in the sum decay slowly with p, so permitting us to replace the sum with an integral. Also we may average the $\cos^2$ function over monomers (n) to get a factor of $1/2$. We then find:

$$\begin{aligned}\phi(t) &\simeq 6D_{CM}t + \frac{2k_BT}{k_1}\int_0^{\infty}\frac{1}{p^2}\left(1 - e^{-p^2 t/\tau_R}\right)dp \qquad (39)\\ &= 6D_{CM}t + \frac{Nb^2}{3\pi^2}\left(\frac{t}{\tau_R}\right)^{1/2}\alpha\,. \qquad (40)\end{aligned}$$

Here $\alpha = \int z^{-1/2}(1-e^{-z})dz$ is a coefficient of order unity. The result is remarkable: each monomer executes an 'anomalous' or sub-Fickian diffusion, such that its mean square displacement goes as $t^{1/2}$ rather than t (as for ordinary diffusion). This behaviour persists until times of order the Rouse time, after which each monomer is carried by the (faster) centre of mass motion of the whole molecule.

2.3.4 So what is going on?

In some ways, the structure of the Rouse modes is misleading: they seem to imply longer range correlations along the chain than in fact exist. The sub-Fickian diffusion arises physically from the *absence* of such correlations. To diffuse a distance ΔR_n, the n-th monomer requires its motion to be correlated with the $(\Delta R_n)^2/b^2$ other monomers in the region spanned by ΔR_n. This arises as a straightforward consequence of their *connectivity*. All other monomers have motions *uncorrelated* with it, so cannot contribute to the effective drag for that motion, which is $\zeta_{\text{eff}} = \zeta_0(\Delta R_n)^2/b^2$. Thus, from the law of normal diffusion and the Einstein relation $D_{\text{eff}} = k_B T/\zeta_{\text{eff}}$, we obtain

$$\phi(t) \equiv \langle|\Delta\mathbf{R}_n|^2\rangle \simeq \frac{k_B T b^2}{\zeta_0\langle|\Delta\mathbf{R}_n|^2\rangle} t \Longrightarrow \phi(t) \simeq \sqrt{\left(\frac{k_B T b^2}{\zeta_0} t\right)} \,. \tag{41}$$

The 'extra-drag' effect runs out of new monomers when the chain has diffused its own radius of gyration, which is at the Rouse time. After this the drag saturates at a level set by the entire chain, and all monomer motions become correlated as ordinary (centre-of-mass) diffusion takes over. So a log-log plot of the monomeric displacement looks like Figure 12.

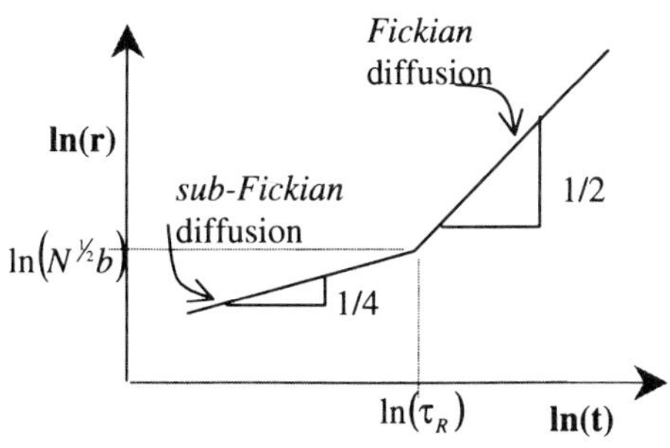

Figure 12. *Monomer diffusion with time in the Rouse model.*

2.3.5 Stress relaxation in the Rouse model

The (deviatoric) stress formula we derived above (Equation 16) has a very simple representation in terms of the Rouse modes:

$$\sigma_{ij} = \frac{\mathbb{C}}{N}\sum_p k_p \langle X_{pi}(t)X_{pj}(t)\rangle \,, \tag{42}$$

where, on X_{pi}, the first suffix is the mode index and the second a Cartesian one. To find $G(t)$, we consider a step strain in shear of size γ. As the step is applied, all the vector mode amplitudes $\mathbf{X}_p(t)$ deform affinely: in a fast enough strain, the chain co-ordinates follow the applied shear. (For a proof, see [7].) Hence $X_{px}(0^+) = X_{px}(0^-) + \gamma X_{py}(0^-)$, giving

$$\sigma_{xy}(0^+) = \gamma \frac{\mathbb{C}}{N} \sum_p k_p \left\langle X_{py}(0^-) X_{py}(0^-) \right\rangle = \gamma \frac{\mathbb{C}}{N} \sum_p k_B T . \tag{43}$$

Each mode then decays back to equilibrium with its own time constant $\tau_p = \tau_1/p^2$, giving for the time dependent modulus:

$$G(t) = \frac{\sigma_{xy}(t)}{\gamma} = \frac{\mathbb{C} k_B T}{N} \sum_p e^{-2p^2 t/\tau_1} . \tag{44}$$

Again, for times $t \ll \tau_R$, the modes are effectively continuous and the sum is approximated well by the integral $\int dp \exp(-2p^2 t/\tau_1) \simeq (t/\tau_1)^{-1/2}$. So we find that, until a final crossover to exponential decay beyond the Rouse time, the Rouse model has a relaxation modulus which is a power-law ($G(t) \sim t^{-1/2}$). From Equation 11, we then have also $G'(\omega) \sim G''(\omega) \sim \omega^{1/2}$. This form can be seen, for example in the high-frequency parts of the polyisoprene linear and star rheology data we saw in Figures 7 and 9.

In conjunction with the physics we used to understand the scaling of monomer diffusion, this behaviour follows from our argument (Section 1.2) for the modulus in soft matter. We estimated this as $k_B T$ per effective degree of freedom (one that couples to strain on the relevant time scale). In this case, after a time t, we allot $k_B T$ of modulus to each unrelaxed subchain. Such a chain contains $n(t)$ monomers where $n(t) \sim (\Delta R_n(t))^2/b^2 \sim t^{1/2}$. The number of such subchains thus decays as $t^{-1/2}$, giving $G \sim t^{-1/2}$, until the sub-Fickian regime ceases. The expected behaviour on a log-log plot is therefore as shown in Figure 13. Note that the longest relaxation time scales with molecular weight as N^2, but the viscosity scales as $\mathbb{C} k_B T N$. This is because at the longest relaxation time, the remaining stress is carried only by the lowest Rouse mode: the density of these modes is one per chain, or $\mathbb{C}/N$.

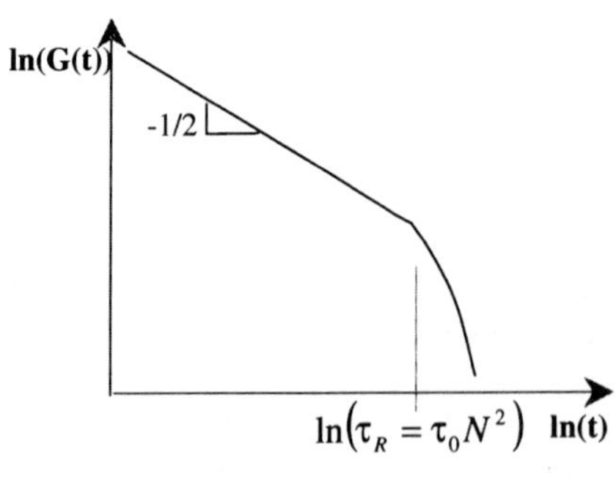

Figure 13. *Stress relaxation in the Rouse model.*

2.4 Entangled chains: reptation dynamics

Now we consider the motion of a chain in a forest of topological constraints arising from its neighbours. The chain behaves as if it were confined to a tubelike region along its contour. The tube diameter a will depend on the concentration of polymer (in a way that is still not clarified theoretically, but experimentally goes like $a \sim \mathbb{C}^{-1/2}$). Only the chain ends are free to explore the melt without the constriction of the tube, as shown in Figure 14.

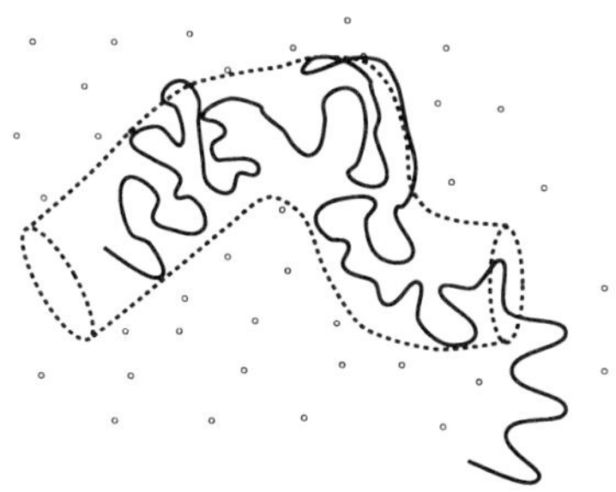

Figure 14. *Reptation of a linear polymer chain in a tube arising from topological constraints with its neighbours.*

At small times and small distances, the presence of the tube will not be felt, so $G(t)$ and monomer displacements at early times will be unchanged. This is true only until the Rouse time τ_e of pieces of chain that just span the tube—these are called 'entanglement segments' and have N_e monomers with $a^2 \simeq N_e b^2$. (This N_e is directly proportional to the 'entanglement molecular weight' M_e encountered in Figure 7.) After that, each section of tube *will constrain the enclosed piece of chain to the orientation of the piece originally present there, until the tube section is traversed by a free end.* So a good way to understand $G(t)$ is to view the stress as carried by tube segments.

There is a typical waiting time for the arrival of a free end, because Brownian motion for $t > \tau_e$ causes 1-dimensional curvilinear diffusion of the chain along the tube contour, termed *reptation.* To evacuate all the tube occupied at $t = 0$ (when a step-strain might be applied), the waiting time is $\tau_d \simeq L^2/D_c$ where $L = Nb^2/a$ (the length of the random walk coarse-grained on the scale of a), and where $D_c = k_B T/N\zeta_0$, is the curvilinear diffusion constant. (This coincides with the centre-of-mass diffusion constant for the whole Rouse chain in free space [7].) So we expect $\tau_d \simeq N^3 b^4 \zeta_0 / k_B T a^2$. For time scales between τ_e and τ_d (a range that grows as $(N/N_e)^3$), we expect a near-plateau in $G(t)$, with an amplitude of $G_0 \simeq \mathbb{C} k_B T/N_e$.

To calculate $G(t)$ more precisely within the tube model, we calculate the survival probabilities for segments of original tube during stress relaxation. (Although the chain is constrained by new tube as it moves out of the old one, the new tube segments are isotropically oriented and do not contribute to the deviatoric stress.) In the frame of the *chain,* a given tube segment behaves as a particle diffusing on the curvilinear coordinate x with diffusion constant D_c that is absorbed by boundaries at 0 and L. So if $\Psi(x, x', t)$ is the probability that a tube segment initially at position x' on the chain has diffused to x at time t without encountering a chain end, it will obey the diffusion equation,

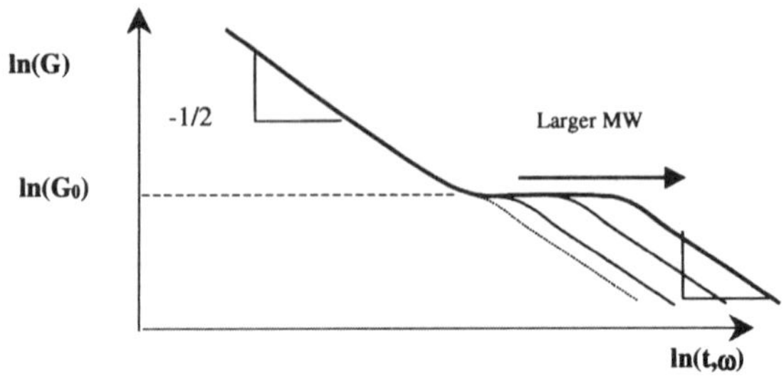

Figure 15. *Stress relaxation in the tube model.*

$D_c\partial^2\Psi/\partial x^2 = \partial\Psi/\partial t$, with $\Psi(x,x',0) = \delta(x-x')$ as initial condition and $\Psi(0,x',t) = \Psi(L,x',t) = 0$ as boundary conditions.

This is equivalent to the Fourier problem of heat diffusion through a slab with cooled faces. To solve it, we expand in the normalised eigenfunctions that obey the boundary conditions, $\phi_p(x) = \sqrt{2/L}\sin(p\pi x/L)$ (for integer p):

$$\Psi(x,x',t) = \sum_{p=1}^{\infty} u_p(t,x')\phi_p(x)\,. \tag{45}$$

Substituting into the diffusion equation gives the time dependence of the coefficients $u_p(t) = u_p(0)\exp(-p^2t/\tau_d)$. This gives a precise meaning to the disengagement (reptation) time τ_d:

$$\tau_d = \frac{L^2}{\pi^2 D_c} = \frac{N^3 b^4 \zeta_0}{\pi^2 k_B T a^2}\,. \tag{46}$$

The final ingredient is to find the initial amplitudes $u_p(0)$:

$$u_p(0) = \int_0^L \phi_p(s)\Psi(s,x',0)ds = \sqrt{\frac{2}{L}}\sin\left(\frac{p\pi x'}{L}\right)\,. \tag{47}$$

So, using Equation 45 the final solution for the survival probability is

$$\Psi(x,x',t) = \sum_{p=1}^{\infty}\frac{2}{L}\sin\left(\frac{p\pi x'}{L}\right)\sin\left(\frac{p\pi x}{L}\right)\exp\left(\frac{-p^2t}{\tau_d}\right)\,. \tag{48}$$

To find the stress remaining we must calculate the mean survival probability of all tube segments (regardless of their initial and final values x, x'):

$$G(t) = G_0\int_0^L dx\int_0^L dx'\,\Psi(x,x',t) = G_0\sum_{p\ \mathrm{odd}}\frac{8}{\pi^2p^2}e^{-p^2t/\tau_d}\,. \tag{49}$$

As expected, the result is nearly single-exponential, and certainly in qualitative accord with the data in Figure 7. However, the prediction for the density of higher modes (decaying rapidly, as p^{-2}) is less than that experimentally seen. Approximating the sum in Equation 49 with an integral, we find $G''(\omega) \sim \omega^{-1/2}$ at frequencies higher than $1/\tau_d$, whereas Figure 7 shows a flatter decay.

2.5 Some comments on Rouse and reptation dynamics

The existence of modes with shorter relaxation times than the reptation time τ_d, arises from the greater rapidity with which tube segments originally near a chain end are evacuated, compared with those near the middle of the chain. The calculation above ignores all fluctuations of the total path length of the molecule along the tube (the chain is assumed to translate with its centre-of-mass) so would not be expected to estimate this contribution accurately. It turns out that the approximation of fixed path length is directly related to the difference between the cubic dependence of τ_d on N predicted above (Equation 46) and the experimentally observed result, $\tau_d \sim M^{3.4}$ [2, 7]. Moreover, an accurate calculation of these fluctuations becomes necessary to make any progress at all with entangled branched polymers (Section 3 below).

The two types of polymer dynamics presented above have established themselves as rather fundamental. For example, at long enough length scales, the dynamics of an unconstrained random walk with *any* local rule for its motion becomes equivalent to the Rouse description (see also Kremer, this volume). Similarly, reptation arises quite generally in the constrained case (another example is in polyethylene crystals!). Moreover the two dynamics are 'orthogonal' in their natural mathematical representations: Rouse modes do not diagonalise reptation dyanamics nor *vice versa* (despite our use of Fourier modes, which superficially resemble the Rouse modes, en route to Equation 49). However, in real polymer melts and entangled solutions both dynamics co-exist. This is because the tubes themselves are not permanent objects, but are subject to local rearrangement as constraints from neighbouring chains are released. Without this additional relaxation mechanism, the tube model overpredicts the alignment of chains in a strong shear flow, and severely underestimates the shear stress. Various authors have developed a formalism in which the chains *reptate* within tubes that behave as *Rouse* objects. While it is clear how to do this for weak flows (*e.g.* [11]), the nonlinear case remains the subject of current work, which it is hoped will provide a powerful formalism for melts at high shear rates.

3 Branched entangled polymers

In this section, we bring together several themes of this volume that converge in the study of entangled branched polymers. (For a fuller overview, see the recent review article [2].) They will furnish us with an attractive example of 'slow dynamics' with the following characteristics:

- Configurations relax by activated diffusion from entropic traps.
- Dynamics are exponentially slow in a tunable parameter (in this case the molecular weight of dangling arms).
- The relaxation is highly co-operative.
- The barrier distributions can be tailored accurately by chemistry in real experiments.
- The activated dynamics has hierarchical features dependent on polymer branching.

- The resulting theory is quantitative for linear response, with just two parameters required from experiment. These depend on local chemistry; results for polymers of all architectures then follow.
- There are natural extensions to non-linear rheology (not described here).

3.1 A preliminary exercise: diffusive barrier crossing

In many physical problems featured in this volume (see *e.g.* the lectures of Bouchaud and of Frenkel) we need to know the mean waiting time for escape of a single degree of freedom, such as a diffusing particle, over a barrier. This is usually written $\tau \simeq \tau_0 \exp[U/k_BT]$, where U is the barrier height and τ_0 some intra-trap diffusion time. This is often good enough, but there are instances where an accurate expression for τ_0 is needed, and branched polymers are one such case.

Consider a potential well $U(x)$ with a single minimum at $x = 0$. We want to calculate the average first passage time of a particle through a position $s > 0$, given that it is introduced at $x = 0$ at $t = 0$. This is equivalent to the mean lifetime of the particle if an absorbing wall is placed at $x = s$, or, equivalently, if the potential $U(x)$ is replaced by one which drops abruptly to $-\infty$ at $x = s$ (so a particle crossing this point never returns). In this language, s marks the top of the barrier over which particles escape (Figure 16).

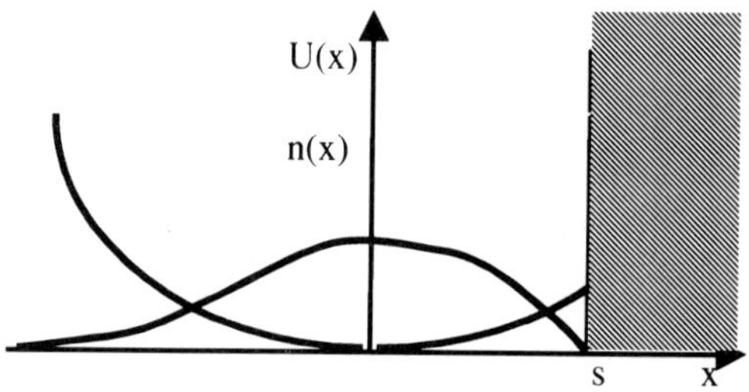

Figure 16. *Potential and steady-state distribution function for particles diffusing over the barrier at $x = s$.*

To solve this problem we imagine introducing a steady current $j\delta(x)$ of diffusers at the origin, and wait until a steady-state number density $n(x)$ of diffusers has been established. Then the total number of particles in the distribution is just the 'supply' current multiplied by the mean survival time τ. So

$$\tau(s) = \frac{1}{j}\int_{-\infty}^{s} n(x)dx\,. \tag{50}$$

We assume that the diffusion constant is D and work in units of k_BT for the energy. Then in the diffusive limit, $n(t,x)$ satisfies

$$\frac{\partial n}{\partial t} = -\frac{\partial}{\partial x}D\left(-\frac{\partial n}{\partial x} - n\frac{\partial U}{\partial x}\right) + j\delta(x) = 0\,, \tag{51}$$

at steady state. In $x > 0$ we may integrate this once directly to give

$$\frac{\partial n}{\partial x} + n\frac{\partial U}{\partial x} = -\frac{j}{D}\,, \tag{52}$$

and once more by using an integrating factor e^U to give (setting $U(0) = 0$)

$$n(x) = \frac{j}{D} e^{-U(x)} \int_x^s e^{U(x')} dx' \quad \text{for} \quad x > 0 \,. \tag{53}$$

For negative x, there is no net current in steady state: the mean current solely transports material from the origin to the absorber at $x = s > 0$. Hence the density for $x < 0$ obeys the equilibrium (Boltzmann) distribution, with the prefactor chosen to match the solution in $x > 0$. So

$$n(x) = \frac{j}{D} e^{-U(x)} \int_0^s e^{U(x')} dx' \quad \text{for} \quad x < 0 \,. \tag{54}$$

Now integrating $n(x)$ over all x (where it helps to reverse x, x' in the order of integration), and using Equation 50, we find

$$\tau(s) = \frac{1}{D} \int_0^s dx' e^{U(x')} \int_{-\infty}^{x'} dx \, e^{-U(x)} \,. \tag{55}$$

This exact solution for the mean lifetime can be further approximated when the barrier is high ($U(s) \gg 1$). For now the inner (x) integral in Equation 55 is completely dominated by the contribution near the origin where U is at its minimum, and a good approximation is the Gaussian integral found by expanding U to second order in x. The outer (x') integral is likewise dominated by the contribution near the upper limit. This may be expanded in terms of $U'(s)$ (or $U''(s)$ if the first derivative is zero at s) to give an exponential integral. The final result is

$$\tau(s) \simeq \frac{k_B T}{D U'(s)} \sqrt{\frac{k_B T \pi}{2 U''(0)}} \exp[U(s)/k_B T] \,. \tag{56}$$

Here factors of $k_B T$ have been restored. This shows the prefactor $\tau_0(s)$ of the dominant exponential activation factor is not necessarily close to a naive estimate of the diffusion time ($\tau_0 \simeq s^2/D$), especially when the potential barrier is large. We will find below that in the case of branched polymers escaping from topological traps, the full dependence of the pre-exponential factor on $U(x)$ is essential to producing quantitative results from theoretical models that can be compared to experimental data.

3.2 Experimental rheology of star polymers

We have already seen the large effect of introducing a branch point into the molecules of a polymer melt, in the relaxation modulus shown in Figure 9, where star polymers and linear chains are compared. There are equally remarkable differences in the way the viscosity η varies with molecular weight. Instead of the $N^{3.4}$ dependence of linear polymers, the viscosity increase for stars is dominated by an exponential growth. Comparison between different chemistries indicates that it is always the number of entanglements along the star polymer arms that matters ($\eta \sim \exp[\upsilon N_a/N_e]$, with N_a the size of an arm and υ a constant). More remarkably, the number of arms (providing that this at least 3 and not more than 30 or so) affects neither the viscosity (Figure 17) nor the relaxation spectrum.

The relaxation spectrum is vastly broader for star polymer melts than linear chains (Figure 18). As the molecular weight of the arms is increased, so the near-plateau in

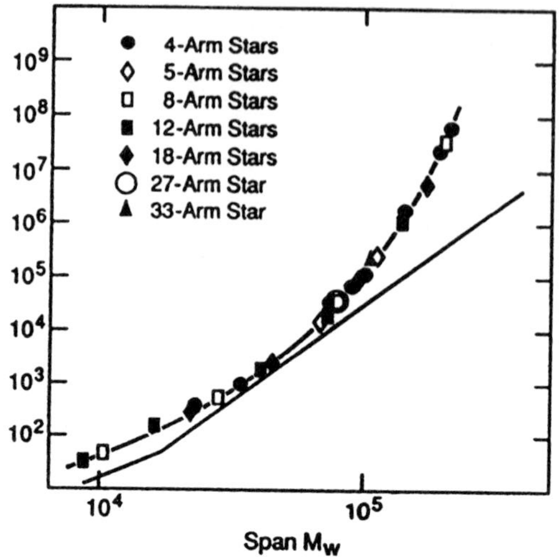

Figure 17. *Log-log plot of the viscosities of star polyisoprenes with molecular weight of the arms. The line represents accumulated data from linear polymers[3]. Stars of different numbers of arms fall onto the same plot.*

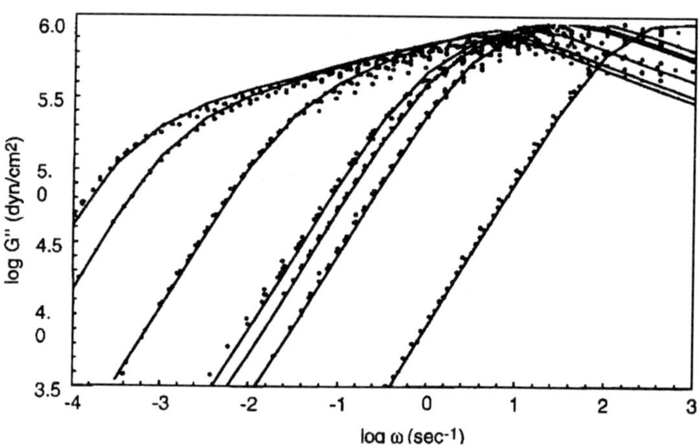

Figure 18. *Data on series of PI star polymers from [3] and corresponding theoretical predictions using the theory of [4].*

$G''(\omega)$ grows towards lower frequencies with the same exponential dependence as the viscosity. Other qualitative effects emerge in the rheology of more general branched polymers. The important industrial material 'low density polyethylene', which has a tree-branched structure, exhibits severe extensional hardening (compare Figure 6) in constrast to melts of linear chains, and often sets up quite different patterns of flow in complex geometries [2].

3.3 A tube model for star polymers

The qualitative prediction of the tube model for star polymers is clear to see, and was identified long ago by de Gennes [5]. The branch point suppresses reptation because double-occupation of a single tube by two arms is entropically unfavourable (one of the arms sacrifices its configurational entropy at length scales larger than a). However, the basic mechanism for configurational (and stress) relaxation is unchanged—tube segments must be visited by chain ends. In entangled star polymers this can only happen by 'path length fluctuations' in which a free end retraces its way back down the tube contour before re-emerging again into new tube. To make way for the retracting end, the arm must effectively shorten, which it does by creating unentangled loops within, or emerging from, the tube. In so doing all previously occupied tube from the original point occupied

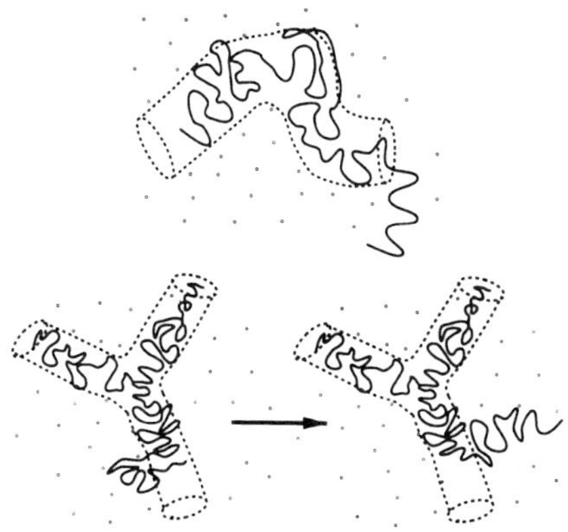

Figure 19. *Dominant relaxation processes in (top) linear and (bottom) star entangled polymers showing arm retraction and tube reconfiguration.*

by the free end, to its point of deepest retraction, is reconfigured (and corresponding stresses relaxed). Clearly, shallow retractions will happen much more frequently than deep retractions—this is the origin of the huge spread in relaxation times observed in star polymers (Figure 18). The deepest retractions themselves will become exponentially rare as the molecular weight of the arm increases—this is the origin of the molecular weight dependence of the retraction times in Figure 17. In other words, there is an effective potential well against which the free end moves along the tube. Moreover, we can see that the stress relaxation occurs without diffusion of the branch point itself, and occurs independently in each arm. The picture therefore gives also a qualitative understanding of the observed independence of the number of arms.

To make all this quantitative we need to calculate the effective potential in which the entangled path length of the arm executes its random motion. One way to construct the potential proceeds by noting that the entropy loss ($\simeq k_B$ per tube segment) of constraining an entangled polymer to its tube is quantitatively equivalent to applying a (constant) tension of $3k_BT/a$ along its length [7]. The free energy change associated with withdrawing an end is the work done against this tension. The entropic chain tension arises in a physical way: at time scales short enough for the tube constraints to be effectively permanent, each chain end is subject to random Brownian motion (at the scale of an entanglement strand) such that it may make a random choice of exploration of possible paths into the surrounding melt. Only one of these choices corresponds to retracing the chain back along its tube, shortening the so-called 'primitive path' (or curvilinear tube contour). Far more choices correspond to extending the primitive path. The net effect is as though to pull on the chain end with the stated tension, which is sustained by the free ends. The equilibrium length of the primitive path is the one that balances this tension against the entropic spring force.

We can, then, write a potential $U(z)$ for the length of the primitive path z by including both the (quadratic) curvilinear entropic-elastic term and the (linear) end-tension term as follows:

$$U(z) = \frac{3k_BT}{2Nb^2}z^2 - \frac{3k_BT}{a}z = \frac{3k_BT}{2Nb^2}s^2 + \text{const.}\,, \tag{57}$$

where $L = Nb^2/a$ is the equilibrium primitive path length of the chain and $s = L - z$ is the co-ordinate that measures the retraction of the free end from its equilibrium position. This quadratic potential will determine the fluctuation dynamics of an arm of an entangled star polymer: it gives the free energy paid for a retraction of the free end a distance s along the tube. Whenever this happens, the tube orientation is relaxed for all tube segments whose primitive path distance from the branch point is between $L - s$ and L.

The observations above can be turned into a semi-quantitative theory for star-polymer stress-relaxation [5] which is amenable to more quantitative refinement [6]. The key observation is that the diffusion equation for stress release, which arises in linear polymers via the passage of free ends out of oriented tube segments, is modified in star polymers by the potential $U(s)$. Each position along the arm, s, will possess its own characteristic stress relaxation time $\tau(s)$ given by the average first passage time of the diffusing free end to s.

But this is just the problem we addressed in the 'preliminary exercise' above! In Figure 16, the curve for $U(s)$ is now the quadratic potential (Equation 57) given by the tube model, and simple substitution of the arc-length potential into the general result (Equation 56), using $a^2 = (4/5)N_eb^2$ [7], gives for the longest relaxation time of a star with arm size N_a:

$$\tau(L) = \tau_e\left(\frac{N_a}{N_e}\right)^3\frac{\pi^3}{4}\exp\left(\frac{15}{8}\frac{N_a}{N_e}\right) \tag{58}$$

where τ_e is the Rouse time of an entanglement segment. (Note: the prefactor in Equation 58 is not insignificant!)

The relaxation modulus in this case can be written

$$G(t) = \frac{G_0}{L}\int_0^L p(s,t)ds\,, \tag{59}$$

where $p(s, t)$ is the survival probablity of the tube segment at s (the probability that it has not been visited by the free end before time t). To a good approximation this is just $\exp[-t/\tau(s)]$, and for highly entangled arms we can approximate it further by a step function in s. For consider the state of relaxation at any time t intermediate between $\tau(a)$ (the relaxation time of the first tube segments near the end of the arm) and $\tau(L)$ (that of the core-segments of the star). At t, some internal tube segment will typically be just in the process of reconfiguration via its first 'visit' by the free end. This segment will have an arclength co-ordinate s given roughly by $\tau(s) = t$. All segments exterior to the segment $s(t)$ (given by the inverse of the function $\tau(s)$) are almost certain to have relaxed, because their relaxation time scales are *exponentially* shorter than the current time t, while segments nearer to the core are conversely almost certainly unrelaxed.

Unfortunately, although qualitatively promising, this version of the theory fails disastrously at the quantitative level. A glance at the polyisoprene ($M_e = 5000$) star data (Figure 18) will suffice: the sample with the longest arm molecular weight (105,000) is predictied to carry an exponential factor in $\tau(L)$ (Equation 58) of approximately 10^{18}. Yet this factor must describe roughly the width of the 'relaxation shoulder' in $G''(\omega)$ in Figure 18, which is only 6 decades broad. The problem lies with the need to account for 'constraint release'—a small correction in the case of linear polymers (Section 2.5), but which in the case of star polymers becomes quite dominant. Fortunately it is also much simpler to treat in the case of star polymers, as we see next.

3.4 Hierarchical constraint release in star polymer melts

The great significance of constraint release to the dynamics of entangled star polymers arises from the very broad distribution of relaxation time scales we have discussed above. Fortunately, the same breadth of time scales provides a simple way of calculating the effect [4, 10]. As a consequence of the exponential separation of relaxation time scales along a star arm, by the time the population of tube segments of some s is relaxing, all segments at $s' < s$ (nearer a chain end) have renewed their configurations, typically many times. So chain segments at s (and those of $s' > s$) effectively *do not entangle* with these fast segments at the time scale $\tau(s)$ and beyond. Alternatively we can say that the tube is widened due to this effective dilution of the entanglement network: fast-relaxing segments act as *solvent* for the slower ones [2].

The new information necessary to make this approach quantitative is the dependence of the entanglement parameter N_e on the concentration Φ of unrelaxed segments. This is known from experiments on dilution of polymer melts (by theta-solvents: see the lectures by Khokhlov, this volume) to be approximately $N_e(\Phi) = N_{e0}/\Phi$, which corresponds to the approximately quadratic concentration dependence of $G_0 \sim \Phi^2$. (See [4] for a more general treatment.) At any stage in the relaxation dynamics of a melt of identical star polymers, therefore, when a segment s is currently relaxing for the first time, the effective N_e is $N_e(s) = N_{e0}/(1 - s/L)$. To recompute the relaxation times $\tau(s)$ with the dynamic dilution assumption, we consider the activated diffusion in a *hierarchical* way: to retract from s to $s + ds$, the *attempt frequency* is $\tau(s)^{-1}$ (the rate of relaxation events at level s), and the barrier height for progressing from s to $s + ds$ is $[U(s + ds; N_e(s)) - U(s; N_e(s))]$ where the notation for U indicates the dependence (through the tube diameter) on the 'running value' of N_e. Taking the limit of ds small gives the differential equation

$$\frac{dU}{ds} = \frac{\partial U}{\partial s}(s; N_e(s)) , \tag{60}$$

where N_e is held constant for the partial derivative on the right hand side. (The latter is found by differentiating Equation 57 with respect to $s = L - z$ and substituting for a in terms of N_e.) Integration of the result for dU/ds gives a renormalised potential $U(s)$

which is now (in units of $k_B T$) a cubic in s :

$$U_{\text{eff}}(x) = \frac{15}{8}\frac{N}{N_e}(x^2 - \frac{2}{3}x^3), \tag{61}$$

where we write $x = s/L$ for the fractional arm length retracted. The terminal time and viscosity are dominated, as we saw above, by the potential at complete retraction $U_{\text{eff}}(1)$. In units of $k_B T$ this was previously $(15/8)(N_a/N_e)$ (see Equation 58) but is now given by $(15/24)(N_a/N_e)$, in far closer agreement with experiments: the 18 decade shoulder for G'' in Figure 18 becomes 6 decades, as observed.

Equation 59 for $G(t)$ also needs modifying since each element of chain ds contributing to the stress relaxation now does so in an environment diluted by $(1 - s/L)$, so Equation 59 picks up this factor within the integrand as a coefficient of $p(s,t)$. The shape of the relaxation spectrum predicted by this procedure does indeed fit rheological data on pure star melts better than the previous theory [4], especially when corrected at high frequencies by a crossover to nonactivated tube loss very near the free end. The curves through the experimental points in the data for $G''(\omega)$ on the PI stars in Figure 18 were calculated via this scheme, using literature values of the two fitting parameters required, τ_e (a horizontal shift on the figure) and G_0 (a vertical shift). They fit the data without further adjustment. Indeed, a great strength of this remarkably powerful theoretical framework is that in principle, *only* these two parameters are required for all molecular weights and architectures of a given chemistry. For example, there are straightforward generalisations of the above to bimodal blends of two star polymer fractions of different arm molecular weight. Without any change of parameters these account well for data on carefully-synthesised samples [12].

3.5 General chain architectures

The picture of hierarchical retraction dynamics with dynamic dilution can be generalised in a straightforward way to arbitrarily branched polymer 'trees'. For structures with many branch points a simplification is to treat the relaxation in discrete stages, calculating the time scales at which arm retraction has penetrated to each layer of the tree. At each stage the effective geometry of the molecule simplifies, as faster relaxing (outer) segments cease to be part entangled network, but instead dilute the current value of N_e.

For example, the Cayley tree of n layers and functionality f [13], contains f^n segments in its outermost layer and $(f^{n+1} - f)/(f - 1)$ segments altogether. (In such a tree, each stem branches into f stems, with no dead ends until the nth layer is reached, at which point all stems terminate.) The effective concentration of unrelaxed segments after m levels have relaxed is $C(m) = (f^{n-m+1} - f)/(f^{n+1} - f) \simeq f^{-m}$ when n is large. Solving the retraction problem from level m to level $m + 1$ (with the approximation that the effective concentration at level m is valid throughout that stage of the hierarchy) gives the recurrence relation

$$\tau_{m+1} = \tau_m \exp[v(N_x/N_e)f^{-m}], \tag{62}$$

with solution

$$\tau_m = \tau_0 \exp\left(v(N_x/N_e)\left(\frac{1 - f^{-m}}{1 - f}\right)\right). \tag{63}$$

Here $v = 15/8$ and N_x the number of monomers between branch points on the tree. At this level of approximation we may also take $G(t) \simeq G_0[C(m(t))]^2$. This leads to the logarithmic form of stress relaxation

$$G(t) = G_0 \ln \left(\frac{\tau_{\max}}{t} \right)^{\theta}, \tag{64}$$

where $\tau_{\max}$ is a (finite) limiting relaxation time (τ_m as $m \to \infty$), and θ is a branching-dependent exponent with a value of 2 in this case. It turns out that other, less regular, tree-like structures also have this form of $G(t)$ but with different values for θ. For example, the ensemble of randomly-branched trees predicted to occur at the classical mean-field gelation point has $\theta = 4$ [14].

Experimental verification for more general architectures with well-controlled materials has so far only proceeded to two-level branching. A number of groups have studied polymers shaped like the letter H in the melt; extra arms can be added at the same two junctions to make a 'pom-pom' polymer [15][16]. In the H-polymer case the frequency-dependent rheology directly reflects the chain structure (once the polymers are well-entangled) with features in $G''(\omega)$ arising from both outer arms and the central 'cross-bar' (at lower frequencies). These are shown in Figure 20. The theory (solid curves) does

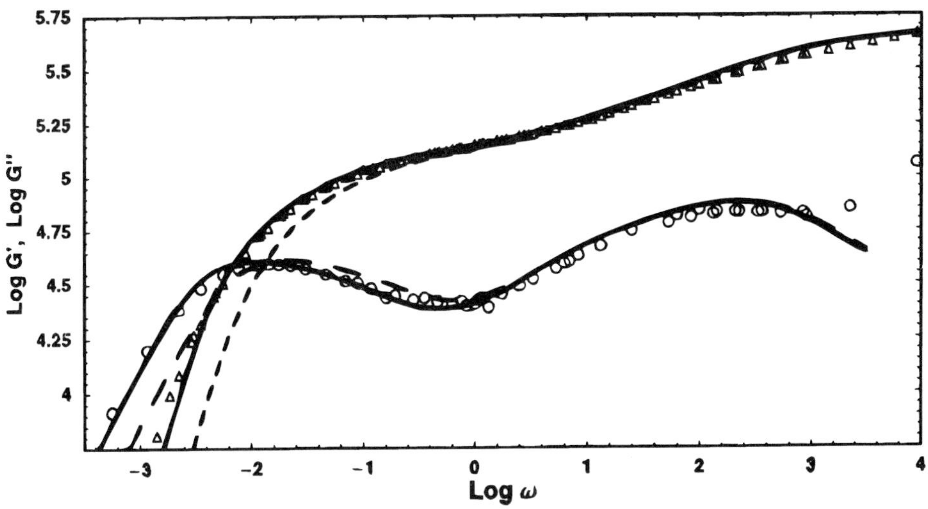

Figure 20. *Linear viscoelastic data for an H-polyisoprene melt with molecular weights for arms of $M_a = 20000$ and the cross-bar of $M_c = 111000$ (synthesised by J. Allgaier [15]). Solid and dashed lines are the theory with and without polydispersity respectively.*

indeed grasp the quantitative form of the rheology, once it is realised that the cross-bar motion is actually *reptation*, in spite of the branched nature of the polymer! For at long time scales when the outer arms have completely relaxed, only the cross-bar sections of the molecules remain topologically 'active' and so behave as linear polymers in tubes with a diameter set by their mutual entanglements only (see Figure 21).

The sharpness of the peak in $G''(\omega)$ at low frequencies in the H-polymer data arises precisely from the narrowness of the mode distribution in reptation of a linear chain (Equa-

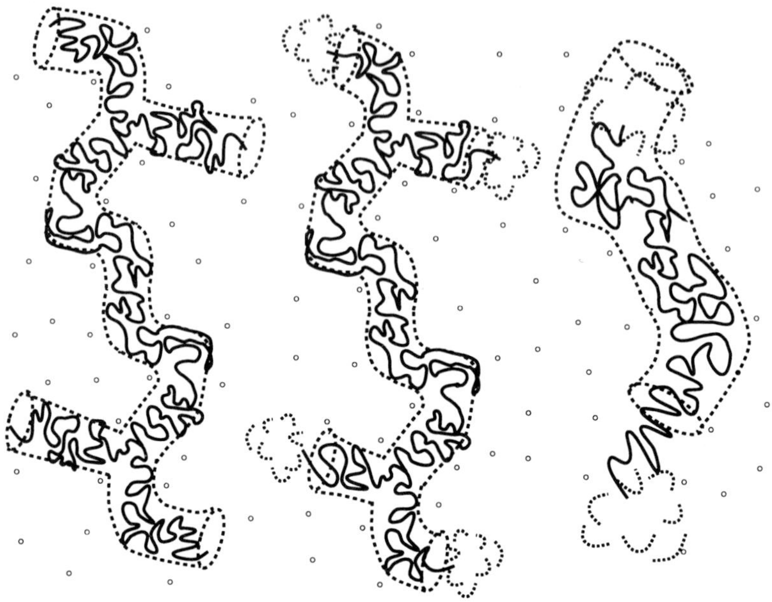

Figure 21. *Successive stages in the configurational relaxation of an entangled H-polymer. In the final stage it reptates as a linear chain.*

tion 49). One additional important insight arose from studies on these highly monodisperse model materials—the role of residual polydispersity. Even the small (10% level) variation in molecular weights of the arms in the specially polymerised polyisoprene H-molecules had a significant quantitative effect on the linear relaxation spectrum. This is because the diffusion constants of the branch points depend exponentially on the arm molecular weight. This greatly amplifies the contribution of the few arms that are significantly longer than the average, incresing relaxation times overall. The calculation is straightforward for small polydispersity, and produces the full curves in Figure 20 (dashed curves are for the purely monodisperse case).

3.6 Conclusions

Although at first sight very complex, the system of entangled flexible branched polymers seems to give rise to a rather simple picture of hierarchical dynamics for configurational relaxation, needing only two parameters for each chemistry of polymer. The remarkable slowness of trapped dynamics results in molecular relaxation times on a time scales of seconds and even hours, for moderate molecular weights. Hierarchical dynamics can produce features very well separated in time scale even though they arise from relatively close molecular 'neighbours' along the chain backbone. An exciting challenge for the next few years lies in the application of this branch of soft condensed matter physics to the full complexity of industrial materials.

References

[1] Larson R G, 1999, *The Structure and Dynamics of Complex Fluids* (Clarendon Press, Oxford).

[2] McLeish T C B and Milner S T, 1999, *Adv Polym Sci* **143**, 195.

[3] Fetters L J, Kiss A D, Pearson D S, Quack G F, Vitus F J, 1993, *Macromolecules* **26**, 647.

[4] Milner S T and McLeish T C B, 1997, *Macromolecules* **30**, 2159.

[5] de Gennes P G, 1975, *J Phys (Paris)* **36**, 1199.

[6] Pearson D S and Helfand E, 1984, *Macromolecules* **19**, 888.

[7] Doi M and Edwards S F, 1986, *The Theory of Polymer Dynamics* (Clarendon Press, Oxford).

[8] Rouse P E, 1953, *J Chem Phys* **21**, 1272.

[9] Zimm B H, 1954, *J Chem Phys* **24**, 269.

[10] Ball R C and McLeish T C B, 1989, *Macromolecules* **22**, 1911.

[11] Viovy J L, Rubinstein M and Colby R H, 1991, *Macromolecules* **24**, 3587.

[12] Blottière B, McLeish T C B, Hakiki A, Young, R N and Milner S T, 1998, *Macromolecules* **31**, 9295.

[13] McLeish T C B, 1988, *Europhys Lett* **6**, 511.

[14] Rubinstein M, Zurek S, McLeish T C B and Ball R C, 1990, *J Phys (Paris)* **51**, 757.

[15] McLeish T C B, Allgaier J, Bick D K, Bishko G, Biswas P, Blackwell R, Blottière B, Clarke N, Gibbs B, Groves D J, Hakiki A, Heenan R, Johnson, Kant R, Read D J and Young R N, 1999, *Macromolecules* **32**, 6734.

[16] Archer L A and Varshney S K, 1998, *Macromolecules* **31**, 6348.

Introduction to colloidal systems

Daan Frenkel

FOM Institute for Atomic and Molecular Physics, The Netherlands

1 Introduction

Karl Marx has said that, in history, things always happen twice: the first time as a tragedy, the second time as a farce. This comment of Marx applied to Napoleon I and III. However, if we strip the subjective interpretation (tragedy or farce) from this sentence, it could apply to many phenomena in physics. In physics, there often appears to be a similarity between phenomena on very different length and time-scales but, on closer inspection, there are important, even qualitative differences. Examples abound: in some respects, light waves resemble ripples on a pond but, in most respects, they are totally different. The Bohr model of the atom resembled a planetary system but, of course, the differences are so important that, in the end, they led to the demise of the Bohr model. These two examples illustrate an important point: in physics, analogies are very useful in formulating an approximate description of a phenomenon—but even more interesting than the analogy itself, is its breakdown.

In many ways, colloids behave like giant atoms, and quite a bit of the colloid physics can be understood in this way. However, much of the interesting behaviour of colloids is related to the fact that they are, in many respects, *not* like atoms. In these lectures, I shall start from the picture of colloids as oversized atoms or molecules, and then I shall selectively discuss some features of colloids that are different. My presentation of the subject might seem a bit strange, because I am a computer simulator, rather than a colloid scientist. Colloids are the computer simulator's dream, because many of them can be represented quite well by models—such as the hard-sphere and Yukawa models—that are far too simple to represent molecular systems. On the other hand, colloids are also the simulator's nightmare, or at least challenge, because if we look more closely, simple models do not work: this is sometimes true for the static properties of colloids (*e.g.* in the case of charged colloids) and even more often, in the case of colloid dynamics.

What are colloids? Usually, we refer to a substance as a colloidal suspension if it is a dispersion of more-or-less compact particles with sizes within a certain range (typically, 1nm–1μm). However, it would be more logical to classify colloids according to some physical criterion. To this end, we should compare colloidal particles with their 'neighbours': small molecules on one end of the scale, and bricks on the other. What distinguishes colloids from small molecules? I would propose that the important difference is that for the description of colloids, a detailed knowledge of the 'internal' degrees of freedom is not needed—in particular, the discrete, atomic nature of matter should be irrelevant. That is *not* to say that the chemical nature of the constituent atoms or molecules is irrelevant—simply that, in order to describe a colloid, we do not need to know the detailed microscopic arrangement of these constituents. This definition has the advantage that it allows for the fact that particles may behave like colloids in some respects, and like 'molecules' in others. For instance, we cannot hope to understand the biological function of proteins if we do not know their atomic structure. However, we can understand a lot about the phase behaviour of proteins without such knowledge. This ambiguous nature of macromolecules may persist even at length scales that are usually considered colloidal. For instance, for the biological function of the Tobacco Mosaic Virus, the precise sequence of its genetic material is important. But its tendency to form colloidal liquid crystals depends only on coarse-grained properties, such as shape, flexibility and charge.

Let us next consider the other side of the scale. What is the difference between a colloidal particle and a brick? The behaviour of colloids is governed by the laws of statistical mechanics. In equilibrium, colloidal suspensions occur in the phase with the lowest free energy, and the dynamics of colloids in equilibrium is due to thermal (Brownian) motion. In principle, this should also be true for bricks. But in practice, it is not. In order for bricks to behave like colloids, they should be able to evolve due to Brownian motion. There are two reasons why bricks do not. First of all, on earth, all particles are subject to gravity. The probability of finding a particle of mass m at a height h above the surface of the earth is given by the barometric height distribution:

$$P(h) = \exp(-mgh/k_BT) \ , \tag{1}$$

where m is the effective mass of the colloidal particle (*i.e.* the mass, minus the mass of the displaced solvent), T is the temperature and k_B is Boltzmann's constant. The average height of the colloid above the surface is equal to $\langle h \rangle = k_BT/(mg)$. For a 1kg brick at room temperature, $\langle h \rangle = O(10^{-20})$ cm. This tells us something that we all know: bricks don't float around due to thermal motion. One way to delimit the colloidal regime is to require that $\langle h \rangle$ is larger than the particle diameter. Suppose we have a spherical particle with diameter σ and (excess) mass density ρ, then our criterion implies

$$\frac{\pi g \rho \sigma^4}{6} = k_BT \ . \tag{2}$$

For a particle with an excess density of 1g/cm^3, the above equality is satisfied for a value of $\sigma \approx 1\mu$m, *i.e.* on earth. In the microgravity environment that prevails in space, much larger particles would behave like colloids (not bricks though, because it is virtually impossible to reduce all accelerations to less than $10^{-20}g$). Another way to make large particles behave like colloids on earth, is to match the density of the solvent to that of the particle. Yet, even if we could succeed in doing all this for a brick, it would still not behave like a colloid. Colloidal particles should be able to move due to diffusion (*i.e.* thermal motion). How long does it take for a particle to move a distance equal to its own diameter? In a time t, a particle typically diffuses a distance $\sqrt{2Dt}$. For a spherical particle, the diffusion constant is given by the Stokes-Einstein relation $D = k_BT/(3\pi\eta\sigma)$, where η is the viscosity of the solution. Hence, a particle diffuses a distance comparable to its own diameter in a time

$$\tau = O(\eta\sigma^3/k_BT) . \tag{3}$$

For a $1\mu m$-colloid in water, this time is of the order of one second. For a brick, it is of the order of ten million years. Hence, even though bricks in zero-gravity may behave like colloids, they will not do so on a human time-scale. Clearly, what we define as a colloid, also depends on the observation time. Again, 1 micron comes out as a natural upper limit to the colloidal domain.

In summary, a colloid is defined by its behaviour. For practical purposes, the colloidal regime is between 1 nanometre and 1 micrometre. But these boundaries are not sharp. And the lower boundary is ambiguous: a particle may behave like a colloid in some respects, but not in others.

2 Forces between colloids

Most colloidal suspensions are solutions of relatively large particles in a simple molecular solvent. Yet, the description of the static properties of such a solution resembles that of a system of atoms in vacuum—somehow, the solvent does not appear explicitly. At first sight, this seems like a gross omission. However, as pointed out by Onsager [1], we can eliminate the degrees of freedom of the solvent in a colloidal dispersion. What results is the description that only involves the colloidal particles, interacting through some *effective* potential (the 'potential of mean force') that accounts for all solvent effects. Below, I briefly sketch how this works. Consider a system of N_c colloids in a volume V at temperature T. The solvent is held at constant chemical potential μ_s, but the number of solvent molecules N_s is fluctuating. The 'semi-grand' partition function of such a system is (with $\beta = 1/k_BT$)

$$\Xi(N_c, \mu_s, V, T) \equiv \sum_{N_s=0}^{\infty} \exp(\beta\mu_s N_s) Q(N_c, N_s, V, T) . \tag{4}$$

The canonical partition function $Q(N_c, N_s, V, T)$ is given by the classical expression for a mixture

$$Q(N_c, N_s, V, T) = \frac{q_{\mathrm{id},c}(T)^{N_c} q_{\mathrm{id},s}(T)^{N_s}}{N_c! N_s!} \int dr^{N_c} dr^{N_s} \exp[-\beta U(r^{N_c}, r^{N_s})] . \tag{5}$$

where $q_{\mathrm{id},\alpha}$ is the kinetic and intra-molecular part of the partition function of a particle of species α, and r^{N_c} (r^{N_s}) denotes a $3N_c$ ($3N_s$) dimensional vector specifying a complete set of colloid (solvent) coordinates. The $q_{\mathrm{id},\alpha}$ terms are assumed to depend only on temperature, and not on the inter-molecular interactions (sometimes this is not true, *e.g.* in the case of polymers—I shall come back to that point later). In what follows, I shall usually drop the factors $q_{\mathrm{id},\alpha}$ (more precisely, I shall account for them in the definition of the chemical potential: *i.e.* $\mu_\alpha \to \mu_\alpha + k_BT \ln q_{\mathrm{id},\alpha}$). The interaction potential $U(r^{N_c}, r^{N_s})$ can always be written as $U_{cc} + U_{ss} + U_{sc}$, where U_{cc} is the direct colloid-colloid interaction (*i.e.* $U(r^{N_c}, r^{N_s})$ for $N_s = 0$), U_{ss} is the solvent-solvent interaction (*i.e.* $U(r^{N_c}, r^{N_s})$ for $N_c = 0$), and U_{sc} is the solvent-colloid interaction $U(r^{N_c}, r^{N_s}) - U_{cc}(r^{N_c}) - U_{ss}(r^{N_s})$. With these definitions, we can write

$$Q(N_c, N_s, V, T) = \frac{1}{N_c!} \int dr^{N_c} \exp[-\beta U_{cc}] \left\{ \frac{1}{N_s!} \int dr^{N_s} \exp[-\beta(U_{ss} + U_{sc})] \right\} \tag{6}$$

and hence

$$\begin{aligned} \Xi(N_c, \mu_s, V, T) = & \frac{1}{N_c!} \int dr^{N_c} \exp[-\beta U_{cc}] \\ & \times \left\{ \sum_{N_s=0}^{\infty} \frac{\exp(\beta \mu_s N_s)}{N_s!} \int dr^{N_s} \exp[-\beta(U_{ss} + U_{sc})] \right\} . \end{aligned} \tag{7}$$

We can rewrite this in a slightly more suggestive form. If we define the usual canonical and grand-canonical partition functions for solvent alone as

$$Q_s(N_s, V, T) \equiv \frac{1}{N_s!} \int dr^{N_s} \exp[-\beta U_{ss}] , \tag{8}$$

$$\Xi(\mu_s, V, T) \equiv \sum_{N_s=0}^{\infty} \exp(\beta \mu_s N_s) Q_s(N_s, V, T) , \tag{9}$$

then

$$\begin{aligned} \Xi(N_c, \mu_s, V, T) = & \frac{1}{N_c!} \int dr^{N_c} \exp[-\beta U_{cc}] \\ & \times \left\{ \sum_{N_s=0}^{\infty} \exp(\beta \mu_s N_s) Q_s(N_s, V, T) \left\langle \exp[-\beta U_{sc}] \right\rangle_{N_c, N_s, V, T} \right\} \\ = & \frac{\Xi(\mu_s, V, T)}{N_c!} \int dr^{N_c} \exp[-\beta U_{cc}] \left\langle \exp[-\beta U_{sc}] \right\rangle_{\mu_s, T} , \end{aligned} \tag{10}$$

where

$$\left\langle \exp[-\beta U_{sc}] \right\rangle_{\mu_s, V, T} \equiv \frac{\sum_{N_s=0}^{\infty} \exp(\beta \mu_s N_s) Q_s(N_s, V, T) \left\langle \exp[-\beta U_{sc}] \right\rangle_{N_c, N_s, V, T}}{\Xi(\mu_s, V, T)} . \tag{11}$$

Note that this quantity still depends on all the colloid coordinates, r^{N_c}: it is the average over solvent coordinates of the Boltzmann factor for the solvent-colloid interaction. We now define the *effective* colloid-colloid interaction as

$$U_{cc}^{\text{eff}}(r^{N_c}) \equiv U_{cc}(r^{N_c}) - k_B T \ln \left\langle \exp[-\beta U_{sc}(r^{N_c})] \right\rangle_{\mu_s, V, T} . \tag{12}$$

We refer to $U_{cc}^{\text{eff}}(r^{N_c})$ as the *potential of mean force.* Note that the potential of mean force depends explicitly on the temperature and on the chemical potential of the solvent. In the case where we study colloidal suspensions in *mixed* solvents, the potential of mean force depends on the chemical potential of all components in the solvent (an important example is a colloidal dispersed in a polymer solution).

At first sight, it looks as if the potential of mean force is a totally intractable object. For instance, even when the colloid-solvent and solvent-solvent interactions are pairwise additive, the potential of mean force is not. (Note that we have, thus far, not even assumed pairwise additivity). However, we should bear in mind that even the 'normal' potential energy function that we all think we know and love, is also not pairwise additive—that is why we can hardly ever use the pair potentials that describe the intermolecular interactions in the gas phase to model simple liquids. In fact, in many cases, we can make very reasonable estimates of the potential of mean force. It also turns out that the dependence of the potential of mean force on the chemical potential of the solvent molecules is a great advantage: it will allow us to *tune* the effective forces between colloids *simply by changing the composition of the solvent.* (You all know this: simply add some vinegar to milk, and the colloidal fat globules in the milk start to aggregate.) In contrast, in order to change the forces between atoms in the gas phase, we would have to change Planck's constant or the mass or charge of an electron. Hence, colloids are not simply giant atoms, they are *tunable* giant atoms.

We shall now briefly review the nature of inter-colloidal interactions. It will turn out that, almost all colloid-colloid interactions depend on the nature of the solvent and are, therefore, potentials of mean force.

2.1 Hard-core repulsion

Colloidal particles tend to have a well-defined size and shape. They behave like solid bodies—in fact, many colloidal particles *are* fairly solid (*e.g.* the colloids that Perrin used to determine Avogadro's number were small rubber balls, silica colloids are small glass spheres and PMMA colloids are made out of plastic). Solid bodies cannot interpenetrate. This property can be related to the fact that, at short range, the interaction between (non-reactive) atoms is harshly repulsive. This is due to the Pauli exclusion principle. This hard-core repulsion is about the only colloid-colloid interaction that is essentially independent of the solvent. In fact, colloidal crystals can be dried and studied in the electron microscope because the Pauli exclusion principle works just as well in vacuum as in solution. However, there are also other mechanisms that lead to 'hard-core' repulsion in colloids: for instance, short-ranged Coulomb-repulsion between like-charged colloids, or entropic repulsion between colloids that have a polymer 'fur', or even solvent-induced repulsion effects. All these repulsion mechanisms are sensitive to the nature of the solvent. We shall come back to them later.

2.2 Coulomb interaction

The Coulomb interaction would seem to be the prototype of a simple, pairwise additive interaction. In fact, it is. However, for every charge carried by the colloidal particles, there is a compensating charge in the solvent. These counter charges 'screen' the direct Coulomb repulsion between the colloids. I put the word 'screen' in quotes because it is too passive a word to describe what the counterions do: even in the presence of counterions and added salt ions the direct, long-ranged Coulomb repulsion between the colloids exists—but it is almost completely compensated by a net attractive interaction due to the counterions. The net result is an *effective* interaction between the colloids that is short-ranged *i.e.* it decays asymptotically as $\exp(-\kappa r)/r$, with κ the inverse screening length ($\kappa = 1/r_D$) that appears in the Debye-Hückel theory of electrolytes:

$$\kappa = \sqrt{\frac{4\pi}{\epsilon k_B T} \sum_i \rho_i q_i^2} \ ,$$

where ϵ is the dielectric constant of the solvent and ρ_i is the number density of ionic species i with charge q_i. (Here, and below, we used rationalised units for electrostatics, rather than SI units).

The first expression for the effective electrostatic interaction between two charged colloids was proposed Derjaguin, Landau, Verweij and Overbeek (DLVO) [2]:

$$V_{\text{Coulomb}} = \left(\frac{Q \exp(\kappa R)}{1 + \kappa R} \right)^2 \frac{\exp(-\kappa r)}{\epsilon r} \ , \tag{13}$$

where r is the distance between the two charged colloids, Q is the (bare) charge of the colloid and R is its 'hard-core' radius. Ever since, there have been attempts to improve on the DLVO theory. However, the theory of the effective electrostatic interaction between colloids is subtle and full of pitfalls. Usually, the electrostatic interaction between like-charged colloids is repulsive. However, under certain conditions it can be attractive. Sogami and Ise [3] have reported many experiments that provide evidence for such attraction. These authors suggested that this attraction should even be present at the level of the effective pair interaction. Recently, however, detailed experimental information has become available [4] that suggests that the Coulomb attraction between like-charged colloids is *not* present in the interaction between an isolated pair of colloids in the bulk solvent. At present, experiment and theory both suggest that all attractive interactions are either mediated by the presence of confining walls [5–7], (but see, however, [8]) or, in the bulk, they are due to many-body effects [9]. In addition, fluctuations in the charge distribution on the colloids may lead to dispersion-like attractive interactions (see *e.g.* [10]) that are also non-pairwise additive. Having said all this, the old DLVO theory usually yields an excellent first approximation for the electrostatic interaction between charged colloids.

2.3 Dispersion forces

Dispersion forces are due to the correlated zero-point fluctuations of the dipole moments on atoms or molecules. As colloids consist of many atoms, dispersion forces act between colloids. However, it would wrong to conclude that the solvent has no effect on the dispersion forces acting between colloids. After all, there are also dispersion forces acting between the colloids and the solvent, and between the solvent molecules themselves. In fact, for a pair of polarisable molecules, the dispersion interaction depends on the polarisabilities (α_1 and α_2) of the individual particles

$$u_{\text{disp}}(r) \sim -\frac{3\alpha_1 \alpha_2 h \sqrt{\nu_1 \nu_2}}{4\pi r^6} \equiv -\frac{C_{\text{disp}}(12)}{r^6} \ , \tag{14}$$

where $h\nu_i$ is a characteristic energy associated with the optical transition responsible for the dipole fluctuations in molecule i (in what follows, we shall assume the frequency ν_i to be the same for all molecules). The net dispersion force between colloidal particles in suspension depends on the difference in polarisability per unit volume of the solvent and the colloid. The reason is easy to understand: if we insert two colloidal particles in a polarisable solvent, we replace solvent with polarisability density $\rho_s\alpha_s$ by colloid with polarisability density $\rho_c\alpha_c$. If the two colloidal particles are far apart, each colloid contributes a constant amount proportional to $-\rho_s\alpha_s(\rho_c\alpha_c-\rho_s\alpha_s)$ to the dispersion energy. However, if at short inter-colloidal distances there is an additional *effective* colloid-colloid interaction that is proportional to $-(\rho_c\alpha_c - \rho_s\alpha_s)^2/r_{cc}^6$, then this leads to an *attractive interaction* irrespective of whether the polarisability density of the colloids is higher or lower than that of the solvent. On the other hand, in a colloid mixture, the dispersion force need not be attractive: if the polarisability density of one colloid (denoted by $c1$) is *higher* than that of the solvent, and the polarisability density of the other (denoted by $c2$) is *lower*, then the positive-definite square $(\rho_c\alpha_c - \rho_s\alpha_s)^2$ is replaced by the negative product $(\rho_{c1}\alpha_{c1} - \rho_s\alpha_s)(\rho_{c2}\alpha_{c2} - \rho_s\alpha_s)$ and hence the effective dispersion forces between these two colloids are *repulsive*.

The polarisability density of bulk phases is directly related to the refractive index. For instance, the Clausius-Mosotti expression for the refractive index is

$$\frac{n^2-1}{n^2+2} = \frac{4\pi\rho\alpha}{3} . \tag{15}$$

Hence, if the refractive index of the solvent is equal to that of the colloidal particles, then the effective dispersion forces vanish! This procedure to switch off the effective dispersion forces is called *refractive index matching*. In light-scattering experiments on dense colloidal suspensions, it is common to match the refractive indices of solvent and colloid in order to reduce multiple scattering. Thus, precisely the conditions that minimise the dispersion forces are optimal for light-scattering experiments.

Colloids are not point particles, therefore Equation 14 has to be integrated over the volumes of the interacting colloids, to yield the total dispersion interaction

$$V_{\text{disp}}(r) = -\frac{A}{6}\left\{\frac{2R^2}{r^2-4R^2} + \frac{2R^2}{r^2} + \ln\frac{r^2-4R^2}{r^2}\right\} , \tag{16}$$

where A is the so-called *Hamaker constant.* In the simple picture sketched above, A would be proportional to $(\rho_c\alpha_c-\rho_s\alpha_s)^2$. However, in a more sophisticated theoretical description of the dispersion forces between macroscopic bodies (see *e.g.* the book by Israelachvili [11]), the Hamaker constant can be related explicitly to the (frequency-dependent) dielectric constants of the colloidal particles and the solvent. This analysis affects the value of the constant A but, to a first approximation, not the functional form of Equation 16.

2.4 DLVO potential

Combining Equations 13 and 16, we obtain the DLVO potential that describes the interaction between charged colloids

$$V_{DLVO}(r) = \left(\frac{Q\exp(\kappa R)}{1+\kappa R}\right)^2 \frac{\exp(-\kappa r)}{\epsilon r} - \frac{A}{6}\left\{\frac{2R^2}{r^2-4R^2} + \frac{2R^2}{r^2} + \ln\frac{r^2-4R^2}{r^2}\right\} . \tag{17}$$

This potential is shown in Figure 1. Note that, at short distances, the dispersion forces always win. This suggests that the dispersion interaction will always lead to colloidal

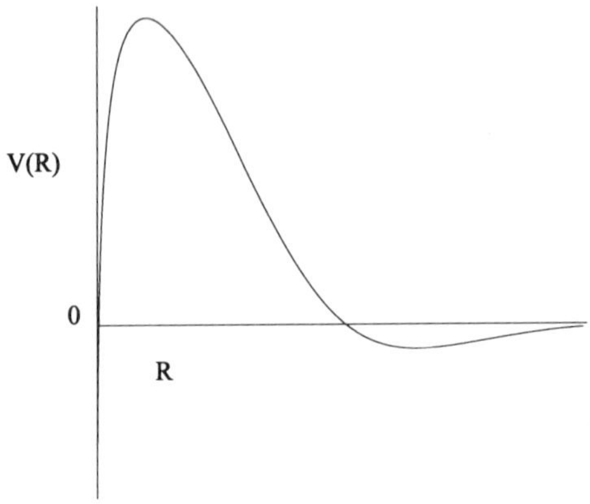

Figure 1. *The DLVO potential has a deep minimum at short distances. At larger distances, the Coulomb repulsion dominates. This leads to the local maximum in the curve. At still larger distances, the dispersion interaction may lead to a secondary minimum.*

aggregation. However, the electrostatic repulsion usually prevents colloids from getting close enough to fall into the primary minimum of the DLVO potential. The height of this stabilising barrier depends (through κ) on the salt concentration. Adding more salt will lower the barrier and, eventually, the colloids will be able to cross the barrier and aggregate.

Density matching—an intermezzo

In addition to refractive index matching, it is useful to try to match the density of the solvent to that of the colloid. This has an utterly negligible effect on the interaction between colloids. But, as far as gravity is concerned, density-matched colloidal particles are neutrally buoyant—that is they behave as if they have a very small (ideally zero) positive or negative excess mass. This is the mass that enters into the barometric height distribution (Equation 1). Hence, by density-matching, we can study bulk suspensions of colloids that would otherwise quickly settle on the bottom of the container.

2.5 Depletion interaction

One of the most surprising effects of the solvent on the interaction between colloids, is the so-called depletion interaction. Unlike the forces that we have discussed up to this point, the depletion force is not a solvent-induced modification of some pre-existing force between the colloids. It is a pure solvent effect. It is a consequence of the fact that the colloidal particles exclude space from the solvent molecules. To understand it, return to Equation 12:

$$U_{cc}^{\text{eff}}(r^{N_c}) \equiv U_{cc}(r^{N_c}) - k_B T \ln \left\langle \exp[-\beta U_{sc}(r^{N_c})] \right\rangle_{\mu_s,V,T} \quad .$$

Let us consider a system of hard particles with no additional attractive or repulsive interaction. In that case, all the contributions to the second term of the *effective* potential in Equation 12 are depletion interactions. These interactions can be attractive, even though all direct interactions in the system are repulsive.

To illustrate this, consider a trivial model system, namely a 2-dimensional square lattice with at most one particle allowed per square [12].

Figure 2. *Two-dimensional lattice model of a hard-core mixture of large colloidal particles (grey squares) and small solvent particles (black squares). Averaging over the solvent degrees of freedom results in a net attractive interaction (depletion interaction) between the 'colloids'.*

Apart from the fact that no two particles can occupy the same square cell, there is no interaction between the particles. For a lattice of N sites, the grand-canonical partition function is:

$$\Xi = \sum_{\{n_i\}} \exp[\beta\mu_c \sum_i n_i] \,. \tag{18}$$

The sum is over all allowed sets of occupation numbers $\{n_i\}$ and μ_c is the chemical potential of the 'colloidal' particles. Next, we include small 'solvent' particles that are allowed to sit on the links of the lattice (see Figure 2). These small particles are excluded from the edges of a cell that is occupied by a large particle. For a given configuration $\{n_i\}$ of the large particles, one can then calculate exactly the grand canonical partition function of the small particles. Let $M = M(\{n_i\})$ be the number of free spaces accessible to the small particles. Then clearly:

$$\Xi_{\text{small}}(\{n_i\}) = \sum_{l=0}^{M} \frac{M! z_s^l}{l!(M-l)!} = (1+z_s)^{M(\{n_i\})}, \tag{19}$$

where $z_s \equiv \exp(\beta\mu_s)$ is the fugacity of the small particles. M can be written as

$$M(\{n_i\}) = Nd - 2d \sum_i n_i + \sum_{\langle ij \rangle} n_i n_j \,, \tag{20}$$

where we have given the result for general space dimension d; Nd is the number of links on the lattice and the second sum is over nearest-neighbour pairs and comes from the fact that when two large particles touch, the number of sites excluded for the small particles is $4d-1$, not $4d$. Whenever two large particles touch, we have to correct for this overcounting of excluded sites. The total grand-partition function for the mixture is:

$$\Xi_{\text{mixture}} = \sum_{\{n_i\}} \exp\left[(\beta\mu_c - 2d\log(1+z_s))\sum_i n_i + [\log(1+z_s)]\sum_{\langle ij\rangle} n_i n_j\right] , \tag{21}$$

where we have omitted a constant factor $(1+z_s)^{Nd}$. Now we can bring this equation into a more familiar form by using a standard procedure to translate a lattice-gas model into a spin model. We define spins s_i such that $2n_i - 1 = s_i$ or $n_i = (s_i+1)/2$. Then we can write Equation 21 as

$$\Xi_{\text{mixture}} = \sum_{\{n_i\}} \exp\left[\frac{\beta\mu_c - d\log(1+z_s)}{2}\sum_i s_i + \frac{\log(1+z_s)}{4}\sum_{\langle ij\rangle} s_i s_j + \text{const.}\right] . \tag{22}$$

This is simply the expression for the partition function of an Ising model in a magnetic field with strength $H = (\mu_c - d\log(1+z_s)/\beta)$ and an effective nearest neighbour *attraction* with an interaction strength $J = \log(1+z_s)/(4\beta)$.

There is hardly a model in physics that has been studied more than the Ising model. In two dimensions, the partition function can be computed analytically in the zero field case [13]. In the language of our mixture model, no external magnetic field means:

$$(1+z_s)^d = z_c, \tag{23}$$

where $z_c = \exp\beta\mu_c$, the large particle fugacity.

Several points should be noted. First of all, in this simple lattice model, summing over all solvent degrees of freedom resulted in effective *attractive* nearest neighbour interaction between the hard-core colloids. Secondly, below its critical temperature, the Ising model (for $d > 1$) exhibits spontaneous magnetisation. In the mixture model, this means that, above a critical value of the fugacity of the solvent, there will be phase transition in which a phase with low $\langle n_c\rangle$ (a dilute colloidal suspension) coexists with a phase with high $\langle n_c\rangle$ (concentrated suspension). Hence, this model system with purely repulsive hard-core interaction can undergo a demixing transition. This demixing is purely entropic.

2.6 Depletion flocculation

Let us next consider a slightly more realistic example of an entropy-driven phase separation in a binary mixture, namely polymer-induced flocculation of colloids. Experimentally, it is well known that the addition of a small amount of free, non-adsorbing polymer to a colloidal suspension induces an effective attraction between the colloidal particles and may even lead to coagulation. This effect has been studied extensively and is theoretically well understood [14–17]. As in the example discussed above, the polymer-induced attraction between colloids is an *entropic* effect: when the colloidal particles are close together, the total number of accessible polymer conformations is larger than when the colloidal particles are far apart.

To understand the depletion interaction due to polymers, let us again consider a system of hard-core colloids. To this system, we add a number of ideal polymers. Ideal, in this case means that, in the absence of the colloids, the polymers behave like an ideal gas. The configurational integral of a single polymer contains a translational part (V) and an intramolecular part, Z_{int}, which, for an ideal (non-interacting) polymer, is simply the sum over all distinct polymer configurations. In the presence of hard colloidal particles, only part of the volume of the system is accessible to the polymer. How much, depends on the conformational state of the polymer. This fact complicates the description of the polymer-colloid mixture, although numerically, the problem is tractable [18] .

To simplify matters, Asakura and Oosawa [14] introduced the assumption that, as far as the polymer-colloid interaction is concerned, the polymer behaves like a hard sphere with radius R_G. (Here R_G is the radius of gyration, which is comparable to other characteristic measures of polymer size, such as the RMS end-to-end distance; see Khokhlov, this volume.) What this means is that, as the polymer-colloid distance becomes less than R_G, most polymer conformations will result in an overlap with the colloid, but when the polymer-colloid distance is larger, most polymer conformations are permitted (this assumption has been tested numerically [18], and turns out to be quite good). As the polymers are assumed to be ideal, it is straightforward to write down the expression for the configurational integral of N_p polymers, in the presence of N_c colloids at fixed positions r^{N_c} :

$$\int dr^{N_p} \exp[-\beta(U_{ss}+U_{sc})] = \left\{\int d\mathbf{r}_p \exp[-\beta U_{sc}(r^{N_c};\mathbf{r}_p)]\right\}^{N_p} = V_{\text{eff}}^{N_p}(r^{N_c}) \, ,$$

where V_{eff} is the effective volume that is available to the polymers. Equation 10 then becomes

$$\begin{aligned}\Xi(N_c,\mu_s,V,T) &= \frac{1}{N_c!}\int dr^{N_c} \exp[-\beta U_{cc}(r^{N_c})] \sum_{N_p=0}^{\infty} \exp(\beta\mu_p N_p)\frac{V_{\text{eff}}^{N_p}(r^{N_c})}{N_p!} \\ &= \frac{1}{N_c!}\int dr^{N_c} \exp[-\beta U_{cc}(r^{N_c})] \exp(z_p V_{\text{eff}}(r^{N_c})) \, , \end{aligned} \tag{24}$$

where $z_p \equiv \exp(\beta\mu_p)$. Clearly, the effective colloid-colloid potential is now

$$U_{\text{eff}}(r^{N_c}) = U_{cc}(r^{N_c}) - \beta^{-1} z_p V_{\text{eff}}(r^{N_c}) \, . \tag{25}$$

This equation shows that the correction to the colloid-colloid interaction is due to the fact that the volume available to the polymers depends on the configuration of the colloids. The reason why this should be so is easy to understand. Consider two colloids of radius R at distance $r_1 \gg 2(R+R_G)$. In that case, every colloid excluded a spherical volume with radius $R+R_G$ to the polymers (see Figure 3).

Equation 25 shows that the depletion attraction increases with the polymer fugacity or, what amounts to the same thing, with the osmotic pressure of the polymers in solution. The more polymer we add to the suspension, the stronger the attraction. The range of the attraction depends on the size R_G of the polymers. The larger R_G, the longer the range of the attraction. If we model polymers as mutually interpenetrable spheres with radius R_G, then the explicit expression for the depletion interaction between a pair of

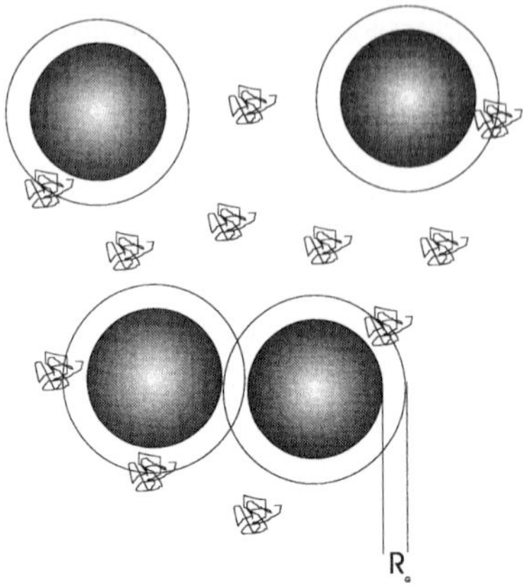

Figure 3. *Hard-core colloids exclude a shell with thickness R_G to the ideal polymers in the solution. When the colloids are far apart, the excluded volumes simply add up. At shorter distances, the excluded volumes overlap and the total volume available to the polymers increases.*

colloids is, for $2R < r < 2(R + R_G)$,

$$V_{\text{dep}}(r) = -\frac{4\pi(R+R_G)^3 z_p k_B T}{3}\left\{1 - \frac{3r}{4(R+R_G)} + \frac{1}{16}\left(\frac{r}{R+R_G}\right)^3\right\}, \tag{26}$$

where we have subtracted a constant term from the potential (namely the contribution of two colloids at a distance $r \gg 2(R + R_G)$). Equation 26 shows clearly that, by changing the size of the added polymers and their concentration, we can change both the range and the strength of the attractive interaction between the colloids. In Section 3, I shall discuss the effect of this tunable attraction on the phase behaviour of polymer-colloid mixtures.

One final comment is in place: the true depletion interaction is *not* pairwise additive. This is clear if we consider three colloidal spheres: if the three exclusion zones overlap, the total excluded volume is larger than would be estimated on basis of the pair-terms alone. Hence, three-body forces yield a *repulsive* correction to the depletion interaction. Note that three-body forces are only important if R_G/R is large enough to get the three exclusion zones to overlap. This holds *a fortiori* for the 4-body forces (that are, again, attractive), *etc.* This feature of the depletion interaction does not depend on the details of the Asakura-Oosawa model. In fact, direct simulations of hard colloids and (lattice) polymers [18] show exactly the same effect.

2.7 Why colloidal materials are soft

Let me return to the picture of colloids as giant atoms. We now know that this is an oversimplification—the origins of the effective interaction between colloids often have no counterpart in atomic physics. Yet, if we ignore all these subtleties, there are similarities. Both atoms and colloids have an effective hard-core diameter: σ_a for atoms, σ_c for colloids. Typically, $\sigma_c/\sigma_a = \mathcal{O}(10^3)$. The characteristic interaction energies between colloids ε_c are of the order of the thermal energy k_BT. For atomic solids, the interaction energy ε_a depends on the nature of the interatomic interaction: it may vary from a value comparable to k_BT for van der Waals crystals, to a value of the order of electron-volts for covalently bonded materials (*e.g.* diamond). Knowing the characteristic sizes and the characteristic interaction energies of the particles, is enough to give an order-of-magnitude estimate of various physical properties (basically, this is simply an over-extension of van der Waals' Law of Corresponding States). For instance, the elastic constants of a solid have the dimensions [energy/volume]. That means that the elastic constants of a dense colloidal suspension are of the order k_BT/σ_c^3. For an atomic van der Waals solid, the elastic constants are of the order k_BT/σ_a^3. In other words: the force needed to deform a colloidal crystal is a factor $\sigma_c^3/\sigma_a^3 \approx 10^9$ smaller than for an atomic crystal held together by dispersion forces (and these are the softest atomic crystals). Clearly, colloidal matter is very easily deformable; it is indeed 'soft matter'.

2.8 Polydispersity

All atoms of a given type are identical. They have the same size, weight and interaction strength. This is usually not true for colloids. In fact, all synthetic colloids are to some degree polydisperse, *i.e.* they do not all have the same size (or mass, or shape, or refractive index). This polydispersity is usually a complicating factor: it makes it more difficult to interpret experimental data (*e.g.* X-ray or neutron scattering, or dynamic light-scattering). In addition, it may broaden phase coexistence regions and, in some cases even completely wipe out certain phases. However, polydispersity is not all bad: it also leads to interesting new physics. For instance, sometimes polydispersity may induce a new phase that is not stable in the monodisperse limit [19]. In general, the effect of polydispersity on the stability of phases is most pronounced in the high-density limit. In that limit, polydispersity may lead to a frustration of the local packing.

3 Colloidal phase behaviour

In Section 2, I explained that the interactions between colloids can often be *tuned.* It is possible to make (uncharged, refractive-index matched, sterically stabilised) colloids that have a steep repulsive interaction and no attraction. These colloids behave like the hard-core models that have been studied extensively in computer simulation of simple fluids. But it is also possible to make (charged) colloids with smooth, long-ranged repulsion. And, using for instance, added polymer to induce a depletion interaction, colloids can be made with variable ranged attractions. Finally, colloids need not be spherical. It is possible to make colloidal rods and disks. Below, I briefly discuss some of the interesting consequences that this freedom to design the colloid-colloid interaction has for the phase behaviour.

3.1 Entropic phase transitions

The second law of thermodynamics tells us that any spontaneous change in a closed system results in an increase of the entropy, S. In this sense, all spontaneous transformations of one phase into another are entropy driven. However, this is not what the term 'entropic phase transitions' is meant to describe. It is more common to consider the behaviour of a system that is not isolated, but can exchange energy with its surroundings. In that case, the second law of thermodynamics implies that the system will tend to minimise its Helmholtz free energy $F = E - TS$, where E is the internal energy of the system and T the temperature. Clearly, a system at constant temperature can lower its free energy in two ways: either by *increasing* the entropy S, or by *decreasing* the internal energy E.

In order to gain a better understanding of the factors that influence phase transitions, we must look at the statistical mechanical expressions for entropy. The simplest starting point is to use Boltzmann's expression for the entropy of an isolated system of N particles in volume V at an energy E,

$$S = k_B \ln \Omega \,, \tag{27}$$

where k_B, the Boltzmann constant, is simply a constant of proportionality. Ω is the total number of (quantum) states that is accessible to the system. In the remainder of these lecture notes, I shall often choose my units such that k_B=1. The usual interpretation of Equation 27 is that Ω, the number of accessible states of a system, is a measure for the disorder in that system. The larger the disorder, the larger the entropy. This interpretation of entropy suggests that a phase transition from a disordered to a more ordered phase can only take place if the loss in entropy is compensated by the decrease in internal energy. This statement is completely correct, provided that we use Equation 27 to *define* the amount of disorder in a system. However, we also have an *intuitive* idea of order and disorder: we consider crystalline solids ordered, and isotropic liquids disordered. This intuitive picture suggests that a spontaneous phase transition from the fluid to the crystalline state can only take place if the freezing lowers the internal energy of the system sufficiently to outweigh the loss in entropy: *i.e.* the ordering transition is 'energy driven'. In many cases, this is precisely what happens. It would, however, be a mistake to assume that our intuitive definition of order always coincides with the one based on Equation 27. In fact, the aim of this section is to show that many 'ordering' transitions that are usually considered to be energy-driven may, in fact, be entropy driven. I stress that the idea of entropy-driven phase transitions is an old one. However, it has only become clear during the past few years that such phase transformations may not be interesting exceptions, but the rule!

In order to observe 'pure' entropic phase transitions, we should consider systems for which the internal energy is a function of the temperature, but not of the density. Using elementary statistical mechanics, it is easy to show that this condition is satisfied for classical hard-core systems. Whenever these systems order at a fixed density and temperature, they can only do so by increasing their entropy (because, at constant temperature, their internal energy is fixed). Such systems are conveniently studied in computer simulations. But, increasingly, experimentalists—in particular, colloid scientists, have succeeded in making real systems that behave very nearly as ideal hard-core systems [24]. Hence, the phase transitions discussed below can, and in many cases, do occur in nature. Below I list examples of entropic ordering in hard-core systems. But I stress that the list is far from complete.

3.2 Computer simulation of (liquid) crystals

The earliest example of an entropy-driven ordering transition is described in a classic paper of Onsager [1], on the isotropic-nematic transition in a (three-dimensional) system of thin hard rods. Onsager showed that, on compression, a fluid of thin hard rods of length L and diameter D *must* undergo a transition from the isotropic fluid phase, where the molecules are translationally and orientationally disordered, to the nematic phase. In the latter phase, the molecules are translationally disordered, but their orientations are, on average, aligned. This transitions takes place at a density such that $(N/V)L^2D$ is of order unity. Onsager considered the limit $L/D \to \infty$. In this case, the phase transition of the hard-rod model can be found exactly [33]. At first sight it may seem strange that the hard rod system can *increase* its entropy by going from a disordered fluid phase to an orientationally ordered phase. Indeed, due to the orientational ordering of the system, the orientational entropy of the system decreases. However, this loss in entropy is more than offset by the increase in translational entropy of the system: the available space for the centre of any one rod increases as the rods become more aligned. In fact, we shall see this mechanism returning time and again in ordering transitions of hard-core systems: the entropy *decreases* because the density is no longer uniform in orientation or position, but the entropy *increases* because the free-volume per particle is larger in the ordered than in the disordered phase.

The most famous, and for a long time controversial, example of an entropy-driven ordering transition is the freezing transition in a system of hard spheres. This transition had been predicted by Kirkwood in the early fifties [20] on the basis of an approximate theoretical description of the hard-sphere model. As this prediction was quite counter-intuitive and not based on any rigorous theoretical results, it met with wide-spread skepticism until Alder and Wainwright [21] and Wood and Jacobson [22] performed numerical simulations of the hard-sphere system that showed direct evidence for this freezing transition. Even then, the acceptance of the idea that freezing could be an *entropy* driven transition, came only slowly [23]. However, by now, the idea that hard spheres undergo a first-order freezing transition is generally accepted.

Since the work of Hoover and Ree [25], we have known the location of the thermodynamic freezing transition. We now also know that the face-centered cubic phase is more stable than the hexagonal close-packed phase [26], but by only $10^{-3}k_BT$ per particle. To understand how little this is, consider the following: if we used calorimetric techniques to determine the relative stability of the *fcc* and *hcp* phases, we would find that the free-energy difference amounts to some 10^{-11} cal/cm^3! Moreover, computer simulations allow us to estimate the equilibrium concentration of point defects (in particular, vacancies) in hard-sphere crystals [27]. At melting, this concentration is small, but not very small (of the order of one vacancy per four-thousand particles).

The next surprise in the history of ordering due to entropy came in the mid-eighties when computer simulations [28] showed that hard-core interactions alone could also explain the formation of more complex liquid crystals. In particular, it was found that a system of hard sphero-cylinders (*i.e.* cylinders with hemi-spherical caps, see Figure 4) can form a smectic liquid crystal, in addition to the isotropic liquid, the nematic phase and the crystalline solid [29]. In the smectic (A) phase, the molecules are orientationally ordered but, in addition, the translational symmetry is broken: the system exhibits a

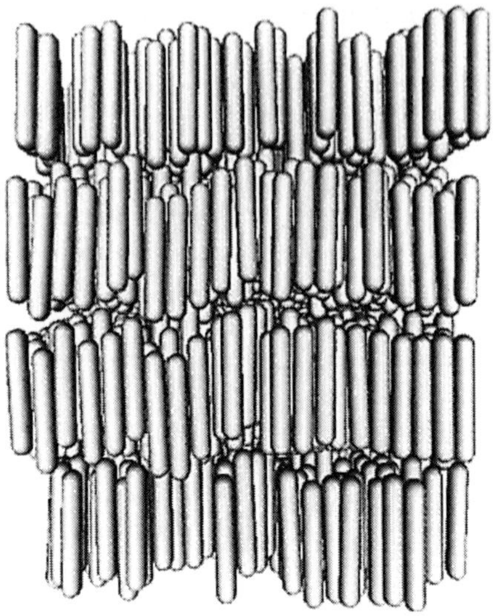

Figure 4. *Snapshot of a hard-core smectic liquid crystal.*

one-dimensional density-modulation. (See also Roux, this volume.) Subsequently, it was found that some hard-core models could also exhibit columnar ordering [30]. In the latter case, the molecules assemble in liquid-like stacks, but these stacks order to form a two-dimensional crystal. In summary, hard-core interaction can induce orientational ordering and one-, two- and three-dimensional positional ordering.

3.3 To boil—or not to boil...

Why do liquids exist? We are so used to the occurrence of phenomena such as boiling and freezing that we rarely pause to ask ourselves if things could have been different. Yet the fact that liquids must exist is not obvious *a priori.* This point is eloquently made in an essay by Weisskopf [31]:

> *The existence and general properties of solids and gases are relatively easy to understand once it is realised that atoms or molecules have certain typical properties and interactions that follow from quantum mechanics. Liquids are harder to understand. Assume that a group of intelligent theoretical physicists had lived in closed buildings from birth such that they never had occasion to see any natural structures. Let us forget that it may be impossible to prevent them to see their own bodies and their inputs and outputs. What would they be able to predict from a fundamental knowledge of quantum mechanics? They probably would predict the existence of atoms, of molecules, of solid crystals, both metals and insulators, of gases, but most likely not the existence of liquids.*

Weisskopf's statement may seem a bit bold. Surely, the liquid-vapour transition could have been predicted *a priori.* This is a hypothetical question that can never be answered. But, as I shall discuss below, in colloidal systems there may exist an analogous phase transition that has not yet been observed experimentally and that was found in simulation before it had been predicted. To set the stage, let us first consider the question of the liquid-vapour transition. In his 1873 thesis, van der Waals gave the correct explanation for a well known, yet puzzling feature of liquids and gases, namely that there is no essential distinction between the two: above a critical temperature T_c, a vapour can be compressed continuously all the way to the freezing point. Yet below T_c, a first-order phase transition separates the dilute fluid (vapour) from the dense fluid (liquid) [32]. It is due to a the competition between short-ranged repulsion and longer-ranged attraction. From the work of Longuet-Higgins and Widom [35], we now know that the van der Waals model (in which molecules are described as hard spheres with an infinitely weak, infinitely long-ranged attraction [34]) is even richer than originally expected: it exhibits not only the liquid-vapour transition but also crystallisation (Figure 5).

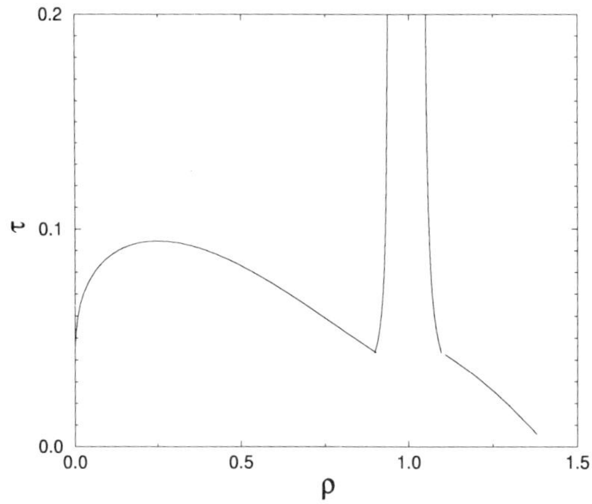

Figure 5. *Phase diagram of a system of hard spheres with a weak, long-range attraction (the 'true' van der Waals model). The density is expressed in units* σ^{-3}, *where* σ *is the hard-core diameter. The 'temperature'* τ *is expressed in terms of the van der Waals a-term:* $\tau = k_B T v_0/a$, *where* v_0 *is the volume of the hard spheres. (Hence the van der Waals mean-field equation of state reads* $(p + aN/V)(V - Nv_0) = Nk_BT$*). Plotted is the coexistence line: below this, vapour-liquid or fluid-crystal coexistence occurs.*

The liquid-vapour transition is possible between the critical point and the triple point, and in the van der Waals model, the temperature of the critical point is about a factor two large than that of the triple point. There is, however, no fundamental reason why this transition should occur in every atomic or molecular substance, nor is there any rule that forbids the existence of more than one fluid-fluid transition. Whether a given compound will have a liquid phase, depends sensitively on the range of the intermolecular potential: as this range is decreased, the critical temperature approaches the triple-point temperature, and when T_c drops below the latter, only a single stable fluid phase remains. In mixtures of spherical colloidal particles and non-adsorbing polymer, the range of the attractive part of the effective colloid-colloid interaction can be varied by changing the size of the polymers (see Section 2.6). Experiment, theory and simulation all suggest that when the width of the attractive well becomes less than approximately one third of the diameter of the colloidal spheres, the colloidal 'liquid' phase disappears.

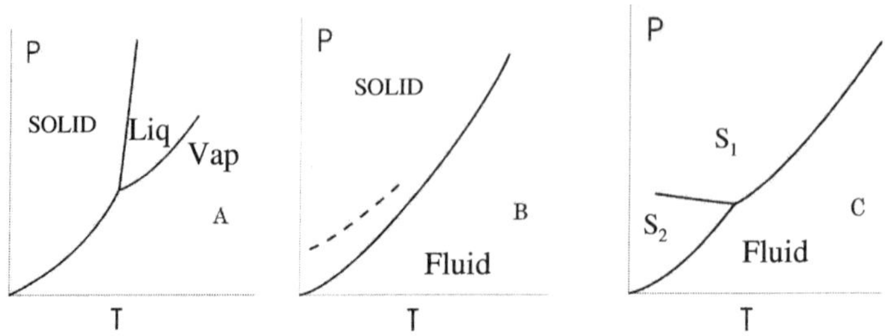

Figure 6. *Sequence of phase diagram for a system of spherical particles as the interaction range is varied. The range of the interaction decreases from left to right in the sequence. Diagram A is normal for a simple molecular substance (the liquid-vapour line ends in a critical point while the liquid-solid line continues indefinitely). In diagram B, the liquid-vapour line is metastable only, but can have dynamical consequences (see Section 5) (dotted). In diagram C, there is an isostructural solid-solid coexistence line.*

Figure 6 shows schematically the evolution of the phase-diagram of a system of spherical particles with a variable ranged attraction. As the range of attraction decreases, the liquid-vapour curve moves into the metastable regime. For very short-ranged attraction (less than 5% of the hard-core diameter), a first-order iso-structural solid-solid transition appears in the solid phase [36]. It should be stressed that phase diagrams of type **B** in figure 6 are common for colloidal systems, but rare for simple molecular systems. A possible exception is C_{60} [37]. Phase diagrams of type **C** have, thus far, not been observed in colloidal systems. Nor had they been predicted before the simulations appeared (this suggests that Weisskopf was right).

4 Colloid dynamics

For the computer simulator, the study of colloid dynamics is a challenge. The reason is that colloid dynamics spans a wide range of time-scales. No single simulation can cover all time-scales simultaneously. Below, I shall discuss two aspects of colloid dynamics that clearly illustrate the time-scale problem. The first is colloidal *hydrodynamics*. The second is homogeneous nucleation of a new phase from a metastable phase.

4.1 Hydrodynamic effects in colloidal suspensions

Colloid dynamics is a research field in its own right (see *e.g.* [38]). Clearly, I cannot cover this field in a few pages. I therefore wish to focus on a few simple concepts that are useful when thinking about the dynamics of colloidal particles. The analogy between colloids and atoms that is useful when discussing the static properties of colloidal matter, breaks down completely when discussing the dynamics. The reason is that atoms in a dilute gas phase move *ballistically*, colloids in a dilute suspension move *diffusively*. In order to understand the motion of colloids, we have to consider the hydrodynamic properties of the surrounding solvent. Just imagine what would happen if kinetic gas theory applied to the motion of colloids: then the frictional force acting on a spherical colloid would be caused by independent collisions with the solvent molecules, and we would find that the frictional force is proportional to the velocity of the colloid, $\mathbf{v}$ (which is correct) and the effective area of the colloid (πa^2) (which is wrong). In fact, the true frictional force on a colloid moving at a constant velocity $\mathbf{v}$ is given by the Stokes expression

$$\mathbf{F}_{\text{frict}} = -6\pi\eta a\mathbf{v} , \tag{28}$$

where η is the viscosity of the solvent and a the radius of the colloid.

The Stokes relation can be derived from hydrodynamics, however this derivation does not make it intuitively obvious why the friction is proportional to a rather than to a^2. Below, I shall give a hand-waving derivation that is more intuitively appealing (although the answer is not quite right). We start with the assumption that the time evolution of any flow field $\mathbf{u}(\mathbf{r}, t)$ in the solvent obeys the Navier-Stokes equation for an incompressible fluid

$$d_s \left(\frac{\partial \mathbf{u}(\mathbf{r}, t)}{\partial t} + \mathbf{u}(\mathbf{r}, t) \cdot \nabla \mathbf{u}(\mathbf{r}, t) \right) = \eta \nabla^2 \mathbf{u}(\mathbf{r}, t) - \nabla p(\mathbf{r}, t) ,$$

where $\mathbf{u}(\mathbf{r}, t)$ is the flow velocity at point $\mathbf{r}$ and time t, d_s is the mass density of the solvent and $p(\mathbf{r}, t)$ is the hydrostatic pressure. I shall consider the case that $\mathbf{u}(\mathbf{r}, t)$ is 'small' (low Reynolds-number regime, see [38]). Then we can neglect the $\mathbf{u} \cdot \nabla \mathbf{u}$ term. Let us now consider the situation where the solvent is in contact with a flat surface (see Figure 7). Initially, both fluid and wall are at rest. At time $t = 0$, the wall is given a tangential

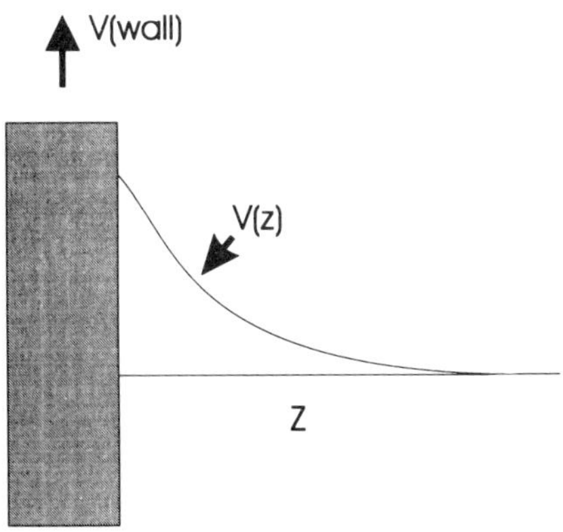

Figure 7. *When a wall is suddenly given a tangential velocity* $\mathbf{v}_{\text{wall}}$, *the transverse velocity field penetrates* diffusively *into the bulk fluid.*

velocity $\mathbf{v}_{\text{wall}}$. We assume that this velocity is parallel to the y-direction. The normal to the surface defines the z-direction. In this geometry, the equation of motion for the flow field reduces to

$$d_s \frac{\partial u_y(z, t)}{\partial t} = \eta \nabla_z^2 u_y(z, t) .$$

But this is effectively a *diffusion* equation for the transverse velocity. The 'diffusion coefficient' is equal to $(\eta/d_s) \equiv \nu$. This diffusion coefficient for transverse momentum is called the *kinematic viscosity*. The larger ν is, the faster transverse momentum diffuses away from its source. Diffusion equations typically show up when we consider the transport of a quantity that is conserved, such as mass, energy or (in this case) momentum.

Let us now use this concept of diffusing momentum to estimate the frictional drag on a sphere. To simplify matters, I shall pretend that the transverse momentum is a *scalar* rather than a *vector*. Clearly, this is wrong, but it will not affect the qualitative answer. A moving sphere acts as a source of transverse momentum. The transverse momentum flux j_T is related to the gradient in the transverse velocity field (v_T) by

$$j_T = -\eta \nabla v_T \,.$$

In steady state, $\nabla^2 v_T(r) = 0$. If the transverse velocity were a scalar, the solution to this equation would be

$$v_T(r) = v_0 \frac{a}{r} \,, \tag{29}$$

where v_0 is the velocity of the colloidal sphere. The transverse momentum current density is then

$$j_T = \eta v_0 \frac{a}{r^2} \,.$$

The frictional force on the sphere is equal to minus the total rate at which momentum flows into the fluid

$$F_{\text{frict}} = -4\pi r^2 j_T = -4\pi \eta a v_0 \,, \tag{30}$$

which is almost Stokes' law (the factor 4π instead of 6π is due to our cavalier treatment of the vectorial character of the velocity).

This trivial example shows that the conservation of momentum is absolutely crucial for the understanding of colloid dynamics. A second result that follows almost immediately from Equation 29 is that the flow velocity at a distance r from a moving colloid, decays as $1/r$. Through this velocity field, one colloid can exert a drag force on another colloid. This is the so-called *hydrodynamic interaction*, and is very long ranged. Again, for a correct derivation, I refer the reader to [38].

Having established a simple language for the discussion of colloid dynamics, we can make estimates of the relevant time-scales that govern the time evolution of a colloidal system. The shortest time-scale τ_s, is usually not even considered. It is the time-scale on which the solvent behaves as a *compressible* fluid. If we set a colloid in motion, this will set up a density disturbance. This density modulation will be propagated away as a sound wave (carrying with it one third of the momentum of the colloid [38]). This sound wave will have moved away after a time $\tau_s = a/c_s$ (where c_s is the velocity of sound). Typically, $\tau_s = \mathcal{O}(10^{-10}\text{s})$. The next time-scale is the one associated with the propagation of hydrodynamic interactions: τ_H. It is of the order of the time it takes transverse momentum to diffuse a typical interparticle distance: $\tau_H = \mathcal{O}(\rho^{-2/3}/\nu)$, where ρ is the number density of the colloids. In dense suspensions, the typical inter-particle distance is comparable to the diameter of the colloids, and then $\tau_H = \mathcal{O}(a^2/\nu)$. Usually, this time-scale is of the order of 10^{-8}s. Next, we get the time-scale for the decay of the initial velocity of a colloid. If we assume (somewhat inconsistently, as it will turn out) that this decay is determined by Stokes' law, we find that the decay of the velocity of a colloid occurs on a time-scale $\tau_v = \mathcal{O}(M_c/\eta a)$, where M_c is the mass of a colloid. Since $M_c = (4\pi a^3 d_c/3)$, where d_c is the mass density of the colloid, then we can write $\tau_v = \mathcal{O}(d_c a^2/\eta)$. In a dense suspension, $\tau_v = (d_c/d_s)\tau_H$. This means that, for a neutrally buoyant colloid, there is no separation in time-scales between τ_v and τ_H.

The final time-scale in colloid dynamics is the one associated with appreciable displacements of the colloids. As the colloids move diffusively, and as the diffusion constant is related to the Stokes friction constant by $D = k_B T/(6\pi\eta a)$, the time it takes a colloid to diffuse over a distance comparable to its own radius is $\tau_R = \mathcal{O}(a^2/D) \sim \mathcal{O}(\eta a^3)$. This τ_R is of the order of milliseconds to seconds. Clearly, there is a wide time-scale separation between τ_R and the other times. For times that are much longer than τ_v and τ_H, we can pretend that the colloids perform uncorrelated Brownian motion. However, this is not quite correct: even though the hydrodynamic interactions have long decayed, they render the effective diffusion constant of every colloid dependent on the instantaneous configuration of its neighbours. This is one of the reasons why the theory of colloid dynamics is not simple [38].

Let me, however, conclude this section on colloid dynamics with something that can easily be understood on the basis of diffusion of transverse momentum. The Stokes-Einstein relation provides an expression for the frictional force acting on a colloidal particle that moves at velocity v_c through the solvent: $F_{\text{frict}} = -6\pi\eta a v_c$. At first sight, it appears that this equation allows us to compute the rate at which the initial velocity of a particle decays

$$M_c \frac{\partial v_c}{\partial t} = -6\pi\eta a v_c \ , \tag{31}$$

and the solution to this equation is

$$v_c(t) = v_c(0) \exp\left(-\frac{6\pi\eta a}{M_c} t\right) \ . \tag{32}$$

This answers looks reasonable. It even yields the correct expression for the diffusion constant. Indeed, using the Green-Kubo relation between the self-diffusion constant and the velocity-autocorrelation function $\langle v_x(0)v_x(t)\rangle$:

$$D = \int_0^\infty dt \langle v_x(0)v_x(t)\rangle \ , \tag{33}$$

we find $D = k_B T/(6\pi\eta a)$, as it should. Still, Equation 32 is wrong. Velocity fluctuations of colloidal particles (or, for that matter, even atoms [39]) do not decay exponentially, but with a power law.

In terms of the diffusive transport of transverse momentum, this is easy to understand. Consider a colloidal particle of mass M_c having an initial velocity v_c. Part of the initial momentum of the particle is carried away by sound waves (in fact, one third of the initial momentum). The remainder is converted to transverse momentum and is transported away diffusively. After a time t, the transverse momentum has diffused over a typical distance $\sqrt{2\nu t}$. That means that (two-thirds of) the initial momentum of the particle, $M_c v_c$ is now contained in a spherical volume with radius $\sqrt{2\nu t}$. This volume has a total mass proportional to $d_s(\nu t)^{3/2}$. The average flow velocity of the fluid in this volume is equal to its momentum divided by its mass:

$$v_{av} \simeq \frac{M_c v_c}{d_s(\nu t)^{3/2}} . \tag{34}$$

The velocity of the colloidal particle is equal to this average flow velocity. Hence, for long times, the velocity-autocorrelation function of the colloidal particles decays as

$$\langle v_x(0)v_x(t)\rangle \simeq \frac{k_B T}{d_s(\nu t)^{3/2}} , \tag{35}$$

where we have used $\langle v_x^2(0)\rangle = k_B T/M_c$. As is clear from Equation 35, the velocity correlation function of a colloidal particle decays as $t^{-3/2}$. Several points should be noted: first of all, the decay described by Equation 35 can be, and has been, observed experimentally (see e.g. [40]). The power-law decay of the velocity auto-correlation function only describes the asymptotic behaviour. A similar analysis can be applied to the decay of the angular momentum of a rotating colloidal particle (in that case, the decay goes as $t^{-5/2}$ [41]). The presence of a wall perturbs the diffusion of transverse momentum. Using arguments very similar to the one above, one may then derive new power-law exponents for the decay of rotational and translational velocity correlation functions [42].

5 Metastability and nonequilibrium dynamics

It is well known that liquids can be supercooled before they freeze and vapours can be supersaturated before they condense: the resulting phases are *metastable*. Similar phenomena arise in colloids too. In what follows, we discuss the escape from such a metastable phase.

5.1 Homogeneous nucleation in colloidal suspensions

A homogeneous phase can be supercooled because the only route to the more stable state is via the formation of small *nuclei*. The free energy of such nuclei is determined not only by the difference in chemical potential between vapour and liquid, which drives the nucleation process, but also by their surface free energy. In classical nucleation theory (CNT) [43][44] it is assumed that the nuclei are compact, spherical objects, that behave like small droplets of bulk phase. The surface free energy term is always positive, because of the work that must be done to create an interface. Moreover, for small droplets this term dominates and hence the free energy of a nucleus increases with size. Only when the droplet has reached a certain critical size, does the volume term takes over, and the free energy decrease. It is only from here on that the nucleus grows spontaneously into a bulk liquid. The free energy of a spherical liquid droplet of radius R in a vapour is then given by

$$\Delta G = 4\pi R^2\gamma + \frac{4}{3}\pi R^3\rho\Delta\mu, \tag{36}$$

where γ is the surface free energy, ρ is the particle number density in the bulk liquid, and $\Delta\mu$ is the difference in chemical potential between bulk liquid and bulk vapour. Clearly, the first term on the right hand side of Equation 36 is the surface term, which is positive, and the second term is the volume term, which is negative; the difference in chemical potential is the driving force for the nucleation process. The height of the nucleation barrier can easily be obtained from the above expression, yielding

$$\Delta G^* = \frac{16\pi\gamma^3}{3\rho^2\Delta\mu^2}. \tag{37}$$

This equation shows that the barrier height depends not only on the surface free energy γ (and the density ρ), but also on the difference in chemical potential $\Delta\mu$. The difference in chemical potential is related to the supersaturation. Hence, the height of the free-energy barrier that separates the stable from the metastable phase depends on the degree of supersaturation. At coexistence, the difference in chemical potential is zero, and the height of the barrier is infinite. Although equally likely to be in the liquid or vapour phase, once the system is one state or the other, it will remain in this state; the system simply cannot transform into the other state.

Macroscopic thermodynamics dictates that the phase that is formed in a supersaturated system is the one that has the lowest free energy. However, nucleation is an essentially dynamic process, and therefore one cannot expect *a priori* that on supersaturating the system the thermodynamically most stable phase will be formed. In 1897, Ostwald [45] formulated his step rule, stating that the crystal phase that is nucleated from the melt need not be the one that is thermodynamically most stable, but the one that is closest in free energy to the fluid phase. Stranski and Totomanow [46] re-examined this rule and argued that the nucleated phase is the phase that has the lowest free-energy barrier of formation, under the conditions prevailing. The simulation results discussed below suggest that, even on a *microscopic* scale, something similar to Ostwald's step rule seems to hold.

5.2 Coil-globule transition in condensation of dipolar colloids?

The formation of a droplet of water from the vapour is probably the best known example of homogeneous nucleation of a polar fluid. However, the nucleation behaviour of polar fluids (including polar colloids) is still poorly understood. In fact, while classical nucleation theory gives a reasonable prediction of the nucleation rate of nonpolar substances, it seriously overestimates the rate of nucleation of highly polar compounds, such as acetonitrile, benzonitrile and nitrobenzene [47, 48]. In order to explain the discrepancy between theory and experiment, several nucleation theories have been proposed. It has been suggested that in the critical nuclei the dipoles are arranged in an anti-parallel head-to-tail configuration [47, 48], giving the clusters a non-spherical, prolate shape, which increases the surface-to-volume ratio and thereby the height of the nucleation barrier. In the oriented dipole model introduced by Abraham [49], it is assumed that the dipoles are perpendicular to the interface, yielding a size dependent surface tension due to the effect of curvature of the surface on the dipole-dipole interaction. However, in a density-functional study of a weakly polar Stockmayer fluid, it was found that on the liquid (core) side of the interface of critical nuclei, the dipoles are not oriented perpendicular to the surface, but parallel [50].

We have studied the structure and free energy of critical nuclei, as well as pre- and postcritical nuclei, of a highly polar Stockmayer fluid [51]. In the Stockmayer system, the particles interact via a Lennard-Jones pair potential plus a dipole-dipole interaction potential

$$v(\mathbf{r}_{ij}, \boldsymbol{\mu}_i, \boldsymbol{\mu}_j) = 4\epsilon\left[\left(\frac{\sigma}{r_{ij}}\right)^{12} - \left(\frac{\sigma}{r_{ij}}\right)^{6}\right] - \frac{3(\boldsymbol{\mu}_i\cdot\mathbf{r}_{ij})(\boldsymbol{\mu}_j\cdot\mathbf{r}_{ij})}{r_{ij}^5} + \frac{\boldsymbol{\mu}_i\cdot\boldsymbol{\mu}_j}{r_{ij}^3}. \tag{38}$$

Here ϵ is the Lennard-Jones well depth, σ is the Lennard-Jones diameter, $\boldsymbol{\mu}_i$ denotes the dipole moment of particle i and $\mathbf{r}_{ij}$ is the vector joining particle i and j. We have studied the nucleation behaviour for reduced dipole moment $\mu^* = |\boldsymbol{\mu}|/\sqrt{\epsilon\sigma^3} = 4$, which is close to the value for water. We have computed [51] the excess free energy $\Delta\Omega$ of a cluster of size n in a volume V, at chemical potential μ and at temperature T, from the probability distribution function $P(n)$

$$\beta\Delta\Omega(n,\mu,V,T) \equiv -\ln[P(n)] = -\ln[N_n/N]. \tag{39}$$

Here β is the reciprocal temperature; N_n is the average number of clusters of size n and N is the average total number of particles. As the density of clusters in the vapour is low, the interactions between them can be neglected. As a consequence, we can obtain the free-energy barrier at any desired chemical potential μ' from the nucleation barrier measured at a given chemical potential μ via

$$\beta\Delta\Omega(n,\mu',V,T) = \beta\Delta\Omega(n,\mu,V,T) - \beta(\mu'-\mu)n + \ln\left[\rho(\mu')/\rho(\mu)\right], \tag{40}$$

where $\rho = N/V$ is the total number density in the system.

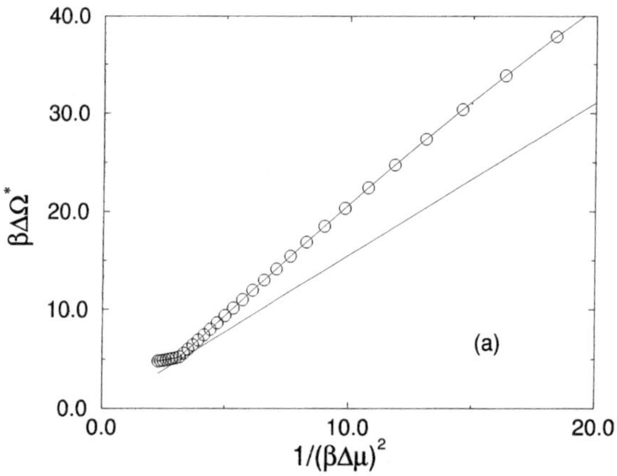

Figure 8. *Comparison of the barrier height between the simulation results (open circles) and classical nucleation theory (straight solid line) for a Stockmayer fluid with reduced dipole moment $\mu^* = |\boldsymbol{\mu}|/\sqrt{\epsilon\sigma}^3 = 4$ and reduced temperature $k_BT/\epsilon = 3.5$. The chemical potential difference between the liquid and the vapour is $\Delta\mu$.*

Figure 8 shows the comparison between the simulation results and CNT for the height of the barrier. Clearly, the theory underestimates the barrier height. As the nucleation rate is dominated by the height of the barrier, our results are in qualitative agreement with the experiments on strongly polar fluids [47, 48], in which it was found that CNT overestimates the nucleation rate. But, unlike the experiments, the simulations allow us to investigate the microscopic origins of the breakdown of classical nucleation theory.

In classical nucleation theory it is assumed that already the smallest clusters are compact, more or less spherical objects. In a previous simulation study on a typical nonpolar fluid, the Lennard-Jones fluid, we found that this is a reasonable assumption [52], even for nuclei as small as ten particles. However, the interaction potential of the Lennard-Jones system is isotropic, whereas the dipolar interaction potential is anisotropic. (On the other hand, the bulk liquid of this polar fluid is isotropic.) We find that the smallest clusters, that initiate the nucleation process, are not compact spherical objects, but chains, in which the dipoles align head-to-tail (Figure 9). In fact, we find a whole variety of differently

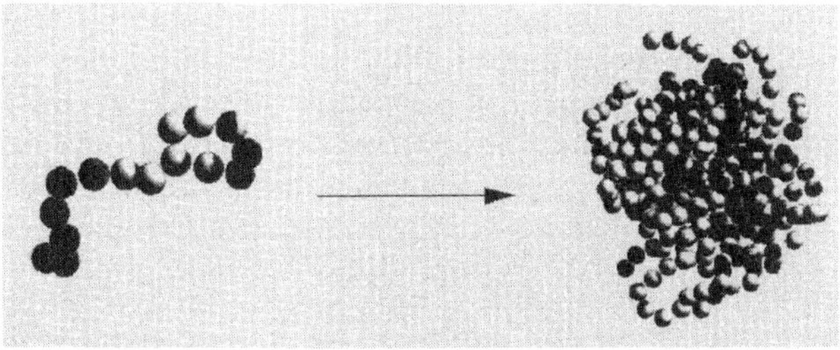

Figure 9. *Left: sub-critical nucleus in a supercooled vapour of dipolar spheres. The dipolar particles align head-to-tail. Right: critical nucleus. The chain has collapsed to form a more-or-less compact, globular cluster.*

shaped sub-critical clusters in dynamical equilibrium: linear chains, branched-chains, and 'ring-polymers'. Initially, as the cluster size is increased, the chains become longer. But, beyond a certain size, the clusters collapse to form a compact globule. The Stockmayer fluid is a simple model system for polar fluids and the mechanism that we describe here might not be applicable for all fluids that have a strong dipole moment. However, it is probably not a bad model for colloids with an embedded electrical or magnetic dipole. The simulations show that the presence of a sufficiently strong permanent dipole may drastically change the pathway for condensation.

5.3 Crystallisation near a metastable critical point

Proteins are notoriously difficult to crystallise. The experiments indicate that most proteins only crystallise under very specific conditions [54–56], otherwise remaining indefinitely as metastable, fluid suspensions. Moreover, the conditions are often not known beforehand. As a result, growing good protein crystals is a time-consuming business. Interestingly, there seems to exist a similarity between the phase diagram of globular proteins and of colloids with short-range attractive interactions [57]. In fact, a series of studies [58–61] show that the phase diagram of a wide variety of proteins is of the kind shown in Figure 6B. Rosenbaum and Zukoski [57, 62] observed that the conditions under which a large number of globular proteins can be made to crystallise, map onto a narrow temperature range of the computed fluid-solid coexistence curve of colloids with short-ranged attraction [63]. If the temperature is too high, crystallisation is hardly observed at all, whereas if the temperature is too low, amorphous precipitation rather than crystallisation occurs. Only in a narrow window around the metastable liquid-vapour critical point, can high-quality crystals be formed. In order to grow high-quality protein crystals, the quench should be relatively shallow, and the system should not be close to a glass transition. Under these conditions, the rate-limiting step in crystal nucleation is the crossing of the free-energy barrier. Using simulation, it is possible to study the nucleation barrier, and the structure of the critical nucleus in the vicinity of this metastable critical point [64].

We performed simulations on a model system for particles with a short-ranged attraction, for a number of state points near the metastable critical point. These state-points were chosen such that on the basis of classical nucleation theory the same height of the barrier could be expected. In order to find the free-energy barrier, we have computed the free energy of a nucleus as a function of its size. However, we first have to define what we mean by a 'nucleus'. As we are interested in crystallisation, it might seem natural to use a crystallinity criterion. However, we expect that crystallisation near the critical point is influenced by critical density fluctuations within the metastable fluid. We therefore used not only a crystallinity criterion, but also a density criterion. We define the size of a high-density cluster (be it solid- or liquidlike) as the number of particles, N_ρ, within a connected region of significantly higher local density than the particles in the remainder of the system. The number of these particles that is also in a crystalline environment is denoted by N_{crys}. In our simulations, we have computed the free-energy 'landscape' of a nucleus as a function of the two coordinates N_ρ and N_{crys}.

Figure 10 shows the free-energy landscape for $T = 0.89T_c$ and $T = T_c$. We find that away from T_c (both above and below), the path of lowest free energy is one where the increase in N_ρ is proportional to the increase in N_{crys} (Figure 10A). Such behaviour is expected if the incipient nucleus is simply a small crystallite. However, around T_c, critical density fluctuations lead to a striking change in the free-energy landscape (Figure 10B). First, the route to the critical nucleus leads through a region where N_ρ increases while N_{crys} is still essentially zero. In other words: the first step towards the critical nucleus is the formation of a liquidlike droplet. Then, beyond a certain critical size, the increase in N_ρ is proportional to N_{crys}, that is, a crystalline nucleus forms inside the liquidlike droplet.

Clearly, the presence of large density fluctuations close to a fluid-fluid critical point has a pronounced effect on the route to crystal nucleation. But, more importantly, the nucleation barrier close to T_c is much lower than at either higher or lower temperatures (Figure 11). The observed reduction in ΔG^* near T_c by some $30k_BT$ corresponds to an increase in nucleation rate by a factor 10^{13}. Finally, let us consider the implications of this reduction of the crystal nucleation barrier near T_c. An alternative way to lower the crystal nucleation barrier would be to quench the solution deeper into the metastable region below the solid-liquid coexistence curve. However, such deep quenches often result in the formation of amorphous aggregates [57, 61, 62, 65–68]. Moreover, in a deep quench, the thermodynamic driving force for crystallisation ($\mu_{\text{liq}} - \mu_{\text{cryst}}$) is also enhanced. As a consequence, the crystallites that nucleate will grow rapidly and far from perfectly [55]. Thus the nice feature of crystal nucleation in the vicinity of the metastable critical point is, that crystals can be formed at a relatively small degree of undercooling. It should be

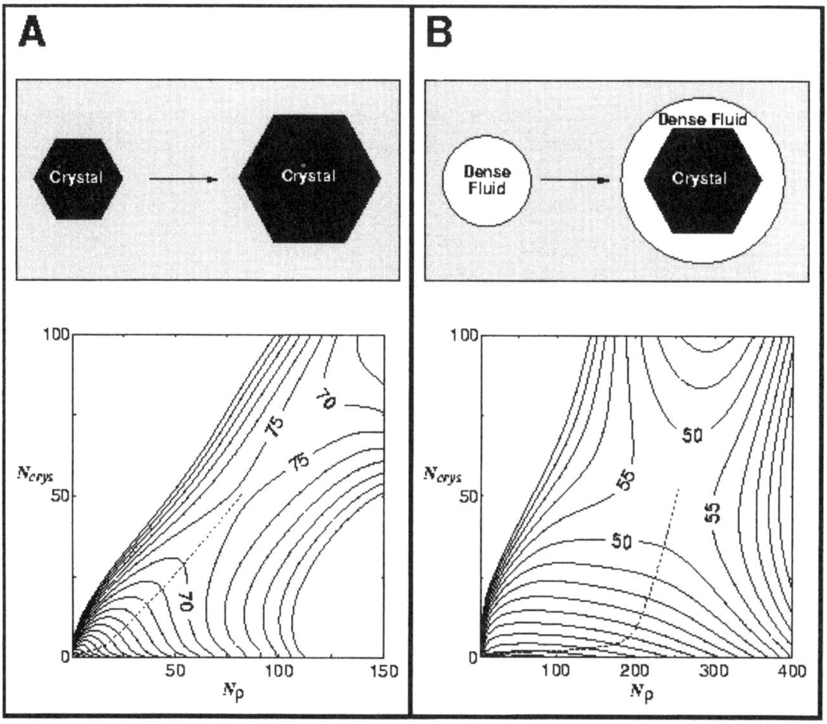

Figure 10. *Contour plots of the free-energy landscape along the path from the metastable fluid to the critical crystal nucleus, for our system of spherical particles with short-ranged attraction. The curves of constant free energy are drawn as a function of N_ρ and N_{crys} (see text) and are separated by $5k_BT$. If a liquidlike droplet forms in the system, we expect N_ρ to become large, while N_{crys} remains essentially zero. In contrast, for a normal crystallite, we expect that N_ρ is proportional to N_{crys}. Panel A shows the free energy landscape well below the critical temperature ($T/T_c = 0.89$). The lowest free-energy path to the critical nucleus is indicated by a dashed curve. Note that this curve corresponds to the formation and growth of a highly crystalline cluster. Panel B: The same, but now for $T = T_c$. In this case, the free-energy valley (dashed curve) first runs parallel to the N_ρ axis (formation of a liquid-like droplet), and moves towards a structure with a higher crystallinity (crystallite embedded in a liquid-like droplet). The free energy barrier for this route is much lower than the one shown in A.*

stressed that nucleation will also be enhanced in the vicinity of the fluid-fluid *spinodal*. Hence, there is more freedom in choosing the optimal crystallisation conditions. Finally, I note that in colloidal (as opposed to protein) systems, the system tends to form a gel before the metastable fluid-fluid branch is reached. A possible explanation for the difference in behaviour of proteins and colloids is discussed in [69].

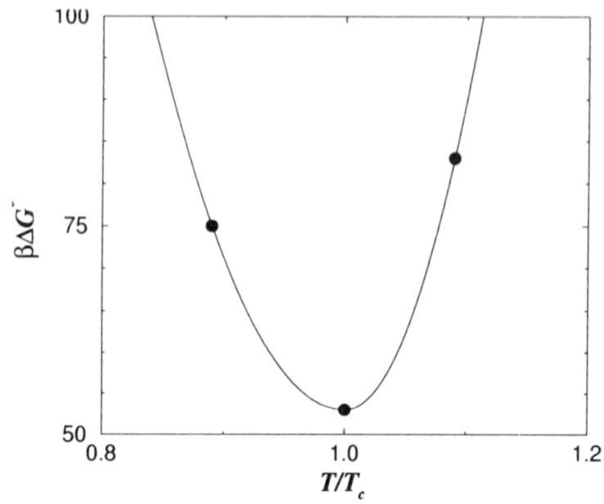

Figure 11. *Variation of the free-energy barrier for homogeneous crystal nucleation, as a function of T/T_c, in the vicinity of the critical temperature. The solid curve is a guide to the eye. The simulations show that the nucleation barrier goes through a minimum around the metastable critical point (see text).*

5.3.1 Microscopic step rule

Ostwald formulated his step rule more than a century ago [45] on the basis of macroscopic studies of phase transitions. The simulations suggest that also on a microscopic level, a 'step rule' may apply and that metastable phases may play an important role in nucleation. We find that the structure of the pre-critical nuclei is that of a metastable phase (chains/liquid). As the nuclei grow, the structure in the core transforms into that of the stable phase (liquid/fcc-crystal). Interestingly, in the interface of the larger nuclei traces of the structure of the smaller nuclei are retained.

5.4 Concluding remarks on nucleation dynamics

The reader may have noticed that I have discussed the subject of homogeneous nucleation without ever discussing the actual *dynamics* of the barrier-crossing process. The reason is that usually (well away from the gelation point) the barrier height completely dominates the variation of the nucleation rate. However, in a full description of nucleation in colloids, the actual dynamics of the barrier crossing process should be taken into account. (See McLeish, this volume, for similar remarks in a different, polymeric context.) Computationally, this is feasible, but non-trivial—after all, the dynamics of colloids in suspension is itself quite complex. But the techniques to study this problem exist.

6 Conclusion

Finally: I realise that my introduction to colloid physics has been biased and superficial. Biased because, as a simulator, I tend to focus on idealised models. Superficial because, wherever I could, I gave quick-and-dirty explanations instead of decent derivations. As much as possible, I have tried to refer the reader to the 'correct' literature. But, as these lectures are not meant to be an exhaustive review, I have surely omitted many more relevant references than I have quoted. I hope that both the reader and the offended authors will forgive me.

Acknowledgments

The work of the FOM Institute is part of the research program of Stichting Fundamenteel Onderzoek der Materie (FOM) and is supported by NWO. I gratefully acknowledge the contributions of numerous students postdocs and colleagues, in particular: Alain Stroobants, Alfons van Blaaderen, Ard Louis, Bela Mulder, Chris Lowe, David Kofke, David Oxtoby, Emmanuel Trizac, Evert Jan Meijer, Henk Lekkerkerker, Ignacio Pagonabarraga, James Polson, Jan Veerman, Jeroen van Duijneveldt, Jose Cuesta, Maarten Hagen, Maria-Jose Ruiz-Montero, Marjolein Dijkstra, Martin van der Hoef, Martin Bates, Massimo Noro, Mike Allen, Norbert Kern, Peter Bladon, Peter Bolhuis, Pieter Rein ten Wolde, Richard Sear, Sander Pronk, Thierry Biben and Tony Ladd.

References

[1] L Onsager, *Ann NY Acad Sci* **51** 627 (1949)

[2] E J Verwey and J Th G Overbeek, *Theory of the Stability of Lyophobic Colloids* (Elsevier, Amsterdam, 1948)

[3] I Sogami and N Ise, *J Chem Phys* **81** 6320 (1984)

[4] A E Larsen and D G Grier, *Nature* **385** 230–233 (1997)

[5] G M Kepler and S Fraden, *Phys Rev Lett* **73** 356 (1994)

[6] W R Bowen and A O Sharif, *Nature* **393** 663–665 (1998)

[7] D Goulding and J P Hansen, *Europhys Lett* **46** 407 (1999)

[8] J C Neu, *Phys Rev Lett* **82** 1072 (1999)

[9] R vanRoij and J P Hansen, *Phys Rev Lett* **79** 3082 (1997),
R van Roij, M Dijkstra and J P Hansen, *Phys Rev E* **59** 2010 (1999)

[10] J Ray and G S Manning, *Langmuir* **10** 2450 (1994),
J L Barrat and J F Joanny, *Adv Chem Phys* **94** 1 (1996),
N Grønbech-Jensen, R J Mashl, R F Bruinsma, and W M Gelbart, *Phys Rev Lett* **78** 2477 (1997),
B-Y Ha and Andrea J Liu, *Phys Rev Lett* **79** 1289 (1997) and *Phys Rev Lett* **81** 1011 (1998)

[11] J Israelachvili, *Intermolecular and Surface Forces*, 2nd edn (Academic Press, London, 1992)

[12] D Frenkel and A A Louis, *Phys Rev Lett* **68** 3363 (1992). In fact, in a different context, this model had been introduced by B Widom, *J Chem Phys* **46** 3324 (1967)

[13] L Onsager, *Phys Rev* **65** 117 (1944)

[14] S Asakura and F Oosawa, *J Chem Phys* **22** 1255 (1954); *J Polym Sci* **33** 183 (1958)

[15] A Vrij, *Pure Appl Chem* **48** 471 (1976)

[16] A P Gast, C K Hall and W B Russel, *J Colloid Interface Sci* **96** 251 (1983)

[17] H N W Lekkerkerker, W C Poon, P N Pusey, A Stroobants and P B Warren, *Europhysics Lett* **20** 559 (1992)

[18] E J Meijer and D Frenkel, *Phys Rev Lett* **67** 1110 (1991), *J Chem Phys* **100** 6873–6887 (1994), *Physica A* **213** 130–137 (1995)

[19] M A Bates and D Frenkel, *J Chem Phys* **109** 6193 (1998)

[20] J E Kirkwood in *Phase Transformations in Solids*, R Smoluchowski, J E Mayer and W A Weyl (eds),Wiley, New York (1951), p 67

[21] B J Alder and T E Wainwright, *J Chem Phys* **27** 1208 (1957)

[22] W W Wood and J D Jacobson, *J Chem Phys* **27** 1207 (1957)

[23] *The Many-Body Problem*, J K Percus, editor, Interscience, New York, 1963

[24] P N Pusey, in *Liquids, freezing and glass transition*, J P Hansen, D Levesque and J Zinn-Justin (editors),North-Holland, Amsterdam, 1991, p 763

[25] W G Hoover and F H Ree, *J Chem Phys* **49** 3609–3617 (1968)

[26] P G Bolhuis, D Frenkel, S-C Mau and D A Huse, *Nature* **388** 235 (1997)

[27] C H Bennett and B J Alder, *J Chem Phys* **54** 4796 (1970)

[28] D Frenkel, H N W Lekkerkerker and A Stroobants, *Nature* **332** 822 (1988)

[29] Peter Bolhuis and Daan Frenkel, *J Chem Phys* **106** 666 (1997)

[30] J A C Veerman and D Frenkel, *Phys Rev* **A45** 5633 (1992)

[31] V F Weissskopf, *Trans NY Acad Sci II*, **38**, 202 (1977)

[32] J S Rowlinson, *Studies in Statistical Mechanics* Vol XIV, J L Lebowitz (editor), North Holland, Amsterdam, 1988

[33] R F Kayser and H J Raveche, *Phys Rev* **A17** 2067 (1978)

[34] P C Hemmer and J L Lebowitz, in *Critical Phenomena and Phase Transitions* **5b**, C Domb and M Green (editors), Academic Press, New York, (1976)

[35] H C Longuet-Higgins and B Widom, *Mol Phys* **8** 549 (1964)

[36] P G Bolhuis and D Frenkel, *Phys Rev Lett* **72**, 221 (1994)
P G Bolhuis, M H J Hagen and D Frenkel, *Phys Rev* **E50** 4880 (1994)

[37] M H J Hagen, E J Meijer, G C A M Mooij, D Frenkel and H N W Lekkerkerker, *Nature*, **365** 425 (1993)

[38] J K G Dhont, *An Introduction to Dynamics of Colloids*, Elsevier, Amsterdam, 1996

[39] B J Alder and T E Wainwright, *Phys Rev A* **1** 18 (1970)

[40] J X Zhu, D J Durian, J Muller, D A Weitz and D J Pine, *Phys Rev Lett* **68** (16) 2559 (1992)

[41] C P Lowe, D Frenkel and A J Masters, *J Chem Phys* **103** 1582 (1995)

[42] I Pagonabarraga, M H J Hagen, C P Lowe and D Frenkel, *Phys Rev E* **58** 7288 (1998)

[43] M Volmer and A Weber, *Z Phys Chem* **119** 227 (1926)

[44] R Becker and W Döring, *Ann Phys* **24** 719 (1935)

[45] W Ostwald, *Z Phys Chem* **22** 289 (1897)

[46] I N Stranski and D Totomanow, *Z Physikal Chem* **163** 399 (1933)

[47] D Wright, R Caldwell, C Moxeley, and M S El-Shall, *J Chem Phys* **98** 3356 (1993)

[48] D Wright and M S El-Shall, *J Chem Phys* **98** 3369 (1993)

[49] F F Abraham, *Science* **168** 833 (1970)

[50] V Talanquer and D W Oxtoby, *J Chem Phys* **99** 4670 (1993)

[51] P R ten Wolde, D W Oxtoby and D Frenkel, *Phys Rev Lett* **81** 3695 (1998)

[52] P R ten Wolde and D Frenkel, *J Chem Phys* **109** 9901 (1998)

[53] I Kusaka, Z-G Wang, and J H Seinfeld, *J Chem Phys* **108** 3446 (1998)

[54] A McPherson, *Preparation and Analysis of Protein Crystals*, Krieger Publishing, Malabar (1982)

[55] S D Durbin and G Feher, *Ann Rev Phys Chem* **47** 171 (1996)

[56] F Rosenberger, *J Crystal Growth* **166** 40 (1996)

[57] D Rosenbaum, P C Zamora, and C F Zukoski, *Phys Rev Lett* **76** 150 (1996)

[58] C R Berland, G M Thurston, M Kondo, M L Broide, J Pande, O O Ogun, and G B Benedek, *Proc Natl Acad Sci USA* **89** 1214 (1992)

[59] N Asherie, A Lomakin, and G B Benedek, *Phys Rev Lett* **77** 4832 (1996)

[60] M L Broide, T M Tominc, and M D Saxowsky, *Phys Rev E* **53** 6325 (1996)

[61] M Muschol and F Rosenberger, *J Chem Phys* **107** 1953 (1997)

[62] D Rosenbaum and C F Zukoski, *J Crystal Growth* **169** 752 (1996)

[63] M H J Hagen and D Frenkel, *J Chem Phys* **101** 4093 (1994)

[64] P R ten Wolde and D Frenkel, *Science* **277** 1975 (1997)

[65] A George and W W Wilson, *Acta Crytallogr* **D50** 361 (1994)

[66] S M Ilett, A Orrock, W C K Poon and P N Pusey, *Phys Rev E* **51** 1344 (1995)

[67] W C K Poon, A D Pirie, and P N Pusey, *Faraday Discuss* **101** 65 (1995)

[68] W C K Poon, *Phys Rev* **E55** 3762 (1997)

[69] M Noro, N Kern and D Frenkel, *Europhysics Lett* (in press, 1999)

Computer simulations in soft matter science

Kurt Kremer

Max-Planck-Institute für Polymerforschung, Mainz, Germany

1 Introduction

This school covers a huge area of modern chemical physics. For all the questions considered, computer simulation methods are becoming, or are already, important scientific tools. To give an introduction one can choose a number of different approaches ranging from a somewhat complete, but certainly superficial overview to the deep discussion of a few examples. For the present contribution I try to combine both approaches, accepting the necessary shortcomings. After a general introduction on simulation strategy and characteristic length scales, the basics of Monte Carlo and molecular dynamics will be introduced. Later on, as examples, I will discuss the simulation of polymer melts and networks as well as first attempts towards a multiscale modelling of polymeric materials. A number of nice reviews and books, some extensive, exist [1–11]. Thus, the present article can also be viewed as a guide to the literature.

There are two basic concepts that are used in computer simulations of materials. A conceptually direct approach is the molecular dynamics method. One numerically solves Newton's equations of motion for a collection of particles, which interact via a suitable interaction potential. Through the equations of motion, a natural time scale is built in. The simulation samples phase space *deterministically*. Though this sounds very simple, there are many technical and conceptual complications, which I will come back to later. The second approach, the Monte Carlo method, samples phase space *stochastically*. Monte Carlo is intrinsically stable but has no inbuilt time scale. There are ways to rectify this (discussed below).

These two very different basic approaches are both used to explore the statistical properties of systems and materials. Nowadays many applications employ hybrid methods, where combinations of both are used. Before going into details we first ask when each kind of model is appropriate. To do this we must discuss the different time and length scales relevant for the question under consideration. Figure 1 illustrates this for the example of polymers.

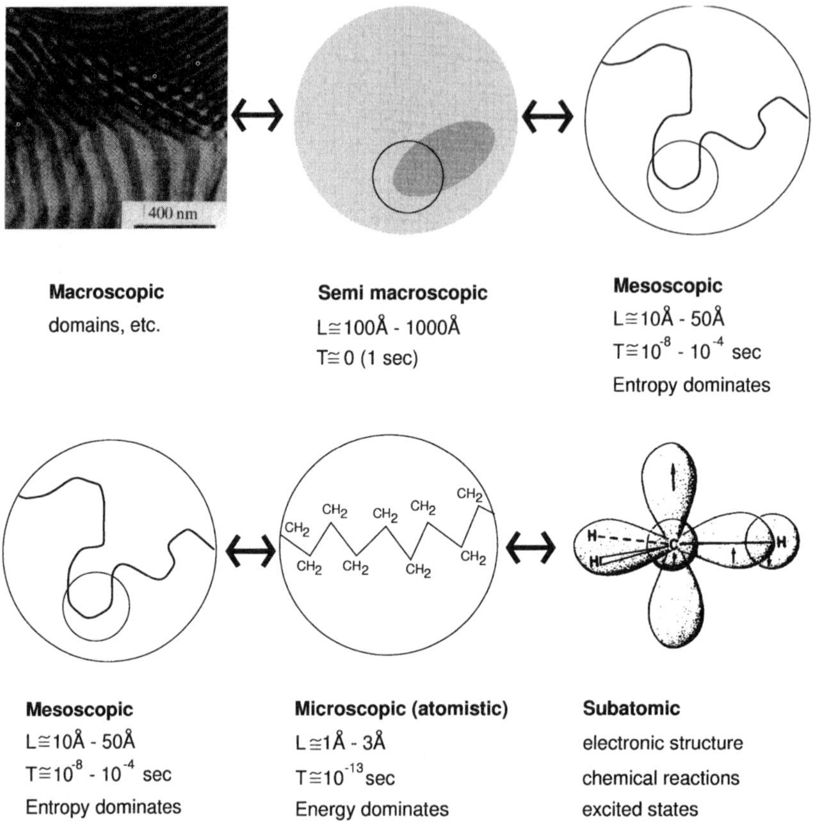

Figure 1. *The different time and length scales in polymer problems.*

Starting from the top of Figure 1 one first observes domain structures. On this level a continuum description is appropriate and one should not use 'particle based' simulation techniques. Looking more closely one could observe a marked chain in a melt or dense solution of otherwise identical chains as a very pale shadow. The typical extension of the shadow is given by the diameter of the overall coil, as indicated. Looking more closely more of the chain structure is revealed. This is the universal entropy-dominated coil regime. Typical experimental time scales are indicated. Only if one again looks much more closely can the chemical details of the polymers be identified. There the behaviour is governed by the local chemical details and is energy-dominated. The lower time boundary is given by the highest frequency, which usually is from the C-C bond oscillations. To study reactions, or excited states, details of the electronic structure have to be considered. Here we deal with the three 'middle levels' of the above scheme. Typical coarse-grained simulations are situated somewhere in between the coil and the microscopic regime.

At a first sight it is tempting to simulate a melt of polymers with all details of the chemical structure of the monomers included. For instance, the chain diffusion constant D could be measured by monitoring the mean square displacement of the monomers or of the centre-of-mass of the chains. This, however, is tempting only at the very first glance. Even with the largest computers, one would need an exceedingly large amount of computer time. As is true for many disordered, complex materials, polymers can be characterised by a hierarchy of different length and time scales, spanning an extremely wide range [12]. Figure 1 illustrates that the typical range easily can exceed ten decades in time. On the microscopic level the properties are dominated by the local oscillations of bond angles and lengths. (To study chemical reactions or excited states, one requires quantum chemistry methods that are beyond the scope of the present paper.) Their typical time constant of about 10^{-13} sec requires a simulation time step of 10^{-15}s. On the semi-macroscopic level the behaviour is dominated by the overall relaxations of conformation of the chains or even larger units (domains *etc.*). These times, depending on chain length and temperature, can easily reach seconds and, if one approaches the glass transition, even longer. To cover such a range within a conventional computer simulation is certainly impossible. On the other hand, it is important to relate the chemical structure of a system to its macroscopic properties. Thus, a long standing challenge within the modelling of complex materials is to apply 'multi-scale' methods to cover the range from microscopic to the semi-macroscopic regime [8, 11, 12].

2 Basics of computer simulations

2.1 Molecular dynamics (MD)

MD simulations date back to the early fifties. (For a rather complete overview see [1].) Consider a cubic box of volume $V = L^3$ containing N identical particles of mass m. In order to avoid surface effects and (as much as possible) finite size effects, one typically uses periodic boundary conditions; the particle number density is $\rho = N/L^3$. The first simulations employed hard spheres of radius R_0, which have volume fraction $\rho_V = (4/3)\pi R_0^3\rho$. The 'excluded volume' interaction potential $U(\mathbf{r}_{ij})$, with $\mathbf{r}_{ij}$ the distance vector between two particles i and j, is given by

$$U(\mathbf{r}_{ij})/k_BT = \begin{cases} 0 \quad , & |\,\mathbf{r}_{ij}\,| \;\geq\; 2R_0 \\ \infty \quad , & |\,\mathbf{r}_{ij}\,| \;<\; 2R_0 \,. \end{cases}$$

Since the interactions are athermal (U/k_BT is independent of T) the only relevant variable is the volume fraction ρ_V. The particles are assigned random initial velocities with the condition that their total momentum is zero. Then one performs on the computer a (conceptually) simple collision dynamics. The temperature defines the time scale via the equipartition theorem: $m\langle|\dot{\mathbf{r}}|_i^2\rangle/2 = 3k_BT/2$. The simulation progresses the system to the point where two particles meet. At each such collision the necessary momentum and energy exchange is performed, and the run continues. (Since the hard sphere potential is the most anharmonic one possible, the system equilibrates very fast.) It is clear that for large systems this dynamics requires very small time steps, as very many collisions occur. Since all the sequential collisions have to be taken care of, such a programme cannot easily be parallelised, an important feature for modern large-scale applications. For some time, the hard sphere problem was mainly of historical interest. However, with the development of modern colloid science (see *e.g.* Frenkel, Lekkerkerker, and Chaikin, this volume) hard sphere simulations became very important again. On the computer, 'microgravity experiments' are a little easier than in space; however, our systems are still somewhat smaller!

Many of the technical problems discussed above can be avoided if one replaces the hard sphere potential by a soft sphere potential, such as the Lennard Jones potential $U^{LJ}(\mathbf{r}_{ij})$, dervied originally for interactions of noble gases. In its simplest form, for two identical particles

$$U^{LJ}(\mathbf{r}_{ij}) = 4\epsilon\left[\left(\frac{\sigma}{r_{ij}}\right)^{12} - \left(\frac{\sigma}{r_{ij}}\right)^{6}\right] \tag{1}$$

Usually, a cutoff r_c is introduced for the range of the interaction. This typically varies between 2.5 σ (a classical LJ interaction with an attractive well) and $2^{1/6}\sigma$ where only the repulsive part of the potential remains. The unit of energy can be chosen as ϵ, the unit of length σ and of mass m. This defines the 'LJ-units' of temperature $[T] = \epsilon/k_B$, time $[t] = \sqrt{\sigma^2 m/\epsilon}$ and number density $[\rho] = \sigma^{-3}$. In many practical programs σ, m, ϵ are set to one.

The straightforward MD approach is now to integrate Newton's equations of motion:

$$m_i\ddot{\mathbf{r}}_i = -\nabla_i \sum_{j\neq i} U(\mathbf{r}_{ij})\,. \tag{2}$$

Since energy in such a simulation is conserved, we generate the microcanonical ensemble. Again the average kinetic energy defines the temperature T via $\Sigma_i m_i|\dot{\mathbf{r}}_i|^2/2 = 3Nk_BT/2$.

A simple but very efficient and stable integration scheme for Equation 2 is the Verlet algorithm. With a simulation time step $\delta t \ll 2\pi/\omega_{max}$ where ω_{max} is the highest characteristic frequency of the system (*e.g.* the Einstein frequency of an LJ crystal) we have (with $m = 1$ in one dimension)

$$r_i(t+\delta t) = r_i(t) + \delta t\dot{r}_i(t) + \frac{\delta t^2}{2}\ddot{r}_i(t) + \frac{\delta t^3}{6}\dddot{r}_i(t) \tag{3}$$

$$r_i(t-\delta t) = r_i(t) - \delta t\dot{r}_i(t) + \frac{\delta t^2}{2}\ddot{r}_i(t) - \frac{\delta t^3}{6}\dddot{r}_i(t)\,. \tag{4}$$

Addition and subtraction of these two equations yields

$$r_i(t+\delta t) = 2r_i(t) - r_i(t-\delta t) + \delta t^2\ddot{r}_i(t) + 0(\delta t^4) \tag{5}$$

$$v_i(t) = \dot{r}_i(t) = \frac{1}{2\delta t}[r_i(t+\delta t) - r_i(t-\delta t)] + 0(\delta t^3)\,. \tag{6}$$

Thus, the position and velocity calculations have an algorithmic error of $0(\delta t^4), 0(\delta t^3)$ respectively. There are many variants of this basic method used throughout the literature: one can follow the realistic time evolution of a system, as long as the forces/potentials are realistic and as long as classical mechanics is sufficent. If the system is ergodic, which is not trivial and requires 'mixing of modes' (there is the famous Fermi-Pasta-Ulam problem, where one asks how anharmonic a potential has to be in order to equilibrate a one dimensional chain of particles [24]) one can equate ensemble averages to averages over time steps:

$$\langle A\rangle = \frac{1}{M}\sum_{i=1}^{M} A(t_i) \tag{7}$$

for any physical quantity A of interest. This is the basis for simulating a microcanonical ensemble [7], also called the NVE ensemble, where all extensive thermodynamics variables of the system, namely N, V, E are kept constant. Nowadays, most applications employ other ensembles such as the canonical (NVT), the isobaric-isothermal (NPT) or even the grand canonical (μPT) ensemble. As a general rule, in order to avoid two-phase coexistence and equilibration problems one should choose an ensemble which has as many intensive variables kept constant as possible.

So far only equilibrium MD simulations have been mentioned. In addition there is a huge literature on different nonequilibrium simulations to study the shear viscosity of liquids [13, 14] or other transport properties like heat conductivity [15].

2.2 Monte Carlo simulations

The classical version of MD simulation, just outlined, is a fully deterministic simulation technique. While there are many variants of the classical MD which add stochastic terms (see below) the other extreme, namely purely stochastic sampling, corresponds to the classical Monte Carlo (MC) approach.

Let's come back to the basic problem of hard spheres. We have seen that the only relevant parameter is the volume fraction in this athermal case (if one is only concerned with structural information). Then one can replace the ballistic motion of spheres by a stochastic one. Starting from a particular configuration, randomly a sphere is selected and displaced by a random jump. If this new configuration does not comply with excluded volume constraints the move is rejected, if it does, it is accepted. Then one starts the whole procedure again. Once every particle had a chance to move once (on average), our Monte Carlo step is complete. This is the most basic Monte Carlo simulation (see *e.g.* [1, 16, 17, 18]). Since for hard spheres there is no energy involved, all states have the very same probability $P_{\rm eq}$. Also if the system is in state x, the probability $W(x \to y)$ of it jumping into y is the same as that from y to x. So the dynamics trivially fulfils the detailed balance condition

$$W(x \to y)P_{\rm eq}(x) = W(y \to x)P_{\rm eq}(y). \tag{8}$$

To discuss a few more basic aspects of MC simulation let's go back to the 'fruit fly of MC simulations', the 2-d Ising model [1, 17, 18]. Consider a two dimensional square lattice, with spins on each lattice site. Each spin can assume two states $S_i = \pm 1$, corresponding to the 'spin up' and the 'spin down' state. The energy function (Hamiltonian) is

$$H(\{S_i\}) = -J\sum_{\langle i,j\rangle} S_iS_j + h\sum_i S_i$$

where $\langle i,j\rangle$ denotes all nearest neighbour pairs, J is the coupling constant, and h is an applied external field. Any observable A, *e.g.* the magnetisation, is given on average by

$$\begin{aligned}
\langle A\rangle &= \sum_{\{S_i\}} A(\{S_i\})P_{\rm eq}(\{S_i\}), && (9)\\
P_{\rm eq}(\{S_i\}) &= \exp(-H/k_BT)/Z, && (10)\\
Z &= \sum_{\{S_i\}} \exp(-H/k_BT).
\end{aligned}$$

A direct method of calculation would be to sample all possible states. There are however 2^{L^2} states, L being the lattice linear dimension, making this impossible for large systems. (For small systems such enumeration can lead to very good results!) Instead we sample phase space stochastically. Taking a spin at random, we flip the spin and calculate the energy change. With $P(x)$ the Boltzmann probability of the orginal state and $P(y)$ that of the new state, detailed balance is obeyed if

$$\frac{W(x \to y)}{W(y \to x)} = \exp\{-(H(x) - H(y))/k_B T\}. \tag{11}$$

This is normally a sufficient condition for a MC simulation to relax into thermal equilibrium, though this may take a very long time. (However, we will see later that in some cases improper choice of 'moves' can lead to a non-ergodic behaviour with equilibrium only attained on a subspace!) Algorithms without detailed balance will not be discussed here (but see Mukamel, this volume).

The *Metropolis criterion* is the one most frequently used to accept or reject a move:

$$W(x \to y) = \Gamma \begin{cases} \exp\{-(H(x) - H(y))/k_B T\} & , \quad \triangle H > 0 \\ 1 & , \quad \triangle H < 0\,. \end{cases} \tag{12}$$

Since only the ratio of the rates W is relevant, Γ is an arbitrary constant between zero and one. A random number r, equally distributed between 0 and 1, is used to decide upon the acceptance of a move. If $r < W(x \to y)$ the move is accepted, otherwise rejected. (Usually $\Gamma = 1$ is chosen so that any move that lowers the energy is accepted.) This is the basic MC procedure, which allows the sampling of phase space, and equilibrium averages to be found from Equation 7.

In many cases, however, one also would like to gain information on the dynamics of a (model) system. How can one use MC simulations, without an intrinsic time scale, to obtain information on the dynamics? The method described before evolves a system from one state to another by a local spin flip. By this local stochastic flips the magnetisation changes with 'time'. This can also be seen as a dynamic MC method based on a stochastic Markov process, where subsequent configurations $x \to x' \to x'' \to ...$ are generated with transition probabilities $W(x \to x'), W(x' \to x'')$... To a large extent the choice of the move is arbitrary, as long as one can interpret it as a local basic unit of motion. As seen before, the actual choice of W is somewhat arbitrary as well: the prefactor Γ in Equation 12 can vary. Γ actually can be interpreted as fixing an attempt rate, $\Gamma = \tau_0^{-1}$, for the moves and so introduces a time scale. Thus we reinterpret the transition probability $W(x \to y)$ as a transition probability per unit time [3, 19]. To compare the simulated dynamics with an experiment, the basic task is to properly determine τ_0 (*e.g.* from the diffusion of chains in a polymer melt). It is obvious, however, that this (overdamped) simulation does *not* include any hydrodynamic effects (see Frenkel, this volume) since there is no momentum involved. (There are very interesting more advanced methods like 'dissipative particle dynamics' and 'lattice Boltzmann MC' which are currently being developed to repair this [20, 21, 22].)

Using this interpretation, ensemble averages can again be written as time averages:

$$\langle A \rangle = \frac{1}{M - M_0} \sum_{i=M_0+1}^{M} A\left(x(t_i)\right) \cong \frac{1}{t - t_0} \int_{t_0}^{t} dt' A\left(x(t')\right) \tag{13}$$

where we can interpret one attempted move per particle as one time step. Typically, the first configurations in a simulation are not characteristic equilibrium configurations. One first has to 'relax' the system into equilibrium, meaning the data for the first M_0 steps are omitted. In this interpretation the dynamic Monte Carlo procedure is nothing but a numerical realisation of a Markov process described by a master equation

$$\frac{d}{dt}P(x,t) = -\sum_y W(x \to y)P(x,t) + \sum_y W(y \to x)P(y,t)$$

with $P(x,t)$ the probability of being in microstate x at time t. The condition of detailed balance is sufficient to ensure that $P_{\text{eq}}(x) = \exp[-H(x)/k_BT]/Z$ is the unique steady-state solution of the master equation (so long as all states are mutually accessible). Thus $P(x,t)$ must relax towards $P_{\text{eq}}(x)$ as $t \to \infty$ irrespective of the starting state. Note however that the choice of a 'good' starting state can save enormous amounts of CPU time (reducing M_0).

So far, the two extreme cases for classical, particle-based computer simulations were discussed: microcanonical MD and MC. There are many approaches in between. Here I just want to mention them. The techniques range from

- **pure MD** where Newton's equations of motion are solved: $m\ddot{x} = -\nabla U$, via
- **MD coupled to a heat bath and a friction** ('Langevin MD', 'Noisy MD'), where one solves $m\ddot{x} = -\nabla U - \zeta\dot{x} + f(t)$ with ζ a friction and f a random force, via
- **Brownian Dynamics (BD)** where one solves $\zeta\dot{x} = -\nabla U + f$, via
- **Force biased MC**, where attempted moves are selected from the beginning according to local forces, to
- **plain MC** as described above.

3 Polymer simulations

3.1 General considerations

Now let us turn to some specific questions related to polymers. Compared to simulations of small molecules, polymers (like many other forms of soft matter) require special attention due to the huge number of intra-molecular degrees of freedom. This causes both computational advantages and disadvantages. Polymers are, of course, chain molecules built of repeat units called monomers or (by physicists) 'beads'. Examples range from the simple, widely used, PE to the technically very important but more complicated BPA-PC, which is used, among other things, for compact discs:

PE	$(CH_2)_N$	polyethylene
PS	$(CH_2(CH(C_6H_5))_N$	polystyrene
PEO	$((CH_2)_2O)_N$	polyethylene oxide
BPA-PC	$(C_6H_4C(CH_3)_2C_6H_4CO_3)_N$	bisphenyl A polycarbonate

While three out of the above examples are only soluble in organic solvents, PEO is also water soluble. Chemistry and biology of course provide many more complex examples.

The molecular weight of a single polymer molecule can easily reach $\sim 10^5$ Daltons (several thousands of monomers). Full scale MD simulations are not suitable (Figure 1), but in any case, would be restricted to the study of a very few specific systems. Even then it is very difficult to determine proper intra- *and* inter-molecular interaction potentials, and one would also need an effective and computationally convenient parametrisation for them. In particular the complicated inter-chain or polymer-solvent interactions are only poorly understood at present.

Despite these seemingly tremendous complications, the situation is not that bad. To what extent do we really need the chemical details? Figure 1 gives a caricature of the various relevant scaling regimes. It also illustrates universality. Properties, which are governed by length scales larger than a few Kuhn lengths are independent of the chemistry. For example in the limit of long chains of N monomers the mean squared end-to-end distance $\langle R^2(N) \rangle = AN^{2\nu}$. The exponent ν is universal, and takes values $0.588 \simeq 6/5$ in a good solvent (for $d = 3$) and 0.5 in the melt or in a Theta solvent (see Khokhlov, this volume). The chemical details are hidden in the prefactor A. This suggests a relatively satisfactory situation, namely for most physics questions one can confine the simulation to the simplest and, for computational purposes, fastest models. In many cases we just need the monomer-monomer excluded volume and the chain connectivity. These models are often called coarse-grained models. Later on, we will face the problem of 'mapping properties back to a given chemical system' in order to determine amplitudes (like A) as well.

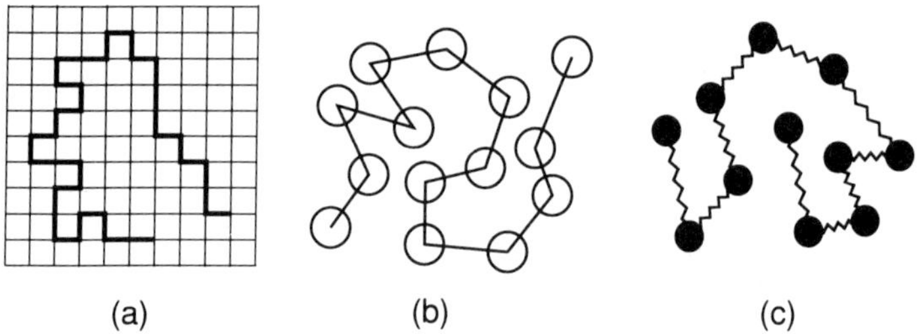

Figure 2. *Typical models for polymers used in simulations; (a) SAW on lattice, (b) pearl necklace, (c) bead-spring model.*

We now apply the previously described techniques to coarse-grained polymer models. Typically, three classes of models are used for computer simulations of polymers, see Figure 2. The first and historically most widely used model of a single chain is the *self-avoiding walk* (SAW) on a lattice [3]. In such a walk, each lattice point can be occupied only once; it is trivial to introduce nearest-neighbour energies and other generalisations. The second model is the direct generalisation of this for continuous space, the *pearl necklace model*. The chain consists of hard spheres of diameter σ_0 and a fixed bond length ℓ_0. The third variant is a *bead-spring model*, mainly used in MD simulations: the monomers are particles which interact with, in most cases, a purely repulsive Lennard Jones interaction (Equation 1). For the bonded nearest-neighbours along the chain, an additional spring potential is added which, together with the repulsion, determines the bond length ℓ.

The purpose of a simulation is to generate statistically independent conformations (equilibrium) or to follow the time evolution of a given global conformation (dynamics). Again, the most natural ansatz would be to simply perform a MD simulation, where Newton's equations of motion (Equation 2) for a bead spring model are solved numerically. However, this direct approach cannot be used for an isolated chain. The reason lies in its structural properties: a linear-chain polymer without excluded volume interactions has the structure of a random walk (harmonic chain). Its structure and dynamics can be described by eigenmodes, the Rouse modes, which decouple [23]. Thus solving Equation 2 exactly, without excluded volume or external noise, would never equilibrate the chain.

For a chain with excluded volume, the Rouse modes are no longer eigenmodes. However, the deviations, which affect the large length scales, come mainly from 'long range' contacts between monomers which are far apart along the chain but close by in space. For long chains the internal monomer density decays with a power law $N^{1-d\nu}$. (The chains are fractal objects with a fractal dimension $d_f = 1/\nu$; see Pine, this volume). Consequently, these 'long range' collisions, which cause the swelling of the chains, are very infrequent: equilibration is not guaranteed. The single SAW is an example of the direct relevance of the aforementioned Fermi-Pasta-Ulam problem, which to date is not yet solved [24].

Thus the natural MD approach can only be used for a chain interacting directly with solvent or other chains, or for chains with long range interaction potentials. To avoid this problem one has to couple the motion of chain beads to a stochastic process, for example a heat bath comprising a friction and a random force. In the overdamped limit this results in Brownian dynamics. The other approach is the dynamic Monte Carlo method, where again the conformation changes come from local stochastic jumps. Both approaches follow realistically the (Langevin) dynamics of a chain. Though this leads to information about chain dynamics as well as statics, this should be avoided if the dynamic information is not needed. To illustrate this, we compare in Table 1 a simple liquid (or a lattice gas) and a polymer melt of the same density and total number of particles. The comparison in the

	Liquid	Polymer melt
particles	$N_{\rm tot}$	$N_{\rm tot}$
density	ρ_0	ρ_0
CPU time per timestep	$\propto N_{\rm tot}$	$\propto N_{\rm tot}$
equilibration distance	$\propto \rho_0^{-1/d} \propto N_{\rm tot}^0$	$\propto \langle R^2(N)\rangle^{1/2} \propto N^{1/2}$
physical relaxation time	$\tau \propto N_{\rm tot}^0$	$\tau_N \propto N^z, \;\; 2 \le z \le 3.4$ (varies with chain length)
CPU time for relaxation	$\propto N_{\rm tot}$	$\propto N_{\rm tot} \cdot N^z$

Table 1. *An illustration of the time scales in polymer simulations*

Table assumes that we are not close to a critical point of the liquid, where critical slowing down occurs. But in fact, the exponent $z = 2 \ldots 3.4$ in the relaxation time $\tau \sim N^z$ of polymers is analagous to that for critical slowing down [23, 25]. (In the language of critical phenomena the inverse chain length corresponds to $(T - T_c)/T_c$.) This N-dependent slowing causes significant problems since N values of interest start at around 50 to 100. As a result there is one important rule for polymer simulation: if possible, avoid algorithms with *realistically slow* physical dynamics. (However, if one wants to study chain dynamics itself, there is no alternative: see Section 4.)

3.2 How to generate conformations

Having demonstrated that the methods which follow the realistic dynamics are not very useful for single isolated chains, we now try to generate the conformations by a purely stochastic process.

3.2.1 Simple sampling techniques

First let us consider SAWs on a lattice. To fulfil the excluded volume requirement each lattice site can only be occupied once; but otherwise each conformation of an N-step walk has the same probability. If we fix the first step, then each new step is taken with probability $1/(q-1)$, where q is the coordination number of the lattice, and we account for the fact that backward steps are ruled out.

This most naive way of generating conformations with equal probability is called simple sampling. Each time an attempted new bond hits a site which is already occupied, one has to start again at the beginning. Otherwise, different SAW conformations will receive different probabilities, generating an (attractive) effective interaction among the monomers. Each sampled conformation is therefore taken randomly out of the $q(q-1)^{N-1}$ possible random walk paths, which do not include direct back-folding, whereas the total number of SAWs on a lattice is given by [3]

$$Z(N) = c_0\, q_{\text{eff}}^N\, N^{\gamma-1} \qquad N \gg 1 \tag{14}$$

with $q_{\text{eff}} < q-1$ and c_0 a number of order unity. The critical exponent γ is dimension- but not lattice-dependent ($\gamma \approx 7/6$ for $d=3$, and $4/3$ for $d=2$). Typical numbers for q_{eff} are 2.6385 (square, $q=4$), 2.879 (diamond, $q=4$) and 4.6835 (simple cubic, $q=6$) [3]. Thus the success rate of the sampling process, $A(N) = A_0(q_{\text{eff}}/(q-1))^N N^{\gamma-1}$, decays exponentially with N. A typical value for A is *e.g.* $A(100) = 0.03$ on the diamond lattice. This illustrates that simple sampling is only useful to get a rough estimate for chain properties like q_{eff} and γ.

The first improvement was suggested as early as 1955 by Rosenbluth and Rosenbluth [26]. Their idea of *biased sampling* is to look ahead for at least one step in order to overcome the attrition. More modern approaches look several steps ahead or implement this idea within in a dynamical scheme [8, 10]. Here I explain two alternatives, which can mainly be used for isolated chains.

3.2.2 Dimerisation

The *dimerisation* approach (Figure 3) [27, 28] allows a simultanous determination of both the exponents γ and ν. The idea is to begin by generating many short SAWs of

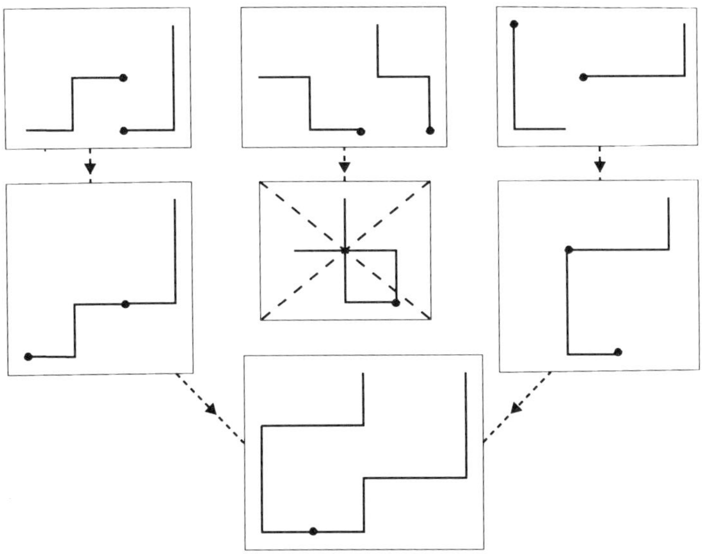

Figure 3. *Illustration of the dimerisation method.*

length N_0. These walks can then be combined randomly. For every step the success rate P_{bin} for a binary assembly of walks of length N_1, and $N_2 = N - N_1$ is $P_{\text{bin}}(N_1, N_2) = Z(N)/(Z(N_1)Z(N_2))$. Using $Z \propto q^N N^{\gamma-1}$ we get $P_{\text{bin}}(N_1, N_2) = N^{\gamma-1}/(N_1 N_2)^{\gamma-1}$; obviously P_{bin} is optimal for $N_1 = N_2 = N/2$. The original procedure, where short walks were generated and then stored, is computationally not very efficient. It is more direct to generate each sub-walk by simple sampling, as follows [3, 29]. After a SAW of N_1 monomers is generated we continue. If the next N_1 steps violate the excluded volume condition *within the second piece*, we start at N_1 again. Only an overlap with the first N_1 monomers makes it necessary to start at the very beginning. This defines a very efficient hierarchical procedure.

Batoulis and Kremer generalised this approach to study good-solvent properties of star polymers [29]. In particular, the dependence of γ on the number f of arms of the star is interesting and very difficult to estimate analytically. Using this method we were able to give precise results for $\gamma(f)$ and also size statistics such as the mean radius of gyration, and the mean hydrodynamic radius $\langle R_H \rangle = N^{-2} \sum_i \sum_{i \neq j} \langle 1/r_{ij} \rangle$. For R_H, very strong corrections to scaling are observed, which explain deviations of experiments such as light scattering from the asymptotic power law behaviour. Figure 4 shows the extrapolation of the hydrodynamic radii as a function of $N^{-1/2}$, which was thought to be the correction to the scaling for this quantity. The correct value is $1 - \nu$ [30].

3.2.3 Pivot algorithm

For this algorithm [31, 32], as illustrated in Figure 5, a point on the chain is chosen randomly and one part of the chain is rotated at random. As with dimerisation, this can easily be done for both lattice and off-lattice systems. Given a chain of length N and a

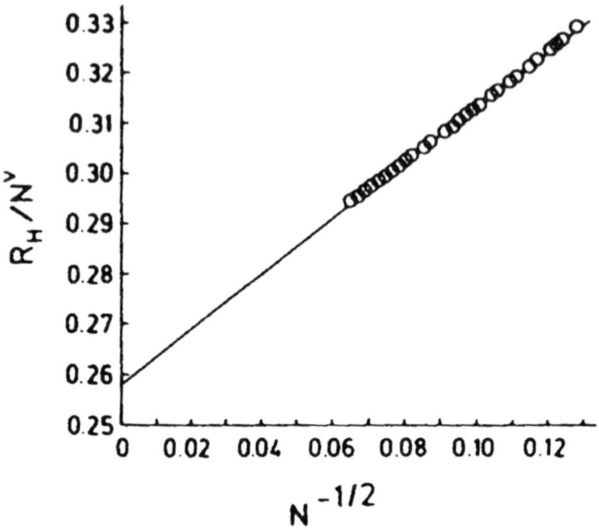

Figure 4. *The extrapolated hydrodynamic radii as a function of $N^{1/2}$ [29].*

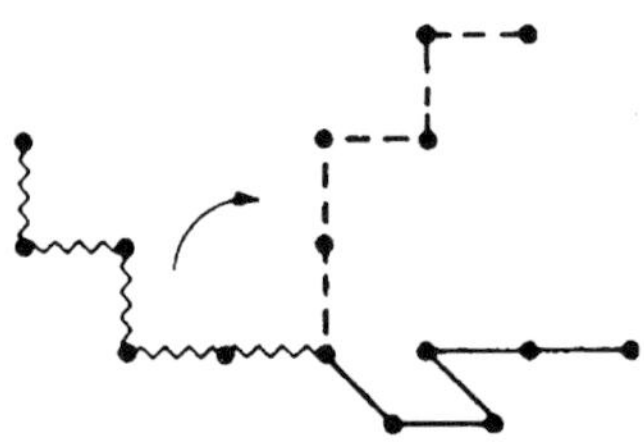

Figure 5. *Illustration of the pivot algorithm.*

pivot point at position N_1, then the acceptance rate p is given by the probability that the new conformation has no overlaps. On a lattice this needs a simple $\mathcal{O}(N)$ check, while in continuous space this is more difficult. On lattices, especially for $d = 2$, one has to take care that the choice of moves does not contain hidden conservation rules. The approach is then ergodic as shown by Madras and Sokal [31] who claim an $N^{0.2}$ power law for the 'relaxation time' of the mean-square end-to-end distance. This method was used to obtain very precise estimates of the exponent ν, $\nu = 0.7496 \pm 0.0007$ ($d = 2$: exact value 3/4) and $\nu = 0.592 \pm 0.003$ ($d = 3$), which is about as accurate as the results coming from dimerisation. For the single isolated SAW there is probably no better way to generate very quickly many conformations that are globally different.

However there is no 'free lunch'. As soon as the concentration is increased, either due to other chains or monomer-monomer attraction along the same chain, the acceptance rate is dramatically reduced. A more subtle aspect is that this method relaxes large length scales very fast, while short ones need a longer time. For example, nearest-neighbour bond correlations need up to $\mathcal{O}(N^2)$ moves to relax completely. For many problems it is essential to cover the equilibration of the short as well as the long distances. This can be overcome in two ways (depending on the model). One is a hybrid method of pivot moves and molecular dynamics simulations (for bead spring models) [33]. For fixed bond lengths it is sufficient instead to generalise the pivot approach so that not only tails of the chain, but also internal pieces of arbitrary length, are rotated. The choice of lengths is then adjusted to the density, so that the acceptance rate stays above 10%. This version can be applied to ring polymers as well.

3.2.4 Many-chain methods: generalised reptation algorithms

In semidilute solutions or melts the generalised pivot algorithm will only produce acceptable success rates for moves involving a small number of monomers. Eventually it joins the class of dynamic algorithms, which follow realistically the very slow physical dynamics of the system. This is certainly not what we are looking for in the present context.

The generalised reptation algorithm is an effective way to simulate dense solutions, and also for single chains, in the collapsed regime for $T < \Theta$, the Theta temperature (see Khokhlov, this volume). The method is explained here for a single lattice chain. The original idea of the 'slithering snake' goes back to Kron in 1965 and Wall and Mandel 1975 [34]. One randomly takes an end monomer and tries to add it at the other chain-end with a random orientation. If the new chain fulfils the SAW condition, the move is accepted, otherwise rejected. In this way the chain moves (forward and backward) like a snake along its own contour. (The algorithm resembles, but should not be confused with, the *physically realistic* reptation dynamics of polymer melts: see McLeish, this volume, and below.) For this method detailed balance is fulfilled trivially, but on a lattice, only a subspace of all conformations is reached. (This subspace excludes conformations where both chain ends are completely surrounded by other monomers.) In continuous space this problem is absent. The extension to many chains as well as interacting chains is trivial. One can also introduce a grand canonical version by allowing the chain length to fluctuate. The use of pointers allows for very fast and efficient codes.

How fast is the relaxation of a chain conformation in CPU time? To estimate this, let us look at the position of the middle monomer of the chain. At first this does not move. A measure of the relaxation time T_N^{CPU} is the time until this space point is not on the chain for the first time. For random walks this first passage time follows the same power law as a one-dimensional diffusion along the chains. Since each move only requires a constant N-independent number of operations we get $T_N^{CPU} \propto N^2$ for ideal chains. (To simulate this is a good check on the program.) This method is $\mathcal{O}(N)$ faster than algorithms with 'realistic dynamics' such as the Rouse model (see below), whose *real* relaxation time is $T_N \sim N^2$ giving N^3 in CPU time.

Somewhat more complicated is the situation for SAWs. In physical time one expects for the diffusion constant $D \propto N^{-1}$. With $R^2 \propto N^{2\nu}$ and $DT_N \approx R^2$ we expect $T_N^{CPU} \sim T_N \sim N^{1+2\nu}$ for this algorithm. This is, however, not correct. Numerically one finds for the autocorrelation function of $\langle R^2 \rangle$ a relaxation time of $T_N \propto N^{2.15} \quad (d=2)$ and $N^{2.02} \quad (d=3)$ instead of $N^{2.5}$ and $N^{2.18}$ respectively. This result, although of advantage for simulations, is not completely understood. The explanation is probably in the difference between diffusion times and first passage times [35]. In addition, it should be kept in mind that because of the SAW condition, SAWs are correlated objects!

3.3 More complex single polymer problems: two examples

The above methods are the basis for many applications to more complicated chemical designs or molecular architectures. Branched polymers were already mentioned. There is however a whole zoo of other systems available: H-shaped objects, comb polymers, double stranded polymers *etc.*. Two other interesting classes are copolymers, especially random copolymers, often dicussed in the context of protein folding, and charged polymers (see Khokhlov, this volume). For both, simulation studies are becoming more and more important. Here I want to discuss two applications on the level of single chains: polyampholytes, and the collapse transition of PEO in water under pressure. In both cases a somewhat specialised model of a single chain is needed, but the simulation methods are essentially the ones described above.

Many biopolymers are on average neutral, however they contain, from a physicist's view, randomly distributed charges, of either sign, along the backbone. The simplest model of such a *polyampholyte* chain [36] treats electrostatics and excluded volume only. A chain with an extensive number of excess charges of one sign, is certainly stretched, while one might expect the opposite from an overall-neutral chain. Analytical theories give contradictory results and are only approximate.

To approach the problem Kantor and coworkers [36] looked at a random copolymer, where the comonomers are either positively or negatively charged. With N monomers, on the average there is an excess charge of $\sqrt{N}$. The authors used a simple cubic lattice simulation to study chain conformations as a function of temperature (with fixed dielectric constant of the background 'solvent'). They found that in the limit of very long chains (and/or low temperatures) those chains with less than $\mathcal{O}(\sqrt{N})$ excess charges will collapse, while the others will be expanded (Figure 6).

For standard, uncharged polymers, chain conformation changes continously as a function of temperature and/or solvent quality. But there are contrasting examples as well. PEO is soluble in water under normal atmospheric pressure but separates out under high pressure. There is a Flory type theory, which relates this phase separation/collapse transition to the destruction of the hydration shell around the oxygen under pressure [39]. If this hydration shell is destroyed, one is left with the strong hydrophobic interaction of hydrocarbon backbone and the water. Within a simulation this can be taken account of by a simple model [40], where the monomers have an internal degree of freedom. The situation with a hydration shell corresponds to the good solvent case and a fairly large excluded volume. If the hydration shell is destroyed, the excluded volume is smaller and the monomers attract each other, since water is a very poor solvent for the hydrocarbon part of PEO. We now can study the conformations as a function of temperature for a given (pressure-dependent) activation barrier for the hydration shell. It turns out that already on the single chain level the collapse transition becomes discontinuous.

Figure 7 illustrates this for the chain form factor $S(q)$. Right at the phase transition point one gets an energy distribution function with two maxima. Dividing the configuration space at the minimum yields two coexisting sets of conformations. The figure shows $S(q)$ averaged separately for the two subsets and clearly displays a collapsed and an expanded structure. Most of these runs were performed with the slithering snake algorithm combined with a Metropolis update of the hydration shell.

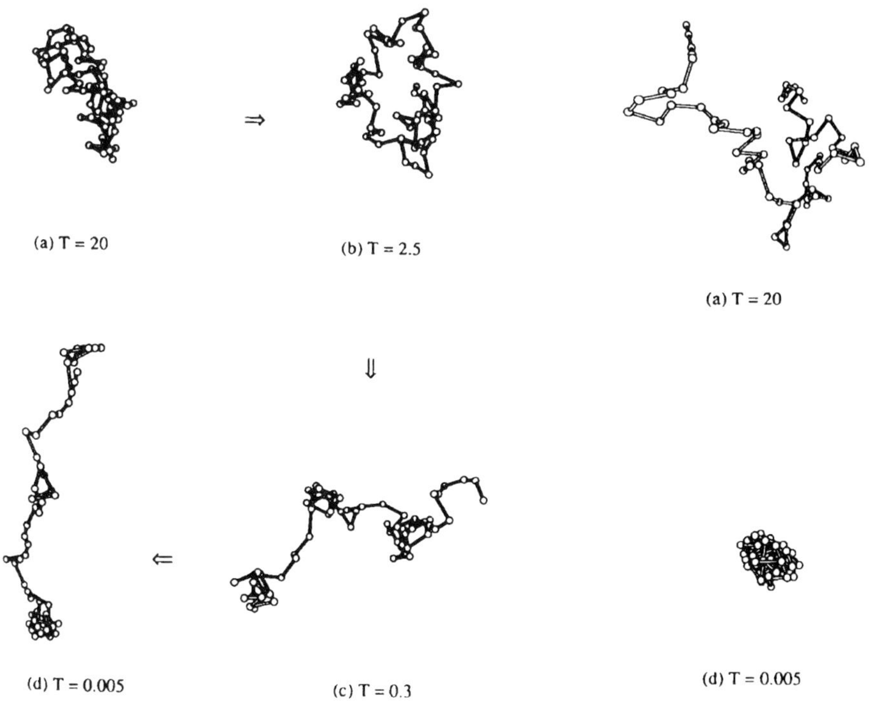

Figure 6. *Left, (a-d): Typical conformations of a polyampholyte chain with an asymmetric charge distribution as a function of interaction strength (temperature), from [36]. Note that this picture holds for isolated, finite length chains only. At finite concentration, due to the overall neutrality of the whole system, oppositely charged chains and agglomerates will cluster step by step [37, 38]. Right: (a,d) conformations of a chain that is neutral overall.*

4 Polymer dynamics

To obtain information about dynamics, we have to follow the slow, physically realistic, simulation path. There are many tricks to vectorise or simplify the algorithms, but these only influence the *prefactors* of the power laws shown in Table 1. Faced with a naturally very slow process, we are forced to employ extremely simple models.

To test the applicability of any chosen method we need a basic model for dynamics, the Rouse model [23]. This is still the only model on a 'molecular level', which can be solved analytically. The Rouse model treats the dynamics of a Gaussian random walk in the overdamped (Brownian) limit. All the complicated inter-chain and intra-chain interactions are summarised in the viscous and random forces from the heat bath. Thus, we totally disregard excluded volume and topological constraints beyond the plain chain connectivity. This model, which represents the dynamics of molten but unentangled (short) chains can readily be solved analytically (see McLeish, this volume).

Here we recap the essentials. With $\mathbf{r}_i$ the position of the i-th monomer, the Rouse

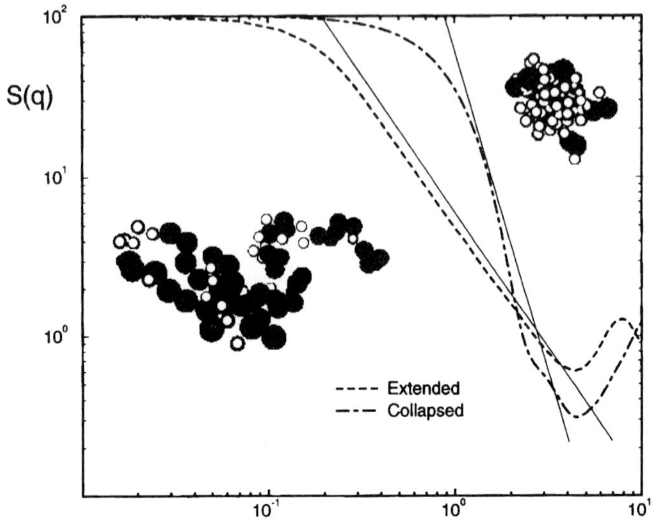

Figure 7. *Static structure function for $N = 100$ at the collapse transition point for the simple PEO model. The bigger spheres denote the hydrated monomers. For the expanded configurations these are equally distributed all over the chain, while for the collapsed ones only a few surface sites remain available to build up a solvation shell. The asymptotic slopes for the two cases of $q^{-1/0.588}$ (swollen coil) and q^{-4} (collapsed globule) are indicated by straight lines; from [40].*

equation of motion is

$$\zeta \frac{\mathrm{d}\mathbf{r}_i}{\mathrm{d}t} = -\kappa(2\mathbf{r}_i - \mathbf{r}_{i+1} - \mathbf{r}_{i-1}) + \mathbf{f}_i(t) \,. \tag{15}$$

Here ζ is the monomeric friction constant, κ the bond spring constant and $\mathbf{f}_i(t)$ the random (heat bath) force with $\langle \mathbf{f}_i \rangle = 0$, and (via the fluctuation dissipation theorem) $\langle \mathbf{f}_i(t)\mathbf{f}_j(0)\rangle = 2\zeta T \delta(t) \underline{\mathbf{I}} \delta_{ij}$. In the discrete version, the Rouse amplitudes $\mathbf{X}_p$ are given by

$$\mathbf{X}_p(t) = \frac{1}{N-1}\left\{\sum_{i=1}^{N-1} \mathbf{r}_i(t) \cos\left(\frac{p\pi(i+1/2)}{N-1}\right)\right\} \tag{16}$$

with relaxation times (for large N)

$$\tau_p = \tau_1/p^2 \quad ; \quad \tau_1 \equiv \tau_{\text{Rouse}} = \frac{\zeta N \langle R^2 \rangle}{3\pi^2 k_B T} \,. \tag{17}$$

Eigenmodes decay exponentially, $\langle \mathbf{X}_p(t)\mathbf{X}_p(0)\rangle / \langle |\mathbf{X}|_p^2 \rangle = \exp(-t/\tau_p)$, resulting in an overall chain diffusion constant $D = k_B T/\zeta N$ and a single chain dynamic structure function $S(q,t)$ scaling as $\ln(S(q,t)/S(q,0)) \propto -q^2\sqrt{t}/6$.

The most natural quantity to measure from a simulation is the mean square displacement of the individual monomers, $g_1(t) = \langle |\mathbf{r}_i(t) - \mathbf{r}_i(0)|^2 \rangle$. One finds (see Figure 8)

$$g_1(t) \sim \begin{cases} t^{1/2} & t < \tau_1 \\ t^1 & t \geq \tau_1, \end{cases}$$

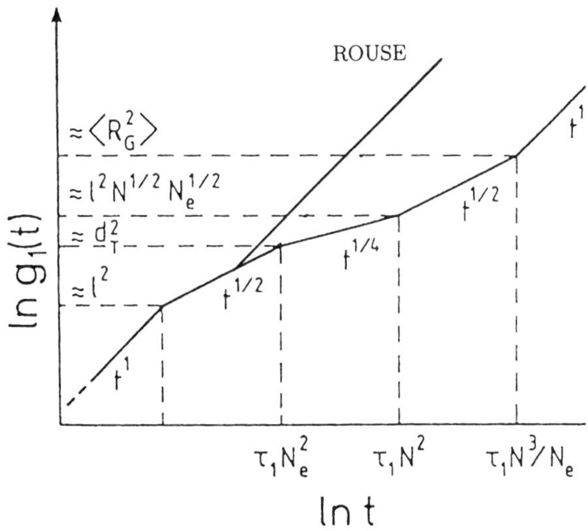

Figure 8. *A sketch of the monomer mean square displacements for the standard Rouse model and for the reptation model. The time and length scales are indicated for the reptation case.*

while the centre of gravity of the chain always follows $r_{\rm cm}^2(t) \propto t$.

This model is a basis for the dynamic interpretation of stochastic algorithms. Besides the standard requirements for MC procedures (detailed balance *etc.*) we need the following property (for an extensive discussion of details see [3]): to simulate polymer dynamics, any algorithm with stochastic (or other artificial) dynamics must involve only local moves, and must yield Rouse dynamics for Gaussian random walks. Below we concentrate on lattice models; however, the generalisation to continuous space is obvious.

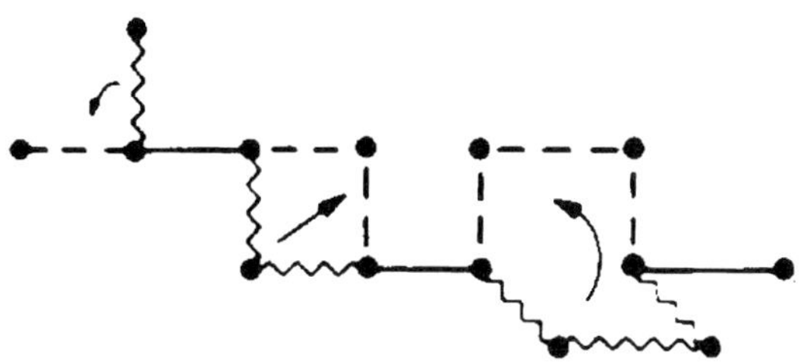

Figure 9. *Typical kink jump moves for a dynamic Monte Carlo simulation.*

The standard procedure is the kink jump method, Figure 9, as follows: (i) select monomer at random; (ii) select trial local move at random; (iii) check the move is topologically permitted; (iv) implement a Metropolis check; (v) decide on acceptance; (vi) start over again.

In a simulation we are mainly interested in SAWs, but to check the validity of a method for dynamics we need to recover Rouse behaviour for ideal, Gaussian chains. To check a given set of local move rules for the SAW, the test for Rouse behaviour is actually performed on non-reversing random walks (NRRWs). NRRWs are walks which do not allow direct back-folding but do not have any long range excluded volume interaction. (Such walks are ideal Gaussian coils on large length scales, but have the same local conformations as SAWs and consequently the same set of local moves.)

To achieve Rouse dynamics, the set of local moves must contain moves which create new bond vectors $\{\mathbf{b}_i\}$ inside the chain. In Figure 9, there are 2-bond moves and 3-bond moves. The 2-bond moves only exchange positions of bonds along the chain but do not change the set of bond vectors. For rings, an algorithm only containing such moves would never equilibrate, while for open chains new bond vectors could only diffuse in from the freely rotating chain ends. This would increase the relaxation time at least by a factor of N. Local rotations, such as the 3-bond move, are needed, so that $\{\mathbf{b}_i\}$ is not conserved and Rouse behaviour recovered. The minimal number of bonds involved in $d = 3$ varies from 2 in continuum or on the fcc lattice to 4 on the diamond lattice. But in $d = 2$, for the 180° rotation of the 3-bond move we again only exchange bonds. To circumvent this difficulty, one can use off-lattice models, where both the bond lengths and angles vary. For such a model, the excluded volume constraints are usually time consuming to evaluate. Distances have to be calculated at every time step between several monomers. This results in the loss of the main advantage of the MC methods compared with the MD.

A now frequently used alternative approach is the *bond fluctuation* (BF) method of Carmesin and Kremer [41], which combines advantages of lattice simulations and continuous space. Figure 10 illustrates the method for $d = 2$ and $d = 3$. Each monomer consists

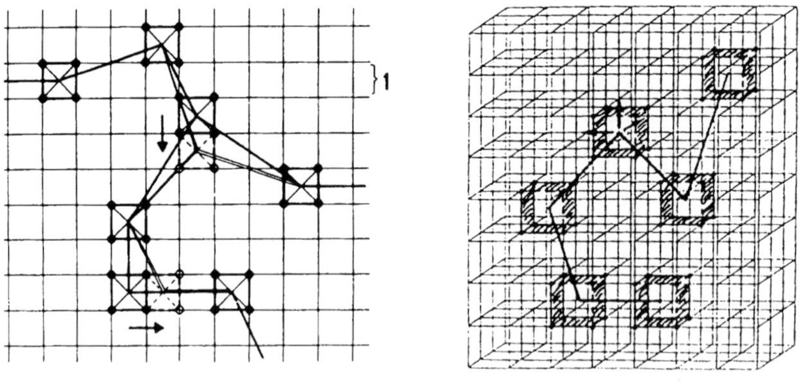

Figure 10. *An illustration of the bond fluctuation model for a 2-d branched polymer and for a 3-d linear polymer. Typical elementary moves are indicated. Each monomer is represented by a square (cube) of four (eight) occupied sites on a 2-d (3-d) unit lattice.*

of 2^d lattice sites. In addition to the excluded volume interaction, the bond length l is restricted to a maximum extension to avoid bond crossing. On the square lattice, one has the constraint that $1 < l < \sqrt{16}$. For $d = 3$, the situation is slightly more complicated. In this case a set of 108 different bonds are allowed [42, 43, 44]. Since each monomer occupies 2^d sites, but every jump only requires 2^{d-1} empty sites, the method works effectively at high densities. (It also suffers less from non-ergodicity problems than the standard methods do.) For the 3-d study of the dynamics of polymer melts, densities as large as 0.5 were used, although higher densities are no problem. This corresponds to a very high physical chain density. Skolnick *et al.* [45] find for SAWs on a simple cubic lattice at $\phi = 0.5$ a static screening length for the excluded volume interaction of about 12 monomers, indicating that they are in a semi-dilute regime, whereas Paul *et al.* [46] find for BFM, and $\phi = 0.5$ ($d = 3$) a value of about 2. On a scale larger than a trimer the excluded volume interaction is already screened as it is in a polymer melt (see McLeish, this volume).

A first serious application was the study of a 2-d melt of chains [47]. There one would expect the chains to segregate for entropic reasons [25]. In a Monte Carlo study using the bond fluctuation model chains of up to $N = 100$ monomers at a density of up to 80% occupied lattice sites were investigated. Figure 11 shows the segregation procedure clearly.

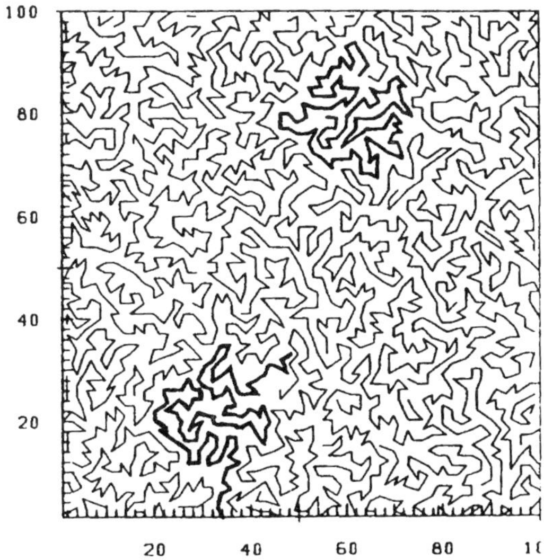

Figure 11. *A 2-d polymer melt with* $N = 100$ *and* $\Phi = 0.8$ *for an almost completely equilibrated sample [47].*

More interesting, however, is the *dynamics* of the $d = 2$ melt. The chains cannot cut through each other; on the other hand, they are compact objects with $\langle R^2(N)\rangle \propto N \propto R^d$. One finds a typical 2-d soft sphere liquid with an average of 6 chains surrounding a given chain. The Rouse model does not take into account any constraints on motion besides the connectivity. A first test is to measure the Rouse mode relaxation spectrum. The autocorrelation function for each mode shows an almost perfect single exponential decay. On the other hand we can also measure the diffusion constant D_0 of the chains directly from the mean square displacements. Each test yields an independent estimate of the monomeric friction constant ζ, which should be consistent, if the Rouse model describes the motion. For $d = 2$, however, one finds that $\zeta_{\rm mode} > \zeta_D$, and that $\zeta_{\rm mode}$ is governed by the conformational relaxation of the chain, while D_0 is controlled by motion of the chain without really reorganising its internal conformational structure. Recently this picture was impressively supported by experiments on 2-d confined DNA molecules [48]. An extension, which puts these results into a much wider perspective, is 2-d polymer glasses [49]. It could be shown, for this case, that time temperature superposition (see McLeish, this volume) directly followed from the Rouse dynamics.

More relevant to most actual experiments, of course, is the study of 3-d melts, where one wants to test the concept of chain reptation. Besides standard kink-jump and BF MC, another alternative is to perform a MD simulation. Simulations using the MD method usually employ the bead-spring model [50, 9], with, in addition, each monomer weakly coupled to a heat bath. (Technically, this is a hybrid method or 'noisy MD'.) Each polymer chain consists of N monomers of mass m connected by an anharmonic spring. The monomers interact through a repulsive Lennard-Jones potential given by Equation 1 with $r_c = 2^{1/6}\sigma$. For connected monomers we add an attractive interaction

$$U^{\rm bond}(r) = \begin{cases} -0.5\kappa R_0^2 \ln\left[1 - \left(\frac{r}{R_0}\right)^2\right] & r \leq R_0 \\ \infty & r > R_0 \end{cases} . \tag{18}$$

For melts the parameters $\kappa = 30\epsilon/\sigma^2$ and $R_0 = 1.5\sigma$ usually are chosen, while in solution softer potentials are mostly used [9].

In this algorithm, the equation of motion for monomer i is

$$m_i \ddot{\mathbf{r}}_i = -\boldsymbol{\nabla}_i \sum_{j\neq i} U(\mathbf{r}_{ij}) - \Gamma \frac{\mathrm{d}\mathbf{r}_i}{\mathrm{d}t} + \mathbf{W}_i(t) . \tag{19}$$

Here Γ is the bead friction which acts to couple the monomers to the heat bath, and $\mathbf{W}_i(t)$ describes the random force acting on each bead, with $\langle \mathbf{W}_i(t)\mathbf{W}_j(t')\rangle = \delta_{ij}\delta(t-t')\underline{\mathbf{I}}2k_BT\,\Gamma$. Strict MD runs for polymers typically exceed the stability limits of a microcanonical simulation; these small damping and noise terms extend stability and define a canonical ensemble. We have used $\Gamma = 0.5\tau^{-1}$ and $T = 1.0\epsilon$ in most cases and $m = 1$. This value of Γ is large enough to stabilise the run but small enough *not* to produce in itself Rouse-like behaviour on length and time scales of the order of a few bond lengths. Thus Γ and $\mathbf{W}$ are not be confused with ζ and $\mathbf{f}$ in the Rouse Equation 15 for a single chain; in this simulation, the Rouse friction is dominated instead by collisions with other chains. The program can be vectorised [51, 52] or parallelised [53]. Most older runs used a predictor-corrector scheme and Gaussian random numbers. One can however use equally distributed random numbers, so long as these have the correct mean value and second moment. The use of a Verlet algorithm then allows for time steps as large as 0.012τ yielding up to about 300000 particle timesteps per second for a typical vector processor or more than 100000 particle steps per processor or on a Cray T3E. With this method a huge variety of systems have been studied. Most recently a melt of chains of length N = 10000 was simulated [54, 55].

4.1 MC versus MD for melts

Which method is best to use for studying the dynamics of dense polymeric systems? I think that the choice should be between the BF Monte Carlo method and MD, or variants of these two approaches. For the athermal case (hard core repulsion only) we can compare the CPU time to reach the crossover time for entanglements τ_e (see Section 4.2 below) as a measure of the relative speed of the algorithms. By doing this we estimate that BF Monte Carlo is somewhat faster than MD on vector computers. However the inclusion of soft interactions has a much stronger effect on slowing down BF than it does for MD. (The main speed advantage of the MC comes from the simple acceptance test for the moves compared to the more complicated force calculations of MD.) Which method to use therefore depends on the particular question under consideration, whether it is better to work in the continuum (for example, to study shear flow), and whether it is acceptable to have stochastic dynamics on all time scales: using MD, the Rouse behaviour for short chains (or early times) is a consequence of the interactions, while it is built in explicitly in MC. If one is interested in the behaviour of gels or polymer networks under swelling or elongation, or the forces between polymer brushes [56, 57], then a continuum simulation using MD is probably more appropriate; generally speaking, the MD method is more flexible. Note also that with increasing computer power, workstations are becoming more and more important. They typically do not take advantage of especially fast integer arithmetic, reducing even further the advantage of the BF MC approach compared to continuum methods.

4.2 Dynamics and flow of melts: reptation

The dynamics of polymer melts is observed experimentally to change from an apparent Rouse-like behaviour to a dramatically slower dynamics for chains exceeding the characteristic length N_e. (For longer chains, one observes a much slower diffusion, $D \propto N^{-2}$, and an increased viscosity, $\eta \propto N^{3.4}$.) There are several theoretical models which try to explain this behaviour. However, only the reptation concept of Edwards and de Gennes [23] and variants of this approach take the non-crossing of the chains explicitly into account. This approach is the only one which, at least qualitatively, can account for a wide variety of different experimental results, such as neutron spin echo scattering, diffusion and viscosity. While it cannot explain all experimental data it does remarkably well, particularly considering its conceptual simplicity.

The idea of reptation is that the topological constraints imposed by the surrounding on each chain cause a motion along the polymers own coarse-grained contour. The diameter of the tube, to which the chain is constrained, is the diameter of a subchain of length N_e, namely $d_T \propto N_e^{1/2}$. The chains follow the Rouse relaxation up to the time $\tau_e \propto N_e^2$. For longer times the constraints become dominant and the chain moves along its own contour. To leave the tube, the chain has to diffuse along the tube a distance of the order of its own contour length, $d_T N/N_e$. In order to leave this original tube the chain needs a time $\tau_d \propto N^2(N/N_e)$, giving $D \propto N^{-2}$ and $\eta \propto N^3$. The difference between the predicted and the measured exponent for η is still not completely understood. For the mean square displacements of the monomers, $g_1(t)$, the model (Figure 8) predicts (a) the standard Rouse behaviour $g_1(t) \propto t^{1/2}$ for $t \leq \tau_e \propto N_e^2$; (b) the Rouse relaxation along the tube with $g_1(t) \propto t^{1/4}$ for $t \leq \tau_N \propto N^2$; (c) the diffusion along the tube with $g_1(t) \propto t^{1/2}$ for

$t \leq \tau_d \propto N^3$ and (d) the free diffusion of the chains. These four distinct regimes are a direct consequence of the reptation model. (For a more general discussion of this and the competing but less successful concepts, and a rather complete list of references see [9].) Below we review a few results, which have been obtained recently by both MD and MC simulations.

The transition from Rouse to reptation can be identified from the centre-of-mass diffusion constant,

$$D(N) = \frac{k_B T}{N\zeta}, \quad N < N_e, \tag{20}$$

$$D(N) \propto N^{-2}, \quad N \gg N_e. \tag{21}$$

Here ζ is the monomeric friction coefficient. Several forms exist in the literature for the prefactor of the second expression and the crossover. To compare results from different simulations and also experiment, a plot of $D(N)/D_{\mathrm{Rouse}}(N)$ versus N/N_e or M/M_e respectively should give one universal curve since N_e is thought to be the only characteristic scale at the crossover. Here $D_{\mathrm{Rouse}} = k_B T/\zeta N$ and M_e is the experimental entanglement molecular weight. This mapping is important for our understanding, since experiment and simulation use different methods to estimate M_e or N_e. In simulations N_e is determined by the crossover towards the $t^{1/4}$ regime in $g_1(t)$, while normally, experimentally, the plateau modulus from the stress relaxation function is used (McLeish, this volume). The scaling of the different data onto one curve shows that N_e resulting from the plateau modulus is typically about a factor of 2.3 larger than that from the mean square displacements. Figure 12 shows the results for simulation and experiment [9, 54, 55]. The MD simulations were performed at a density of $\rho = 0.85\sigma^{-3}$ and the MC simulations [46] at two different volume fractions $\phi = 0.4, 0.5$. The experimental data are NMR measurements of Pearson *et al.* [58]. Both simulations and the experiment show a clear crossover from the Rouse $D(N) \propto N^{-1}$ behaviour towards the $D(N) \propto N^{-2}$ regime. From these numbers it is clear that the simulations are in a position to analyse the crossover towards the reptation regime in some detail and, increasingly, to reach the fully entangled regime.

As mentioned earlier, a signature of reptation is the mean square displacement of the *monomers*, not just that of the chain centre of mass. (Experimentally the monomeric motion can either be observed by various NMR techniques [59, 60] or by neutron scattering [61, 62].) Figure 13 shows data on both quantities from the bead-spring MD simulations of Pütz *et al.* [55] for a variety of chain lengths. For short times, all the data collapse onto one single curve, indicating that initially the monomers do not feel the constraints acting on the full chain to which they belong. Only for larger times does one find the crossover to a slower motion. It is important to notice that this crossover occurs, within the error bars, at the same times and monomer displacements, independent of chain length, at least for the longer chains, which shows that N_e and τ_e are independent of N as expected. For longer times the slope in Figure 13 is around 0.26 for the longest chain, a little larger than the $t^{1/4}$ expected from the reptation model. Using for the crossover $g_1(\tau_e) = 2\langle R_g^2(N_e)\rangle$ the data of Figure 13 give $N_e \approx 35$ and $\tau_e \approx 1800\tau$ from the MD data. Within the reptation model one would expect the $t^{1/4}$ regime to hold up to the Rouse time of the chain. (For a detailed discussion see [9, 23].) For the earlier MD results [63, 64] on chains of $N \leq 150$ the chains were too short to observe a clean indication of a crossover from Rouse to reptation, since $N/N_e \approx 4.3$.

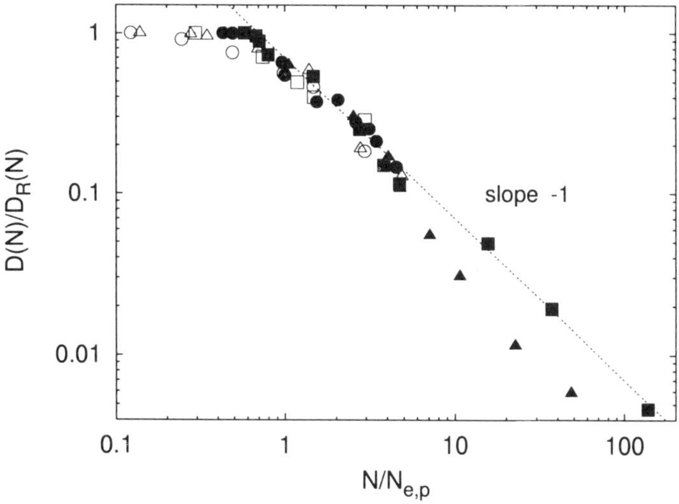

Figure 12. *Scaled diffusion constant $D(N)/D_R(N)$ versus scaled chain length $N/N_{e,p}$ for polystyrene (•) ($M_{e,p} = 14600$, $T = 485$ K), polyethylene (■) ($M_{e,p} = 870$, $T = 448$ K), PEB2 (filled △), ($M_{e,p} = 992$, $T = 448$K), our bead spring model (△) ($N_{e,p} = 72$), the bond-fluctuation model for $\Phi = 0.5$ (□) and tangent hard spheres at $\Phi = 0.45$ (○). All data are scaled with $N_{e,p}$ from the plateau modulus or with 2.2 N_e from $g_1(t)$, from [55].*

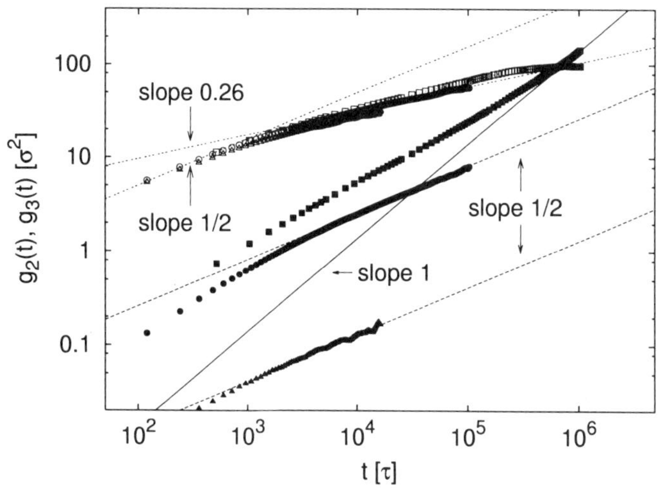

Figure 13. *Open symbols: mean square monomeric displacements $g_2(t)$. (This is defined as for $g_1(t)$ but is measured in centre of mass frame of each chain and averaged over the five middle monomers of the chain). Closed symbols: mean square centre of mass displacement $g_3(t)$. Chain lengths are $N = 350$ (■), $N = 700$ (○) and $N = 10000$ (△). The straight lines show some power laws to guide the eye. The local reptation power laws $g_2(t) \propto t^{1/4}$ and $g_3(t) \propto t^{1/2}$ are verified with remarkable clarity. From Ref. [55].*

Above we have discussed a variety of different equilibrium simulations, which try to follow the motion of individual chains. Kröger and co-workers [65] used the MD simulation model to perform NEMD (nonequilibrium MD) simulations, where the sample is subject to a steady shear or stress. They were able to cover chains of up to $N = 100$ monomers or equivalently about $3N_e$. While this is too short to study *e.g.* the viscosity or the shear-alignment of highly entangled chains, changes from Rouse to the entangled regime were observed. Such methods will probably become much more important with the coming generation of faster computers, allowing realistic simulation of melt flow behaviour.

Despite considerable effort it is still not known what an entanglement means physically, and how it can be derived from first principles. However, the reptation model does appear to work quite well.

5 Network structures

There has been extensive research on polymer networks for many years. A major problem is that it is experimentally not feasible to collect the necessary structural information in order to allow an unambiguous test of theoretical predictions. By simulation it is possible to analyse in detail the structural properties. This allows for a more thorough investigation of the effects of the non-crossing of the chains. One expects the consequences of conserved topology to be even stronger in crosslinked systems than in uncrosslinked ones. (This should hold for dilute polymer gels as well as for dense networks; however due to the importance of hydrodynamic effects, gels are even more complicated to simulate.) There have been various simulations on the properties of dense polymer networks. These simulations range from rather rigid model systems, where the crosslinks cannot move, to highly complicated, fully mobile randomly crosslinked polymer melts [56, 66]. The simulation techniques used are both MD and (lattice) MC. More recently also the swelling behaviour of networks has been investigated [54].

Duering et al [67] and Everaers and Kremer [68, 69, 70] and Pütz [54, 71, 72] performed MD simulations on a variety of different model networks with two different interaction potentials. These different simulations were used to study the influence of the possibility of chains crossing each other, the influence of different kinds of disorder (strand length, connectivity, topology) as well as the elasticity and relaxation as a function of the strand length.

Figure 14 illustrates the three different classes of networks. The first system, an equilibrated polymer melt with randomly placed crosslinks, certainly resembles most closely many experimental systems such as those produced by radiation crosslinking or vulcanisation. These systems have several kinds of disorder. The strands between subsequent crosslinks are extremely polydisperse, with an exponential distribution of strand lengths. Only a few crosslinks per chain are needed to give a gel-fraction which is nearly 100%, though the elastically active part of the network is significantly smaller. In this case there are many dangling chain ends or even some dangling clusters, like little trees which are connected to the network by only one strand. These dangling pieces are expected to be responsible for the extremely slow decay (in the time-dependent modulus $G(t)$, for example) seen in most experiments. In addition, trapped loops and knots may play an important role in both linear and nonlinear elasticity, while the length of the shortest

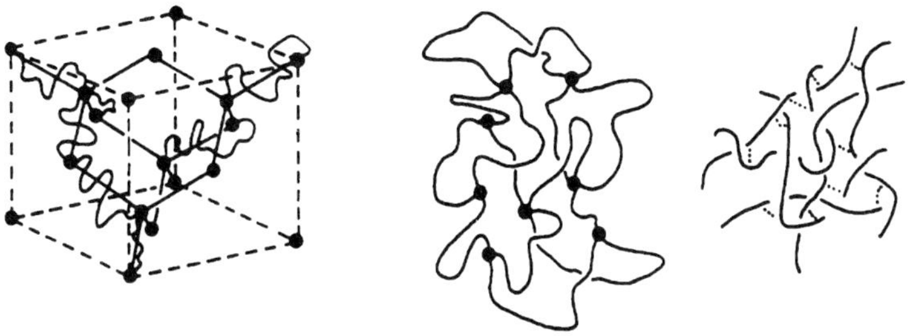

Figure 14. *Sketches of the typical network topologies considered in simulations. Left: interpenetrating networks; centre: end-linked chains; right: random cross-links.*

chemical pathway through the network cluster is expected to be of special relevance in nonlinear elasticity (since extension of a sample beyond this length forces breakage of covalent bonds). Although it is possible to identify the elastically active part of the cluster unambiguously [67], such disordered systems are extremely difficult to analyse in a systematic manner.

A somewhat more idealised case is an end-crosslinked polymer melt [67, 71]. Such systems have been considered experimentally as well [73]. There one starts with a monodisperse melt. After equilibrating, the chains are kinetically crosslinked at the ends. By this route, all strands in the network have the same length. (In a different type of end-linking, each free end binds instead to a randomly selected neighbour bead.) Now there are only two sources of quenched disorder in the system. One involves permanent topological links, the second, the distribution of chemical pathways through the system, which again is highly polydisperse, either at the level of individual network strands, or that of a percolating chemical pathways across the system.

Finally one can consider highly idealised networks, which certainly are somewhat remote from experiments, where the *only* source of disorder is the occurrence of knotted loops. These are the interpenetrating lattice networks, in which each subnet is an ordered diamond lattice that interpenetrates other, similar nets [68, 69, 70]. They can be analysed and investigated in a variety of ways in order to shed some light on the role of entanglements and conserved topology. The effects of disorder in networks are especially important under strong elongational stress. There they dominate the nonlinear regime, as Figure 15 illustrates.

6 Further reading

There are many other interesting polymer simulation problems, which were not discussed above. An extensive overview can be found in a recent book edited by K. Binder [74]. For a more microscopic approach the recent book edited by Monnerie and Suter [11] provides an interesting set of papers. Here we summarise these fields briefly.

Polyelectrolytes. The simplest charged polymers are not polyampholytes (see Section 3.3) but weakly charged polyelectrolytes. The whole field of macromolecules with long range interactions is still not well developed in spite of many years of research [75, 76, 77]

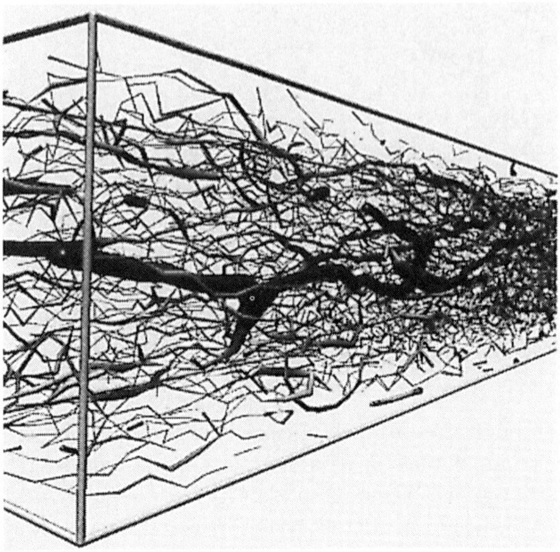

Figure 15. *Visualisation of the stress distribution in a highly extended polymer network [70] from a simulation of randomly interpenetrating polymer networks with diamond topology. The only source of disorder in this particular system are random knots. The 'shortest topological paths' (thick lines) carry most of the stress.*

(see also Khokhlov, this volume). Simulations start to give some clear answers, which await a better theoretical understanding. Until now we have been able to handle up to about 1000 chain monomers and 1000 counterions [78, 79].

Tethered chains. There is a huge literature on polymers that stick with either one or both ends to another object. The most common classes, polymer networks and star polymers, were discussed above. For details of simulations on the other classes of tethered chains see [80]. If the polymers are not connected to a centre molecule, but stuck to a surface/interface, one speaks of a polymer brush. Such brushes range from amphiphilic monolayers, where the chains are very much stretched (*e.g.* layers of a membrane) to rather soft systems. In addition a huge body of recent work considers tethered surfaces or membranes (two dimensional polymer sheets). A discussion of these topics would, however, require a review of its own.

Phase separation. Polymers in melts and dense solutions interpenetrate each other strongly. However, they are 'poor in translational entropy' (Khokhlov, this volume), so that even a very small repulsion between unlike polymers leads to phase separation. Even deuteration is sufficient to derive phase separation from similar protonated chains. Phase separation has attracted much simulation activity over the years; for a review see [74].

Block co-polymers. A block co-polymer consists of two or more stretches of different polymer linked together in a single chain. As a function of composition they can show lamellar phases, crystalline ordered phases of spheres, bicontinous phases and so on. (See Roux, this volume, for the surfactant analogues.) Most simulations so far were done using a quasi grand canonical approach, derived from blend simulations. Recently, also the dynamics was studied by MD simulation [81].

7 Linking the scales

A simulation eventually should provide complete information about the properties of the system under investigation. This goes beyond the generic problems discussed so far. Thus we have to link different scales: we want to relate the chemical structure of a system to the overall behaviour of the material [12]. One of the long-standing challenges within the modelling of complex materials is to apply methods which allow coverage of the range from the microscopic to the semi-macroscopic regime. Recently, methods were developed which map polymers to a mesoscopic level and then reintroduce the atomistic structure [82, 83, 84].

Below I first describe a mapping procedure to go from a microscopic description of a polymer chain to the mesoscopic description, which allows fairly effective simulation on a coarse grained level. The study of three modifications of one polymer structure, namely polycarbonates, approaching the glass transition from above, allows a test of the sensitivity. To check the quality of this approach, the chemical details of the chains will then be reintroduced into the coarse-grained conformations and results compared to neutron scattering results.

Thereafter, first steps toward the next level of description are discussed. Starting from the conformations of polymer chains on the coarse-grained level, each chain is mapped onto an extended soft particle with only three internal degrees of freedom. The aim is to provide a general approach to simulating specific polymers without losing the essential parts. Unlike other approaches all methods will be within continuous space and not confined to a lattice structure.

7.1 Coarse graining

We describe a systematic approach to renormalise the intra-chain interactions towards a coarser level for three different polycarbonates [82]. The three modifications of the basic polycarbonate structure are BPA-PC, BPZ-PC, and TMC-PC (Figure 16). Although the backbone sequence is the same these have remarkably different physical properties. For the first two (BPA and BPZ) the glass transition temperature T_G is roughly the same ($T_G \approx 420$K), while the third one has a glass transition temperature which is about 80 to 100 K higher, around 500 K. On the other hand, BPA-PC is ductile while BPZ-PC is brittle; TMC-PC is less ductile than BPA but much less brittle than BPZ. This is also reflected in the difference of the generalised activation energy within a Vogel-Fulcher fit (see Kob, this volume, and Equation 24 below): BPA and TMC have roughly the same activation energy while BPZ has a significantly higher one. Not only are the glass transition properties different, but also the entanglement chain lengths N_e in the melt are significantly different. For BPA-PC an extremely short entanglement length of $N_e = 7$ monomers is reported. This length increases through $N_e = 9$ monomers (BPZ) to $N_e = 14$ monomers for TMC-PC. In particular, the extremely short entanglement length for BPA-PC is not understood. Considering other well-studied polymers like polyethylene or PDMS (polydimethylsiloxane), one would expect it to be larger by a factor of at least 5 to 10. Whether this is the result of a special local chain structure (banana shaped repeat units joined by almost pivot-like junctions) is beyond the scope of the present discussion, but is a matter of current studies.

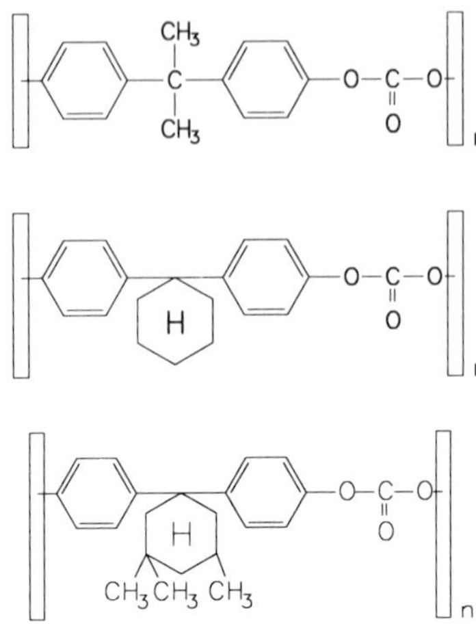

Figure 16. *Three modifications of BPA repeat units tested for the coarse graining procedures. From top to bottom: BPA-PC, BPZ-PC, TMC-PC.*

The coarse graining procedure is explained in detail here for BPA-PC. Ideally, the method is parameter free and as simple as possible. In addition we would like to stay as close as possible to the chemical structure in order to be able to reintroduce the chemical details later without too many problems. The coarse-grained monomers have to be designed to be easily be identified with specific chemical groups of the polymer itself. Considering the chemical structure of the three different polycarbonates a 2:1 mapping onto spherical beads seems to be a first reasonable choice, as illustrated in Figure 17. The resulting coarse-grained structure then only has four relevant internal degrees of freedom: the bond length ℓ between carbonate and isopropylidene group, α, the carbonate-isopropylidene-carbonate bond angle, β, the isopropylidene-carbonate-isopropylidene bond angle and ϑ, the torsion angle.

To arrive at the coarse-grained interactions from the microscopic model one can imagine a number of empirical fitting procedures. Here, we follow a different route. The coarse-grained potentials not only have to include energetic aspects of the microscopic model but also entropic terms from the different possibilities of local conformations. We first use intra-chain distribution functions to construct the bonded potentials in the coarse-grained model. Knowing the potential functions of the detailed chemical system, it is rather straightforward to perform an MC simulation to a very high accuracy at a given temperature of the conformations of individual free random walks. The probability distribution functions of conformations of such a model system are only dependent on temperature and originate from the bonded interactions along the backbone of a chain. The potentials for the microscopic models are derived from *ab initio* quantum chemistry calculations. Using the microscopic model to generate configurations, we sample the probability distribution function $P(\ell, \alpha, \beta, \vartheta)$ for the coarse-grained model in the limit of single isolated random walks.

The coarse-grained distribution function is temperature dependent via the Boltzmann weights of the different states of the microscopic model. The most crucial assumption now is, that the distribution function of the set of variables factorises into independent

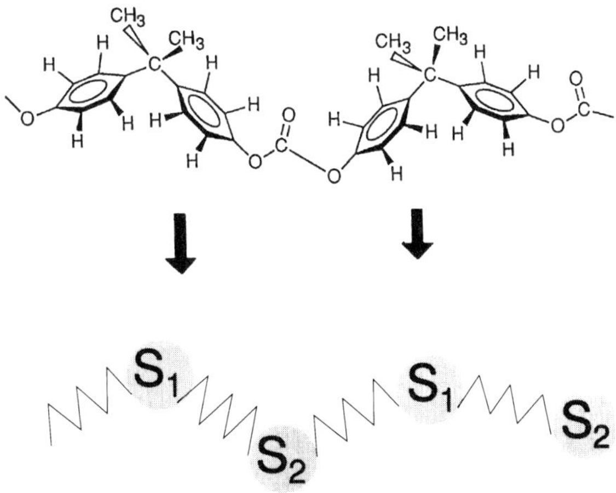

Figure 17. *Illustration of the mapping procedure for a 2:1 mapping where the repeat unit of a BPA-PC chain is replaced by two monomers of a generalised bead spring chain. The geometrical centres of the carbonate group and the geometrical centre of the isopropylidene group respectively are mapped onto the centres of the new spherical bead [82].*

distribution functions of the individual variables:

$$P(\ell, \alpha, \beta, \vartheta) = P(\ell)P(\alpha)P(\beta)P(\vartheta).$$

The distribution function P is determined at each temperature separately, and may be written

$$\begin{aligned} P(\ell) &\propto \exp(-U(\ell)) \\ P(\alpha) &\propto \exp(-U(\alpha))\ldots \end{aligned} \tag{22}$$

U is a generalised potential function at that temperature, already expressed in units of k_BT. (In effect, this allows one to keep the simulation temperature at $k_BT = 1$, which is of technical advantage for molecular dynamics simulations.) From Equation 22 we get the forces

$$\begin{aligned} F_\ell &= -\frac{d}{d\ell}\ln P(\ell) \\ F_\alpha &= -\frac{d}{d\alpha}\ln P(\alpha)\ldots \end{aligned} \tag{23}$$

as they originate from the conformations of the coarse-grained model. This approach avoids the fitting of a functional form of the coarse-grained potential functions to the microscopic parameters. The only fitting procedure is a smoothing for potentials in order to get rid of the scatter in the sampling and to stabilise the resulting force. There is no need to determine the partition function explicitly, since it only shows up as a constant in the potential and thus does not alter the resulting forces.

Using this model we can now simulate dense polymer systems. The volume of the effective hard-sphere beads of the coarse-grained model is adjusted to give the same Van-der-Waals volumes as in the experimental case (normalised to the simulation density). In the case of BPA-PC the carbonate group and the isopropylidene group are represented by spheres with a radius of 3.02Å and 3.11Å respectively. No further specific excluded volume interactions, nor any directional interaction is taken into account. The simulation density is adjusted to the experimental mass density in every case, and then the volume held constant; after this, there is no freedom left to adjust parameters.

If this procedure is to reproduce the essential aspects of the different chemical species, then not only should the static structure which comes out of this simulation compare well to the experimental systems, but also the dynamics. A detailed discussion of the dynamic properties is given elsewhere, but note that for the range of temperatures which we are investigating here, it seems reasonable to assume that the simulation time scales linearly with the physical time. (Possible deviations originating from the different shape of the potentials instead of taking different temperatures are neglected at this stage.)

MD melt simulations were performed as described before (Equation 19). The excluded volume interactions of the monomers are taken into account through a repulsive Lennard-Jones interaction. For the present system the background friction Γ is about 100 times weaker than the monomer-monomer friction. For static properties, the mapping between simulation and physical units requires a length, fixed by equating the the mass density in simulation and experiment. Starting from a mass density of $\rho_{pc} = 1.05\mathrm{g/cm}^3$ (BPA-PC at 500 K) and the simulation number density of $\rho_{MD} = 0.85\sigma^{-3}$, we arrive at a length scaling of $\sigma = 5.56$Å for the present case.

To compare the dynamic properties, a time scale is also required. This is found by using the Rouse model to calculate the melt viscosity in terms of the centre-of-mass diffusion constant of the chains, and equating this to the observed viscosity at some reference temperature. This is possible as we have one case where the highest experimental temperature and the lowest simulation temperature coincide. For the present example one gets $\tau = 2.21 \times 10^{-10}$ seconds, where τ is the simulation (Lennard-Jones) time unit. The simulation time step is typically $\delta t = 0.01\tau$. However, the absolute comparison of dynamic quantities should only be taken as indicative since the experimental systems and the simulations systems comprise different chain lengths, and also the effect of polydispersity might alter this absolute scale by some prefactor.

Compared to other molecular dynamics simulations of microscopic models, the simulation time step is roughly three orders of magnitude larger than usual. Taking the simplicity of the potentials and the short range nature of the interactions into account the resulting speed-up is of order 10^4. The simulated systems typically comprised between 1000 and 10000 model monomers on chains of 20 or 60 model monomers. For our cubic simulation box this means that one can easily simulate systems of up to 125Å^3. As it turns out, the inter-chain interactions strongly modify the angular distribution functions compared to the isolated chain. These and other static properties are discussed in Section 7.2 below, where chemical detail is reintroduced.

For the coarse-grained model we first check the dynamical properties as a function of temperature, especially approaching the glass transition temperature. The properties of many materials when approaching the glass transition temperature are well described by the so-called Vogel-Fulcher behaviour

$$D = D_0 \exp\left(-\frac{A_0}{T - T_{VF}}\right) \tag{24}$$

given here for the tracer diffusion constant D of the chains: A_0 is a generalised activation energy and T_{VF}, the so called Vogel-Fulcher temperature, typically is about 80 degrees below the calorimetric glass transition temperature. The prefactor D_0 is a hypothetical high temperature diffusion constant. For the present situation D_0 is easy to determine, because it simply corresponds to the freely jointed polymer melt with athermal excluded volume, and all the chemistry dependent intra-molecular interactions set to zero. (Extensive computer simulations are available for that case.)

Figure 18 gives a Vogel-Fulcher plot of the three polycarbonate modifications. The

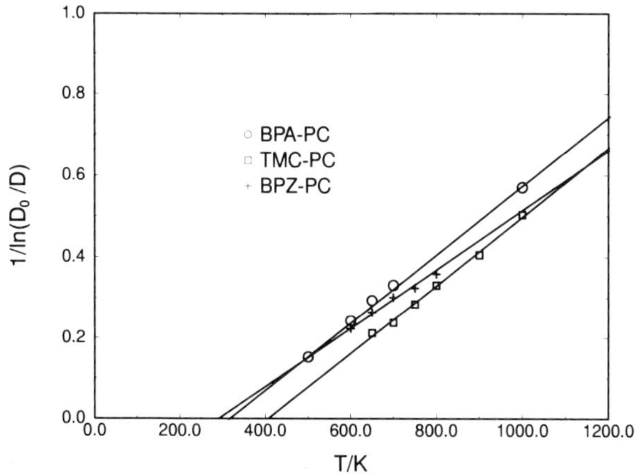

Figure 18. *Vogel-Fulcher plot of the chain diffusion constants D for the three different polycarbonate modifications, as indicated in the figure, for $N = 20$ model monomers [82].*

results qualitatively match the experimental situation, namely that the Vogel-Fulcher temperature for TMC-PC is about 80–100 degrees above the Vogel-Fulcher temperature of BPZ-PC and BPA-PC while the generalised activation energy, which in Figure 18 is the slope of the lines, is roughly the same for BPA and TMC but is different for BPZ-PC. Even quantitatively the results are not that different from the typical experimental value as Table 2 shows.

T_{VF}	BPA-PC	TMC-PC	BPZ-PC
Simulation (N=20)	322	407	292
Experiment	387	477	392
A_0	BPA-PC	TMC-PC	BPZ-PC
Simulation (N=20)	1305	1363	1443
Experiment	1012	1073	1534

Table 2. *Activation energies A_0 (below) and Vogel-Fulcher temperatures T_{VF} (above) for experiment and simulation. While the shift for T_{VF} (simulated) is consistent with expectations, the deviations (about 30%) for the experimental determination of A_0 are probably due to the large polydispersity of the typical commercial samples.*

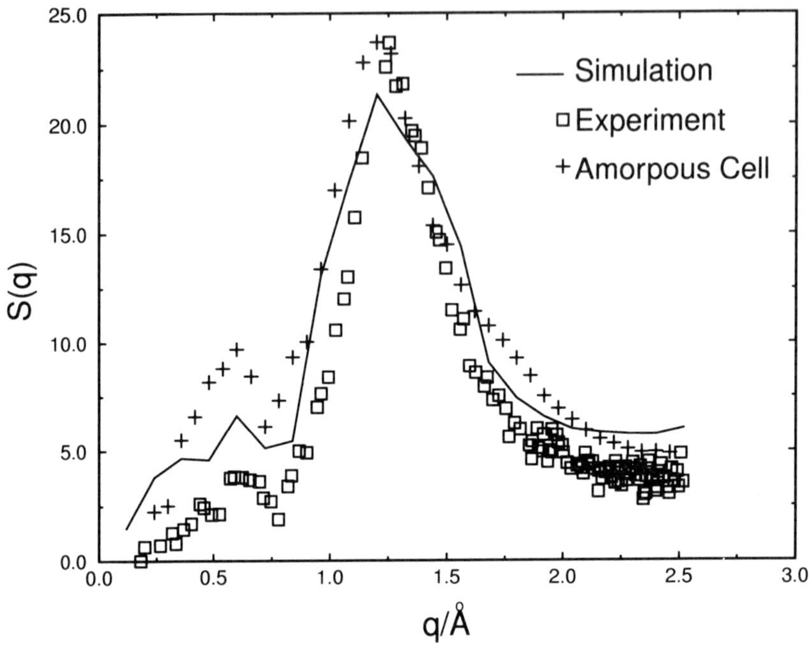

Figure 19. *Coherent structure function $S(q)$ in absolute units in comparison to amorphous cell simulation and neutron scattering data [82, 84].*

7.2 Restoring chemical detail

There are various ways to check the quality of the resulting structures with respect to experiment. A typical test would be to compare the mean square end-to-end distance $\langle R^2 \rangle$ to results from scattering experiments. However, since the experimental samples are highly polydisperse, the resulting answers from scattering experiments are somewhat questionable [85]. Furthermore, a crucial check is the direct comparison of conformations of systems. To compare the conformations resulting from the simulations unambiguously to experiment, we reintroduce the chemical details into the coarse-grained chain [83, 84]. This is one of the reasons why it was important to devise a mapping procedure which stays near to the chemical structure of the objects. We have a one-to-one correspondence of the model monomers to the different parts of the chemical repeat unit of the chains.

To reintroduce the details we use a commercial package since the details of the force fields are no longer essential. We start out with a chemically detailed chain with the correct bond angles and bond lengths, but free torsion angles. Then the chain is placed in the system and rotated via the torsional degree of freedom along the path of the coarse-grained chain. The structure is then optimised by a steepest descent method in energy space. By this approach, all coarse-grained chains are mapped onto their chemically detailed counterparts individually. The deviations of the minimised structure from the starting structure are extremely small: $\langle \Delta r^2 \rangle = 0.01 \text{Å}^2$. In a second step, all chains are combined and, to allow for the introduced Van-der-Waals interaction, are locally

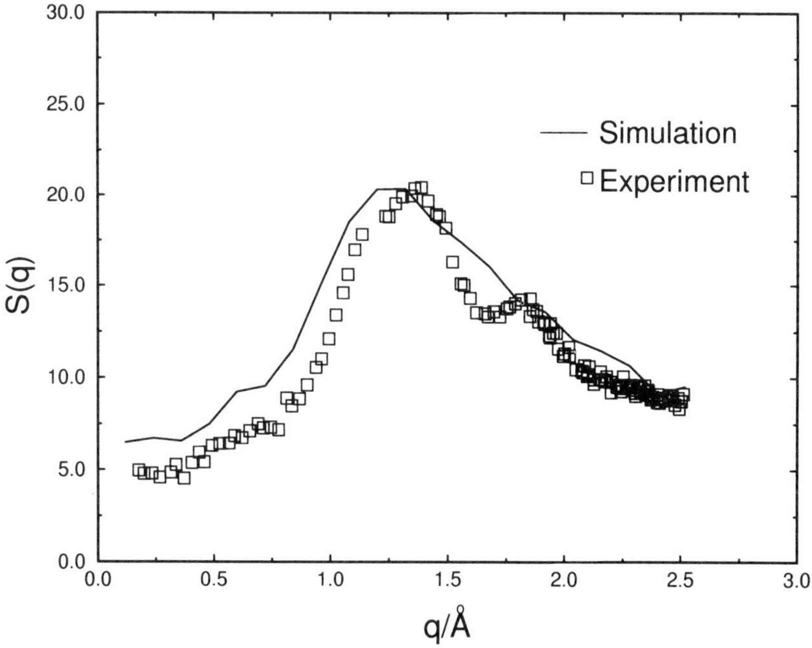

Figure 20. *Comparison of experiment and simulation for the same case as above (see Figure 19), but now for a system with deuterated methyl groups [82, 84].*

equilibrated within a short MD run. This local equilibration of course only includes motion on a very small distance.

Since for the resulting local packing structures the polydispersity effects of the experimental melts should not be that important, we now can calculate the structure function of the whole system and compare it directly to neutron scattering. The neutron scattering structure function $S(q)$ is given by (Pine, this volume)

$$S(q) = \left\langle \left| \sum_j b_j \exp(i\mathbf{q} \cdot \mathbf{r}_j) \right|^2 \right\rangle \quad . \tag{25}$$

For this comparison all atoms of the systems are explicitly included with their corresponding scattering length b.

Figures 19 and 20 give two typical examples for BPA-PC. In Figure 19 we show the results for a fully protonated system and compare the data to both neutron scattering and a previous (amorphous cell) simulation. Figure 20 gives the example where the methyl groups are deuterated. Other structure functions show the same agreement with experiments. For TMC-PC the density fluctuations are greater than for BPA-PC which leads to smeared out peaks for the structure factor. The partially deuterated examples show that the simulations are able to reproduce details of the scattering curves down to the smallest values of q allowed by the box size.

8 An even coarser view on polymers

So far, all models considered allow the correspondence of a bead of a polymer to one or to a few repeat units of a given chemical species. This still means that the number of degrees of freedom that have to be considered is proportional to the number of monomers of a given chain, causing enormous problems if one wants to try to simulate big systems (many chains instead of many monomers). To arrive at a situation where we can simulate many chains we go back to Figure 1. There, three levels of description were illustrated. I have discussed the microscopic and the mesoscopic regime and, in one case, the link between the two. Now we want to consider another step, namely to try to map the chains from the mesoscopic system up to the semi-macroscopic regime where we replace the chains by soft ellipsoidal particles which can strongly overlap in the melt [86, 12]. For such a model each chain is represented by a soft ellipsoid which varies its size and shape.

We separate the free energy of a system into an intra-chain part and an inter-chain part. For the total free energy F we make the ansatz:

$$F = \sum_{i=1}^{M} F^{i}_{\text{intra}} + F_{\text{inter}} \,. \tag{26}$$

The first sum runs over all M chains of the system. First let us consider the intra-chain part of the free energy. In a melt, the allowed conformations of a polymer chain are the same as a self-avoiding walk in 'vacuum'. (The change from self-avoiding to ideal random walk statistics in a melt is a result of the inter-chain interactions causing reweighting of configurations, but not of the intra-chain conformational distribution.) Thus, we characterise the intra-chain contribution to the free energy by the number of states which correspond to a specified moment of inertia tensor of the chain. Denoting this $\underline{\underline{\mathbf{R}}}$ and its eigenvalues R_1, R_2, R_3 (with $R_1 > R_2 > R_3$), we generate microscopically, in a similar manner to that described in the previous sections, a probability distribution $P(\underline{\underline{\mathbf{R}}})$. To each $\underline{\underline{\mathbf{R}}}$ belongs an average intra-chain monomer density distribution $\rho(\mathbf{r}, \underline{\underline{\mathbf{R}}})$ which is sampled as well. Here $\mathbf{r}$ is the position vector from the centre-of-mass in the principal frame. The averaging for ρ is carried out over all conformations with a given $\underline{\underline{\mathbf{R}}}$. Taking into account that the set of allowed conformations of individual chains in the melt and for the isolated chain are identical, the intra-chain contribution to the free energy from chain i is simply given by

$$F^{i}_{\text{intra}} = -k_B T \ln P(\underline{\underline{\mathbf{R}}}^{i}) \,. \tag{27}$$

Now we assume that the inter-chain interaction is given by the pairwise overlap of the ellipsoids of the different chains. Since each inertia tensor corresponds to a density distribution, we can write for the inter-chain free energy contribution of the pair ij

$$F^{(ij)}_{\text{inter}} = \epsilon(N) \int \rho_i(\mathbf{r} - \mathbf{r}^{\text{cm}}_i, \underline{\underline{\mathbf{R}}}^{i}) \rho_j(\mathbf{r} - \mathbf{r}^{\text{cm}}_j, \underline{\underline{\mathbf{R}}}^{j}) d^3 r \tag{28}$$

Here each of the two density distributions is centered on the centre-of-mass of the corresponding chain, and $\epsilon(N)$ is an adjustable parameter accounting for the binary excluded volume as well as the overlapping contribution of the probability distributions. For technical details refer to [86].

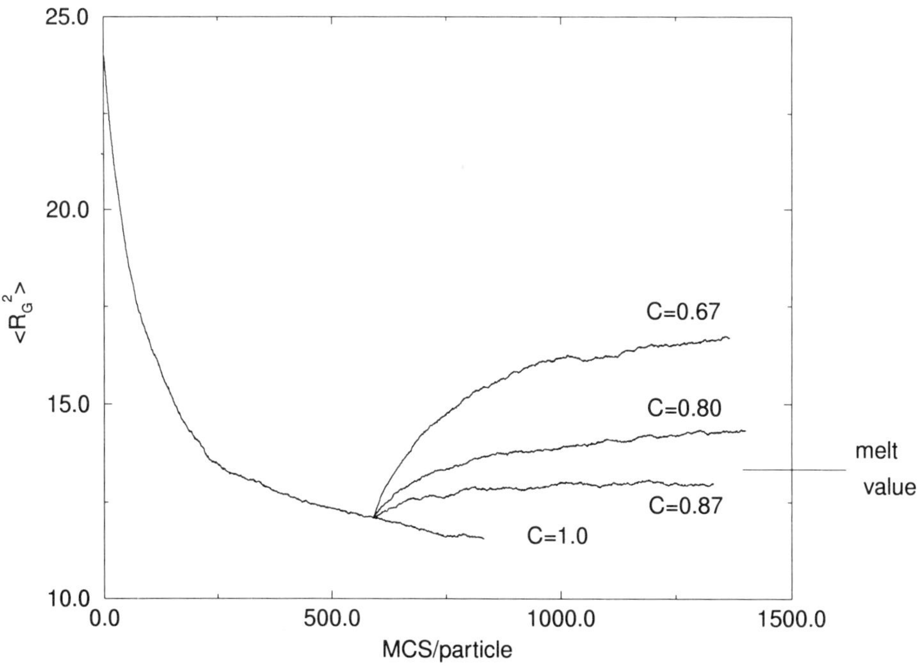

Figure 21. *Evolution of the sample averaged $\langle R_G^2 \rangle$ as a function of Monte Carlo time (4000 particles of N=50). The initial value of $\epsilon(N) = C = 1.0$ was changed to values indicated after 600 MC steps. The indicated melt value corresponds to a comparable system with explicit chains with repulsive Lennard Jones interactions and a number density of 0.85 σ^{-3} [86].*

Here we test this idea for simple coarse-grained polymer models. An extension to a more refined coarse-grained model for *e.g.* polycarbonate should be straightforward and is an objective of future work. Typical systems consist of 10000 chains of N = 100 monomers. The simulation procedure is a standard Metropolis Monte Carlo procedure as described in Section 3. The ellipsoidal particles are first randomly distributed in the system, with a distribution of shapes corresponding to isolated SAWs. Then MC simulation is performed such that the ellipsoids can move in space (translation) and can change both the length and the orientation of their principal axes (shape deformation). Figure 21 gives a typical evolution plot of the ensemble averaged-squared radius of gyration of our ellipsoids, as a function of Monte Carlo time steps, for different parameters $\epsilon(N)$.

The adjustment of ϵ allows a precise mapping of the ellipsoidal model onto the explicit chain models at a given density. To show not only that the end-to-end distance of the ellipsoidal system in the melt agrees with the explicit chain simulation, but also that the chain statistics correspond to Gaussian statistics, we scale the resulting probability distribution function of the radius of gyration for different chain lengths within the random walk scaling scheme. Figure 22 shows this for chain lengths between 25 and 100. Various other control investigations such as the scaling of the correlation hole (the locally reduced density of other chains produced by the self-density of the chain under consideration)

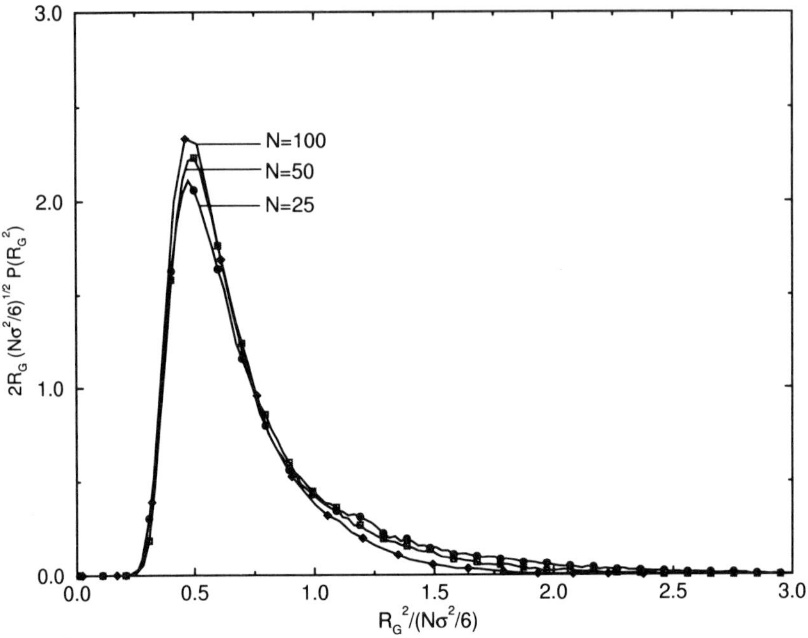

Figure 22. *Scaled distribution functions of $R_G^2(N)$ versus the Gaussian normalised value of R_G for chain lengths as indicated [86].*

support the conclusion that our chains are now Gaussian. In a very similar way as in earlier studies on phase separations of polymers, one can introduce an $\epsilon(N)$ which is able to distinguish between two different species. By doing this one is able to investigate phase separation kinetics and morphology development of huge polymer samples. The next step will be the reintroduction of the explicit chains in order to complete the scheme as indicated in Figure 1.

Acknowledgments

The present chapter is a result of longstanding collaborations with many colleagues through the last years. The work on coarse-grained models especially benefitted from a longstanding collaboration with G. S. Grest, B. Dünweg, K. Binder, R. Everaers, and M. Pütz. CPU time was granted by the German Supercomputer Centre, HLRZ Jülich. The multiscale simulations were supported by the BMBF under Grant No. 03N8008E0 and the Bayer Corp.. We especially thank the other project partners (group of K. Binder, group of D. W. Heermann, group of D. Richter, group of U. Suter and the Bayer polymer physics group) and M. Murat for many helpful disuccions and fruitful collaboration over the years.

References

[1] Binder K, 1996, *Monte Carlo and Molecular Dynamics of Condensed Matter Systems*, Como Conference Proceeding (Bologna: Società Italiana di Fisica)

[2] Baumgärtner and Binder K, 1984, *Applications of the Monte Carlo Method in Statistical Physics* (Berlin: Springer)

[3] Kremer K and Binder K, 1988, Monte Carlo Simulations of lattice models for macromolecules, *Comp Phys Reports* **7** 259

[4] Binder K, 1992, in *Computational Modeling of Polymers*, ed. J Bicerano (Berlin, Heidelberg, New York: Springer)

[5] Roe R J, ed., 1991, *Computer Simulation of Polymers* (Englewood Cliffs, NJ: Prentice Hall)

[6] Mark H E and Erman B, eds., 1992, *Elastomeric Polymer Networks* (Englewood Cliffs: Prentice Hall)

[7] Allen M P and Tildesley D J, 1989, *Computer Simulations of Liquids* (London: Oxford Univ Press) second edn.

[8] Kremer K, 1996, Computer Simulation methods for polymer physics in *Monte Carlo and Molecular Dynamics of Condensed Matter Systems*, eds. K Binder and G Ciccotti Como Conf Proceedings, 1995, (Bologna: Societa Italiana di Fisica)

[9] Whittington S G, 1998, *Numerical Methods for Polymeric Systems* The IMA Volumes in Mathematics and its Applications (New York: Springer) 0-387-98557-3

[10] Frenkel D and Smit B, 1996, *Understanding Molecular Simulation: From Basic Algorithms to Applications* (San Diego, CA: Academic Press)

[11] Monnerie L and Suter U W, 1994, *Atomistic Modelling of Physical Properties*, volume 116 of *Advances in Polymer Science* (Heidelberg: Springer)

[12] Baschnagel J, Binder K, Doruker P, Gusev A A, Hahn O, Kremer K, Mattice W L, Müller-Plathe F, Murat M, Paul W, Santos S, Suter U W and Tries V, 1998, Bridging the Gap Between Atomistic and Coarse Grained Model of Polymers: Status and Perspectives, *Advances in Polymer Science* **152** (Berlin: Springer, 2000)

[13] Evans D J and Morris G P, 1990, *Statistical Mechanics of Non-Equilibrium Liquids* (London: Academic Press)

[14] Hess S, 1995, *Monte Carlo and Molecular Dynamic Simulations in Polymer Science* (New York: Oxford Univ Press)

[15] Müller-Plathe F, 1997, *J Chem Phys.* **106** 6082

[16] Binder K and Heermann D W, 1992, *Monte Carlo Simulation in Statistical Physics* (Berlin: Springer)

[17] Binder K, ed., 1984, *Applications of the Monte Carlo Method in Statistical Physics* (Heidelberg, New York: Springer Verlag)

[18] Binder K, ed., 1992, *Monte Carlo Methods in Condensed Matter Physics* (Berlin, Heidelberg, New York: Springer Verlag)

[19] Binder K, ed., 1979, *Monte Carlo Methods in Statistical Physics* (Berlin, Heidelberg, New York: Springer Verlag)

[20] Ahlrichs P and Dünweg B, 1999, *J Chem Phys* **111** 8225

[21] Hoogerbrugge P J and Kroelman J M V A, 1992, *Europhys Lett* **19** 155

[22] Chen S and Doolen G D, 1998, *Annu Rev Fluid Mech* **30** 329

[23] Doi M and Edwards S F, 1986, *The Theory of Polymer Dynamics* (Oxford: Clarendon Press)

[24] Bennetin G, 1986, in *Proc of Intern School of Physics 'Enrico Fermi'*, eds. G Ciccoti and W W Hoover (Amsterdam: North Holland)

[25] de Gennes P G, 1979, *Scaling Concepts in Polymer Physics* (New York: Cornell Univ Press)

[26] Rosenbluth A W and Rosenbluth M N, 1955, *J Chem Phys* **23** 356

[27] Alexandrowicz Z, 1969, *J Chem Phys* **51** 561

[28] Suzuki K, 1968, *Bull Chem Soc Japan* **41** 538

[29] Batoulis J and Kremer K, 1989, Thermodynamic properties of star polymers: good solvents, *Macromol* **22** 4277

[30] Dünweg B, 1994, *J Chem Phys* **99** 6977

[31] Madras N and Sokal A, 1988, *J Stat Phys* **50** 109

[32] Sokal A D, 1995, Monte Carlo and Molecular Dynamic Simulations in Polymer Science (New York: Oxford Univ Press)

[33] Micka U and Kremer K, 1996, Persistence Length of the Debye-Hückel model of weakly charged flexible polyelectolyte chains, *Phys Rev E* **54 (3)** 2653

[34] Wall F T and Madel F, 1975, *J Chem Phys* **63** 4562

[35] Jagodzsinski O, Eisenriegler E and Kremer K, 1992, Universal shape properties of open and closed polymer chains: Renormalisation group and analysis and Monte Carlo experiments, *J Phys I France* **2** 2243

[36] Kantor Y, Kardar M and Li H, 1994, *Phys Rev E* **49** 1383

[37] Everaers R, Johner A and Joanny J F, 1997, *Europhys Lett* **37** 275

[38] Everaers R, Johner A and Joanny J F, 1997, *Macromol* **30** 8478

[39] Bekiranov S, Bruinsma R and Pincus P A, 1993, *Europhys Lett* **24** 183

[40] Jeppesen C and Kremer K, 1996, Single chain collapse as first order transition: Model for PEO in water, *Europhys Lett* **34** 563

[41] Carmesin I and Kremer K, 1988, THe bond fluctuation method: A new effective algorithm for the dynamics of polymers in all spatial dimensions, *Macromol* **21** 2819

[42] Wittmann H P and Kremer K, 1990, Vectorised version of the bond fluctuation method for lattice polymers, *Comp Phys Comm* **61** 309

[43] Wittmann H P and Kremer K, 1992, Vectorised version of the bond fluctuation method for lattice polymers, *Comp Phys Comm* **71** 343

[44] Deutsch H P and Binder K, 1990, *J Chem Phys* **94** 2294

[45] Skolnick J and Yaris R, 1988, *J Chem Phys* **881** 1418

[46] Paul W, Binder K, Heermann D W and Kremer K, 1991, Dynamics of polymer solutions and melts: Reptation prediction and scaling of relaxation times, *J Chem Phys* **95** 7726

[47] Carmesin I and Kremer K, 1990, Static and Dynamic Properties of 2-d Polyer Melts, *J Phys (Paris)* **51** 915

[48] Maier B and Rädler J, 1999, *Phys Rev Lett* **82** 1911

[49] Wittmann H P, Kremer K and Binder K, 1992, Evidence for time temperature superposition from MC simulations of 2-d polymer melts, *Macrom Theory Simul* **1** 275

[50] Grest G S and Kremer K, 1986, Molecular dynamics simulation for polymers in the presence of a heat bath, *Phys Rev* **A33** 3628

[51] Grest G S, Dünweg B and Kremer K, 1989, Vectorised link cell Fortran code for molecular dynamics simulations for a large number of particles *Comp Phys Comm* **55** 269

[52] Everaers R and Kremer K, 1994, A fast grid search algorithm for molecular dynamics simulations with short range interactions, *Comp Phys Comm* **81** 19

[53] Pütz M and Kolb A, 1998, *Comp Phys Comm* **113** 145

[54] Pütz M, 1999, *PhD Thesis* (Mainz: University of Mainz)

[55] Pütz M, Kremer K and Grest G S, 1999, *Europhys Lett* (in press)

[56] Kremer K and Grest G S, 1995, Entanglement Effects in Polymer Melts and Networks in *Monte Carlo and Molecular Dynamics Simulations in Polymer Science*, ed. K Binder (Oxford: Oxford Univ Press)

[57] Grest G and Murat M, 1995, in *Monte Carlo and Molecular Dynamic Simulations in Polymer Science*, K. Binder, Ed. (New York: Oxford Univ Press)

[58] Pearson D S, Strate G V, von Meerwall E and Schilling F C, 1987, *Macromol* **20** 1133

[59] Fleischer G and Fujara F, 1992, *Macromol* **25** 4210

[60] Graf R, Heuer A and Spiess H W, 1998, *Phys Rev Lett* **80** 26

[61] Ewen B and Richter D, 1997, Neutron spin echo investigations on the segmental dynamics of polymers in melts, *Adv Pol Sc* **134** 1

[62] Schleger P, Farago B, Lartigue C, Kollmar A and Richter D, 1998, *Phys Rev Lett* **81** 124

[63] Kremer K and Grest G S, 1990, Dynamics of entangled linear polymer melts, a molecular dynamics simulation, *J Chem Phys* **92** 5057

[64] Kremer K, 1991, Dynamics of entangled linear polymer melts, a molecular dynamics simulation (correction), *J Chem Phys* **94** 4103

[65] Kröger M, Loose W and Hess S, 1993, *J of Rheology* **37** 1057

[66] Everaers R, 1999, *New J of Phys* **1** 12

[67] Duering E R, Kremer K and Grest G S, 1993, Dynamics of medel networks: the role of the melt entanglement length, *Macromol* **26** 3241

[68] Everaers R and Kremer K, 1996, Topological interactions in model polymer networks, *Phys Rev E* **53** R37

[69] Everaers R and Kremer K, 1995, Test of the foundations of classical ruber elasticity, *Macromol* **28** 7291

[70] Everaers R and Kremer K, 1996, Elastic properties of polymer networks, *J Mol Model* **2** 293–299

[71] Pütz M, Everaers R and Kremer K, 2000, *Phys Rev Lett* in press

[72] Grest G S, Pütz M, Everaers R and Kremer K, 1999, preprint

[73] Bastide J and Candau C, 1995, in *Physical Properties of Gels*, ed. A Cohen-Addad (Chichester: J Wiley)

[74] Binder K, ed., 1995, *Monte Carlo and Molecular Dynamic Simulations in Polymer Science* (Oxford: Oxford Univ Press)

[75] Barrat J L and Joanny J F, 1996, *Adv Chem Phys* **94** 1–66

[76] Förster S and Schmidt M, 1995, *Adv Polym Sci* **120** 51–133

[77] Micka U, Holm C and Kremer K, 1999, Strongly charged, flexible polyelectrolytes in poor solvents - A Molecular Dynamics Study, *Langmuir* **15** 4033

[78] Deserno M and Holm C, 1998, How to mesh up Ewald sums (I), *J Chem. Phys* **109** 7678

[79] Deserno M and Holm C, 1998, How to mesh Up Ewald sums (II), *J Chem. Phys* **109** 7694

[80] Grest G, 1995, in *Monte Carlo and molecular dynamic simulations in polymer science*, K. Binder, Ed. (Oxford: Oxford Univ Press)

[81] Murat M, Grest G S and Kremer K, 1999, Statics and dynamics of symmetric diblock copolymers, *Macromol* **32** 595

[82] Tschöp W, Kremer K, Batoulis J, Bürger T and Hahn O, 1998, Simulation of polymer melts I: coarse graining prcedure for polycarbonates, *Acta Pol* **49** 61

[83] Tschöp W, Kremer K, Hahn O, Batoulis J and Bürger T, 1998, Simulation of polymer melts II: from coarse grained models back to atomistic description, *Acta Pol* **49** 75

[84] Eilhard J, Zirkel A, Tschöp W, Hahn O, Kremer K, Schärpf O, Richter D and Buchenau U, 1999, Spatial correlations in polycarbonates: neutronl scattering and simulation, *J Chem Phys* **110** 1819

[85] Tschöp W, 1996, *Coarse-grained polymer-modelle und die Wiedereinführung der chemischen Details* (University of Bonn: Dissertation)

[86] Murat M and Kremer K, 1998, From many monomers to many polymers: soft ellipsoid model for polymer melts and mixtures, *J Chem Phys* **108** 4340

Equilibrium and flow properties of surfactants in solution

Didier Roux[†]

Centre de Recherche Paul-Pascal, CNRS, Talence, France

1 Introduction

In what follows we give an overview of the basic properties of surfactants in solution and present some recent developments concerning the effect of flow on the lyotropic lamellar phase. In Section 2, we will briefly show how some simple ideas can lead to a qualitative understanding of the phase behaviour of surfactant solutions. Section 3 describes the effect of flow on two types of lamellar phase and details the shear-induced structures that result (the so-called 'onion' textures). We present in Section 4 the viscoelastic properties of these structures and conclude with some theoretical models describing the dynamics of a lamellar phase under flow (Section 5).

2 Phase diagrams and membrane elasticity

A classical binary phase diagram (surfactant + water) exhibits a succession of isotropic, liquid crystalline, and crystalline phases as a function of temperature and composition [1]. Figure 1 shows such a phase diagram for the binary mixture of sodium laurate and water. Besides complicated phases at very low water concentration (which correspond to hydrated solids), the phase diagram contains three main phases: the lamellar phase which is a periodic stack of fluid membranes made of the surfactant molecules (lyotropic liquid crystal), the hexagonal phase made of infinite tubes placed on a two-dimensional triangular lattice (lyotropic liquid crystal) and the isotropic liquid phase made of spherical micelles. (Note that liquid crystal phases are called 'lyotropic' when controlled by concentration as well as temperature.) Although these are the most common phases found in surfactant solutions, many other structures have also been described, certain of them very recently [2–5]. most interesting ones, we should note the cubic phases [2] corresponding either to a crystal of spherical micelles or to more complex structures (*e.g.* bilayers wrapped on a triply periodic 'minimal surface'). The most intriguing structures, and the

[†]Co-authors: Annie Colin, Jacques Leng and Anne-Sophie Wunenburger.

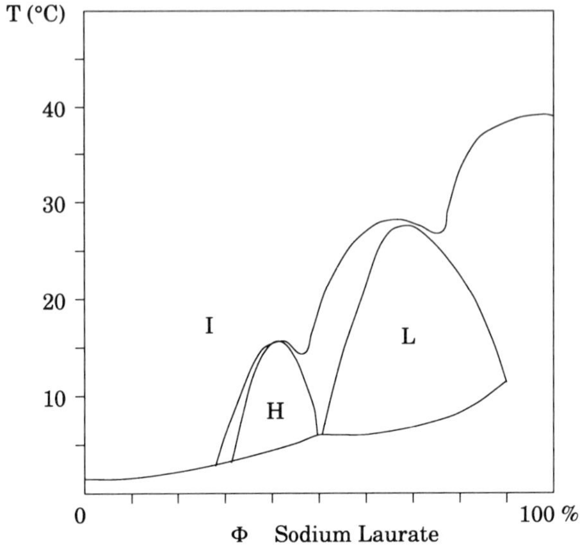

Figure 1. *A typical phase diagram where, as a function of the surfactant concentration, one has successively the isotropic micellar phase (I), the hexagonal phase (H) and the lamellar phase (L). Other regions are biphasic domains.*

ones which were the most difficult to characterise, are complex isotropic liquid phases such as the microemulsion phases [3], the sponge phase [4] and the phase of giant micelles [5].

The complexity of the structure and phase behaviour of these systems has been an experimental and a theoretical challenge. In the last 30 years, a tremendous amount of work has been done leading to a rather unified picture of the way these systems behave. From a theoretical point of view, the direct relation between the microscopic properties of the surfactant and the phase diagram is not accessible. However, it has been very useful to introduce, as proposed by Canham [6] and Helfrich [7], an intermediate step in the statistical physics description of the properties of surfactants in solution. This intermediate description corresponds to the idea that the physics is dominated by the interfacial properties of the hydrophobic/hydrophilic domains. The microscopic structure of the various phases was then attributed to a competition between the curvature energy of the microscopic interface and the entropy (thermal fluctuations). The importance of the curvature energy arises because, for self-assembled structures which are at thermal equilibrium, there is no surface tension at the hydrophobic/hydrophilic interface. (Surface tension is instead the signature of bulk phase separation.) Despite this, the interfacial area is almost fixed: the forces opposing stretching and compression would rapidly become large if the area was changed. Consequently, the first term that matters in the small deformation of the interface is the curvature energy, whose energy scale is most often close to k_BT. Therefore this term in practice controls most of the deformations of the interface.

The concept of curvature energy applied to surfactants in solution turns out to be remarkably efficient. While better adapted to describe dilute phases (mainly because, in that case, the interface becomes very thin compared to the characteristic length of the structure), this concept remains qualitatively valid for more concentrated phases. In all cases, it is a good guideline in trying to understand the phase behaviour and general properties of surfactants in solution.

The main problem with this model is that it relates the stability and the structure of the different phases to three phenomenological parameters which do not have an obvious relationship with the surfactant molecules. These parameters are the spontaneous radius of curvature c_0, the mean curvature rigidity κ and the Gaussian curvature rigidity $\bar{\kappa}$. The energy cost of deforming a surface $\mathcal{S}$, describing the surfactant film, therefore reads:

$$F = \int_{\mathcal{S}} \left[\frac{1}{2}\kappa(c_1 + c_2 - c_0)^2 + \bar{\kappa}c_1c_2\right] dS, \tag{1}$$

where c_1 and c_2 are the two principal radii of curvature of the deformation. Some effort has been devoted to working out microscopic models that link the phenomenological elastic constants to the geometry of the surfactant molecules and these attempts are qualitatively successful. Mechanical models [8], models based on the microscopic description of the molecules [9], and simulations [10] have all been quite useful in understanding the role of the surfactant geometry. Since the most common interfacial shapes are spheres (micelles), cylinders (hexagonal phase) and planes (lamellar phase), it is useful to calculate the elastic energy for these simple cases. One can also quite easily calculate the curvature energy for the unit cell of a cubic minimal surface, which we call a 'cubic element' below.

General properties emerge from such calculations. In particular, we can notice that if the spontaneous radius of curvature is zero (for surfactant *bilayers*, this is the case by symmetry), the elastic energy of a sphere is not a function of its size: $F_{\text{sph.}} = 4\pi(2\kappa + \bar{\kappa})$. This property can easily be generalised (for example to the cubic element) and in general, the bending energy of a finite object is invariant under a change of scale, whenever $c_0 = 0$. Moreover, it has been known since the XIX century that the Gaussian curvature term, $\int_{\mathcal{S}} c_1c_2dS$, is a function of the global topology of the surface and not directly dependent on the local curvatures (this is the Gauss-Bonnet theorem). Since for $c_0 = 0$ the curvature energy, Equation 1, is based on a quadratic expansion around a flat surface, one expects the energy of a sphere and of a cubic element both to be positive. This fixes a range of stability for the values of κ and $\bar{\kappa}$, namely $2\kappa > -\bar{\kappa} > 0$. If $\bar{\kappa}$ becomes less than -2κ, an instability towards very small spheres will develop. Otherwise, if $\bar{\kappa}$ becomes larger than zero, there is instead an instability towards small cubic elements: a periodic surface of very small lattice constant will arise.

The effects of thermal fluctuations on the surfactant aggregates are different depending upon the shape of those aggregates. For spherical objects (micelles), the fluctuations will mainly stabilise the isotropic liquid phase of micelles against the 'crystal' of micelles or other more organised phases (liquid crystals). For cylindrical objects, depending upon the value of the elastic constant of these 'rod-like' micelles, one can have either flexible or rigid systems. If the persistence length (Khokhlov, this volume) of the rods is very large, one will only find cylindrical aggregates in a liquid crystalline arrangement such as the hexagonal phase. However, if the persistence length is small enough (typically smaller than 1000 Å), one can find a phase where the cylindrical micelles are disordered and polymer-like. Just as for regular polymers, this isotropic liquid phase can be found either in dilute or semi-dilute regimes. In such cases, the flexible cylindrical micelles can make a random walk in space and most of their static and dynamic properties can be understood within a model of so-called 'living' (i.e. self-assembled) polymers [5].

Similar ideas apply when surfactants instead form membranes (either monolayers or bilayers). Because κ and $\bar{\kappa}$ have the units of energy, it is possible to compare directly their value to k_BT. Indeed, as shown first by de Gennes [11], one can thereby define a persistence length for fluctuating membranes. This persistence length ξ varies exponentially with the ratio κ/k_BT and consequently, only values of κ not much larger than k_BT lead to a microscopic persistence length ($10^2 \leq \xi \leq 10^4$Å). Relatively small changes in κ differentiate between what we term 'rigid' systems, where the persistence length is quite large ($> 10\mu$m), and 'flexible' systems, where the persistence length is typically less than 1μm.

The easiest way to understand qualitatively the phase diagram of a rigid system is to realise that many of the properties come from a competition between the spontaneous radius of curvature and the geometrical length resulting from the choice of the concentration of the species. Indeed, for a general system made of water, oil and surfactant, and making the reasonable assumption that all the surfactant lies at the water/oil interface, it is quite easy to show that in each case (spheres, cylinders and planes) the characteristic length of the structure (radius of the sphere, radius of the cylinder, or thickness of the oil and water layers) is completely determined by the respective concentrations.

The simplest model is then to take into consideration just the bending energy. It is possible to calculate the most stable structure depending upon the concentration (for a given spontaneous radius of curvature c_0). The result is, naturally, found to be the structure whose curvature best matches the spontaneous radius of curvature. Consequently, for this very simple model of ternary mixtures with no other term apart from the bending energy, one can already finds phase transitions from spherical micelles, to a hexagonal phase, and from there to a lamellar phase by changing the respective surfactant/oil concentrations. Taking into account entropy of mixing and interactions can change the phase diagrams, but this simplest behaviour already gives some reasonable results [12].

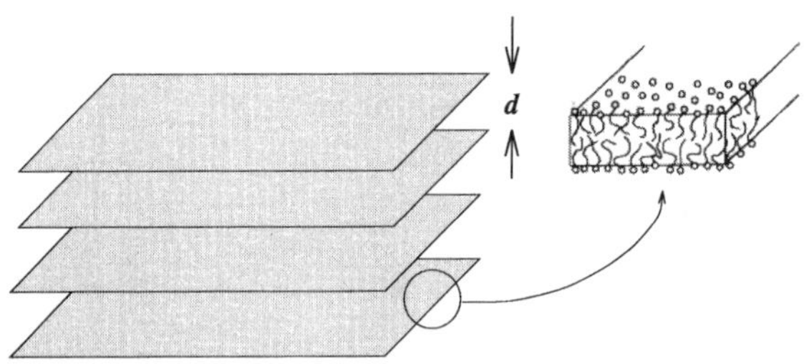

Figure 2. *Schematic drawing of a lyotropic lamellar phase. It consists of a periodic stacking of membranes (repeat distance d), each of a thickness δ, separated by a solvent.*

One of the key questions concerning the lamellar phase is to determine the capacity of this phase to be swollen, *i.e.* the capacity to change the repeat distance d of the lamellar phase (Figure 2) by adding more solvent. Experimentally, depending on the system, the maximum repeat distance can vary from 50 Angstroms to several thousand Angstroms. In order to understand what happens to a lamellar phase when it is swollen with a solvent (either water or an organic solvent), we need a reasonable description of the interactions between the membranes. There are several reviews on membrane-membrane interactions which list all the attractive and repulsive interactions that have been calculated. Most of them have also been measured (electrostatic, Van der Waals, steric...) [13].

One of the most interesting and quite recently discovered long range repulsive interactions comes from the thermal undulations of the membranes. Indeed, Helfrich in 1984 predicted that two membranes subjected to thermal fluctuations should develop a repulsive interaction coming from the multiple collisions they will develop one against the other [16]. This interaction, entropic in origin, has been quantitatively measured by several techniques in sufficiently flexible lamellar phases [13]. Competition between attractive and repulsive interactions can lead to phase transitions, explaining why a lamellar phase cannot be swollen indefinitely. Whereas the competition between electrostatic and Van der Waals interactions can be calculated following the DLVO theory (Frenkel, this volume), the competition between attractive Van der Waals interactions and repulsive undulation forces is much more complex to model, but leads to very interesting behaviours [14, 15].

The phase separation with excess solvent is not the only kind of phase transition that a lamellar phase can experience upon dilution. When a lamellar phase, made of flexible membranes, is swollen with a single solvent, the characteristic repeat distance increases because of the undulation forces acting as a repulsive interaction on the membranes. However, when the d-spacing of the lamellar ordering reaches a length which is comparable to the membrane persistence length ξ, the lamellar phase melts into a sponge phase. (The sponge phase contains a web of bilayer which divides space into two solvent domains. For swelling with equal amounts of oil and water, the analogous phase is the microemulsion.) This phase transition, which corresponds to a change in topology, is also influenced by the Gaussian curvature rigidity ($\bar{\kappa}$) which controls the energy cost of handle formation [4, 17]. Upon adding more solvent the sponge phase itself swells and it eventually undergoes another phase transition to a vesicle phase [4].

The effect of fluctuations on flexible membranes, once analysed in detail, leads to a universal phase diagram where the lamellar, sponge, vesicle phases and their phase transitions can be understood in terms of a competition between curvature energy and entropy [18]. While a lot has been done and understood, some open questions remain and new systems have been studied showing interesting behaviours. We will just cite, as an example, the fact that an extremely dilute microemulsion phase, with a characteristic size of several thousands of Angstroms, has recently been found. This phase *crystallises* upon further swelling into a three dimensional ordered phase [19], unlike most ordered phases which melt upon swelling.

3 Effect of shear on lyotropic lamellar phases

Some years ago, in studying the effect of shear on lyotropic lamellar phases, a new hydrodynamic instability was described [20–24].This instability leads to a phase of multilamellar vesicles compactly packed in space (the so-called *onion texture*). Evidence for the structure of this metastable phase and some dynamics properties such as the viscoelasticity are described in what follows. In the current section, we will briefly describe the basic experimental facts leading to the formation of the onion texture. Then, we will give some details of its viscoelastic behaviour (Section 4) and develop the theoretical description of the lamellar-to-onion transition (Section 5).

As discussed above, a lyotropic lamellar phase is made of surfactant and water (Figure 2), and sometimes contains an additional hydrophobic component (oil). It is a very common phase, most often found for relatively high concentration of the surfactant [1], while in certain special cases, very dilute lamellar phases can be prepared due to long range repulsive interactions between membranes [25]. The symmetry of the phase is that of a 'smectic A' in the nomenclature of the liquid crystalline phases. When present, the hydrophobic compound swells the bilayers. In all the cases, the lamellar phase behaves macroscopically as a viscous liquid whose viscosity varies tremendously depending upon the formulation, and also (because of uncontrolled defects in the packing) on the sample preparation [23, 26]. Upon dilution with extra water, two main behaviours can be described [25]. The dilution of the lamellar phase is limited either by a phase transition to an isotropic liquid phase (micellar phase or sponge phase) or by a phase transition to another liquid crystalline phase. In a limited number of interesting cases the lamellar phase reaches a maximum uptake of water and subsequently phase coexists with excess (virtually pure) water. This arises whenever phospholipids are used as the surfactant.

3.1 Shear diagrams

To understand the effect of flow on such phases, we have studied using rheophysics methods the structure of lyotropic smectic A phases submitted to a simple shear (Figure 3). Using a number of structural probes under shear, such as scattering techniques (light scattering [21], neutron [22], X-ray [27]; see Pine, this volume) or dielectric measurements [28], it was possible to show that the effect of shear can be described using a *shear diagram*. This diagram, which can be considered as a generalisation of the phase diagram for out-of-equilibrium systems, describes the effect of shear as a succession of stationary states of orientation separated by dynamic transitions. Indeed, while always staying thermodynamically within the stable lamellar phase, the sample experiences a series of transitions modifying the orientation of the lamellae with respect of the direction of the shear. These different orientations correspond to differing spatial organisations and densities of the topological defects that are anyway present in most smectic samples. Each transition thus brings a modification of what is named the *texture* of the phase. Consequently, it is not a traditional phase transition but has to be viewed as an instability. It is different, however, from the classical hydrodynamic instabilities observed when a fluid is submitted to shear (convection rolls, *etc.*) because the resulting texture involves no length scale directly related to the size of the shear apparatus. Instead, structure forms on some microscopic (micron) length scale related to the intrinsic properties of the fluid.

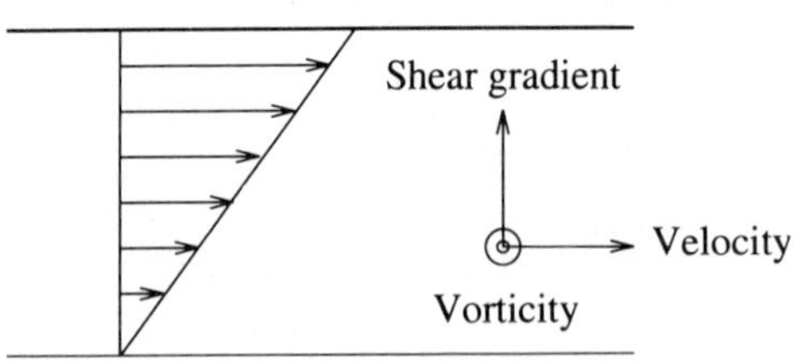

Figure 3. *Characterisation of the geometry of a simple shear flow. The orientations that the membranes of a lamellar phase adopt under shear are described using this geometry.*

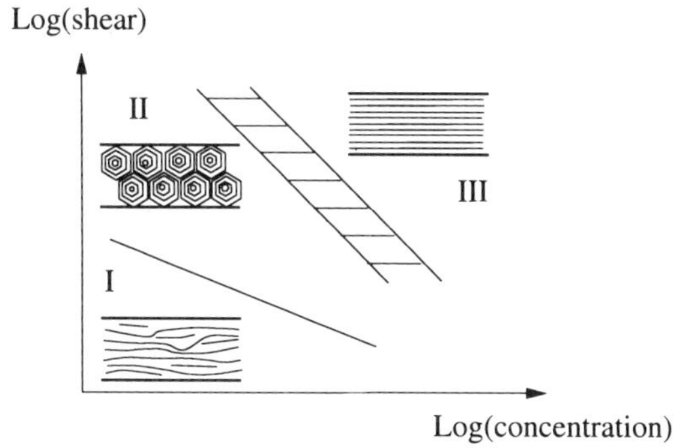

Figure 4. *Representation of the shear diagram of a typical lamellar phase (made of water, dodecane, pentanol, sodium dodecyl sulphate). The horizontal axis represents the concentration of dodecane (the bilayers are swollen upon addition of dodecane), which fixes the characteristic distance d between the lamellae, and the vertical axis represents the shear rate.*

Figure 4 is a schematic representation of the shear diagram obtained in the case of a lamellar phase made of water, dodecane, pentanol and sodium dodecyl sulfate (SDS) [21]. At very low shear rate, the phase is more-or-less oriented with the membranes parallel to the velocity direction. Defects however remain in the velocity direction as well as in the vorticity direction. At high shear rate, the orientation is basically very similar but the defects in the velocity direction have disappeared. In the intermediate regime, a new and interesting orientation appears. The membranes are broken into pieces by the flow and the phase organises itself into a phase of multilamellar vesicles all of the same size. We called these vesicles *onions* because of their multilamellar structure. Figure 5 presents freeze fracture picture of the phase after shearing it [24]. This picture reveals that onions are truly discrete entities and also shows that they adopt a polyhedral shape. While not 'universal', such a phenomenology is quite general and has been encountered in many systems [29, 30].

Figure 5. *Electron-microscopic picture of freeze fracture sample obtained after shearing a lamellar phase. The size of the onions is typically 1μm [24].*

3.2 Techniques used to study structures under flow

Several techniques can be used to demonstrate the existence of the onion texture besides the sophisticated electronic microscopy technique. Direct observation of the texture using a regular optical microscope equipped with crossed polarisers is certainly the easiest one. Figure 6 shows a typical texture of a sheared lamellar phase in the onion state [21]. One easily observes a regular modulation of white and black, with a characteristic length corresponding to the size of the onions. This characteristic length varies with the shear rate until it reaches very small scales below the optical resolution (typically 1μm). A uniform grey colour is then observed.

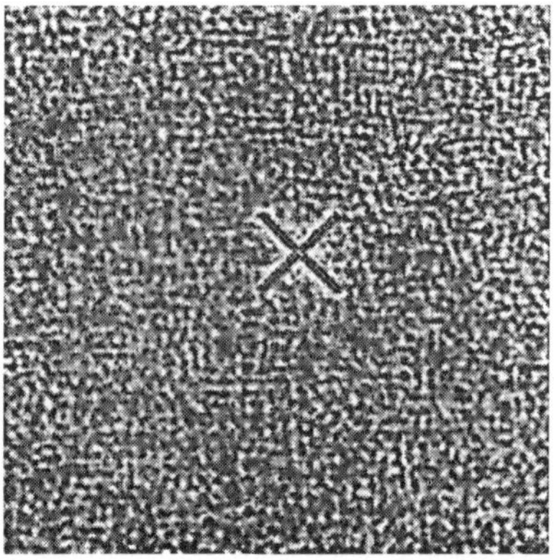

Figure 6. *Typical texture of the onion phase observed using an optical microscope between crossed polarisers.*

The characteristic size is easily detectable using a laser beam and a screen placed at a few centimetres from the sample [21]. Figure 7 presents the small angle patterns obtained when sending a laser beam through the onion texture onto an observation screen [20, 21]. This technique allows the measurement of sizes from 1μm to more than 50μm [20, 21, 30]. Below the micron size scale, a more specialised set-up must be used allowing the experimentalist to access to larger scattering angles.

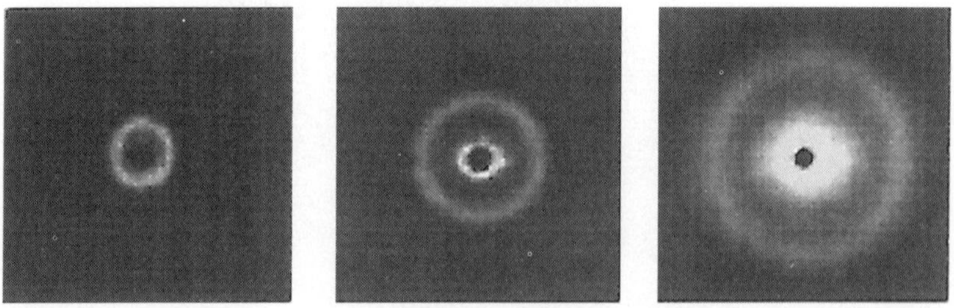

Figure 7. *Small-angle light scattering patterns observed in the onion texture by increasing the shear rate (from left to right). The ring of scattering is directly related to the onion size which obeys $D \sim \dot{\gamma}^{-1/2}$. (see Reference 21 and Figure 8.)*

The narrow ring of scattering observed on the screen shows the presence of a *single* characteristic length scale. Since the sample is birefringent, it is most probably a modulation of the orientation of the layers which is responsible for the contrast, with the index of refraction varying from its ordinary value to its extraordinary one as one traverses the characteristic length. The ring is obtained only if circularly polarised light is sent through the sample. Otherwise, using a linearly polarised beam and crossed analyser, a pattern of four blobs is observed because of the coupling between the polarisation of the incident light and the birefringence of the phase.

The position of the ring in reciprocal space is related to the characteristic size of the modulation. It corresponds directly to the onions diameter D through the classical Debye relation:

$$D = \frac{\lambda}{2n} \frac{1}{\sin(\theta/2)} \tag{2}$$

where n is the index of refraction of the phase (the average one), λ is the light wavelength and θ the scattering angle. The width of the peak is an indication of the uniformity of the size: the narrower it is, the narrower is the size distribution. For the main cases which have been studied, the size within a given sample does not vary more than 20% in radius [21]. The fact that the ring intensity is uniform in all directions is an indication that the onions are disordered. Indeed, they adopt a kind of amorphous arrangement (liquid-like). Because the position of the ring varies with the shear rate, it is possible to measure how the onion size evolves with the shear. Figure 8 represents such an evolution for a liquid-like structure. The size is inversely proportional to the square root of the shear rate : $D \sim \dot{\gamma}^{-1/2}$. This is an easy way to control experimentally the onion size. Depending upon the formulation (the surfactant choice) and the shear rate, sizes ranging from 0.2μm to more than 50μm have been found.

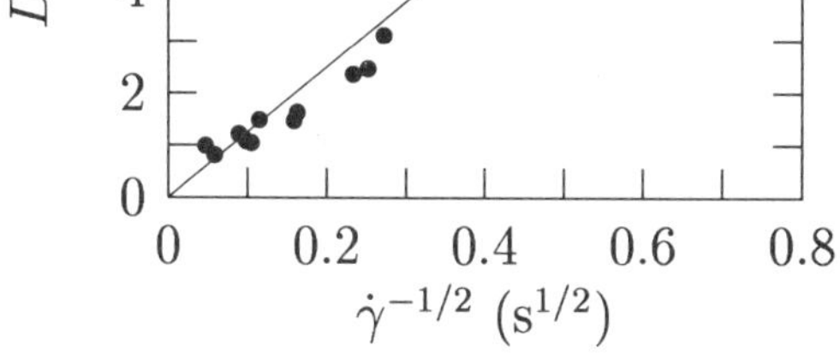

Figure 8. *Evolution of the size of the onions for the SDS–dodecane–water–pentanol system, as a function of the inverse of the square root of the shear rate [21].*

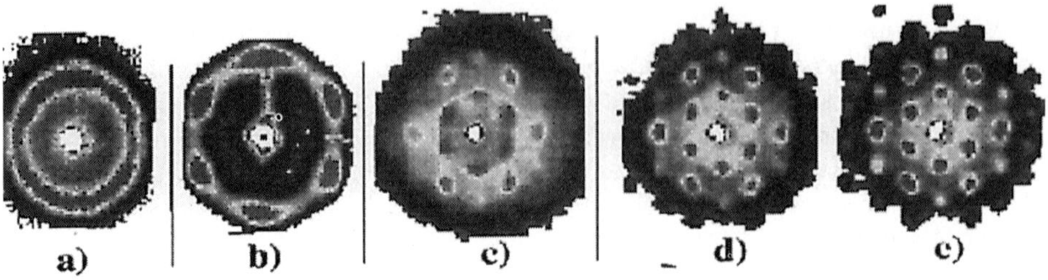

Figure 9. *Evolution of the small angle light scattering patterns as a function of the shear rate. (a) $\dot{\gamma} = 10s^{-1}$, the ring is isotropic and corresponds to an ensemble of monodisperse multilayered vesicles with no long-range order. (b) $\dot{\gamma} = 80s^{-1}$, the ring of scattering is replaced by six dots. The organisation now exhibits a long range order. (c) $\dot{\gamma} = 200s^{-1}$, small angle pattern after the transition of size. The characteristic size of the vesicles is now much bigger and several orders of scattering can be easily seen. (d) $\dot{\gamma} = 0s^{-1}$, same as (c) but after a rapid arrest of the shear. The long range order is kept and even more pronounced. (e) $\dot{\gamma} = 0s^{-1}$, same as (d) but after a few oscillations of small amplitude (made by hand). The long range order is even better, more than 5 orders of diffraction can be seen.*

3.3 Ordered structure of the onion phase under shear

An interesting system made of SDS, octanol and brine exhibits a more complex behaviour [30]. In addition to the steady states previously described, this system exhibits several new transitions. First, a transition between the disordered state described above and an ordered state can be observed. Second, a transition between two states of ordered multilayered vesicles, differentiated by the size of the vesicles, has been found. The transition between these two states is observed as a jump from small to big vesicles when either the shear rate or the temperature is increased. This transition, which is in general discontinuous (the size jumps abruptly), becomes continuous (smooth size evolution) at a critical temperature.

The system studied is a quaternary lyotropic lamellar phase whose phase diagram has been published [31]. A single sample is studied under shear (85.6% of brine at 20g/l of NaCl, 6.5% of SDS, 7.9% of octanol, in weight). This system is studied both as a function of temperature and shear rate. Let us first described what is observed in reciprocal space using small angle light scattering. Above $\dot{\gamma} \approx 1s^{-1}$, an isotropic ring of scattering (Figure 9a) appears characteristic of the multilayered vesicle state. This ring corresponds to the characteristic size of the close packed vesicles. Its radius increases with increasing $\dot{\gamma}$, indicating that the vesicle size decreases with the shear rate. The isotropy of the ring is the signature of no long-range order in the positions of the vesicles. Above a well-defined shear rate of $10s^{-1}$, a modulation in the radial intensity of the ring appears, leading to a well-defined pattern of six spots above $50s^{-1}$ (Figure 9b). This is the so-called *layering transition* which corresponds to the ordering of the multilayered vesicles in planes exhibiting an hexagonal order (in-plane). This transition does not affect the onion size since the dots appear on the ring: the size of the onions before and after the transition is practically the same and is around 3–4μm at the transition. After the transition, the size evolves very slowly with the shear rate, still decreasing when the shear rate increases [30].

When a shear rate of $200s^{-1}$ is reached, a new phenomenon is observed. The previous pattern of six dots evolves toward two rings of scattering: one at the previous position and a new ring appears at smaller angles. With time, the former ring of scattering disappears and a clear new set of dots is seen at a much smaller angle than the previous one. After some time (typically from 20 minutes to two hours), the pattern shown on Figure 9c is observed under shear. This new pattern corresponds to an ordered structure of onions similar to the state previously described, but constituted of much bigger onions (around $10\mu m$ at $T \approx 24°C$). Moreover, even under shear, several orders of diffraction can be observed (up to 3–4). In contrast to the previous ordered state, this pattern persists once the shear is stopped (Figure 9d). The quality of the long range order can even be improved significantly by applying small amplitude oscillations to the shear cell (Figure 9e). We have been able to keep this ordered structure after the shear has been stopped for several days. The exact nature of the structure of this phase has recently been studied in more detail and consists of a so-called *random stacking* structure [32] (see Section 4).

The transition between small and big vesicles can be mapped in the $(\dot{\gamma}, T)$ plane. Figure 10 shows the shear diagram obtained from these measurements. The two regions of ordered vesicles (small and big) are separated by a line corresponding to a discontinuous transition. This line ends on a critical point at $\dot{\gamma} \approx 40s^{-1}$, $T \approx 26°C$. The decrease of the small angle position of the ring or the Bragg peaks is the signature of the large increase in size of the onions. Direct space imaging can also be observed, confirming that the transition does involve a change in the onion size [30].

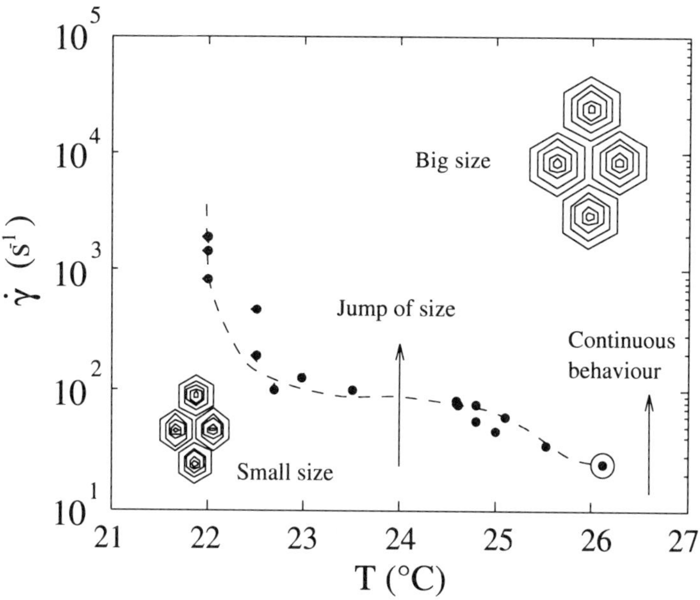

Figure 10. *Shear diagram of the jump-of-size transition. The full circles are the experimental points, the dotted line is a guide for the eye and the circle around the dot is the location of critical point above which the discontinuous transition is replaced by a continuous evolution.*

4 Viscoelasticity of the onion texture

As described above, onions give an opportunity to precisely control the defect texture of a lamellar phase on macroscopic length scales. We present below the viscoelastic behaviour of the onion texture for the two spatial organisations as a function of the radius of the droplets. Whereas the elasticity of the amorphous organisation is like that of concentrated emulsions, the ordered texture exhibits an unusual and still unexplained elastic behaviour. (For an introduction to viscoelasticity, see McLeish, this volume.)

4.1 Viscoelasticity of soft cellular materials

In terms of viscoelasticity, onions can be described as monodisperse soft spheres under pressure. They adopt a polyhedral shape (Figure 5) and belong to a wider class of cellular structures [33]. When the structure is ordered, the shape of the onions is the so-called Wigner-Seitz cell of a crystalline (or random stacking) lattice [24]. The structural analogy with dry foams or concentrated emulsions is obvious, and, indeed, the onion phase can be considered as a foam of lamellar droplets in the dry limit (the volume fraction of the dispersed droplets is close to unity). Linear viscoelastic properties of soft cellular materials have been intensively studied during the past 5 years, and since it is now possible to obtain a good degree of monodispersity, the comparison with theoretical studies is meaningful, though mainly limited to the disordered case. For many soft cellular systems, linear viscoelastic measurements reveal storage and loss moduli in nearly constant ratio ($G'/G'' \approx 10$) with a weak dependence on the frequency within the experimental window (10^{-2}Hz–10Hz). For compressed emulsions or dry foams, the elastic response to an infinitesimal deformation is due to the increase of the surface area of each bubble compared to the unstrained configuration. A calculation for a monodisperse disordered foam, based on Plateau's laws [34], leads to $G' = 0.55\sigma/R$ where σ is the surface tension of the film and R the radius of bubbles [35]. This result is well verified by experiments on fairly monodisperse emulsions [36], where it is found that $G' \approx 0.6\sigma/R$. (Even for a polydisperse sample $G' \approx 0.5\sigma/R$ was found [37].) As we will see in the next sections, onions also can be considered as having an *effective surface tension* σ when subjected to a strain, and a similar scaling law for G' with onion size R is therefore expected.

Whereas the elasticity of soft cellular materials seems to be well understood, their dissipation processes remain unclear. The simple fact that G'' is very often constant at low frequency (and sometimes increases) suggests a broad spectrum of dissipation at lower frequencies, probably due to slow relaxations of the glassy structure. This behaviour may be difficult to incorporate within a linear approximation. Furthermore, each attempt to describe fully the dissipation has led to an underestimation of $G''(\omega)$ compared to experimental values both in the emulsion case [38] or the onion one [39]. Finally, attention has to be paid to the interpretation of experimental data since, on the one hand, the rather elastic behaviour of such materials ($G' \gg G''$) makes it very difficult to determine G'' accurately, and on the other hand, the existence of a linear regime for such disordered systems remains uncertain [40].

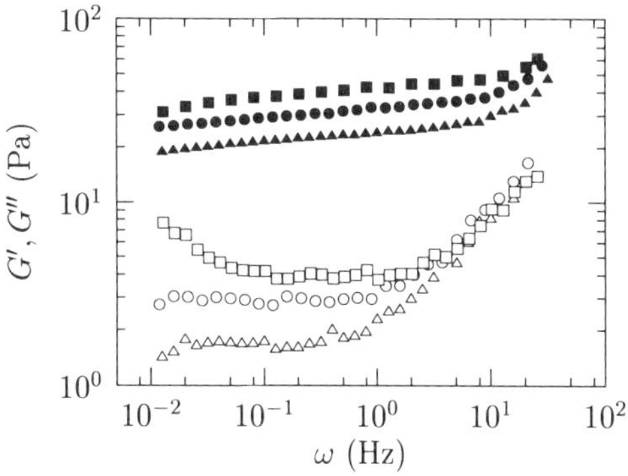

Figure 11. *Viscoelastic behaviour as a function of the frequency for three sizes of the* amorphous *onion texture. Open symbols correspond to G'' and solid ones to G'. (□ 0.37μm, ○ 0.5μm, △ 1.4μm).*

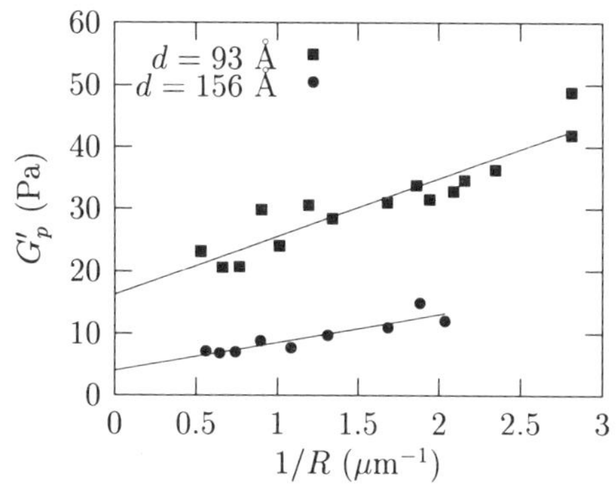

Figure 12. *Evolution of G'_p as a function of $1/R$ for two dilution of the lamellar phase. The behaviour is linear, with $G'_p(1/R \to 0) \neq 0$.*

4.2 The onion texture: results

We have first conducted viscoelastic measurements on the *amorphous* onion texture in the linear regime. For each onion size, the extent of the linear strain regime is measured at different frequencies (ω = 0.1 and 1 Hz). For the biggest onions, we find that the strain should not exceed 0.5% to remain in the linear regime, at least so far as the storage modulus G' is concerned. The typical result is shown in Figure 11, for three different sizes of onions in the disordered case. G' depends only weakly on frequency and the value of the plateau modulus G'_p decreases when the size increases.

We investigated how the onion size influences G'_p by tuning the shear rate of the onion preparation. By this method R varies from 0.3μm to more than 20μm depending on the system. In Figure 12 G'_p is plotted as a function of $1/R$, and a linear behaviour with a non-zero contribution for $1/R \to 0$ is clearly exhibited. As we shall see, this offset can be interpreted as arising from the elastic frustration present within a randomly distorted

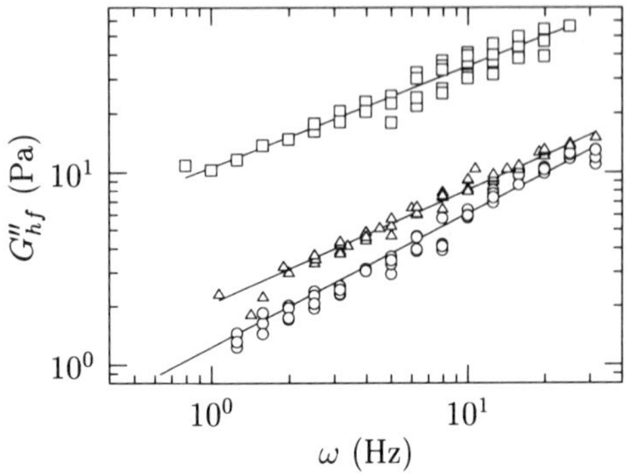

Figure 13. *Loss modulus of the amorphous onion texture as a function of the frequency in the size-independent regime, for three dilutions of the lamellar phase. (□ d=65 Å, △ d=93 Å, ○ d=156 Å).*

packing of lamellar phase onions [44]. The loss modulus G'' exhibits a size dependent shallow minimum followed by a high-frequency rise (Figure 11). The high-frequency behaviour of G'' has been studied carefully for one lamellar phase at two different dilutions (changing the smectic period d and therefore the stiffness of the phase) and is reported in Figure 13. One observes that the behaviour is a power-law $G''_{hf}(\omega) \sim A\omega^b$, where b is a weak function of the dilution ($b \approx 1/2$) and the coefficient scales as $A \sim 1/d^3$.

4.2.1 Dissipation

In the range of accessible frequencies, the behaviour of the loss modulus G'' suggests a cross-over between slow modes of structural relaxation (on the onion scale) at low frequency and fast modes (probably related to thermal fluctuations of the membranes within onions [39]) at higher ones. The structural modes suggest glassy behaviour. Indeed, when the strain amplitude is raised, G'' starts to show an increase, rather than a plateau, at low frequencies, even though the elastic response (G') remains unaltered. In other words, the dissipation increases markedly when the deformation increases, although still in an apparent linear response regime for the elasticity. This behaviour is consistent with some recent experiments on emulsions where it has been shown that some droplets follow irreversible paths when the whole sample is subjected to an oscillatory strain [40]; see also Pine, this volume. The notion of a 'linear regime' has therefore to be questioned. The frequency dependence of the loss modulus is also in qualitative agreement with a recent theory invoking disorder and metastability in 'soft glassy materials' [41].

The high-frequency behaviour does not show the same sort of nonlinearity, but its form remains unexpected. A full calculation of the contribution of thermal fluctuations within lyotropic lamellar phases [39], which should apply in the range of high frequencies, predicts that $G''(\omega) \sim \eta\omega$ (neglecting here some weak logarithmic correction in ω), where η is the viscosity of the solvent (typically 10^{-3} Pa s). Neither the power law nor the order of magnitude are in agreement with our experiments. An alternative approach is offered by Liu *et al.* [42]. They assume that a macroscopically applied shear strain does not produce an affine deformation because droplets are randomly packed. Instead, some regions slip rather than deforming, and this increases the viscous dissipation. They find a contribution which varies as $\omega^{1/2}$, which may dominate the high-frequency behaviour. Nevertheless, we do not expect this contribution to be relevant at the very high volume fraction occupied by the onions ($\phi \simeq 0.99$), so the origin of the observed power law remains unexplained.

4.2.2 Elasticity

The elastic properties of onion textures are probably better understood. Basically, onions are compressed lamellar droplets adopting a polyhedral shape. The energy cost of their deformation can be expressed in terms of the elastic constants $\bar{B}$ and K of the lamellar phase. $\bar{B}$ is the bulk compression modulus (at constant chemical potential) and $K = \kappa/d$ is the bulk bending elastic constant [43] of a smectic stack of fluid layers. A typical length can be extracted from this competition, the so-called 'de Gennes penetration length' λ which is expressed through the ratio between K and $\bar{B}$:

$$\lambda = \sqrt{\frac{K}{\bar{B}}} .$$

In a polyhedral onion, the deformation is concentrated along planes of strong curvature, of thickness λ, radiating from the centre of each onion towards its edges (Figure 5). In such a structure, a rough estimate of the stored energy per unit of volume is:

$$\frac{K}{\lambda^2}\frac{\lambda R^2}{R^3} = \frac{\sqrt{K\bar{B}}}{R} . \tag{3}$$

The first factor in (3) is the curvature cost within the deformed regions, whereas the second is the ratio of the volume in which the deformation is stored to that of a whole onion. The same calculation also gives the modulus scaling as $G' \simeq \sqrt{K\bar{B}}/R$, since this is (roughly) the stored energy density that can couple to an applied strain (see McLeish, this volume). The quantity $\sqrt{K\bar{B}}$ has the dimensions of a surface tension and indeed plays the role of an *effective surface tension* for small deformations of a smectic droplet. So this result is closely analogous to the one given previously for emulsions.

Glassy state

It is argued in Reference [44] that an additional feature contributes to the elasticity. This additional term arises from the glassy aspect of the phase (not included in the previous estimate, Equation 3) and reflects an excess of stored elastic energy quenched in such an organisation. Indeed, the onions of the amorphous texture are randomly deformed by an additional local strain of characteristic magnitude γ_0 compared to an ordered, symmetrical conformation. The main point of the argument is that this quenched strain γ_0 interacts in an intricate manner with a macroscopic deformation, and contributes to the elastic modulus [44]. In simple words, non-symmetric, distorted onions are sensitive to the direction in which they are deformed, in contrast to the symmetric onions in an ordered packing. This model predicts that the disorder contribution to G' is $\gamma_0^2\bar{B}$ which is independent of onion size. (Only in the case of an onion phase without quenched strains does γ_0^2 vanish.) The elastic modulus of an onion phase is finally predicted to be:

$$G'_p = \alpha + \beta \frac{\sqrt{K\bar{B}}}{R}, \tag{4}$$

where $\alpha = \gamma_0^2 \bar{B}$ in the glassy state, $\alpha = 0$ in an ordered state, and β is a phenomenological parameter possibly reflecting the geometry.

In Figure 12, one sees that the agreement with such a description for the glassy state is qualitatively good. Indeed the agreement is even quantitative since experiments have shown that $\alpha \sim 1/d^3$ [44], the same scaling as $\bar{B}$ (assuming that γ_0 does not depend on d). Nevertheless, examining carefully several lamellar phases, some points remain unclear. First, γ_0 depends on the studied lamellar phase, and varies from 3% to more than 10%. The quantitative origin of γ_0 is not known (it is introduced phenomenologically) but it may come from residual polydispersity or other effects.

Second, the parameter β also depends on the studied system. More precisely, it varies from 0.1 to 0.6 depending on the lamellar phase. (Note that the value $\beta = 0.6$ is extremely close to the elasticity of compressed emulsions). Indeed, this result suggests that the simple dimensional analysis of Equation 3 is not accurate enough. This model however catches the main features of the elasticity of the amorphous onion texture.

Ordered phase

Onions may be ordered under shear (Figure 9) in the same way as colloidal suspensions, and the phenomenology is quite similar. At rest, the resulting structure has been fairly well characterised using light and small angle neutron scattering. It consists of hexagonal layers of onions, stacked in one direction in a random ABC sequence without any long-range correlation. This structure is intermediate between the classical face-centred cubic (ABCABC) and hexagonal close packed (ABABAB) crystalline structures. However, the structure contains a certain amount of disorder, firstly via some dislocations (but in a negligible manner compared with a glass), secondly because the structure is indeed *amorphous* in the direction of the stacking of hexagonal layers.

We have conducted viscoelastic measurements on this structure, and typical spectra have exactly the same features as in Figure 12. In Figure 14, G'_p is plotted as a function of $1/R$ for ordered and disordered textures. The result is quite surprising since it shows that G'_{ordered} is only very weakly dependent on R and does not vanish when $1/R \to 0$. The model of Equation 4, which is in reasonable agreement with experiments for the amorphous organisation, predicts exactly the opposite, namely $G'_{\text{ordered}} \sim 1/R$. Although unexpected and certainly not understood, this result suggests that the physical origin of the elasticity may be different in the two cases.

Keeping in mind that the glassy state is correctly described, let us envisage two possible hypotheses. First, the model predicts $\alpha = 0$ only for symmetric onions (the effect of positive and negative strain must be identical for each onion). Our structure, even if crystalline, does seem to be a random sequence of hexagonal layers, but the exact sequence statistics have not been established. It is possible to imagine some sequences which are not symmetric with respect to the strain direction and the symmetric assumption fails. The second hypothesis lies in the elasticity of crystals [45]. The proper expression of

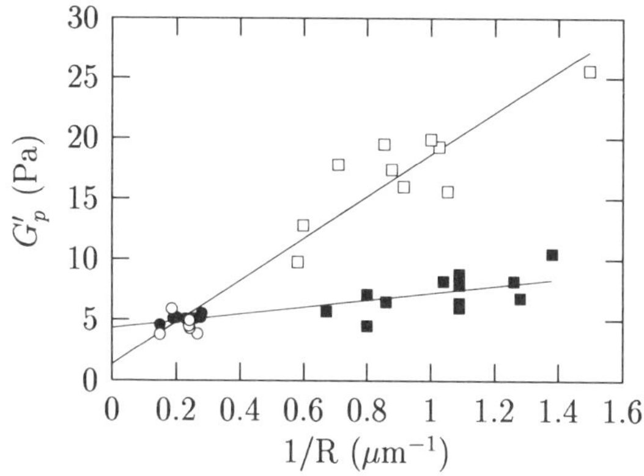

Figure 14. *Elastic modulus of onions as a function of their size and the type of order (full symbols, ordered texture – open symbols, amorphous texture).*

the response to a strain has to be expressed in a tensorial form. On the one hand, the corresponding elastic tensor derives from the symmetry of the crystal and may be strongly altered by the stacking sequence. On the other hand, it is possible to find some particular direction where the response to a strain depends very weakly on the elastic constants, and therefore on the size of the onions. This result however needs to be clarified.

5 Theoretical models of the onion transition

We finally wish to summarise our theoretical understanding of onion formation under shear. Although the formation of the onion texture remains largely unclear, two mechanisms have been proposed to explain its origin. The most recent explanation is described by Zilman and Granek [46]. This mechanism involves the coupling of the short wavelength undulations of the lamellar phase to the flow. These short wavelength undulations give rise to a long range repulsive interaction, the undulation force, and are thus responsible for the stability of the lamellar phase. The authors argue that the shear stress generates an effective force which acts to reduce the excess area of the membranes and suppress the undulations. This decreases the equilibrium layer spacing and is equivalent to applying a dilational strain proportional to the shear rate. At low shear rate, the dilation is balanced by the elastic forces. Above a critical shear rate, the dilation becomes so high that the lamellar phase buckles to give a sinusoidal modulation (see below). To flow, the system then has to find another type of organisation: the onion state appears. This model may explain the lamellar-to-onion transition (I/II) in Figure 4.

This mechanism is attractive but some points remain unclear. The dilational stress induced by the shear is assumed to be proportional to a high effective viscosity, and not to the water viscosity between the smectic layers. This allows the authors to calculate a critical shear rate in good agreement with the experiments. Unfortunately, the exact nature of this high viscosity is not modelled. Moreover, the system is assumed to track the states of minimum free energy (under the effective dilation). As suggested by the authors, a dynamic analysis taking into account hydrodynamic processes is needed to better describe this instability.

The second mechanism is mainly based on the work of Oswald and Kléman [47–50] which show that the *static* undulation instability, due to a dilational strain of the smectic phase, remains when the smectic phase is weakly sheared. (This extends, to the case of shear, the well-known undulation instability which was found in controlled dilation experiments without shear [51–54].) The main hypothesis is that it is this undulation instability which leads directly to the onion state [23]: the dilation is a real, and not an effective, one. In contrast Granek and Zilman [46], Oswald and Ben Abraham have performed a dynamic analysis of this instability but their expansion is valid only at small shear rates. The instability occurs because the gap between the two sliding surfaces in the shear apparatus is not uniform (at least, not at the scale of the lamellar spacing). At equilibrium, the spatial variations of the gap lead to the formation of defects. Dislocations appear to change locally the number of layers. At low shear rates, the dislocations are assumed to move with the flow by permeation. (Permeation is the diffusion of surfactant and/or solvent in a direction normal to the fluid layers.) The lamellar phase exhibits a high viscosity due to the dissipation taking place during this permeation process. At higher shear rates, the permeation process is too slow to allow the dislocations to flow. The lamellae are therefore subjected locally to either a dilational or a compressive strain perpendicular to the layers, leading to the undulation instability. On the one hand, the undulation in the direction of the vorticity is not affected by the flow. On the other hand, in the direction of the flow, the critical dilational strain increases as a function of the shear rate. It is very unlikely that the undulation pattern is able to flow because of the very high density of defects such as parabolic focal conics [55]. At this stage, the onion state appears.

Both mechanisms described above assume the formation of onions to be governed by a dilational strain. According to the work of to Zilman and Granek, this dilatation is inherent and due to the effect of shear on the lamellae. However, as in the case of the second mechanism described above, a dilation can also be due to experimental conditions.

Although these works might be able to describe the I/II transition, to date no description has been given for the second transition between the onion state to the high-shear orientated state: II/III. Following the work of Oswald we tried to answer to the question: what happens to the lamellar phase when the sample is dilated while simultaneously being *strongly* sheared parallel to the layers?

In our model [56] the dilation can either be due to the spatial variations of the gap or induced by an inherent mechanism. We describe the flow using the Navier-Stokes equation where a volume force density due to the smectic elasticity is added [57, 58]. The fluid is incompressible and the flow is coupled to the layer displacement by a permeation equation (the permeability is neglected). We assume that the sample is subjected to a dilational strain perpendicular to the lamellae. Through a linear analysis, we show that undulations may grow in the direction of the flow and of the vorticity, as found in the previous works. This may lead to the appearance of the onion phase. At higher shear rates, we find evidence that the undulation along the flow disappears whereas it persists in the vorticity direction. Such a pattern describes the structure of the lamellar phase in domain III. This work allows us to draw a theoretical phase diagram which describes quantitatively the transition II/III [56].

References

[1] Ekwall P, in *Advances in Liquid Crystals.* Brown G M, Ed., Academic Press, New York 1975

[2] Charvolin J, *J Phys (Paris)*, Colloq 46 C3-173 (1985)

[3] Bellocq A-M, Roux D in *Microemulsions, Structure and Dynamics*, Frieberg S and Bothorel P, Eds., CRC Press, 1987

[4] Porte G, *J Phys: Cond Matt* **4** 8649 (1992)

[5] Cates M E, Candau S J, *J Phys: Cond Matt* **2** 6869 (1990)

[6] Canham P B, *J Theor Biol* **26** 61 (1970)

[7] Helfrich W, *Z Naturforsch* **28** 693 (1973)

[8] Servuss R M, Helfrich W, *J Phys (Paris)* **50** 809 (1989)

[9] Szleifer I, Kramer D, Ben-Shaul A, Gelbart W M and Safran S, *J Chem Phys* **92** 6800 (1990)

[10] Ben-Shaul A and Gelbart W M, *Ann Rev Phys Chem* **36** 179 (1985)

[11] De Gennes P-G, Taupin C, *J Phys Chem* **86** 2294 (1982)

[12] Roux D, Coulon C, *J Phys* **47** 1257 (1986)

[13] Roux D, Nallet F, Safinya C in *Micelles, Membranes, Microemulsions and Monolayers*, Gelbart W M, Ben-Shaul A and Roux D, Eds., Springer Verlag, 1994

[14] Lipowsky R, Leibler S, *Phys Rev Lett* **56** 2541 (1986)

[15] Milner S T, Roux D, *J Phys I France* **2** 1741 (1992)

[16] Helfrich W, *Z Naturforsch* **33a** 305 (1978)

[17] Morse D C, *Phys Rev E* **50** R2423 (1994); Golubović L, *Phys Rev E* **50** R2419 (1994)

[18] Huse D A, Leibler S, *J Phys France* **49** 605 (1988)

[19] Peter U, König S, Roux D and Bellocq A-M, *Phys Rev Lett* **76** 3866 (1996)

[20] Diat O and Roux D, *J Phys II France* **3** 9 (1993)

[21] Diat O, Roux D and Nallet F, *J Phys II* **3** 1427 (1993)

[22] Diat O, Roux D and Nallet F, *J Phys IV* **3** 193 (1993)

[23] Roux D, Nallet F and Diat O, *Europhys Lett* **24** 53 (1993)

[24] Gulik T, Roux D, Degert C and Laversanne R, *Langmuir* **12** 4668 (1996)

[25] Roux D, Nallet F and Safinya C in *Micelles, Membranes, Microemulsions and Monolayers*, Gelbart W , Ben Shaul A and Roux D, Eds., Springer Verlag NY Inc 1993, pp 303-346

[26] Auguste F, Douliez J P, Bellocq A-M, Dufourc E and Gulik-Krzywicki T *Langmuir* **13** 666 (1997)

[27] Roux D in *Theoretical Challenges in the dynamics of complex fluids*, edited by Tom McLeish, Nato ASI, Series E: Applied Sciences 1997, Vol 339 p 203 (Kluwer Academic publishers)

[28] Soubiran L, Roux D and Coulon C *Europhys Lett* **31** 243 (1995)

[29] Lukaschek M, Müller S, Hasenhindl A, Grabowski D A and Schmidt C *Colloid Polym Sci* **274** 1 (1996)

[30] Sierro P, Roux D, *Phys Rev Lett* **78** 1496 (1997)

[31] Hervé P, Roux D, Bellocq A-M, Nallet, F and Gulik-Krzywicki T, *J Phys II France* **3** 1255 (1993)

[32] Leng J, Sierro P, Roux D, Nallet F and Diat O. In preparation.

[33] Verbist G, Weaire D, Phelan R in *Foams and Emulsions*, edited by N Rivier and J-F Sadoc, Nato ASI series Kluwer Academic Publishers, 1999

[34] Plateau J A F, *Statique Expérimentale et Théorique des liquides Soumis aux Seules Forces Moléculaires*, Gauthiers Villard, Paris (1873)

[35] Stamenović D, *J Coll Interface Sci* **145** 255 (1991)

[36] Mason T G, Lacasse M D, Grest G S, Levine D, Bibette J and Weitz D A, *Phys Rev E* **56** 3150 (1997)

[37] Princen H M, *J Coll Interface Sci* **112** 427 (1986)

[38] Buzza D M A, Lu C-Y D and Cates M E, *J Phys II France* **5** 37 (1995)

[39] Lu C-Y D and Cates M E, *J Chem Phys* **101** 5219 (1994)

[40] Hébraud P, Lequeux F, Munch J P and Pine D, *Phys Rev Lett* **78** 4657 (1997)

[41] Sollich P, Lequeux F, Hébraud P and Cates M E, *Phys Rev Lett* **78** 2020 (1997)

[42] Liu A J, Ramaswamy S, Mason T G, Gang H and Weitz D A, *Phys Rev Lett* **76** 3017 (1996)

[43] De Gennes P-G and Prost J, *Physics of Liquid Crystals* Clarendon Press, Oxford 1994

[44] Panizza P, Roux D, Vuillaume V, Lu C-Y D and Cates M E, *Langmuir* **2** 248 (1996)

[45] Wallace D C, *Thermodynamics of Crystals*, John Wiley and Son, 1972

[46] Zilman A.G and Graneck R, *Eur Phys J B* **11** 593 (1998)

[47] Oswald P, and Kléman, M, *J Phys—Lettres* **43** L-411 (1982)

[48] Oswald P, Behar J and Kléman M, *Phil Mag A* **46** 899 (1982)

[49] Oswald P and Ben-Abraham S.I, *J Phys (Paris)* **43** 1193 (1982)

[50] Oswald, P, *J Phys—Lettres* **44** L-303 (1983)

[51] Clark N.A and Pershan P.S, *Phys Rev Lett* **30** 3 (1973)

[52] Delaye M, Ribotta R and Durand G, *Physics Letters* **44A** 139 (1973)

[53] Clark N A and Meyer R B, *Appl Phys Lett* **22** 10 (1973)

[54] Singer S J, *Phys Rev E* **48** 2796 (1993)

[55] Rosenblatt C S, Pindak R, Clark N A and Meyer R B, *J Phys* **38** 1105 (1977)

[56] Wunenberger A-S, Colin A, Colin T and Roux D. Submitted to EPJB

[57] Martin P C, Parodi O, Pershan P S, *Phys Rev A* **6** 2401 (1972)

[58] Brochard F and De Gennes P-G, *Pramāna* sup **1** 1 (1975)

Coarsening dynamics of nonequilibrium phase transitions

Alan J Bray

University of Manchester, UK

1 Introduction

Systems quenched from a disordered phase into an ordered phase do not order instantaneously. Instead, the length scale of ordered regions grows with time ('coarsening') as the different broken symmetry phases compete to select the equilibrium state. To fix our ideas, it is helpful to consider the simplest, and most familiar, system: the ferromagnetic Ising model. Consider a temperature quench, at time $t = 0$, from an initial temperature T_I above the critical point T_C to a final temperature T_F below T_C. At T_F there are two equilibrium phases, with magnetisation $\pm M_0$. Immediately after the quench, however, the system is in an unstable disordered state corresponding to equilibrium at temperature T_I. The theory of phase ordering kinetics is concerned with the dynamical evolution of the system from the initial disordered state to the final equilibrium state.

The first point to note, which underlies the whole field, is that in the thermodynamic limit, final equilibrium is never achieved. This is because the longest relaxation time diverges with the system size in the ordered phase, reflecting the broken ergodicity. Instead, a network of domains of the equilibrium phases develops, and the typical length scale associated with these domains increases with time t. Furthermore, the resulting domain patterns look statistically similar at all (sufficiently long) times, apart from a global change of scale. This is the phenomenon of 'dynamic scaling'.

For pedagogical reasons, we have introduced domain-growth in the context of the Ising model, and will continue to use magnetic language for simplicity. A related phenomenon that has been studied for many decades, however, by metallurgists, is the spinodal decomposition of binary alloys, where the late stages of growth are known as 'Ostwald ripening'. Similar phenomena occur in the phase separation of fluids or binary liquids, although in these cases the phase separation is accelerated by the earth's gravitational field, which severely limits the temporal duration of the scaling regime. The gravitational effect can be moderated by using density-matched binary liquids and/or performing the experiments under microgravity. All of the above systems, however, contain an extra complication not present in the Ising ferromagnet. This is most simply seen by mapping an AB alloy onto an Ising model. If we represent an A atom by an up spin, and a B atom by a down spin, then the *equilibrium* properties of the alloy can be modelled very nicely by the Ising model. There is one important feature of the alloy, however, that is not captured by the Ising model with conventional Monte-Carlo dynamics. Flipping a single spin in the Ising model corresponds to converting an A atom to a B atom (or vice versa), which is inadmissible. The dynamics must conserve the number of A and B atoms separately, *i.e.* the magnetisation (or 'order parameter') of the Ising model should be *conserved*. This will influence the form of the coarse-grained equation of motion, as discussed in Section 2.2 and lead to slower growth than for a non-conserved order parameter.

In all the systems mentioned so far, the order parameter (*e.g.* the magnetisation of the Ising model) is a scalar. In the last few years, however, there has been increasing interest in systems, such as nematic liquid crystals, with more complex order parameters. Consider, for conceptual simplicity, a planar ferromagnet, in which the order parameter is a vector confined to a plane. After a quench into the ordered phase, the magnetisation will point in different directions in different regions of space, and singular lines (vortex lines) will form at which the direction is not well defined. These 'topological defects' are the analogues of the domain walls in the scalar systems. We shall show that, quite generally, an understanding of the relevant topological defects in the system, combined with the scaling hypothesis, will take us a long way towards understanding the forms of the growth laws and scaling functions for phase ordering in a wide variety of systems.

In these lectures I will present an introduction to the theory of phase ordering kinetics. They are based on a review article written a few years ago [1] with some additions to incorporate recent developments. In the first lecture (Section 2) I discuss the dynamical models used to describe phase ordering, or phase separating, systems, and discuss the initial instability that leads to either 'spinodal decomposition' or 'nucleation and growth' when the system is cooled through the equilibrium transition temperature. In the second lecture (Section 3) the scaling phenomenology is discussed in general terms, and dynamical equations are derived for the evolution of the interfaces (or domain walls) in phase-ordering and phase-separating systems. From these equations application of the dynamical scaling hypothesis leads to the growth laws which describe the coarsening of domains in these systems. In the final lecture (Sections 4–6) a more general approach is described which can deal with systems, such as nematic liquid crystals, that are not described by a scalar order parameter. A discussion of the topological defects, analogous to domain walls, which occur in these systems is followed by a general approach to the derivation of coarsening growth laws. The lecture ends with a discussion of spinodal decomposition under a uniform shear flow, illustrated with results from an exactly soluble model.

2 Dynamical models

2.1 Lattice models

The simplest model of order-disorder transitions and phase separation is the Ising spin model on a lattice, with Hamiltonian $H = -J\sum_{\langle ij\rangle} S_i S_j$, consisting of spins, S_i, located on the sites of a lattice and interacting with their nearest neighbours through ferromagnetic ($J > 0$) or antiferromagnetic ($J < 0$) interactions. With the identification of the up and down spins with species A and B of a binary alloy (or binary fluid without hydrodynamics), the case $J < 0$, which leads to a two-sublattice antiferromagnetic ordering, models the order-disorder transition in an alloy, while $J > 0$ models phase separation.

The Ising model does not have an intrinsic dynamics, but a dynamics can be imposed by defining transition rates between spin configurations. Consider two spin configurations, '1' and '2', and let $W(1\to2)$ be the transition rate from 1 to 2. For a system *in equilibrium*, the 'detailed balance' condition, $P_1W(1\to2) = P_2W(2\to1)$, must be satisfied for each pair of states, where P_i $(i = 1, 2)$ is the probability of finding the system in state i. It follows that $W(1\to2)/W(2\to1) = P_2/P_1 = \exp(-\Delta E/k_BT)$, where $\Delta E = E_2 - E_1$ is the energy difference between the states. Two commonly used transition rates are the Metropolis algorithm $W(1\to2) = \max[1, \exp(-\Delta E/k_BT)]$, and the Glauber (or 'heat bath') algorithm $W(1\to2) = [1+\exp(\Delta E/k_BT)]^{-1}$, both of which satisfy detailed balance.

For the non-equilibrium coarsening dynamics of interest here, detailed balance is still necessary in order that the system approach equilibrium locally (*i.e.* within domains) at finite times after the quench, and globally at infinite time. (See Mukamel, this volume, for a discussion of 'driven' systems, where this does not hold.) If the order parameter (*e.g.* magnetisation) is not conserved, a single spin flip is the elementary move between configurations. For a conserved order parameter, 'Kawasaki dynamics' is appropriate, where the elementary move is the interchange of a nearest-neighbour pair of oppositely aligned spins. In the alloy context, Kawasaki dynamics corresponds to interchanging A and B atoms on adjacent sites, while single spin-flip dynamics corresponds to changing an A atom (say) into a B atom. While the latter is obviously unphysical, it leads to no difficulties when (for $J < 0$) the corresponding order parameter is the antiferromagnetic 'staggered magnetisation' which is not a conserved quantity.

For numerical simulations, both the lattice models and the coarse-grained models described below can be used. For analytical purposes, however, working with coarse-grained models is an essential step to furthering one's understanding. Such models are familiar as the Ginzburg-Landau models used to study critical phenomena. Their form is dictated by the required symmetries and conservation laws.

2.2 Coarse-grained models

It is convenient to set up a continuum description in terms of a coarse-grained order-parameter field (*e.g.* the 'magnetisation density') $\phi(\mathbf{x}, t)$, which we will initially take to be a scalar field. A suitable Landau free-energy functional to describe the ordered phase is

$$F[\phi] = \int d^dx \left(\frac{1}{2}(\nabla\phi)^2 + V(\phi)\right) , \tag{1}$$

where the 'potential' $V(\phi)$ has a double-well structure, *e.g.* $V(\phi) = (1-\phi^2)^2$ (see Figure 1). We will take the minima of $V(\phi)$ to occur at $\phi = \pm1$, and adopt the convention that $V(\pm1) = 0$. The two minima of V correspond to the two equilibrium states, while the gradient-squared term in (1) associates an energy cost to an interface between the phases.

In the case where the order parameter is not conserved, an appropriate equation for the time evolution of the field ϕ is

$$\frac{\partial\phi}{\partial t} = -\frac{\delta F}{\delta\phi} = \nabla^2\phi - V'(\phi) , \tag{2}$$

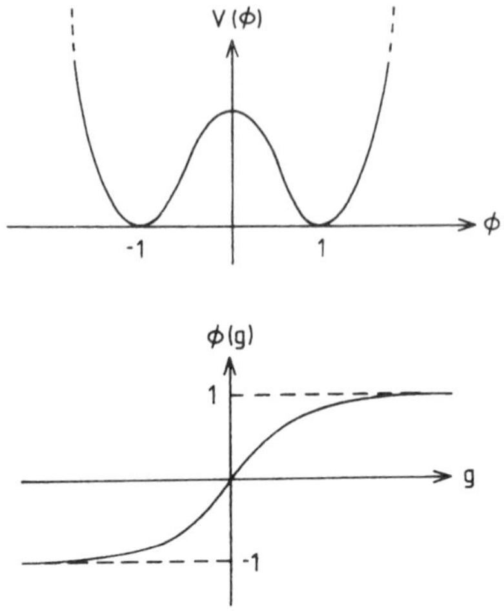

Figure 1. *Top: symmetric double well potential for a scalar order parameter. (For a vector order parameter, an equivalent but rotationally symmetric potential is used; with two components, this resembles a 'mexican hat'.) Bottom: order parameter profile across a domain wall.*

where $V'(\phi) \equiv dV/d\phi$. A kinetic coefficient Γ, which conventionally multiplies the right-hand side of (2), has been absorbed into the timescale. Equation 2, a simple 'reaction-diffusion' equation, corresponds to simple gradient descent, *i.e.* the rate of change of ϕ is proportional to the gradient of the free-energy functional in function space. This equation provides a suitable coarse-grained description of the Ising model, as well as alloys that undergo an order-disorder transition on cooling through T_C, rather than phase separating. Such alloys form a two-sublattice structure, with each sublattice occupied predominantly by atoms of one type. In Ising model language, this corresponds to antiferromagnetic ordering. The magnetisation is no longer the order parameter, but a 'fast mode', whose conservation does not significantly impede the dynamics of the important 'slow modes'.

When the order parameter is conserved, as in phase separation, a different dynamics is required. In the alloy system, for example, it is clear physically that A and B atoms can only exchange locally (not over large distances), leading to diffusive transport of the order parameter, and an equation of motion of the form

$$\frac{\partial \phi}{\partial t} = \nabla^2 \frac{\delta F}{\delta \phi} = -\nabla^2 \left[\nabla^2 \phi - V'(\phi)\right] , \tag{3}$$

which can be written in the form of a continuity equation, $\partial_t \phi = -\nabla \cdot \mathbf{j}$, with current $\mathbf{j} = -\lambda \nabla(\delta F/\delta\phi)$. In (3), we have absorbed the transport coefficient λ into the timescale.

Equations 2 and 3 are sometimes called the Time-Dependent-Ginzburg-Landau (TDGL) equation and the Cahn-Hilliard equation respectively. A more detailed discussion of them in the present context can be found in an article by Langer [2]. The same equations with additional Langevin noise terms on the right-hand sides, incorporating the effects of thermal fluctuations, are familiar from the theory of critical dynamics, where they are 'model A' and 'model B' respectively in the classification of Hohenberg and Halperin [3].

The absence of thermal noise terms in (2) and (3) indicates that we are effectively working at $T = 0$. A schematic Renormalisation Group (RG) flow diagram for T [1] has stable RG fixed points at 0 and ∞, and an unstable fixed point at T_C. Under coarse-graining, temperatures above T_C flow to infinity, while those below T_C flow to zero. We therefore expect the final temperature T_F to be an irrelevant variable (in the scaling regime) for quenches into the ordered phase. This can be shown explicitly for systems with a conserved order parameter [4, 5]. For this case the thermal fluctuations at T_F simply renormalise the bulk order parameter and the surface tension of the domain walls: when the characteristic scale of the domain pattern is large compared to the domain wall thickness (*i.e.* the bulk correlation length in equilibrium), the system behaves *as if* it were $T = 0$, with the temperature dependence entering through T-dependent model parameters.

In a similar way, any short-range correlations present at T_I should be irrelevant in the scaling regime, *i.e.* all initial temperatures T_I are equivalent to $T_I = \infty$. Therefore we will take the *initial conditions* to represent a completely disordered state. For example, one could choose the 'white noise' form

$$\langle \phi(\mathbf{x}, 0)\, \phi(\mathbf{x}', 0) \rangle = \Delta\, \delta(\mathbf{x} - \mathbf{x}') \,, \tag{4}$$

where $\langle \cdots \rangle$ represents an average over an ensemble of initial conditions, and Δ controls the size of the initial fluctuations in ϕ. The equivalent initial condition for lattice models is $\langle S_i(0) S_j(0) \rangle = \delta_{ij}$, again equivalent to $T_I = \infty$. The above discussion, however, indicates that the precise form of the initial conditions should not be important, as long as only short-range spatial correlations are present.

The challenge of understanding phase ordering dynamics, therefore, can be posed as finding the nature of the late-time solutions of deterministic differential equations like (2) and (3), subject to random initial conditions. A physical approach to this formal mathematical problem is based on studying the structure and dynamics of the topological defects in the field ϕ. This is approach that we will adopt. For scalar fields, the topological defects are just domain walls. Before proceeding to a detailed analysis of domain-wall motion, however, we first set the scene by carrying out a linear stability analysis of the coarse-grained equation of motion.

2.3 Linear stability analysis

To first order in ϕ, (2) reads $\partial\phi/\partial t = \nabla^2\phi - V''(0)\phi$, where $V''(0)$ is the curvature at the unstable maximum of $V(\phi)$, *i.e.* $V''(0) < 0$. Fourier transforming this equation, and solving, gives $\phi_{\mathbf{k}}(t) = \exp(\omega_{\mathbf{k}} t)\, \phi_{\mathbf{k}}(0)$, where $\omega_{\mathbf{k}} = |V''(0)| - k^2$. It follows that all Fourier components with $k < k_m = \sqrt{|V''(0)|}$ are *unstable*, *i.e.* grow exponentially with time. (Including thermal noise does not change this conclusion.) The linear analysis will break down, however, at times of order $|V''(0)|^{-1}$. Beyond this time, domains of the equilibrium phases will form and the subsequent dynamics will be controlled by the motion of the domain walls.

The conserved order parameter case is more interesting. In this case, we can take as initial condition a state which, up to small spatial fluctuations, is spatially homogeneous: $\phi(\mathbf{x}, t) = \phi_0$. To determine whether this state is stable against small fluctuations we write $\phi(\mathbf{x},t) = \phi_0 + \tilde{\phi}(\mathbf{x},t)$ in (3) and linearise in $\tilde{\phi}$. After Fourier transforming and solving we obtain once more an exponential dependence of $\tilde{\phi}_{\mathbf{k}}$ on time, but this time with $\omega_{\mathbf{k}} = -k^2[V''(\phi_0) + k^2]$.

If ϕ_0 corresponds to a point in the central part of the potential, between the two inflection points, then $V''(\phi_0) < 0$. In that case we have unstable growth for all k such that $k < k_m = \sqrt{|V''(\phi_0)|}$, with a maximum growth rate at $k = k_m/2$. Phase separation through unstable growth is called 'spinodal decomposition', and the two inflection points of the potential are called 'spinodal points'. If, on the other hand, ϕ_0 lies outside this region, such that $V''(\phi_0) > 0$, small fluctuations in the initial condition decay away to zero. (Again, thermal fluctuations do not change the main conclusion of this linear analysis.)

In the latter case, however, the phase separated state in which a fraction $(1+\phi_0)/2$ of the majority phase, $\phi = 1$ (we are assuming $\phi_0 > 0$), coexists with a fraction $(1-\phi_0)/2$ of the minority phase, $\phi = -1$, has a lower free energy than the homogeneous phase, because the volume free energy gain outweighs the cost of creating an interface between the phases. The uniform state is therefore only metastable. The free energy can be lowered by the nucleation of a 'critical droplet' of the minority phase. The critical droplet is an unstable solution of the time-independent equation $\nabla^2\mu = 0$, where $\mu = \delta F/\delta\phi$ is the chemical potential. Since no current flows, a solution where μ is a constant is required. Furthermore, $\mu = V'(\phi) - \nabla^2\phi$ implies (by considering regions far from the droplet, where $\phi = \phi_0$) that $\mu = V'(\phi_0)$. In the interior of the droplet, far from the boundary, $\mu = V'(\phi_-)$, where ϕ_- is the value inside the droplet.

The radius, R_c, of the critical droplet can be determined as follows. The equation $\mu = V'(\phi) - \nabla^2\phi = V'(\phi_0)$ gives $d^2\phi/dr^2 + (2/r)d\phi/dr - U'(\phi) = 0$, where $U(\phi) = V(\phi) - V'(\phi_0)\phi$ is an effective potential, with minima at ϕ_- and ϕ_0. Multiplying by $d\phi/dr$, integrating from zero to infinity, and recognising that $d\phi/dr$ vanishes at $r = 0$ and $r = \infty$, and is sharply peaked at $r = R_c$, gives $(2/R_c)\int_0^\infty (d\phi/dr)^2 dr = \Delta U$, where $-\Delta U = U(\phi_-) - U(\phi_0)$ is the free energy gain per unit volume in the interior of the droplet. Thus $R_c = 2\sigma/\Delta U$, where $\sigma = \int (d\phi/dr)^2 dr$ is the surface tension (see Section 3.3).

A simpler way to get the same result is to note that the net free energy change to create a droplet of radius R is $\Delta F(R) = 4\pi R^2\sigma - (4\pi R^3/3)\Delta U$, the first and second terms corresponding to surface and volume free energies respectively. This function has a maximum value $\Delta F(R_c) = (16\pi/3)\sigma^3/(\Delta U)^2$ at a radius $R_c = 2\sigma/\Delta U$. Clearly the droplet can freely expand once $R > R_c$. The activation barrier for droplet creation is $\Delta F(R_c)$, so the activation rate per unit volume is proportional to $\exp[-\Delta F(R_c)/k_B T]$. This process is called 'homogeneous nucleation' (see Frenkel, this volume, for an application to colloidal systems).

Since $\Delta U \to 0$ as $\phi_0 \to 1$, nucleation becomes rarer as time progresses, as more and more material of the minority phase get mopped up by the nucleating droplets, pushing the value of ϕ in the background majority phase towards its saturated value. Subsequent growth is by the 'evaporation-condensation mechanism', whereby small drops (whose size is below the current value of R_c defined by the background value of ϕ) evaporate, and material condenses onto the larger drops. This mechanism is discussed in more detail below.

3 Scaling and growth laws

3.1 The scaling hypothesis

Although originally motivated by experimental and simulation results [6–9], for ease of presentation it is convenient to introduce the scaling hypothesis first, and then discuss its implications for observed growth laws and scaling functions. Briefly, the scaling hypothesis states that there exists, at late times, a single characteristic length scale $L(t)$ such that the domain structure is (in a statistical sense) independent of time when lengths are scaled by $L(t)$. It should be stressed that scaling has not been proved, except in some simple models such as the one-dimensional Glauber model [10] and the n-vector model with $n = \infty$ [11]. However, the evidence in its favour is compelling.

We shall find, in Section 5, that the scaling hypothesis, together with a result for the tail of the structure factor, is sufficient to determine the form of $L(t)$ for most cases of interest.

Two commonly used probes of the domain structure are the equal-time pair correlation function

$$C(\mathbf{r}, t) = \langle \phi(\mathbf{x} + \mathbf{r}, t)\, \phi(\mathbf{x}, t) \rangle \ , \tag{5}$$

and its Fourier transform, the equal-time structure factor,

$$S(\mathbf{k}, t) = \langle \phi_{\mathbf{k}}(t)\, \phi_{-\mathbf{k}}(t) \rangle \ . \tag{6}$$

Here angle brackets indicate an average over initial conditions. The structure factor can, of course, be measured in scattering experiments. The existence of a single characteristic length scale, according to the scaling hypothesis, implies that the pair correlation function and the structure factor have the scaling forms

$$\begin{aligned} C(\mathbf{r}, t) &= f(r/L) \ , \\ S(\mathbf{k}, t) &= L^d\, g(kL) \ , \end{aligned} \tag{7}$$

where d is the spatial dimensionality, and $g(y)$ is the Fourier transform of $f(x)$. Note that $f(0) = 1$, since (at $T = 0$) there is perfect order within a domain.

At general temperatures $T < T_c$, $C(\mathbf{0}, t) = M^2$, where M is the equilibrium value of the order parameter. (Note that the *scaling limit* is defined by $r \gg \xi$, $L \gg \xi$, with r/L arbitrary, where ξ is the equilibrium correlation length). We can extract the factor M^2 explicitly by writing $C(\mathbf{r}, t) = M^2 f(r/L)$. The statement that T is irrelevant then amounts to asserting that any remaining temperature dependence can be absorbed into the domain scale L, such that the function $f(x)$ is independent of T. The scaling forms (7) are well supported by simulation data and experiment.

3.2 Two-time scaling and aging

For future reference, we note that the different-time correlation function, defined by $C(\mathbf{r}, t, t') = \langle \phi(\mathbf{x} + \mathbf{r}, t)\, \phi(\mathbf{x}, t') \rangle$, can also be written in scaling form. A simple generalisation of (7) gives [12, 13]

$$C(\mathbf{r}, t, t') = f(r/L, r/L') \ , \tag{8}$$

where L, L' stand for $L(t)$ and $L(t')$. Especially interesting is the limit $L \gg L'$, when (8) takes the form

$$C(\mathbf{r}, t, t') \to (L'/L)^{\bar{\lambda}}\, h(r/L)\;, \qquad L \gg L'\;, \tag{9}$$

where the exponent $\bar{\lambda}$, first introduced by Fisher and Huse in the context of non-equilibrium relaxation in spin glasses [14], is a non-trivial exponent associated with phase ordering kinetics [15]. It has recently been measured in an experiment on twisted nematic liquid crystal films [16]. The *autocorrelation* function, $A(t,t') = C(\mathbf{0}, t, t')$ is therefore a function only of the ratio L'/L, with $A(t) \sim (L'/L)^{\bar{\lambda}}$ for $L \gg L'$.

The fact that the correlation function $A(t,t')$ depends always on *both* times, and not simply on the difference, $t-t'$, is an example of *aging* (see also the lectures of Bouchaud and of Kob, this volume). In particular if one waits a time t_w ('waiting time'), the correlation function $A(t_w, t_w + t)$ depends on the ratio t/t_w, *i.e.* on the age of the system when the measurement was started. In coarsening phenomena, aging is a natural consequence of the fact that the (infinite) system never reaches equilibrium: you can measure the age of the system from the size of the domains.

In the following sections, we explore the forms of the scaling functions in more detail. For example, a linear behaviour of $f(x)$ for small scaling variable, $f(x) = 1 - ax + \cdots$, is a generic feature for scalar fields, both conserved and non-conserved. We shall see that it is a simple consequence of the existence of 'sharp' (in a sense to be clarified), well-defined domain walls in the system. A corollary that we shall demonstrate is that the structure factor scaling function $g(y)$ exhibits a power-law tail, $g(y) \sim y^{-(d+1)}$ for $y \gg 1$, a result known as 'Porod's law' [17, 18]. In Section 5 we shall show that this result, and its generalisation to more complex fields, together with the scaling hypothesis, are sufficient to determine the growth law for $L(t)$.

3.3 Domain walls

It is instructive to first look at the properties of a flat equilibrium domain wall (Figure 1). From (2) the wall profile is the solution of the equation

$$\frac{d^2\phi}{dg^2} = V'(\phi)\;, \tag{10}$$

with boundary conditions $\phi(\pm\infty) = \pm 1$, where g is a coordinate normal to the wall. We can fix the 'centre' of the wall (defined by $\phi = 0$) to be at $g = 0$ by the extra condition $\phi(0) = 0$. Integrating (10) once, and imposing the boundary conditions, gives $(d\phi/dg)^2 = 2V(\phi)$. This result can be used in (1) to give the energy per unit area of wall, *i.e.* the surface tension, as

$$\sigma = \int_{-\infty}^{\infty} dg \left(\frac{d\phi}{dg}\right)^2 = \int_{-1}^{1} d\phi\, \sqrt{2V(\phi)}\;. \tag{11}$$

Note that, for scalar fields, the two terms in (1) contribute equally to the wall energy.

The profile function $\phi(g)$ has a sigmoid form. For $g \to \pm\infty$, linearising (10) around $\phi = \pm 1$ gives

$$1 \mp \phi \sim \exp(-[V''(\pm 1)]^{1/2}|g|)\;, \qquad g \to \pm\infty\;, \tag{12}$$

i.e. the order parameter saturates exponentially fast away from the wall. It follows that the excess energy is localised in the domain walls, and that the driving force for the domain growth is the wall curvature, since the system energy can only decrease through a reduction in the total wall area. The growth mechanism is rather different, however, for conserved and nonconserved fields. We treat nonconserved fields first.

3.4 The Allen-Cahn equation

The existence of a surface tension implies a force per unit area, proportional to the mean curvature, acting at each point on the wall. The calculation is similar to that of the excess pressure inside a bubble. Consider, for example, a spherical domain of radius R, in three dimensions. If the force per unit area is F, the work done by the force in decreasing the radius by dR is $4\pi F R^2 dR$. Equating this to the decrease in surface energy, $8\pi\sigma R dR$, gives $F = 2\sigma/R$. For model A dynamics, this force will cause the walls to move, with a velocity proportional to the local curvature. If the friction constant for domain-wall motion is η, then this argument gives $\eta\, dR/dt = -2\sigma/R$. For general dimension d, the factor '2' on the right is replaced by $(d-1)$.

It is interesting to see how this result arises directly from the equation of motion (2). We consider a single spherical domain of (say) $\phi = -1$ immersed in a sea of $\phi = +1$. Exploiting the spherical symmetry, (2) reads

$$\frac{\partial \phi}{\partial t} = \frac{\partial^2 \phi}{\partial r^2} + \frac{d-1}{r}\frac{\partial \phi}{\partial r} - V'(\phi) \ . \tag{13}$$

Provided the droplet radius R is much larger than the interface width ξ (which could be defined from (12) as $\xi = [V''(1)]^{-1/2}$, say), we expect a solution of the form

$$\phi(r,t) = f\left(r - R(t)\right) \ . \tag{14}$$

Inserting this in (13) gives

$$0 = f'' + \left[\frac{d-1}{r} + \frac{dR}{dt}\right] f' - V'(f) \ . \tag{15}$$

The function $f(x)$ changes from -1 to 1 in a small region of width ξ near $x = 0$. Its derivative is, therefore, sharply peaked near $x = 0$ (*i.e.* near $r = R(t)$). Multiplying (15) by f' and integrating through the interface gives

$$0 = \frac{(d-1)}{R} + \frac{dR}{dt} \ , \tag{16}$$

where we have used $f' = 0$ far from the interface, and that $V(f)$ has the same value on both sides of the interface (in the absence of a bulk driving force, *i.e.* a magnetic field). Integrating (16) gives $R^2(t) = R^2(0) - 2(d-1)t$, *i.e.* the collapse time scales with the initial radius as $t \sim R^2(0)$. Equation (16) is identical to our previous result obtained by considering the surface tension as the driving force, provided the surface tension σ and friction constant η are equal. This we show explicitly below.

The result for general curved surfaces was derived by Allen and Cahn [19], who noted that, close to a domain wall, one can write $\nabla\phi = (\partial\phi/\partial g)_t\, \hat{\mathbf{g}}$, where $\hat{\mathbf{g}}$ is a unit vector normal to the wall (in the direction of increasing ϕ), and so $\nabla^2\phi = (\partial^2\phi/\partial g^2)_t + (\partial\phi/\partial g)_t\, \nabla\cdot\hat{\mathbf{g}}$.

Noting also the relation $(\partial\phi/\partial t)_g = -(\partial\phi/\partial g)_t\,(\partial g/\partial t)_\phi$, (2) can be recast as

$$-\left(\frac{\partial\phi}{\partial g}\right)_t\left(\frac{\partial g}{\partial t}\right)_\phi = \left(\frac{\partial\phi}{\partial g}\right)_t \nabla\cdot\hat{\mathbf{g}} + \left(\frac{\partial^2\phi}{\partial g^2}\right)_t - V'(\phi)\,. \tag{17}$$

Assuming that, for gently curving walls, the wall profile is given by the equilibrium condition (10), the final two terms in (17) cancel. Noting also that $(\partial g/\partial t)_\phi$ is just the wall velocity v (in the direction of increasing ϕ), (17) simplifies to

$$v = -\nabla\cdot\hat{\mathbf{g}} = -K\,, \tag{18}$$

the 'Allen-Cahn equation', where $K \equiv \nabla\cdot\hat{\mathbf{g}}$ is $(d-1)$ times the mean curvature. For brevity, we will call K simply the 'curvature'. An alternative derivation of (18) follows the approach used for the spherical domain, *i.e.* we multiply Equation 17 by $(\partial\phi/\partial g)_t$ and integrate (with respect to g) through the interface. This gives the same result.

Equation 18 is an important result, because it establishes that the motion of the domain walls is determined (for non-conserved fields) purely by the local curvature. In particular, the detailed shape of the potential is not important: the main role of the double-well potential $V(\phi)$ is to establish (and maintain) well-defined domain walls. (Of course, the well depths must be equal, or there would be a volume driving force.)

For a spherical domain, the curvature K is $(d-1)/R$, and (18) reduces to (16). Our explicit treatment of the spherical domain verifies the Allen-Cahn result, and, in particular, the independence from the potential of the interface dynamics.

A second feature of (18) is that the surface tension σ (which *does* depend on the potential) does not explicitly appear. How can this be, if the driving force on the walls contains a factor σ? The reason, as we have already noted, is that one also needs to consider the *friction constant* per unit area of wall, η. The equation of motion for the walls in this dissipative system is $\eta v = -\sigma K$. Consistency with (18) requires $\eta = \sigma$. In fact, η can be calculated independently, as follows. Consider a plane wall moving uniformly (under the influence of some external driving force) at speed v. The rate of energy dissipation per unit area is

$$\frac{dE}{dt} = \int_{-\infty}^{\infty} dg\,\frac{\delta F}{\delta\phi}\frac{\partial\phi}{\partial t} = -\int_{-\infty}^{\infty} dg\left(\frac{\partial\phi}{\partial t}\right)^2\,, \tag{19}$$

using (2). The wall profile has the form $\phi(g,t) = f(g - vt)$, where the profile function f will, in general, depend on v. Putting this form into (19) gives

$$\frac{dE}{dt} = -v^2\int dg\left(\frac{\partial\phi}{\partial g}\right)^2 = -\sigma v^2\,, \tag{20}$$

where the definition (11) of the surface tension σ was used in the final step, and the profile function $f(x)$ replaced by its $v = 0$ form to lowest order in v. By definition, however, the rate of energy dissipation is the product of the frictional force ηv and the velocity; therefore $dE/dt = -\eta v^2$. Comparison with (20) gives $\eta = \sigma$. We conclude that, notwithstanding some contrary suggestions in the literature, the Allen-Cahn equation is completely consistent with the idea that domain growth is driven by the surface tension of the walls.

3.5 Conserved fields

For conserved fields the interfaces cannot move independently. At late times the dominant growth mechanism is the transport of the order parameter from interfaces of high curvature to regions of low curvature by diffusion through the intervening bulk phases. To see how this works, we first linearise (3) in one of the bulk phases, with say $\phi \simeq 1$. Putting $\phi = 1 + \tilde{\phi}$ in (3), and linearising in $\tilde{\phi}$, gives

$$\frac{\partial \tilde{\phi}}{\partial t} = -\nabla^4 \tilde{\phi} + V''(1)\nabla^2 \tilde{\phi} \,. \tag{21}$$

Since the characteristic length scales are large at late times, the ∇^4 term is negligible, and (21) reduces to the diffusion equation, with diffusion constant $D = V''(1)$. The interfaces provide the boundary conditions, as we shall see. However, we can first make a further simplification. Due to the conservation law, the interfaces move little during the time it takes the diffusion field $\tilde{\phi}$ to relax. If the characteristic domain size is L, the diffusion field relaxes on a time scale $t_D \sim L^2$. We shall see below, however, that a typical interface velocity is of order $1/L^2$, so the interfaces only move a distance of order unity (*i.e.* much less than L) in the time t_D. This means that the diffusion field relaxes quickly compared to the rate at which the interfaces move, and is essentially always in equilibrium with the interfaces. The upshot is that the diffusion equation can be replaced by Laplace's equation, $\nabla^2 \tilde{\phi} = 0$, in the bulk.

To derive the boundary conditions at the interfaces, it is convenient to work, not with $\tilde{\phi}$ directly, but with the chemical potential $\mu \equiv \delta F/\delta\phi$. In terms of μ, (3) can be written as a continuity equation,

$$\frac{\partial \phi}{\partial t} = -\nabla \cdot \mathbf{j} \tag{22}$$

$$\mathbf{j} = -\nabla \mu \tag{23}$$

$$\mu = V'(\phi) - \nabla^2 \phi \,. \tag{24}$$

In the bulk, μ and $\tilde{\phi}$ are proportional to each other, because (24) can be linearised to give $\mu = V''(1)\tilde{\phi} - \nabla^2\tilde{\phi}$, and the ∇^2 term is again negligible. Therefore μ also obeys Laplace's equation,

$$\nabla^2 \mu = 0 \,, \tag{25}$$

in the bulk.

The boundary conditions are derived by analysing (24) near an interface. As in the derivation of the Allen-Cahn equation, we consider surfaces of constant ϕ near the interface and introduce a Cartesian coordinate system at each point, with a coordinate g normal to the surface (and increasing with increasing ϕ). Then (24) becomes (compare Equation 17),

$$\mu = V'(\phi) - \left(\frac{\partial \phi}{\partial g}\right)_t K - \left(\frac{\partial^2 \phi}{\partial g^2}\right)_t \tag{26}$$

near the interface, where $K = \nabla \cdot \hat{\mathbf{g}}$ is the curvature. The value of μ at the interface can be obtained (just as in our treatment of the spherical domain in Section 3.4), by multiplying through by $(\partial\phi/\partial g)_t$, which is sharply peaked at the interface, and integrating over g

through the interface. Noting that μ and K vary smoothly through the interface, this gives the completely general relation

$$\mu\Delta\phi = \Delta V - \sigma K \tag{27}$$

at the interface, where $\Delta\phi$ is the change in ϕ across the interface, and ΔV is the difference in the minima of the potential for the two bulk phases. In deriving (27), we have used $(\partial\phi/\partial g)_t \to 0$ far from the interface, and made the identification $\int dg(\partial\phi/\partial g)_t^2 = \sigma$, as in (11), with σ the surface tension. Simplifying to the case where the minima have equal depth (we shall see that the general case introduces no new physics), and taking the minima to be at $\phi = \pm 1$ as usual, gives $\Delta V = 0$ and $\Delta\phi = 2$. Then (27) becomes

$$\mu = -\sigma K/2\ . \tag{28}$$

This (or, more generally, Equation 27) is usually known as the Gibbs-Thomson boundary condition. Note that we have assumed that the order parameter takes its equilibrium value (± 1) in both bulk phases. This is appropriate to the late stages of growth in which we are primarily interested.

To summarise, (28) determines μ on the interfaces in terms of the curvature. Between the interfaces, μ satisfies the Laplace equation (25). The final step is to use (23) to determine the motion of the interfaces. An interface moves with a velocity given by the imbalance between the current flowing into and out of it:

$$v\,\Delta\phi = j_{\mathrm{out}} - j_{\mathrm{in}} = -[\partial\mu/\partial g] = -[\hat{\mathbf{g}}\cdot\nabla\mu]\ , \tag{29}$$

where v is the speed of the interface in the direction of increasing ϕ, g is the usual coordinate normal to interface, $[\cdots]$ indicates the discontinuity across the interface, and we have assumed, as usual, that $\phi \simeq \pm 1$ in the bulk phases.

To illustrate how (25), (28) and (29) are used, we consider again the case of a single spherical domain of negative phase ($\phi = -1$) in an infinite sea of positive phase ($\phi = +1$). We restrict ourselves to $d = 3$ for simplicity. The definition of μ, Equation 24, gives $\mu = 0$ at infinity. Let the domain have radius $R(t)$. The solution of (25) that obeys the boundary conditions $\mu = 0$ at infinity and (28) at $r = R$, and respects the spherical symmetry is (using $K = 2/R$ for $d = 3$) $\mu = -\sigma/r$ for $r \geq R$. Inside the domain, the $1/r$ term must be absent to avoid an unphysical singularity at $r = 0$. The solution of (25) in this region is therefore $\mu = \mathrm{const}$. The boundary condition (28) gives $\mu = -\sigma/R$.

To summarise,

$$\begin{aligned} \mu &= -\sigma/R\ , && r \leq R \\ &= -\sigma/r\ , && r \geq R\ . \end{aligned} \tag{30}$$

Using (29), with $\Delta\phi = 2$, then gives

$$\frac{dR}{dt} = v = -\frac{1}{2}\left[\frac{\partial\mu}{\partial r}\right]_{R-\epsilon}^{R+\epsilon} = -\frac{\sigma}{2R^2}\ , \tag{31}$$

and hence $R^3(t) = R^3(0) - 3\sigma t/2$. We conclude that a domain of initial radius $R(0)$ evaporates in a time proportional to $R^3(0)$. This contrasts with the $R^2(0)$ result obtained for a non-conserved order parameter. In the non-conserved case, of course, the domain simply shrinks under the curvature forces, whereas for the conserved case it evaporates by the diffusion of material to infinity.

We now briefly discuss the case where the potential minima have unequal depths. Consider first a planar interface separating the two equilibrium phases, with order parameter values ϕ_1 and ϕ_2. Since no current flows, $\mathbf{j} = -\nabla\mu = 0$ gives μ =constant. From the definition (24) of μ, and the fact that $\nabla^2\phi$ vanishes far from the interface, it follows that $\mu = V'(\phi_1) = V'(\phi_2)$. On the other hand, the Gibbs-Thomson boundary condition (27) for a flat interface ($K = 0$) gives $\mu = \Delta V/\Delta\phi$. Combining these two results gives

$$V'(\phi_1) = V'(\phi_2) = \frac{\Delta V}{\Delta \phi}\,, \tag{32}$$

leading to the common tangent construction that determines ϕ_1 and ϕ_2 as the points where the common tangent touches the potential. If one now repeats the calculation for a spherical drop, with a domain with $\phi = \phi_1$ immersed in a sea with $\phi = \phi_2$, one obtains the equation of motion for the radius, $dR/dt = -2\sigma/(\Delta\phi)^2R^2$, a simple generalisation of (31). Henceforth, we will consider only the case of degenerate minima.

3.6 Growth laws

The scaling hypothesis suggests a simple intuitive derivation of the 'growth laws' for $L(t)$, which are really just generalisations of the calculations for isolated spherical domains. For model A, we can estimate both sides of the Allen-Cahn equation (18) as follows. If there is a single characteristic scale L, then the wall velocity $v \sim dL/dt$, and the curvature $K \sim 1/L$. Equating and integrating gives $L(t) \sim t^{1/2}$ for non-conserved scalar fields.

For conserved fields (model B), the argument is slightly more subtle. We shall follow the approach of Huse [20]. From (28), the chemical potential has a typical value $\mu \sim \sigma/L$ on interfaces, and varies over a length scale of order L. The current, and therefore the interface velocity v, scale as $\nabla\mu \sim \sigma/L^2$, giving $dL/dt \sim \sigma/L^2$ and $L(t) \sim (\sigma t)^{1/3}$. A more compelling argument for this result will be given in Section 5. We note, however, that the result is supported by evidence from computer simulations [20, 21] (which usually require, however, some extrapolation into the asymptotic scaling regime) as well as a Renormalisation Group (RG) treatment [4, 5]. In the limit that one phase occupies an infinitesimal volume fraction, the original Lifshitz-Slyozov-Wagner theory [22, 23] convincingly demonstrates a $t^{1/3}$ growth.

It is interesting that these growth laws can also be obtained using naive arguments based on the results for single spherical domains [2]. For nonconserved dynamics, we know that a domain of radius R collapses in a time of order R^2. Therefore, crudely speaking, after time t there will be no domains smaller than $t^{1/2}$, so the characteristic domain size is $L(t) \sim t^{1/2}$. Of course, this is an oversimplification, but it captures the essential physics. For conserved dynamics, the same line of argument leads to $t^{1/3}$ growth. In fact, this approach can be used rather generally, for a variety of systems [24], and gives results which agree with the exact growth laws that will be derived in Section 5.

3.7 Binary liquids

The phase separation of binary liquids is a phenomenon of considerable experimental interest. Model B is inappropriate for this system, since it takes no account of the transport of the order parameter by hydrodynamic flow. Here we briefly review the modifications to model B needed to describe binary liquids.

The principal new ingredient is 'advection' of the order parameter by the fluid. The appropriate modification of (3) is

$$\frac{\partial \phi}{\partial t} + \mathbf{v} \cdot \nabla \phi = \lambda \nabla^2 \mu \,, \tag{33}$$

where $\mathbf{v}$ is the (local) fluid velocity, and we have reinstated the transport coefficient λ. The velocity obeys the Navier-Stokes equation which, with the simplification that the fluid is incompressible, reads

$$\rho \left(\frac{\partial \mathbf{v}}{\partial t} + (\mathbf{v} \cdot \nabla)\mathbf{v} \right) = \eta \nabla^2 \mathbf{v} - \nabla p - \phi \nabla \mu \,, \tag{34}$$

where p is the pressure, η the viscosity, and the density ρ is constant. The final term in (34) arises from the free energy change per unit volume $\phi\,\delta\mu$ that accompanies the transport of a fluid region with order parameter ϕ over a distance for which the change in the chemical potential is $\delta\mu$: chemical potential gradients act as a driving force on the fluid.

In the overdamped limit appropriate to most experimental systems, the left side of (34), associated with the inertia of the fluid, can be set to zero. The growth laws can then be obtained by dimensional arguments as follows. From the previous subsection we have $\mu \sim \sigma/L$, and $\nabla\mu \sim \sigma/L^2$. The pressure difference across a curved surface is of order σ/L, so the term ∇p on the right of (34) also scales as σ/L^2. Finally, we can estimate $\nabla^2 v \sim v_f/L^2$, where v_f is the typical fluid velocity. Equating the terms on the right of (34) then gives $v_f \sim \sigma/\eta$.

When the fluid velocity, $v_f \sim \sigma/\eta$, is much smaller than the interface velocity, $v_{\rm int} \sim \lambda\sigma/L^2$, derived from the evaporation-condensation mechanism, *i.e.* when $L \ll (\lambda\eta)^{1/2}$, the fluid motion is irrelevant. Using $v_{\rm int} \sim dL/dt \sim \lambda\sigma/L^2$ gives our previous result, $L \sim (\lambda\sigma t)^{1/3}$ in the 'diffusive regime'.

In the opposite regime, $L \gg (\lambda\eta)^{1/2}$, $v_{\rm int}$ and v_f are of the same order, giving $L(t) \sim \sigma t/\eta$, a result first derived by Siggia [25]. This result has been confirmed by experiments [26] and by numerical simulations [27, 28, 29]. Because the inertial terms are negligible compared to the viscous force here, we will call this the 'viscous hydrodynamic' (or just 'viscous') regime.

Under what conditions is it correct to ignore the 'inertial' terms on the left-hand side of (34)? Using dimensional arguments again, we see that these terms are of order $\rho L/t^2$. Comparing this to the driving term $\phi\nabla\mu \sim \sigma/L^2$ on the right, (the viscous term $\eta\nabla^2 v$ is of the same order in the viscous regime), and using the result derived above, $t \sim \eta L/\sigma$, for this regime, shows that the inertial terms are negligible when $L \ll \eta^2/\sigma\rho$. At sufficiently late times, when this inequality is violated, the inertial terms will therefore be important. In this 'inertial' regime, $L(t)$ is determined by equating the inertial terms, which scale as $\rho L/t^2$, to the driving term $\phi\nabla\mu$, which scales as σ/L^2 (and the viscous term is negligible) to give $L \sim (\sigma t^2/\rho)^{1/3}$. The $t^{2/3}$ growth in the inertial regime was first predicted by Furukawa [30].

To summarise, there are in principle three growth regimes for phase separation in binary liquids, after a deep quench, with the growth laws

$$
\begin{aligned}
L(t) &\sim (\lambda\sigma t)^{1/3} & L &\ll (\lambda\eta)^{1/2} & &(\text{`diffusive'}), & (35)\\
&\sim \sigma t/\eta & (\lambda\eta)^{1/2} \ll L &\ll \eta^2/\rho\sigma & &(\text{`viscous hydrodynamic'}), & (36)\\
&\sim (\sigma t^2/\rho)^{1/3} & L &\gg \eta^2/\rho\sigma & &(\text{`inertial hydrodynamic'}). & (37)
\end{aligned}
$$

These results basically follow from dimensional analysis. The 'inertial hydrodynamic' regime has not, to my knowledge, been observed experimentally. However, a $t^{2/3}$ regime has been observed at late times in simulations of two-dimensional [29, 31] and three-dimensional [32] binary liquids.

Siggia [25] has discussed the physical origin of the linear growth in the 'viscous hydrodynamic' regime. He argues that the essential mechanism is the hydrodynamic transport of fluid along the interface driven by the surface tension. This mechanism, however, can only operate if both phases are continuous. If, by contrast, the minority phase consists of independent droplets (which occurs for volume fractions less than about 15%), this mechanism tends to make the droplets spherical but does not lead to any coarsening. In the absence of thermal fluctuations, therefore, the Lifshitz-Slyozov evaporation-condensation mechanism determines the growth even beyond the nominal crossover length given above. Thermal fluctuations, however, facilitate a second coarsening mechanism, namely droplet coalescence driven by Brownian motion of the droplets. Again, Siggia has given the essential argument. The mobility μ of a droplet of size L is of order $1/\eta L$, so the diffusion constant is given by the Einstein relation as $D = \mu k_B T \sim k_B T/\eta L$, where k_B is Boltzmann's constant. The time for the droplet to diffuse a distance of order L (and coalesce with another droplet) is $t \sim L^2/D \sim \eta L^3/k_B T$, which gives $L \sim (k_B T t/\eta)^{1/3}$. The relative contribution of these two mechanisms (evaporation-condensation and droplet coalescence) to the coarsening rate is has been discussed by Siggia [25].

In recent work, Grant and Elder [33] have considered the possible role of turbulence in the late stages of phase separation for binary liquids. Flow becomes turbulent for large values of the dimensionless Reynolds number, $Re = \rho l_c v_c/\eta$, where l_c and v_c are characteristic length and velocity scales respectively. Here v_c is just the fluid velocity, v_f, but what is the appropriate length scale l_c for the phase separating system? Suppose we tentatively identify l_c with the domain scale $L(t)$. Although $L(t)$ grows with time in the diffusive and viscous regimes, while $v_f \sim \sigma/\eta$ is constant, turbulence does not occur in these regimes since Re becomes of order unity only at the crossover between viscous and inertial regimes. To see this, we put $v_c = \sigma/\eta$ and $l_c = \eta^2/\rho\sigma$, the crossover value from (37). This gives $Re \approx 1$ at the crossover to the inertial regime. In the inertial regime proper, however, using $l_c \sim L(t)$, and $v_c \sim dL/dt$ gives $Re \sim (\rho\sigma^2 t/\eta^3)^{1/3}$, which grows without limit. Grant and Elder argue that the inertial regime cannot be an asymptotic scaling regime since phase separation would be arrested by turbulent remixing of the fluids. Rather, an asymptotic coarsening state would require $L(t) \sim t^{1/2}$ so that Re saturates at late times. These arguments are, however, suggestive rather than rigorous. Possible escape mechanisms from the Grant-Elder scenario, which would allow asymptotic $t^{2/3}$ coarsening, have been suggested by Kendon et al [32, 34], and no indication of departures from $t^{2/3}$ scaling were found in [32] for Reynolds numbers up to 350.

4 Topological defects and Porod tails

4.1 Topological defects

The domain walls discussed in the previous section are the simplest form of 'topological defect', and occur in systems described by scalar fields [35]. They are surfaces, on which the order parameter vanishes, separating domains of the two equilibrium phases. A domain wall is topologically stable: local changes in the order parameter can move the wall, but cannot destroy it. For an isolated flat wall, the wall profile function is given by the solution of (10), with the appropriate boundary conditions, as discussed in Section 3.3. For the curved walls present in the phase ordering process, this will still be an approximate solution locally, provided the typical radius of curvature L is large compared to the intrinsic width (or 'core size'), ξ, of the walls. The same condition, $L \gg \xi$, ensures that typical wall separations are large compared to their width.

Let us now generalise the discussion to vector fields. The '$O(n)$ model' is described by an n-component vector field for which we use the notation $\vec{\phi}(\mathbf{x}, t)$, with a free energy functional $F[\vec{\phi}]$ that is invariant under global rotations of $\vec{\phi}$. A suitable generalisation of (1) is

$$F[\vec{\phi}] = \int d^dx \left(\frac{1}{2} (\nabla\vec{\phi})^2 + V(\vec{\phi}) \right) , \tag{38}$$

where $(\nabla\vec{\phi})^2$ means $\sum_{i=1}^{d} \sum_{a=1}^{n} (\partial_i \phi^a)^2$ (*i.e.* a scalar product over both spatial and 'internal' coordinates), and $V(\vec{\phi})$ is 'mexican hat' (or 'wine bottle') potential, such as $(1 - |\vec{\phi}|^2)^2$ (Figure 1). It is clear that $F[\vec{\phi}]$ is invariant under global rotations of $\vec{\phi}$ (a continuous symmetry), rather than just the inversion symmetry ($\phi \to -\phi$, a discrete symmetry) of the scalar theory. We will adopt the convention that V has its minimum for $|\vec{\phi}|^2 = 1$.

For non-conserved fields, the simplest dynamics (model A) is a straightforward generalisation of (2), namely

$$\frac{\partial\vec{\phi}}{\partial t} = \nabla^2\vec{\phi} - \frac{dV}{d\vec{\phi}} . \tag{39}$$

For conserved fields (model B), we simply add another $(-\nabla^2)$ in front of the right-hand side.

Stable topological defects for vector fields can be generated, in analogy to the scalar case, by seeking stationary solutions of (39) with appropriate boundary conditions. For the $O(n)$ theory in d-dimensional space, the requirement that all n components of $\vec{\phi}$ vanish at the defect core defines a surface of dimension $d - n$ (*e.g.* a domain wall is a surface of dimension $d - 1$: the scalar theory corresponds to $n = 1$). The existence of such defects therefore requires $n \le d$. For $n = 2$ these defects are points ('vortices') for $d = 2$ or lines ('strings', or 'vortex lines') for $d = 3$. For $n = 3$, $d = 3$ they are points ('hedgehogs', or 'monopoles'); see Figure 2. Note that the forms shown are radially symmetric with respect to the defect core: any configuration obtained by a global rotation is also acceptable. For $n < d$, the field $\vec{\phi}$ only varies in the n dimensions 'orthogonal' to the defect core, and is uniform in the remaining $d - n$ dimensions 'parallel' to the core.

For $n < d$, the defects are spatially extended. Coarsening occurs by a 'straightening out' (or reduction in typical radius of curvature) as sharp features are removed, and by the shrinking and disappearance of small domain bubbles or vortex loops. These processes

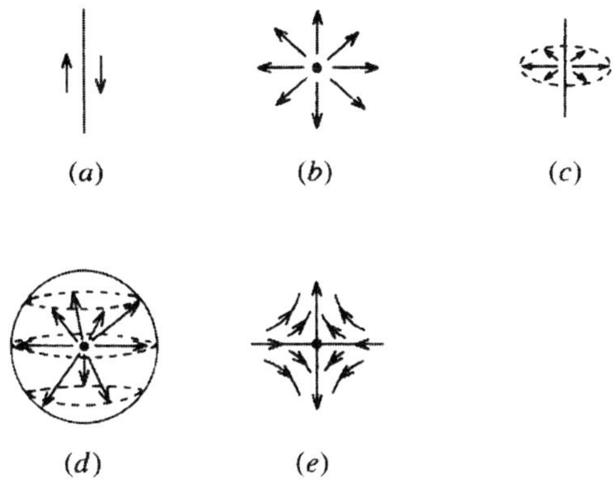

Figure 2. *Topological defects in the O(n) model: (a) domain wall; (b) vortex; (c) string; (d) monopole or hedgehog; (e) antivortex.*

reduce the total area of domain walls, or length of vortex line, in the system. For point defects ($n = d$), coarsening occurs by the mutual annihilation of defect-antidefect pairs. A defect and antidefect have different 'topological charges': *e.g.* for a vortex and antivortex the field rotates by 2π or -2π respectively on encircling the vortex.

For radially symmetric defects, the field $\vec{\phi}$ has the form $\vec{\phi}(\mathbf{r}) = \hat{\mathbf{r}}\, f(r)$, where $\hat{\mathbf{r}}$ is a unit vector in the radial direction, and $f(r)$ is the profile function. Inserting this form into (39), with the time derivative set to zero, gives the equation

$$\frac{d^2 f}{dr^2} + \frac{(n-1)}{r}\frac{df}{dr} - \frac{(n-1)}{r^2} f - V'(f) = 0\ , \tag{40}$$

with boundary conditions $f(0) = 0$, $f(\infty) = 1$. Of special interest is the approach to saturation at large r. Putting $f(r) = 1 - \epsilon(r)$ in (40), and expanding to first order in ϵ, yields

$$\epsilon(r) \simeq \frac{(n-1)}{V''(1)}\frac{1}{r^2}\ , \qquad r \to \infty\ . \tag{41}$$

This should be contrasted with the exponential approach to saturation (12) for scalar fields. A convenient definition of the 'core size' ξ is through $f \simeq 1 - \xi^2/r^2$ for large r. This gives $\xi = [(n-1)/V''(1)]^{1/2}$ for $n > 1$.

4.2 Porod tails

The presence of topological defects, seeded by the initial conditions, in the system undergoing phase ordering has an important effect on the 'short-distance' form of the pair correlation function $C(\mathbf{r}, t)$, and therefore on the 'large-wavenumber' form of the structure factor $S(\mathbf{k}, t)$. To see why this is so, we note that, according to the scaling hypothesis, we would expect a typical field gradient to be of order $|\nabla\vec{\phi}| \sim 1/L$. At a distance r from a defect core, however, with $\xi \ll r \ll L$, the field gradient is much larger, of order $1/r$ (for a vector field), because $\vec{\phi} = \hat{\mathbf{r}}$ implies $(\nabla\vec{\phi})^2 = (n-1)/r^2$. Note that we require $r \gg \xi$ for the field to be saturated, and $r \ll L$ for the defect field to be largely unaffected by other defects (which are typically a distance L away). This gives a meaning to 'short' distances ($\xi \ll r \ll L$), and 'large wavenumbers' ($L^{-1} \ll k \ll \xi^{-1}$). The large field gradients near defects leads to a non-analytic behaviour at $x = 0$ of the scaling function $f(x)$ for pair correlations.

We start by considering scalar fields. Consider two points $\mathbf{x}$ and $\mathbf{x}+\mathbf{r}$, with $\xi \ll r \ll L$. The product $\phi(\mathbf{x})\,\phi(\mathbf{x}+\mathbf{r})$ will be -1 if a wall passes between them, and $+1$ if there is no wall. Since $r \ll L$, the probability to find more than one wall can be neglected. The calculation amounts to finding the probability that a randomly placed rod of length r cuts a domain wall. This probability is of order r/L, so we estimate

$$\begin{aligned} C(\mathbf{r},t) &\simeq (-1)\times(r/L) + (+1)\times(1-r/L) \\ &= 1 - 2r/L\,, \qquad r \ll L\,. \end{aligned} \tag{42}$$

The factor 2 in this result should not be taken seriously.

The important result is that (42) is non-analytic in $\mathbf{r}$ at $\mathbf{r} = 0$, since it is linear in $r \equiv |\mathbf{r}|$. Technically, of course, this form breaks down inside the core region, when $r < \xi$. We are interested, however, in the scaling limit defined by $r \gg \xi$, $L \gg \xi$, with $x = r/L$ arbitrary. The nonanalyticity is really in the scaling variable x.

The nonanalytic form (42) implies a power-law tail in the structure factor, which can be obtained from (42) by simple power-counting:

$$S(\mathbf{k},t) \sim \frac{1}{L\,k^{d+1}}\,, \qquad kL \gg 1\,, \tag{43}$$

a result known universally as 'Porod's law'. It was first written down in the general context of scattering from two-phase media [17]. Again, one requires $k\xi \ll 1$ for the scaling regime. Although the k-dependence of (43) is what is usually referred to as Porod's law, the L-dependence is equally interesting. The factor $1/L$ is simply (up to constants) the total area of domain wall per unit volume, a fact appreciated by Porod, who proposed structure factor measurements as a technique to determine the area of interface in a two-phase medium [17]. On reflection, the factor $1/L$ is not so surprising. For $kL \gg 1$, the scattering function is probing structure on scales much shorter than the typical interwall spacing or radius of curvature. In this regime we would expect the structure factor to scale as the total wall area, since each element of wall with linear dimension large compared to $1/k$ contributes essentially independently to the structure factor.

This observation provides the clue to how to generalise (43) to vector (and other) fields [36, 37]. The idea is that, for $kL \gg 1$, the structure factor should scale as the total volume of defect core. Since the dimension of the defects is $d - n$, the amount of defect per unit volume scales as L^{-n}. Extracting this factor from the general scaling form (7) yields

$$S(\mathbf{k},t) \sim \frac{1}{L^n\,k^{d+n}}\,, \qquad kL \gg 1\,, \tag{44}$$

for the $O(n)$ theory, a 'generalised Porod's law'.

Equation 44 was first derived from approximate treatments of the equation of motion (39) for nonconserved fields [38, 39, 40, 41]. In these derivations, however, the key role of topological defects was far from transparent. The above heuristic derivation suggests that the result is in fact very general (*e.g.* it should hold equally well for conserved fields), with extensions beyond simple $O(n)$ models. Appropriate techniques, which also enable determination of the amplitude of the tail, were developed by Bray and Humayun [37].

5 Growth laws revisited

The Porod tails discussed earlier, together with the scaling hypothesis, provide a basis for deriving exact growth laws for all phase-ordering systems with purely dissipative dynamics.

Although the growth laws for both nonconserved and conserved scalar systems, and conserved fields in general, have been derived by a number of methods, there has up until now been no simple, general technique for obtaining $L(t)$. In particular, the growth laws for non-conserved vector fields have, until recently, been somewhat problematical. Here we describe a very general approach, recently developed by Bray and Rutenberg (BR) [42, 24], to obtain $L(t)$ consistently by comparing the global rate of energy change to the energy dissipation from the local evolution of the order parameter. This method allows the explicit derivation of growth laws for $O(n)$ models, but the results can be also be applied to other systems with similar defect structures, such as liquid crystals.

The BR approach is based on the dissipation of energy that occurs as the system relaxes towards its ground state. The energy dissipation is evaluated by considering the motion of topological defects, when they exist. The defect contribution either dominates the dissipation or gives a contribution that scales with time in the same way as the total dissipation. The global rate of energy change, computed from the time derivative of the total energy, is equated to the energy dissipation from the local evolution of the order parameter. For systems with a single characteristic scale $L(t)$, this approach self-consistently determines the time-dependence of $L(t)$.

5.1 A useful identity

We begin by writing down the equation of motion for the Fourier components $\vec{\phi}_{\mathbf{k}}$:

$$\partial_t \vec{\phi}_{\mathbf{k}} = -k^{\mu}\,(\partial F/\partial \vec{\phi}_{-\mathbf{k}}), \tag{45}$$

The conventional non-conserved (model A) and conserved (model B) cases are $\mu = 0$ and $\mu = 2$, respectively. (Elsewhere in this article, the symbol μ has been used for the chemical potential: the meaning should be clear from the context).

Integrating the rate of energy dissipation from each Fourier mode, and then using the equation of motion (45), we find

$$\frac{d\epsilon}{dt} = \int_{\mathbf{k}} \left\langle \left(\frac{\partial F}{\partial \vec{\phi}_{\mathbf{k}}}\right) \cdot \partial_t \vec{\phi}_{\mathbf{k}} \right\rangle = -\int_{\mathbf{k}} k^{-\mu} \left\langle \partial_t \vec{\phi}_{\mathbf{k}} \cdot \partial_t \vec{\phi}_{-\mathbf{k}} \right\rangle , \tag{46}$$

where $\epsilon = \langle F \rangle / V$ is the mean energy density, and $\int_{\mathbf{k}}$ is the integral $\int d^d k/(2\pi)^d$. We will relate the scaling behaviour of both sides of (46) to that of appropriate integrals over the structure factor, $S(\mathbf{k}, t)$, and its two-time generalisation. Either the integrals converge, and the dependence on the scale $L(t)$ can be extracted using the scaling form (7) (or its two-time generalisation (8)), or the integrals diverge at high k (in the 'ultraviolet', UV) and have to be cut off at $k_{\max} \sim 1/\xi$, corresponding to a dominant contribution from the core scale. It is just this small-scale structure that is responsible for the generalised Porod law (44) for the structure factor, and the time-dependence of any integrals controlled by the core scale can be extracted from a knowledge of the defect structure.

The energy integral. To see how this works, we first calculate the scaling behaviour of the energy density, ϵ, which is captured by that of the gradient term in (1):

$$\epsilon \;\sim\; \left\langle (\nabla\vec{\phi})^2 \right\rangle \;=\; \int_{\mathbf{k}} k^2\, L^d\, g(kL)\ , \tag{47}$$

where we have used the scaling form (7) for the structure factor. For $n > 2$ the integral is UV convergent, and a change of variables yields $\epsilon \sim L^{-2}$. For $n \le 2$, when the integral is UV divergent, we use Porod's law (44) and impose a cutoff at $k \sim 1/\xi$, to obtain [39]

$$\begin{aligned} \epsilon \;&\sim\; L^{-n}\,\xi^{n-2}\ , && n<2\ ,\\ &\sim\; L^{-2}\ln(L/\xi)\ , && n=2\ ,\\ &\sim\; L^{-2}\ , && n>2\ . \end{aligned} \tag{48}$$

We see that the energy is dominated by the defect core density, $\rho_{\rm def} \sim L^{-n}$, for $n < 2$, by the defect field at all length scales between ξ and L for $n = 2$, and by variations of the order parameter at scale $L(t)$ for $n > 2$.

The dissipation integral. We now attempt to evaluate the right hand side of (46) in a similar way. Using the scaling hypothesis for the two-time function,

$$\left\langle \vec{\phi}_{\mathbf{k}}(t)\cdot\vec{\phi}_{-\mathbf{k}}(t') \right\rangle = k^{-d} g(kL(t), kL(t'))\ , \tag{49}$$

which is the spatial Fourier transform of (8), we find

$$\left\langle \partial_t\vec{\phi}_{\mathbf{k}}\cdot\partial_t\vec{\phi}_{-\mathbf{k}} \right\rangle \;=\; \left.\frac{\partial^2}{\partial t\,\partial t'}\right|_{t=t'} \left\langle \vec{\phi}_{\mathbf{k}}(t)\cdot\vec{\phi}_{-\mathbf{k}}(t') \right\rangle \;=\; \dot{L}^2 L^{d-2} h(kL)\ , \tag{50}$$

where $\dot{L} \equiv dL/dt$.

When the k-space integral on the right of (46) is UV convergent we obtain, using (50), $d\epsilon/dt \sim -\dot{L}^2\, L^{\mu-2}$. If, however, the integral is UV divergent, it will be dominated by the behaviour of the integrand near the upper limit, so we need to know the form of the scaling function h in (50) for $kL \gg 1$. It turns out that, in general, the large-kL form is quite complicated, with many different cases to consider [24]. However, we only need the result for those cases where the dissipation integral requires a UV cut-off, otherwise simple power counting is sufficient. For those cases, one additional assumption, which can be verified *a posteriori*, yields a simple and rather general result (Equation 55 below).

5.2 Evaluating the dissipation integral

To see what difficulties arise, and how to circumvent them, it is instructive to consider a scalar field. We want to calculate $\langle \partial_t\phi_{\mathbf{k}}\, \partial_t\phi_{-\mathbf{k}}\rangle$ in the limit $kL \gg 1$. It is clear that $\partial_t\phi$ is appreciably different from zero only near interfaces. In fact, since $d\phi/dt = 0$ in a frame comoving with the interface, we have, near an interface, $\partial_t\phi = -\mathbf{v}\cdot\nabla\phi$, where $\mathbf{v}$ is the interface velocity. In real space, therefore (denoting $\mathbf{r}_1$, $\mathbf{r}_2$ by 1, 2)

$$\langle \partial_t\phi(1)\,\partial_t\phi(2)\rangle = \langle \mathbf{v}(1)\cdot\nabla\phi(1)\,\mathbf{v}(2)\cdot\nabla\phi(2)\rangle \ . \tag{51}$$

The large kL behaviour in Fourier space is obtained from the short-distance ($r \ll L$) behaviour in real space. For $r \ll L$, the points '1' and '2' must be close to the *same* interface. For a typical interface, with radius of curvature of order $1/L$, the speed v is slowly varying along the interface. Furthermore, the interface may be regarded as 'flat' for the calculation of the short-distance correlation, just as in the derivation of Porod's law. It follows that the averages over the interface velocity and position can be carried out independently, giving

$$\langle \partial_t\phi(1)\,\partial_t\phi(2)\rangle = \frac{\langle v^2\rangle}{d}\langle \nabla\phi(1)\cdot\nabla\phi(2)\rangle \ . \tag{52}$$

Fourier transforming this result gives

$$\langle \partial_t\phi_{\mathbf{k}}\,\partial_t\phi_{-\mathbf{k}}\rangle \ = \ \frac{\langle v^2\rangle}{d}\,k^2\,S(\mathbf{k},t) \ \sim \ \frac{\langle v^2\rangle}{Lk^{d-1}} \ , \qquad kL \gg 1 \ , \tag{53}$$

where the Porod result (43) was used in the final step. We will see that Equation 53 requires a careful interpretation.

The next step is to evaluate $\langle v^2\rangle$. Since the characteristic interface velocity is $\dot{L}$, we expect $\langle v^2\rangle \sim \dot{L}^2$. This assumes, however, that the average is dominated by 'typical' values. This, as we shall see, is the key question. To make further progress we introduce the additional assumption, which can be checked *a posteriori*, that the dissipation is dominated by the motion of defect structures of 'characteristic scale' $L(t)$. By the 'characteristic scale' we mean the typical radius of curvature for extended defects ($n < d$), or the typical defect separation for point defects ($n = d$). That is, we are assuming that the dissipation is dominated by the motion of typical defect structures, and not by the disappearance of small domain bubbles, small vortex loops, or by the annihilation of defect-antidefect pairs. If the latter were true, the dissipation would be dominated by structure at the core scale, and the arguments given below would fail. Reference [1] contains a careful discussion of this point.

If the dissipation integral (46) is UV convergent, it can be evaluated by power counting, so the large $kL(t)$ form of (50) is not required. For the required cases where the final integral is UV divergent, the large-$kL(t)$ limit of (50) can be extracted from the physical/geometrical arguments used to obtain the generalised Porod law (44). According to our assumption, we can treat the defects as locally flat (or well separated, for point defects) for $kL \gg 1$. From (50), we are interested in the behaviour of the two-time structure factor, $S(\mathbf{k},t,t') \equiv \left\langle \vec{\phi}_{\mathbf{k}}(t)\cdot\vec{\phi}_{-\mathbf{k}}(t')\right\rangle$, in the limit that the two times are close together. In the limit $kL \gg 1$, this will be proportional to the total density L^{-n} of defect core. Introducing $L = (L(t)+L(t'))/2$ and $\Delta = (L(t)-L(t'))/2$, we obtain

$$\left\langle \vec{\phi}_{\mathbf{k}}(t)\cdot\vec{\phi}_{-\mathbf{k}}(t')\right\rangle \sim \frac{1}{L^n\,k^{d+n}}\,a(k\Delta) \ , \qquad kL \gg 1 \ , \tag{54}$$

where consistency with Porod's law for $t = t'$ requires $a(0) = \text{const}$. Using this in (50) gives

$$\left\langle \partial_t \vec{\phi}_{\mathbf{k}} \cdot \partial_t \vec{\phi}_{-\mathbf{k}} \right\rangle \sim \frac{\dot{L}^2}{L^n\, k^{d+n-2}} \,, \qquad kL \gg 1 \,. \tag{55}$$

This reduces to (53) for $n = 1$ (with $\langle v^2 \rangle \sim \dot{L}^2$). It should be stressed that we are *not* claiming that (55) is a general result, only that it is valid when we need it, *i.e.* when the dissipation integral (46) requires a UV cut-off. There are three possibilities: (i) The integral is UV convergent, its dependence on $L(t)$ can be extracted by a change of variable, and the large-kL behaviour of $\left\langle \partial_t \vec{\phi}_{\mathbf{k}} \cdot \partial_t \vec{\phi}_{-\mathbf{k}} \right\rangle$ is not required. (ii) The integral is UV divergent, but the dissipation is still dominated by structures of scale $L(t)$. Then we can use (55). (iii) The dissipation has significant contributions from structures with local curvature (or spacing) of the order of the core scale. Then one cannot treat the contributions from different defect core elements as independent, (55) no longer holds, and the present approach is not useful. For the moment we will proceed on the assumption that (i) or (ii) obtain. We will show that these possibilities cover nearly all cases. Examples of when (iii) holds will also be given. These include a physically interesting case, $d = n = 2$.

5.3 Results

Putting (55) into the dissipation integral (46) shows that the integral is UV convergent for $kL \gg 1$ when $n + \mu > 2$. Otherwise the integral is dominated by k near the upper cut-off $1/\xi$. This gives

$$\begin{aligned} \int_{\mathbf{k}} k^{-\mu} \left\langle \partial_t \vec{\phi}_{\mathbf{k}} \cdot \partial_t \vec{\phi}_{-\mathbf{k}} \right\rangle &\sim \dot{L}^2\, L^{-n}\, \xi^{n+\mu-2} \,, && n+\mu < 2 \,, \\ &\sim \dot{L}^2\, L^{-n}\, \ln(L/\xi) \,, && n+\mu = 2 \,, \\ &\sim \dot{L}^2\, L^{\mu-2} \,, && n+\mu > 2 \,. \end{aligned} \tag{56}$$

The final step is to equate the dissipation rate (56) to the time derivative of the energy density (48), as required by (46), and solve for $L(t)$. For systems with purely short-ranged interactions, the results, as a function of n and μ, are summarised in Table 1. Note that conservation of the order parameter (which applies in a global sense for any $\mu > 0$) is irrelevant to the growth law for $\mu < 2-n$, where n is treated here as a continuous variable. At the marginal values, logarithmic factors are introduced. The growth laws obtained are independent of the spatial dimension d of the system.

$L(t)$	$n < 2$	$n = 2$	$n > 2$
$n + \mu < 2$	$t^{1/2}$	—	—
$n + \mu = 2$	$(t/\ln t)^{1/2}$	$t^{1/2}$	—
$n + \mu > 2$	$t^{1/(n+\mu)}$	$(t\ln t)^{1/(2+\mu)}$	$t^{1/(2+\mu)}$

Table 1. *The behaviour of $L(t)$ as a function of n and μ.*

For non-conserved fields ($\mu = 0$), we find $L \sim t^{1/2}$ for all systems (with $d > n$ or $n > 2$). Leading corrections in the $n = 2$ case are interesting: the $\ln L$ factors in (48) and (56) will in general have different effective cutoffs, of order the core size ξ. This leads to a logarithmic correction to scaling, $L \sim t^{1/2}(1 + O(1/\ln t))$, and may account for the smaller exponent (~ 0.45) seen in simulations of $O(2)$ systems [43, 44, 45]. Note that for nonconserved scalar fields, the energy (48) and the dissipation (56) have the *same* dependence on the core size ξ (*i.e.* both contain a factor ξ^{-1}), so this dependence cancels from $L(t)$. The fact that the correct $t^{1/2}$ growth is obtained from naive power counting on the linear terms in the equation of motion should therefore be regarded as fortuitous. For example, in systems with long-range interactions, this 'cancellation of errors' no longer occurs, and naive power counting gives an incorrect result for nonconserved scalar fields [46, 42].

For conserved fields ($\mu > 0$) our results agree with an earlier RG analysis [4, 5], with additional logarithmic factors for the marginal cases $n = 2$ and $n + \mu = 2$. Note that the conservation law is only relevant for $n + \mu \geq 2$. Therefore for vector fields ($n \geq 2$), any $\mu > 0$ is sufficient to change the growth law, while for scalar fields ($n = 1$) the conservation law is irrelevant for $\mu < 1$, in agreement with the RG analysis [5] and earlier work of Onuki [47].

Siegert and Rao [48] have performed extensive simulations for $n = 2$, $d = 3$ and $\mu = 2$. In their original paper they fitted $L(t)$ to a simple power, and obtained a growth exponent slightly larger than 1/4. Subsequently, however, Siegert has shown that a very much better fit is obtained when the predicted logarithmic correction is included [49], and this has been confirmed by subsequent work [50].

5.4 Exceptional cases: $n = d \leq 2$

In what cases is our key assumption, that dissipation is dominated by the motion of defect structures of characteristic scale L, correct? Certainly for any $n > 2$, the energy density (48) itself, and hence dissipation, is dominated by variations at scale $L(t)$. Therefore, we limit the discussion to the case $n \leq 2$.

For $n \leq 2$, the energy density is proportional to the defect core volume (with an extra factor $\ln(L/\xi)$ for $n = 2$, see (48)), but we will show that, in general, dissipation is still dominated by defect structures with length scales of order L. To see this, we investigate the contribution to the energy dissipation from small-scale structures (*e.g.* small domains, vortex loops, or defect-antidefect pairs):

$$\begin{aligned} \frac{d\epsilon}{dt} &= \partial_t \int_\xi^\infty dl\, n(l,t)\, \epsilon(l) \\ &= -\int_\xi^\infty dl\, \partial_l j(l,t)\, \epsilon(l) \\ &= j(\xi)\epsilon(\xi) + \int_\xi^\infty dl\, j(l)\, \frac{\partial \epsilon}{\partial l}\,, \end{aligned} \tag{57}$$

where $n(l,t)$ is the number density of defect features of scale l, $\epsilon(l) \sim l^{d-n}$ is the energy of a defect feature (with an extra $\ln(l/\xi)$ factor for $n = 2$), and $j(l,t)$ is the number flux of defect features. We have used the continuity equation, $\partial_t n + \partial_l j = 0$ to obtain the second line of (57), and the t-dependence has been suppressed in the final line. The total number of defect features, N, scales as $N \sim L^{-d}$, and so N does not change significantly over times smaller than $\dot{L}/L$. Since defects only vanish at the core scale, we have $\dot{N} = j(\xi)$.

It follows $j(l)$ has a finite, non-zero, short-distance limit of order $\dot{N} \sim -\dot{L}/L^{d+1}$. We can use this to examine the convergence (at short-distance) of the final integral in (57).

For $d > n$, the integral in (57) is well-behaved at small l, because $\epsilon(l) \sim l^{d-n}$ ($\times \ln(l/\xi)$ for $n = 2$), and the integral dominates the $j(\xi)\epsilon(\xi)$ term. The integral can be estimated by setting $j(l) \simeq j(\xi)$ and introducing a large-distance cut-off at $l \sim L$. This gives $d\epsilon/dt \sim j(\xi)L^{d-n} \sim -\dot{L}/L^{n+1}$ ($\times \ln(L/\xi)$ for $n = 2$). This is just what one gets from differentiating (48) (for the cases $n \leq 2$ considered here), verifying the consistency of the calculation.

For $d = n$ and $n < 2$, however, $\epsilon(l) \sim$ constant, since the (point) defects only interact weakly through the tails of the defect profile. (The one physical example is the $d = 1$ scalar system). The leading contribution to the energy of a defect pair is just the core energy of the individual defects, and dissipation is dominated by the $j(\xi)\epsilon(\xi)$ term in (57), which describes defect pairs annihilating. Since the dissipation occurs at separations $l \sim \xi \ll L$, the derivation of (55) no longer holds. In fact since, the energy of a defect pair depends only weakly on the separation for $l \gg L$, the system will be disordered, with an equilibrium density of defects at any non-zero temperature. At $T = 0$, we expect slow growth that depends on the details of the potential $V(\phi)$ [51]. These cases, including the $d = 1$ scalar system, are at their lower critical dimension, and are beyond the scope of the simplified approach presented here (see [51] for a fuller discussion).

The 2d planar system ($n = d = 2$) is a special case. The logarithm in the energy of a vortex pair, $\epsilon \sim \ln(l/\xi)$, leads to a logarithmically divergent integral in (57), *i.e.* vortex pairs with separations between ξ and L contribute significantly to the energy dissipation. In this case Equation 55, which depends on the $kL \gg 1$ limit being a single defect property, is again questionable. As a result the present method cannot address this case. Indeed, the contributions to the dissipation from all length scales suggest a possible breakdown of scaling.

5.5 Other systems

The strength of this approach is that it can be applied to systems with more complicated order parameters than n-component vectors, provided they have purely dissipative dynamics. Then an equation of the form (46) can be written down. The details of the energy functional (1) are unimportant. The important ingredients are the existence of an 'elastic energy', associated with spatial gradients of the order parameter, the conservation law (if any), characterised by μ (see Eq. 45), and the defect structure if any. The derivation is independent of the initial conditions, and so, *e.g.* applies equally to critical and off-critical quenches *as long as the system scales at late times.* We simply choose a Porod's law (44) to represent the dominant defect type, which is the one responsible for the asymptotic tails of the structure factor scaling function, *i.e.* the one with the smallest 'n'. When the energy density is dominated by defects, *i.e.* when the energy integral (47) is UV divergent, the relation (47) between the energy density and the structure factor, shows that the 'dominant' defects will also be the ones which dominate the energy density. As examples, we consider nematic liquid crystals and Potts models.

In bulk nematic liquid crystals, the 'dominant' defects (in the above sense) are strings, giving a Porod tail of (44) with $n = 2$, which with no conservation law implies $L \sim t^{1/2}$, consistent with experiments [52, 53] and simulations [54].

The q-state Potts model has q equivalent equilibrium phases giving rise to $q(q-1)/2$ different types of domain wall. These can be indexed $\alpha\beta$, where α, $\beta = 1, \ldots q$, are the phases separated by the wall. Three domain walls of type $\alpha\beta$, $\alpha\gamma$ and $\beta\gamma$ can meet at a point ($d = 2$) or line ($d = 3$), which represents a new type of defect. It is clear, however, that the Porod tail and energy density are dominated by the walls, so that the Potts model behaves as an $n = 1$ system. As a result, $L(t) \sim t^{1/2}$ and $t^{1/3}$ for nonconserved and conserved order parameter respectively. Recent numerical results [55, 56] support these predictions, after initial suggestions that the growth was slower.

It should be emphasised that the classification of nematic liquid crystals and Potts models as 'n = 2-like' and 'n = 1-like' respectively, pertain only to the Porod tails and the growth laws. As far as scaling functions (*e.g.* for pair correlations) are concerned, these systems belong to their own universality classes. Similarly, for off-critical quenches of conserved systems, the growth law is independent of the volume fractions of the phases, but the scaling functions are not.

6 Spinodal decomposition under shear flow

The dynamics of phase separation under shear has attracted considerable theoretical [57], experimental [58, 59, 60] and simulational [61] attention in recent years. In the absence of shear, the dynamics of phase separation is, as we have seen, quite well understood. Domains of the two equilibrium phases are formed, and coarsen with time in a manner well-described by the dynamical scaling phenomenology with a single growing length scale $L(t)$ which generally grows as a power law in time, $L(t) \sim t^a$. The structure factor is spherically symmetric, with a maximum at wavevector $k_m \sim L^{-1}$. For binary fluids, the exponent a takes different values depending on the dynamical regime under study. In order of increasing time, there are 'diffusive' ($a = 1/3$), 'viscous hydrodynamic' ($a = 1$) and the 'inertial hydrodynamic' ($a = 2/3$) regimes. The crossover between these regimes is determined by the fluid properties (viscosity, density). Here we will focus on the diffusive regime, in which hydrodynamic effects can be neglected. In the absence of shear, phase separation is described by the Cahn-Hilliard equation for the order-parameter field $\phi(\mathbf{r}, t)$, namely $\partial_t \phi = -\nabla^2(\nabla^2\phi + \phi - \phi^3)$.

If a uniform shear flow is imposed in the x-direction, with shear direction y, the flow velocity is $v_x = \gamma y$, where γ is the shear rate. For an incompressible fluid, the Cahn-Hilliard equation acquires an advective term $\mathbf{v} \cdot \nabla\phi = \gamma y \partial_x \phi$ on the left-hand side. Generalising to an n-component vector order parameter, this gives

$$\partial_t \vec{\phi} + \gamma y \partial_x \vec{\phi} = -\nabla^2 \left(\nabla^2 \phi + [1 - |\vec{\phi}|^2]\vec{\phi} \right). \tag{58}$$

For a system quenched into the two-phase region from the homogeneous phase, an appropriate initial condition is a Gaussian random field with short-range correlations:

$$\langle \phi_i(\mathbf{r}, 0)\phi_j(\mathbf{r}', 0)\rangle = \frac{\Delta}{n}\delta_{ij}\delta(\mathbf{r} - \mathbf{r}').$$

We begin by discussing phase ordering of a vector field in the soluble limit where the number of vector components of the field, n, tends to infinity. To introduce the method, we first consider models A and B before turning to the system with shear.

6.1 The large-n limit: model A

Although not strictly necessary, it is convenient to choose in (39) the familiar 'ϕ^4' potential, in the form $V(\vec{\phi}) = (1 - |\vec{\phi}|^2)^2/4$. With this potential, (39) becomes

$$\frac{\partial \vec{\phi}}{\partial t} = \nabla^2 \vec{\phi} + [1 - |\vec{\phi}|^2]\vec{\phi} \ . \tag{59}$$

The simplest way to take the limit is to recognise that, for $n \to \infty$, $|\vec{\phi}|^2$ can be replaced by its average, to give

$$\frac{\partial \phi}{\partial t} = \nabla^2 \phi + a(t)\, \phi \tag{60}$$

$$a(t) = 1 - \langle |\vec{\phi}|^2 \rangle \ , \tag{61}$$

where ϕ now stands for (any) one of the components of $\vec{\phi}$. Equation 60 can alternatively be derived by standard diagrammatic techniques [15]. Equation 60 can be solved exactly for arbitrary time t after the quench. However, we are mainly interested in late times (*i.e.* the scaling regime), when the solution simplifies. After Fourier transformation, the formal solution of (60) is

$$\phi_{\mathbf{k}}(t) = \phi_{\mathbf{k}}(0)\ \exp[-k^2 t + b(t)] \ , \tag{62}$$

$$b(t) = \int_0^t dt'\, a(t') \ , \tag{63}$$

giving

$$a(t) = \frac{db}{dt} = 1 - \Delta \sum_k \exp[-2k^2 t + 2b(t)] \ , \tag{64}$$

where (4) has been used to eliminate the initial condition. Since we shall find *a posteriori* that $a(t) \ll 1$ at late times, the left side of (64) is negligible for $t \to \infty$. Using $\sum_{\mathbf{k}} \exp(-2k^2 t) = (8\pi t)^{-d/2}$ gives $b(t) \to (d/4) \ln(t/t_0)$, where

$$t_0 = \Delta^{2/d}/8\pi \ . \tag{65}$$

Therefore, $a(t) \to d/4t$ for $t \to \infty$, and the solution of (62), valid at late times, is

$$\phi_{\mathbf{k}}(t) = \phi_{\mathbf{k}}(0)\ (t/t_0)^{d/4}\ \exp(-k^2 t) \ . \tag{66}$$

Using (4) once more, we obtain the structure factor, and its Fourier transform, the pair correlation function as,

$$S(\mathbf{k}, t) = (8\pi t)^{d/2}\ \exp(-2k^2 t) \ , \tag{67}$$

$$C(\mathbf{r}, t) = \exp(-r^2/8t) \ . \tag{68}$$

These results exhibit the expected scaling forms (7), with length scale $L(t) \propto t^{1/2}$. Note that the structure factor has a Gaussian tail, in contrast to the power-law tail (44) found in systems with $n \leq d$.

6.2 The large-n limit: model B

For conserved fields, the calculation proceeds as before, but with an extra $(-\nabla^2)$ on the right-hand side of the equation of motion. Making as before, the replacement $|\vec{\phi}|^2 \to \langle|\vec{\phi}|^2\rangle$ for $n \to \infty$, then with ϕ (any) one component of $\vec{\phi}$, one obtains

$$\frac{\partial \phi}{\partial t} = -\nabla^4 \phi - a(t)\, \nabla^2 \phi \,, \tag{69}$$

with $a(t)$ still given by (61). Transforming to Fourier space, the solution is

$$\phi_{\mathbf{k}}(t) = \phi_{\mathbf{k}}(0)\, \exp[-k^4 t + k^2 b(t)] \,, \tag{70}$$

The function $b(t)$, defined as in (63), satisfies the equation

$$a(t) = db/dt = 1 - \Delta \sum_{\mathbf{k}} \exp[-2k^4 t + 2k^2 b(t)] \,. \tag{71}$$

This equation was solved by Coniglio and Zannetti [11], by first expressing the sum over $\mathbf{k}$ as a parabolic cylinder function, then taking the large-t limit. Here we will take the large-t limit from the outset, and recognise that the sum can then be evaluated using steepest descents. Just as for the nonconserved case, we can show *a posteriori* that $db/dt \ll 1$ at late times, so that this term can be dropped from (71). After the change of variable $\mathbf{k} = [b(t)/t]^{1/2}\mathbf{x}$, we obtain

$$1 = \Delta C_d (b/t)^{d/2} \int_0^\infty x^{d-1} dx\, \exp[2\beta(x^2 - x^4)] \,, \tag{72}$$

where C_d is an uninteresting constant, and

$$\beta(t) = b^2(t)/t \,. \tag{73}$$

Provided $\beta(t) \to \infty$ for $t \to \infty$ (which can be verified *a posteriori*), the integral on the right of (72) can be evaluated by steepest descents. Including the Gaussian fluctuations around the maximum of the integrand at $x = 1/\sqrt{2}$ gives

$$1 = \mathrm{const}\, \Delta\, \beta^{-1/2}\, (\beta/t)^{d/4}\, \exp(\beta/2) \,, \tag{74}$$

with asymptotic solution

$$\beta \simeq (d/2) \ln t \,, \qquad t \to \infty \,, \tag{75}$$

justifying the use of the steepest descents method for large t. Putting this result into (70) gives the final result for the structure factor [11]

$$S(\mathbf{k}, t) \simeq L^{(d/4)\varphi(kL)} \tag{76}$$

$$L \simeq \left(\frac{8}{d}\frac{t}{\ln t}\right)^{1/4} \tag{77}$$

$$\varphi(x) = 1 - (1 - x^2)^2 \,. \tag{78}$$

Here $k_m(t) = 1/L(t)$ is the position of the maximum in $S(\mathbf{k}, t)$. A slightly more careful treatment (retaining the leading subdominant term in (75)), gives an additional logarithmic prefactor such that (asymptotically in time) $\sum_{\mathbf{k}} S(\mathbf{k}, t) = 1$.

Equation 76 is interesting because, in contrast to the nonconserved result (67), it does not have the conventional scaling form. Rather it exhibits 'multiscaling' [11]. For fixed 'scaling variable' kL, the structure factor would vary as L^d, with a *prefactor* depending on the scaling variable. In the multiscaling form (76), for fixed scaling variable, $S(\mathbf{k},t) \sim L^{d\varphi(kL)}$, *i.e.* the *exponent* depends continuously on the scaling variable.

After the discovery of multiscaling in the $n \to \infty$ limit, some effort was devoted to looking for similar phenomena at finite n, notably for scalar systems [62], but also for $n = 2$ [63, 48] and $n = 3$ [64]. However, no evidence was found for any departure from simple scaling for any finite n. At the same time, Bray and Humayun showed, within the context of an approximate calculation based on an idea of Mazenko, that simple scaling is recovered asymptotically for any finite n [65].

6.3 Spinodal decomposition under shear

We now treat equation (58) in the same way. A fuller version of this calculation can be found in [66]. In the limit $n \to \infty$, one can replace $|\vec{\phi}|^2$ in (58) by its mean in the usual way, leading to a self-consistent linear equation. After Fourier transformation this reads

$$\frac{\partial \phi_{\mathbf{k}}}{\partial t} - \gamma k_x \frac{\partial \phi_{\mathbf{k}}}{\partial k_y} = -k^2[k^2 - a(t)]\phi_{\mathbf{k}}, \tag{79}$$

where ϕ is (any) one component of $\vec{\phi}$, and $a(t) = 1 - \langle|\vec{\phi}(t)|^2\rangle$.

The self-consistent determination of $a(t)$ proceeds on the same lines as for model B, but the steepest-descent calculation is technically more difficult than before due to a complete breaking of the rotational symmetry by the imposed shear flow. We refer the interested reader to [66] for the details. The result for the structure factor is

$$S(\mathbf{k},t) = \text{const.}\,(\ln V_s)^{3/2} V_s^{F(\mathbf{q})/F_m},. \tag{80}$$

In this expression $V_s(t) = L_x L_y L_z \sim \gamma t^{7/4}/(\ln t)^{3/4}$ is the 'scale volume' at time t, and $\mathbf{q} = (k_x L_x, k_y L_y, k_z L_z)$ is the scaled wavevector. The scale lengths in the three directions are given by $L_x = \gamma(t^3/b)^{1/2} \sim \gamma(t^5/\ln t)^{1/4}$, and $L_y = L_z = (t/b)^{1/2} \sim (t/\ln t)^{1/4}$, where $b(t) \simeq (7t\ln t/8F_m)^{1/2}$. In equation (80) and below, $F_m = 23/60$ is the maximum value of the function $F(u,v,w)$ [where $\mathbf{q} = (u,v,w)$ in (80)], given by

$$F(u,v,w) = -\frac{1}{5u}\left((u+v)^5 - v^5\right) + \frac{8}{15}u^2 + \frac{4}{3}uv + v^2 - \frac{2}{3}w^2(u^2+3uv+3v^2) + w^2 - w^4. \tag{81}$$

The structure of the function $F(u,v,w)$ is indicated by its turning points, given in Table 2.

For a light scattering experiment (Pine, this volume) with the beam in the z direction, the diffracted intensity is given by the $k_z = 0$ slice of the structure factor, which is determined by $F(u,v,0)$. A contour plot of this function is shown in Figure 3, with the turning points a, b, c, and d of Table 2 indicated. The absolute maxima are at the points d, but these are not much higher than the turning points d, so that $F(u,v,0)$ exhibits two parallel ridges like those observed in experiment [58].

We conclude with some conjectures about the physically realistic case (for binary fluids in the diffusive regime) of a scalar order parameter. These are informed by our

Label	Position	Number	F	Value	Type (3D)
a	$(0,0,0)$	1	0	0	Min
b	$\pm(2/\sqrt{3},0,0)$	2	16/45	.35556	IS
c	$\pm(\sqrt{2}-1/\sqrt{3},-1/\sqrt{2},0)$	2	$(37-12\sqrt{6})/180$	.04225	S2
d	$\pm(\sqrt{2}+1/\sqrt{3},-1/\sqrt{2},0)$	2	$(37+12\sqrt{6})/180$	.36885	S1
e	$\pm(0,0,1/\sqrt{2})$	2	1/4	.25	S1
f	$\pm(\sqrt{3},-1/\sqrt{3},\pm1/\sqrt{6})$	4	23/60	.38333	Max

Table 2. *Stationary points of $F(u,v,w)$: Max = maximum, Min = minimum, Sn = saddle point of type n (the matrix of second derivatives has n positive eigenvalues), IS = 'inflection saddle point' (one positive, one zero, one negative eigenvalue).*

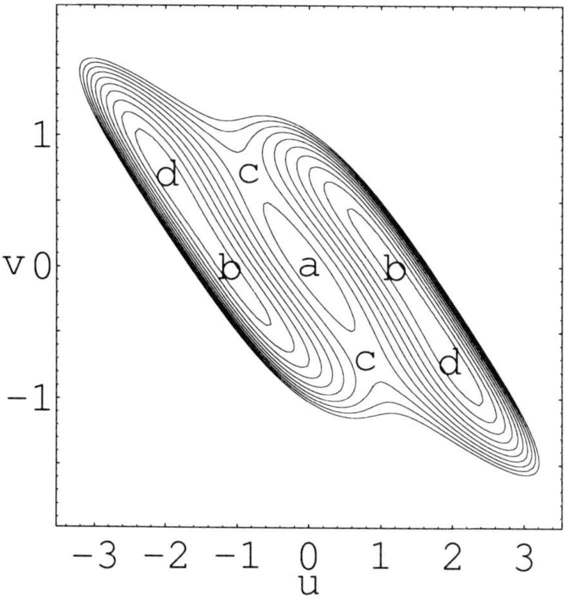

Figure 3. *Contour plot of $F(u,v,0)$ showing approximate locations of the stationary points a,b,c,d from Table 1. Contour lines for $F<-0.1$ are not shown.*

exact solution for $n=\infty$, and by the way the $n=\infty$ solution is known to be modified for scalar fields in the *unsheared* case [65]. First we expect that, for any finite n, the structure factor will exhibit asymptotic scaling of the form $S(\mathbf{k},t) = V_s g(\mathbf{q})$, with $V_s = \Pi_{i=1} L_i$ $(i=x,y,z)$ and $q_i = k_i L_i$, instead of the multiscaling form (80). As in the $n=\infty$ case, we expect the growth of the characteristic scales for directions normal to the flow to obey the same power laws as in the unsheared case, *i.e.* for scalar fields, $L_y \sim L_z \sim t^{1/3}$. The growth in the x-direction can then be deduced from the assumed scaling form for the structure factor: if we multiply the two terms on the left-hand side of (79) by $\phi_{-\mathbf{k}}(t)$, and average, the result $L_x \sim \gamma t L_y$ follows immediately if we insert the scaling form for

$S(\mathbf{k}, t)$ and assume both terms are of the same order in the scaling limit. This leads to the prediction $L_x \sim \gamma t^{4/3}$.

In the viscous hydrodynamic regime, where without shear $L(t) \sim t$, the same heuristics suggest $L_y \sim L_z \sim t$, $L_x \sim \gamma t^2$. The predictions for the length scales are consistent with data on polymer blends [60], though it has been suggested [57] that a stationary state eventually develops for $\gamma t \gg 1$ due to a competition between stretching and breaking of domains. No such stationary state is obtained in the present work, however, and the existence of a stationary (*i.e.* non-coarsening) late-stage limit for more realistic models remains an open question.

References

[1] A J Bray, *Adv Phys* **43**, 357-459 (1994).

[2] J S Langer, in *Solids Far From Equilibrium*, editor C Godrèche (Cambridge University Press, Cambridge 1992).

[3] P C Hohenberg and B I Halperin, *Rev Mod Phys* **49**, 435 (1977).

[4] A J Bray, *Phys Rev Lett* **62**, 2841 (1989).

[5] A J Bray, *Phys Rev B* **41**, 6724 (1990).

[6] K Binder and D Stauffer, *Phys Rev Lett* **33**, 1006 (1974).

[7] J Marro, J L Lebowitz and M H Kalos, *Phys Rev Lett* **43**, 282 (1979)

[8] H Furukawa, *Prog Theor Phys* **59**, 1072 (1978).

[9] H Furukawa, *Phys Rev Lett* **43**, 136 (1979).

[10] A J Bray, *J Phys A* **22**, L67 (1990); J G Amar and F Family, *Phys Rev A* **41**, 3258 (1990). See also B Derrida, C Godrèche and I Yekutieli, *Phys Rev A* **44**, 6241 (1991).

[11] A Coniglio and M Zannetti, *Europhys Lett* **10**, 575 (1989).

[12] H Furukawa, *J Phys Soc Jpn* **58**, 216 (1989).

[13] H Furukawa, *Phys Rev B* **40**, 2341 (1989).

[14] D S Fisher and D A Huse, *Phys Rev B* **38**, 373 (1988). Note that our exponent $\bar{\lambda}$ is called λ in this paper.

[15] T J Newman and A J Bray, *J Phys A* **23**, 4491 (1990).

[16] N Mason, A N Pargellis, and B Yurke, *Phys Rev Lett* **70**, 190 (1993);
H Orihara and Y Ishibashi, *J Phys Soc Jpn* **55**, 2151 (1986);
T Nagaya, H Orihara and Y Ishibashi, *ibid.* **56**, 1898 (1987); **56**, 3086 (1987); **59**, 377 (1990).

[17] G Porod, *Kolloid Z* **124**, 83 (1951); **125**, 51 (1952).

[18] P Debye, H R Anderson and H Brumberger, *J Appl Phys* **28**, 679 (1957);
G Porod, in *Small-Angle X-Ray Scattering*, edited by O Glatter and O Kratky (Academic NY 1982).

[19] S M Allen and J W Cahn, *Acta Metall* **27**, 1085 (1979).

[20] D A Huse, *Phys Rev B* **34**, 7845 (1986).

[21] J Amar, F Sullivan and R Mountain, *Phys Rev B* **37**, 196 (1988);
T M Rogers, K R Elder and R C Desai, *Phys Rev B* **37**, 9638 (1988);
R Toral, A Chakrabarti and J D Gunton, *Phys Rev B* **39**, 4386 (1989);
C Roland and M Grant, *Phys Rev B* **39**, 11971 (1989).

[22] I M Lifshitz and V V Slyozov, *J Phys Chem Solids* **19**, 35 (1961).

[23] C Wagner, *Z Elektrochem* **65**, 581 (1961).

[24] A D Rutenberg and A J Bray, *Phys Rev E* **51**, 5499 (1995).

[25] E D Siggia, *Phys Rev A* **20**, 595 (1979).

[26] N C Wong and C M Knobler, *J. Chem Phys* **69**, 725 (1978); *Phys Rev A* **24**, 3205 (1981); Y C Chou and W I Goldburg, *Phys Rev A* **20**, 2105 (1979); **23**, 858 (1981).

[27] S Puri and B Dunweg, *Phys Rev A* **45**, 6977 (1992); O T Valls and J E Farrell, *Phys Rev E* **47**, 36 (1993); T Koga and K Kawasaki, *Physica* **196A**, 389 (1993).

[28] A Shinozaki and Y Oono, *Phys Rev E* **48**, 2622 (1993).

[29] F J Alexander, S Chen and D W Grunau, *Phys Rev B* **48**, 634 (1993)

[30] H Furukawa, *Phys Rev A* **31**, 1103 (1985).

[31] J E Farrell and O T Valls, *Phys Rev B* **40**, 7027 (1989); **42**, 2353(1990); **43**, 630 (1991).

[32] V M Kendon, J-C Desplat, P Bladon and M E Cates, *Phys Rev Lett* **83**, 576 (1999).

[33] M Grant and K R Elder, *Phys Rev Lett* **82**, 14 (1999).

[34] V M Kendon, PhD Thesis, University of Edinburgh (1999).

[35] For a general discussion of topological defects, see *e.g.* M Kléman, *Points, Lines and Walls, in Liquid Crystals, Magnetic Systems, and Various Ordered Media* (Wiley, New York, 1983).

[36] A J Bray, *Phys Rev E* **47**, 228 (1993).

[37] A J Bray and K Humayun, *Phys Rev E* **47**, R9 (1993).

[38] A J Bray and S Puri, *Phys Rev Lett* **67**, 2670 (1991).

[39] H Toyoki, *Phys Rev B* **45**, 1965 (1992).

[40] F Liu and G F Mazenko, *Phys Rev B* **45**, 6989 (1992).

[41] A J Bray and K Humayun, *J Phys A* **25**, 2191 (1992).

[42] A J Bray and A D Rutenberg, *Phys Rev E* **49**, R27 (1994).

[43] M Mondello and N Goldenfeld, *Phys Rev A* **45**, 657 (1992).

[44] H Toyoki, *J Phys Soc Jpn* **60**, 1433 (1991).

[45] R E Blundell and A J Bray, *Phys Rev E* **49**, 4925 (1994).

[46] A J Bray, *Phys Rev E* **47**, 3191 (1993).

[47] A Onuki, *Prog Theor Phys* **74**, 1155 (1985).

[48] M Siegert and M Rao, *Phys Rev Lett* **70**, 1956 (1993).

[49] M Siegert, private communication.

[50] S Puri, A J Bray and F Rojas, *Phys Rev E* **52**, 4699 (1995).

[51] A D Rutenberg and A J Bray, *Phys Rev E* **51**, 5499 (1995).

[52] A P Y Wong, P Wiltzius and B Yurke, *Phys Rev Lett* **68**, 3583 (1992).

[53] A P Y Wong, P Wiltzius, R G Larson and B Yurke, *Phys Rev E* **47**, 2683 (1993).

[54] R E Blundell and A J Bray, *Phys Rev A* **46**, R6154 (1992).

[55] C Roland and M Grant, *Phys Rev B* **41**, 4663 (1990).

[56] C Jeppesen and O G Mouritsen, *Phys Rev B* **47**, 14724 (1993).

[57] A Onuki, *J Phys.: Condens Matter* **9**, 6119 (1997), and references therein.

[58] C K Chan, F Perrot, and D Beysens, *Phys Rev A* **43**, 1826 (1991).

[59] A H Krall, J V Sengers, and K Hamano, *Phys Rev Lett* **69**, 1963 (1992); T Hashimoto, K Matsuzaka, E Moses, and A Onuki, *Phys Rev Lett* **74**, 126 (1995).

[60] J Läuger, C Laubner, and W Gronski, *Phys Rev Lett* **75**, 3576 (1995).

[61] D H Rothman, *Phys Rev Lett* **65**, 3305 (1990); P Padilla and S Toxvaerd, *J Chem Phys* **106**, 2342 (1997); A J Wagner and J M Yeomans, *Phys Rev E* **59**, 4366 (1999).

[62] A Coniglio, Y Oono, A Shinozaki and M Zannetti, *Europhys Lett* **18**, 59 (1992).

[63] M Mondello and N Goldenfeld, *Phys Rev E* **47**, 2384 (1993).

[64] M Rao and A Chakrabarti, *Phys Rev E* **49**, 3727 (1994).

[65] A J Bray and K Humayun, *Phys Rev Lett* **68**, 1559 (1992).

[66] P N Rapapa and A J Bray, *Phys Rev Lett* **83**, 3856 (2000)—with Erratum in **84**, 791 (2000)

Phase transitions in nonequilibrium systems

David Mukamel

The Weizmann Institute of Science, Rehovot, Israel

1 Introduction

Collective phenomena in systems far from thermal equilibrium have been a subject of extensive studies in recent years. Usually these systems are driven out of equilibrium by external fields, such as electric field in the case of conductors, pressure gradient in the case of fluids, temperature gradient in the case of heat conductors, chemical potential gradient in the case of growth problems and many others [1–4]. These driving fields are very common in nature and are found in a large variety of physical systems such as granular and traffic flow [5–7], gel electrophoresis [8, 9], super-ionic conductors [10, 11] to give a few examples. In many cases these systems reach a steady state, which unlike the equilibrium case, is characterised by non-vanishing currents. Here we consider possible collective phenomena and phase transitions which may take place in such steady states. The main problem in studying nonequilibrium systems is the lack of general theoretical framework within which they could be analysed. As a result they are far less understood as compared with equilibrium systems where the Gibbs picture provides such a theoretical framework.

Before discussing nonequilibrium systems it is useful to consider briefly systems in *thermal equilibrium*. Here decades of studies have yielded a fairly detailed understanding of their thermodynamic behaviour. Many rules which govern phase transitions occurring in these systems have been derived. For example, it has been shown that the critical exponents associated with a phase transition may be classified into *universality classes*. These classes do not depend on the detailed interactions in the system but rather on a few parameters such as the symmetry of the system and of the order parameter associated with the transition, the dimensionality of the system and the range of interactions. Therefore, in order to study theoretically the critical behaviour of a given system it is sufficient to analyse the simplest possible model which belongs to the same universality class. For reviews see, for example, [12, 13]. It has also been shown that phase transitions and spontaneous symmetry breaking do not take place at low dimension. In particular, no phase transition is expected to take place in a one dimensional system at finite temperatures as long as the interactions are short range [14]. Moreover, breaking of continuous symmetry may take place under the same conditions only in dimensions higher than two [15]. Other rules derived by Landau relate the nature of the transition, namely whether it is first order or continuous, to the symmetry of the systems [14]. If the symmetry allows a third order term in the expansion of the free energy in the order parameter, such as in the case of the transition from a liquid to a nematic liquid crystal phase [16], the transition cannot be continuous and is necessarily first order. On the other hand if the symmetry is such that no third order term is allowed, such as in the transition from a paramagnetic to a ferromagnetic phase, the transition may either be first order or continuous, depending on the details of the interactions. The Gibbs phase rule is another very useful example of a rule which governs the phase diagrams of systems in equilibrium [14]. It deals with fluids composed of c components. The thermodynamic phase space of such systems is of $c+1$ dimensions, comprising the temperature, pressure, and $c-1$ chemical potentials. According to the rule, the manifold in this space on which n different phases coexist is of $D = 2+c-n$ dimensions. Another rule deals with the phase diagram near a triple point, where three coexistence lines meet. According to this 180° rule, each of the three angles defined by the intersecting coexistence lines must be less than 180°. This is a direct result of the convexity of the free energy [17].

These rules and many others, some of which are related to disordered systems [18, 19], provide extremely useful tools for analysing and understanding phase diagrams and critical behaviour of models and physical systems in equilibrium. By simply identifying the symmetry of the system and the nature of the order parameter involved in the phase transition one can usually find the universality class of the transition and even obtain a rough idea of the possible phase diagram.

Our degree of understanding of collective behaviour far from thermal equilibrium is at a much more primitive stage. Since a general theoretical framework for studying nonequilibrium phenomena does not exist, one cannot derive similar rules which would be as general as those for equilibrium systems. Rather, one has to resort to studying specific models and probe the resulting types of phase diagrams and phase transitions, with the hope that some general picture might emerge.

In the present lectures we consider stochastic driven systems in one dimension and discuss some interesting collective behaviour which they display. Unlike equilibrium one dimensional systems which do not exhibit phase transitions, nonequilibrium systems exhibit a rich variety of collective phenomena such as first order and continuous phase transitions, spontaneous symmetry breaking (SSB), phase separation, slow coarsening processes and many others. Mechanisms which lead to these phenomena are discussed.

The article is organised as follows: in Section 2 the concept of detailed balance is discussed, the lack of which is characteristic of nonequilibrium systems. A necessary and sufficient condition for the existence of detailed balance is presented. In Section 3 a simple driven model, the totally asymmetric exclusion process, is introduced and its phase diagram for a system with open boundaries is calculated using a mean field approximation. The phase diagram exhibits several phases separated by first order and continuous transitions. The matrix method which enables one to obtain exact results for steady state properties is outlined in Section 4. A model which displays spontaneous symmetry breaking in one dimension is introduced in Section 5 and a model exhibiting phase separation accompanied by slow coarsening processes is described in Section 6. Open problems and perspectives are briefly discussed in Section 7.

2 Detailed balance and driven systems

In this section we make some general considerations concerning the evolution of dynamical systems. Let C be a microscopic configuration, and let $P(C,t)$ be the probability that the system is in the microscopic configuration C at time t. The dynamics of the system is defined in terms of the transition rates $W(C \to C')$ from a configuration C to C'. The equation which governs the evolution of the distribution function $P(C,t)$ takes the form

$$\frac{\partial P(C,t)}{\partial t} = \sum_{C'} W(C' \to C) P(C',t) - \sum_{C'} W(C \to C') P(C,t). \tag{1}$$

The first sum represents the rate of flow, in configuration space, of probability into C while the second sum corresponds to the outgoing flow from this configuration. In a steady state the two terms are equal, yielding zero net flow from any configuration.

Systems in thermal equilibrium are characterised by an energy function, or a Hamiltonian, $E(C)$. The steady state distribution $P(C)$ is proportional to $e^{-E(C)/k_B T}$, where T is the temperature and k_B is the Boltzmann constant. Given an energy function $E(C)$ one can always find transition rates $W(C \to C')$, such as the Metropolis rates, which obey detailed balance. Here the two sums cancel term by term

$$W(C \to C') P(C) = W(C' \to C) P(C'), \tag{2}$$

for any pair of configurations C and C'.

On the other hand dynamical systems are not defined by an energy function but rather by transition rates. When a system is not in thermal equilibrium, the resulting steady state is such that detailed balance (2) is not satisfied. We will basically use this lack of detailed balance as a definition of nonequilibrium.

Given the dynamics of a system, namely the transition rates, it is of interest to know whether or not detailed balance is satisfied. Since, in general, the steady state distribution cannot be calculated, a direct check of the detailed balance condition (2) is not possible. Thus a criterion for existence of detailed balance which is based directly on the transition rates and does not require the knowledge of the steady state is highly desirable. Such a criterion is provided by the following equations. Let $C_1, C_2, \ldots, C_k$ be a set of k microscopic configurations. A necessary and sufficient condition for the existence of detailed balance is that for any such set one has

$$W(1 \to 2) W(2 \to 3) \ldots W(k \to 1) = W(1 \to k) W(k \to k-1) \ldots W(2 \to 1), \tag{3}$$

where for simplicity we have denoted C_i by i. It is easy to check that (3) is a necessary condition. When detailed balance is satisfied one may replace $W(i \to i+1)/W(i+1 \to i)$ by $P(i+1)/P(i)$, where $P(i)$ is the steady state distribution with respect to which detailed balance is satisfied. Using these relations (3) is easily verified.

To demonstrate that this is a sufficient condition as well, we use (3) to derive the steady state distribution. We start with an arbitrary configuration 1 and denote its steady state weight by $P(1)$. The weight of states 2 which are directly connected with 1 (namely, for which $W(1 \to 2) > 0$), may thus be defined using the detailed balance relation, $P(2) = P(1) W(1 \to 2)/W(2 \to 1)$. This process may then be repeated to define the weights of states directly connected with states 2 *etc.*, until all microscopic configurations have been reached. The weight of a microscopic configuration k which may be reached from 1 via intermediate states $2, 3, \ldots, k-1$ is thus given by

$$P(k) = P(1)\frac{W(1\to 2)\ldots W(k-1\to k)}{W(k\to k-1)\ldots W(2\to 1)}. \tag{4}$$

For this procedure to be self-consistent one has to verify that any path between configurations 1 and k yields the same $P(k)$. It is a straightforward matter to show that this follows directly from (3).

Thus, to demonstrate that a dynamical system defined by its transition rates is not in thermal equilibrium it is sufficient to find a single path in configuration space for which (3) is not satisfied. This is usually quite easy to check, making it a very useful criterion. When (3) is not satisfied the system exhibits non-vanishing probability currents between configurations which is an indication of the system not being in thermal equilibrium.

A simple prototypical model of driven systems, termed the 'standard model', was introduced by Katz *et al* [1, 20]. This is a driven lattice gas model defined on a hypercubic lattice with periodic boundary conditions. Each site i is either occupied by a particle or is vacant, with $\sigma_i = 0, 1$ being the occupation number. In the absence of drive, an Ising Hamiltonian is assumed

$$H = -J\sum_{\langle ij\rangle}\sigma_i\sigma_j, \tag{5}$$

where the sum is over nearest neighbour (nn) sites $\langle ij\rangle$. The evolution of the system is defined by Kawasaki dynamics, which allows particles to hop between nearest neighbour sites. Let C and C' be two configurations obtained from each other by an interchange of a single pair of nn occupation numbers σ_i and σ_j. The transition rate between C and C' may be taken as the Metropolis rate $W(C\to C') = w(\beta\Delta H)$, where $\Delta H = H(C') - H(C)$ and $w(x) = \min(1, e^{-x})$. This dynamics leads to the expected Boltzmann equilibrium distribution. In $d > 1$ dimensions the system exhibits the usual Ising transition from a homogeneous phase at high temperatures to a phase separated state at low temperatures.

Introducing a driving field E along one of the axes, the transition rates are modified by adding a term uE to ΔH where

$$u = -1\ ,\ 0\ ,\ +1, \tag{6}$$

for a hop along, transverse or opposite to the field direction, respectively. The transition rates are thus given by

$$W(C\to C') = w(\beta(\Delta H + uE)). \tag{7}$$

Due to the periodic boundary conditions in the direction of the driving field, these rates do not obey detailed balance, and the steady state exhibits non-vanishing currents.

In spite of the simplicity of the model, no exact results for the steady state properties are available (even in one dimension). Extensive numerical studies of this model in two dimensions demonstrate that the phase separation transition which exists in zero drive, persists for non-zero driving fields. At low temperatures the system exhibits stripes of high density and low density regions which are oriented along the field direction. These stripes coarsen with time leading to a phase separated state.

3 Asymmetric exclusion process in one dimension

In this section we consider the phase diagram of the one dimensional 'standard model' defined in the previous section for $J = 0$. Here the only interaction between the particles is the hard-core interaction which prevents more than one particle from occupying the same site. This process is called asymmetric simple exclusion process (ASEP). Furthermore, we consider the limit $E \rightarrow \infty$, called totally asymmetric simple exclusion process (TASEP). In this limit particles are restricted to move only to the right, with no backward moves. This model turns out to be sufficiently simple to allow for exact calculation of some of its steady state properties. In spite of its simplicity, the model with open boundary conditions exhibits a rather rich and complicated phase diagram, displaying both continuous and discontinuous phase transitions (see below). This is clearly a direct consequence of the fact that the dynamics of a nonequilibrium nature. The model and many variants of it have been a subject of extensive studies in recent years [21–30].

The dynamics of the model is defined as follows: at any given time a pair of nn sites is chosen at random. If the occupation numbers of these sites are (+ 0) an exchange is carried out

$$+\,0 \;\rightarrow\; 0\,+, \tag{8}$$

with rate unity. All other configurations, remain unchanged. Here and in the following we interchangeably use 1 or + to denote an occupied site. For periodic boundary conditions the system reaches a trivial steady state in which all microscopic configurations have the same weight. This may be verified by direct inspection of the master equation (1). It is easy to see that the number of configurations to which a given configuration C may flow is equal to the number of configurations flowing into C. To verify that this is the case note that a microscopic configuration C is composed of alternating segments of +'s and 0's. According to the dynamics (8), the system leaves C when the rightmost particle in one of the + segments moves one step to the right. Thus the number of configurations that C may flow into is equal to l, the number of + segments in this configuration. Similarly C may be reached when the leftmost particle in one of the + segments hops into its position. The number of configurations flowing into C is therefore also l. Since all non-vanishing transition rates are 1, the state where all configurations have equal weights is stationary. Therefore in the steady state the system exhibits no correlations, apart from the trivial correlations arising from the fact that the overall density of particles is fixed.

The steady state current J is given by

$$J = \langle \sigma_i (1 - \sigma_{i+1}) \rangle, \tag{9}$$

where the brackets denote a statistical average with respect to the steady state weights of the microscopic configuration. Since in the steady state the system exhibits no correlations the current may be written, in the large system limit, as

$$J = p(1 - p), \tag{10}$$

where $p_i = \langle \sigma_i \rangle$ is the density at site i, and the index i is omitted in (10) since the average density is homogeneous, independent of i. Equation (10), relating the current to the density is known as the fundamental relation (or fundamental diagram). The interesting feature in this relation is that the current is not a monotonic function of the density but rather it exhibits a maximum at $p = 0.5$. This feature is a result of the hard-core interaction between the particles and it affects rather drastically the steady state properties of the system when open boundary conditions are considered.

We now turn to the model with *open* boundary conditions. Here, particles are introduced into the system at the left end, they move through the bulk according to the conserving dynamics (8), and leave the system at the right end. To be more specific, at the left boundary ($i = 1$) the move

$$0 \to +, \tag{11}$$

is carried out with a rate α. Similarly, at the right boundary ($i = N$) one takes

$$+ \to 0, \tag{12}$$

with a rate β. For a schematic representation of the model see Figure 1. The overall

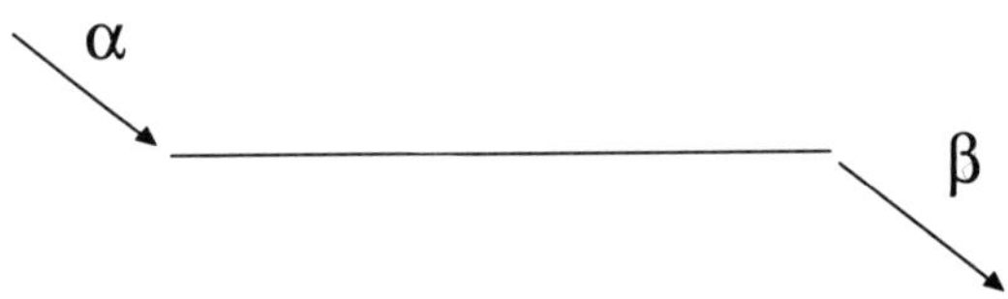

Figure 1. *A schematic representation of the open boundary conditions for the totally asymmetric exclusion process.*

dynamics is non-conserving. Particles are conserved in the bulk but are not conserved at the boundaries. Unlike the case with periodic boundary conditions, the steady state distribution is not trivial, and correlations between the densities at different sites do not vanish. However, far from the boundaries the distribution function is expected to be well approximated by the homogeneous one, suggesting that correlations are small.

We are interested in the steady state of this model for given rates α and β. For large α and small β, namely for a large feeding rate and a small exit rate, the overall density is expected to be high. On the other hand for small α and large β the density is expected to be low. In addition, for a large system, where away from the boundaries the local density is expected to vary very slowly, the fundamental relation (10) is expected to hold locally. Thus the current in the system cannot exceed a maximal current, as suggested by (10). These features yield a rather rich phase diagram, as α and β are varied.

We start by considering the phase diagram in the mean field approximation [21, 22, 24]. Since correlations in this system are expected to be vanishingly small away from the boundaries, this approximation is expected to yield a rather accurate phase diagram. In fact it turns out that the phase diagram obtained in this way is exact.

To derive the mean field equations we note that the current $J_{i,i+1}$ between sites i and $i+1$ is given by $\langle \sigma_i(1-\sigma_{i+1})\rangle$ for $i = 1, \ldots, N-1$. In addition the currents at the two ends are given by $J_0 = \alpha\langle 1-\sigma_1\rangle$ and $J_N = \beta\langle\sigma_N\rangle$. Neglecting correlations one finds that in the steady state, where all currents are equal, the following equations have to be satisfied:

$$J = \alpha(1 - p_1) = p_1(1 - p_2) = \ldots = p_{N-1}(1 - p_N) = \beta p_N. \tag{13}$$

Solving these equations for $J, p_1, \ldots, p_N$ the density profile in the steady state and the current are obtained.

It is instructive to consider these equations in the continuum limit. Replacing p_i by $p(x)$ in (13) with $0 \le x \le L$ yields the bulk current

$$J(x) = p(1 - p) - D\frac{\partial p}{\partial x}, \tag{14}$$

where D is the diffusion constant which, by rescaling x, may be taken as 1. In this expression the first term represents the drive while the second term is the ordinary diffusion current. The evolution of the system is governed by the continuity equation $\partial p/\partial t = -\partial J/\partial x$, together with the boundary conditions $J(0) = \alpha(1 - p(0))$ and $J(L) = \beta p(L)$. In the steady state the current J in (14) is independent of x yielding a density profile which has one of the two following forms

$$\begin{aligned} p(x) &= 0.5 + v \tanh[v(x - x_0)] \\ p(x) &= 0.5 + v \coth[v(x - x_0)], \end{aligned} \tag{15}$$

where $v^2 = 1/4 - J$. The two parameters x_0 and v (or alternatively the current J) are determined by the two boundary conditions, and are thus related to α and β.

By matching the boundary conditions the density profiles and the current are obtained. The resulting phase diagram is given in Figure 2. The system is found to exhibit three distinct phases in the limit of large length L:

- Low density phase for which the bulk density is smaller than 0.5 with $x_0 = O(L)$. The density profile is basically flat, except for a small region near the right end. In this phase $p(0) = \alpha$ and $J = \alpha(1 - \alpha)$. It exists for $\alpha < \beta$ and $\alpha < 1/2$.

- High density phase with bulk density larger than 0.5, and $x_0 = -O(L)$. The density profile is flat except at a small region near the left end. Here $p(L) = 1 - \beta$ and $J = \beta(1 - \beta)$. This phase exist for $\beta < \alpha$ and $\beta < 1/2$. The two phases coexist on the line $\alpha = \beta < 1/2$.

- A maximal current phase in the region $\alpha > 1/2$ and $\beta > 1/2$. Here the bulk density is 1/2 exhibiting structures at both ends of the system. These structures decay algebraically as $1/x$ when moving away from the ends. In this phase the current is maximal, namely $J = 1/4$.

The phase diagram exhibits a first order line on which the high density and the low density phases coexist and two second order lines separating these phases from the maximal current phase. Typical schematic density profiles in the various phases are also given in Figure 2.

In the mean field approximation fluctuations are neglected, and therefore by itself this analysis may not serve as a demonstration that phase transitions do take place in $1d$ away from thermal equilibrium. In fact the mean field approximation yields phase transitions

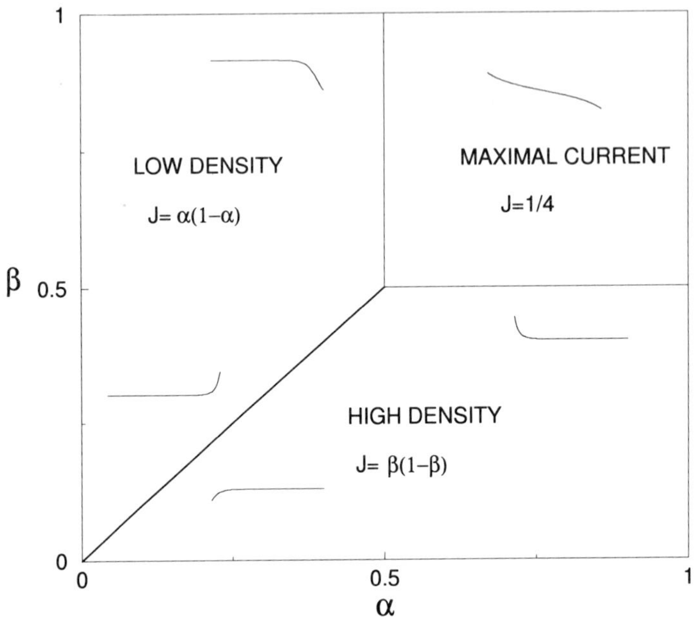

Figure 2. *The (α, β) phase diagram of the totally asymmetric exclusion process exhibiting low density, high density and maximal current phases. The thick line represents a first order transition while thin lines correspond to continuous transitions. Schematic density profiles in the various phases are given.*

in equilibrium $1d$ systems, where they are known not to exist. It is therefore important to examine the role of fluctuations in this driven system and demonstrate that indeed the phase transitions found within the mean field approximation remain when fluctuations are taken into account. This has indeed been demonstrated for the TASEP [24, 25, 26]. A method which goes beyond the mean field approximation and allows exact calculations of steady state properties of some driven $1d$ models is described in the next section.

4 Matrix method

A matrix method for calculating some steady state properties of the TASEP was introduced a few years ago [25]. The method has since then been generalised and applied to other models of driven systems. In the following we briefly outline the method as applied to the TASEP with open boundary conditions described above.

We are interested in calculating the steady state distribution function $P(\sigma_1, \ldots, \sigma_N)$. In the matrix method one tries to express the distribution function by a matrix element of a product of particular matrices. Let D and E be two square matrices and $\langle W|$ and $|V\rangle$ be two vectors. For any given configuration, $(\sigma_1, \ldots, \sigma_N)$ one considers the matrix product in which each occupation number σ_i is replaced by either a matrix D or a matrix

E depending on whether σ_i is 1 or 0, respectively. The key question is whether one can find matrices D and E and vectors $\langle W|$ and $|V\rangle$ such that $P(\sigma_1, \ldots, \sigma_N)$ is proportional to the $\langle W|, |V\rangle$ matrix element of this product. Within this representation the distribution function may be written as

$$P(\sigma_1, \ldots, \sigma_N) \propto \langle W| \prod_{i=1}^{N} [\sigma_i D + (1-\sigma_i)E] |V\rangle. \tag{16}$$

A priori, it is not at all clear that such representation is available. However, if such representation exists, it may yield a straightforward (though sometimes tedious) way for calculating the distribution function. For example, the density at, say, site $i = 1$ may be expressed as

$$p_1 = \frac{1}{Z_N} \sum_{\{X_i\}} \langle W|DX_2 \ldots X_N|V\rangle, \tag{17}$$

where $X_i = D, E$, and the normalisation factor Z_N is given by

$$Z_N = \sum_{\{X_i\}} \langle W|X_1 X_2 \ldots X_N|V\rangle = \langle W|C^N|V\rangle, \tag{18}$$

with

$$C = D + E \quad . \tag{19}$$

Thus, we may rewrite (17) as

$$p_1 = \frac{1}{Z_N} \langle W|DC^{N-1}|V\rangle \ . \tag{20}$$

Densities at other sites and density-density correlation functions may similarly be expressed by other matrix elements.

The main question at this point is how to find matrices and vectors such that (16) holds. To this end we consider the local currents in the system. Using the matrix representation, the current between sites i and $i+1$ ($i = 1, \ldots, N-1$) may be expressed as

$$J_{i,i+1} = \frac{1}{Z_N} \langle W|C^{i-1} DE C^{N-i-1}|V\rangle, \tag{21}$$

and the currents at the two ends

$$\begin{aligned} J_0 &= \frac{\alpha}{Z_N} \langle W|EC^{N-1}|V\rangle \\ J_N &= \frac{\beta}{Z_N} \langle W|C^{N-1}D|V\rangle. \end{aligned} \tag{22}$$

In the steady state all currents are equal. Taking matrices D, E and vectors $\langle W|, |V\rangle$ which satisfy

$$\begin{aligned} &DE = D + E \ \ (= C) \\ &\alpha\langle W|E = \langle W| \\ &\beta D\,|V\rangle = |V\rangle, \end{aligned} \tag{23}$$

guarantees that all currents are equal, with

$$J = \frac{Z_{N-1}}{Z_N}. \tag{24}$$

The question is whether these relations (23) are sufficient to guarantee that the resulting distribution (16) is a steady state. For (16) to be a steady state one has to make sure that

$$\frac{\partial P}{\partial t} = 0, \tag{25}$$

for each of the 2^N microscopic configurations. The relations (23) only directly guarantee that $N+1$ of these 2^N equations are satisfied. This may suggest that (23) may not be sufficient to guarantee that the steady state distribution is given by (16). However it can be shown, by direct inspection of Equation (25) that (23) yields the steady state of the system.

The problem is thus reduced to first finding matrices and vectors which satisfy (16), and then calculating some matrix elements to obtain, for example, the current J. It is straightforward to show that for $\alpha + \beta \neq 1$ the matrices which satisfy (16) have to be of infinite order. Such matrices have been found and the current and density profiles and other correlation functions have been calculated [25]. The resulting phase diagram coincides with that obtained by the mean field approximation, although the density profiles are different. For example, in the algebraic, maximal current, phase the local density decays to the bulk density like $1/\sqrt{x}$ at large distances from the boundary, unlike the mean field result which yields a $1/x$ profile.

The matrix method proved to be very powerful in yielding steady state properties of TASEP dynamics. It has been applied and generalised to study partially asymmetric exclusion processes (ASEP) [31–33] and models with more than one type of particles [34–37]. In addition, replacing matrices by tensors proved to be useful in some cases [38, 39]. However the method is restricted to one dimension. It is not standard in the sense that it cannot be applied to an arbitrary dynamical model. Moreover, there is no simple way to tell *a priori* whether or not it may be applicable for a specific model.

5 Spontaneous symmetry breaking in one dimension

In this section we consider a simple dynamical model which exhibits *spontaneous symmetry breaking* (SSB) in one dimension.

The model may be pictorially described in the following way: consider a narrow bridge connecting two roads. Cars travelling on the bridge in opposite directions do not block each other, although they may slow the traffic flow in both directions. We assume that the two roads leading to the bridge from both sides are statistically identical, so that the arrival rates of cars at the two ends of the bridge are the same. This system clearly has a right-left symmetry. Thus if this symmetry is not spontaneously broken one would expect that the long time average of the current of cars travelling to the right would be the same as that of cars travelling to the left. The question is whether the bridge is capable of exhibiting breaking of the right-left symmetry, and spontaneously turning itself into a 'one-way' street, where the current in one direction is larger than the current in the other direction. It turns out that this may indeed take place in the limit of a long bridge, demonstrating that SSB may take place in 1d nonequilibrium systems.

To model the 'bridge' problem we generalise the TASEP discussed in the previous section [37, 40, 41]. We consider a 1d lattice of length N. Each lattice point may be occupied by either a (+) particle (positive charge) moving to the right, a (−) particle (negative charge) moving to the left or by a vacancy (0). In addition positive (negative) charges are supplied at the left (right) end and are removed at the right (left) end of the system.

The dynamics of the model is defined as follows: at each time step a pair of nearest neighbour sites is chosen and an exchange process is carried out

$$+\,0 \rightarrow 0\,+ \qquad 0\,- \rightarrow -\,0 \qquad +\,- \rightarrow -\,+, \tag{26}$$

with rates 1, 1 and q, respectively. Furthermore, at the two ends particles may be introduced or removed. At the left boundary ($i = 1$) the processes

$$0 \rightarrow + \qquad - \rightarrow 0, \tag{27}$$

take place with rates α and β, respectively. Similarly, at the right boundary ($i = N$), one has the processes

$$0 \rightarrow - \qquad + \rightarrow 0, \tag{28}$$

with rates α and β, respectively (see Figure 3). In the 'bridge' language the boundary terms may be viewed as traffic lights which control the feeding and exit rates at the two ends.

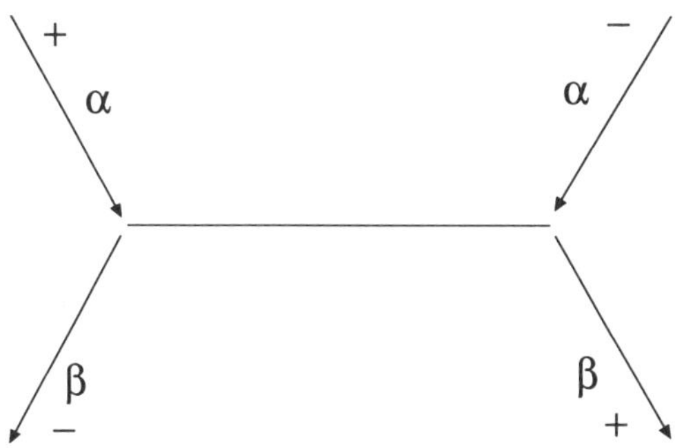

Figure 3. *A schematic representation of the input and output rates of the 'bridge' model.*

Since the parameters α and β are the same on both ends of the systems the dynamics obviously possesses a right-left symmetry. The question of interest is whether or not this symmetry is preserved in the steady state. Clearly, for small α and large β the density of particles in the system is expected to be low, the two types of particles do not block each other, and the steady state is expected to be symmetric. On the other hand for β much smaller than α, particles are blocked in the system, the density is high and it is possible that symmetry breaking takes place.

We start by considering the mean field approximation. It is straightforward to derive the mean field equations for the steady state. They take the form

$$\begin{aligned} J_+ &= p_i[1 - p_{i+1} - (1-q)m_{i+1}] \\ J_- &= m_{i+1}[1 - m_i - (1-q)p_i], \end{aligned} \tag{29}$$

for $i = 1, \ldots, N-1$, where p_i and m_i are the densities of the (+) and (−) particles at site i, respectively, and J_+ and J_- are the currents of the positive and negative particles, respectively. In addition to the bulk equations one has four other equations for the currents at the boundaries

$$\begin{aligned} J_+ &= \alpha(1 - p_1 - m_1) = \beta p_N \\ J_- &= \beta m_1 = \alpha(1 - p_N - m_N). \end{aligned} \tag{30}$$

These $(2N + 2)$ equations may be solved numerically for $(p_1, \ldots, p_N; m_1, \ldots, m_N; J_+, J_-)$ to yield the (α, β) phase diagram of the model. It is found that for large β the steady state is symmetric (with $J_+ = J_-$) while for small β the two currents are unequal in the steady state.

The matrix method discussed in the previous section has been generalised and applied to this model [37]. However it turned out that a self-consistent matrix representation could be found for this model only for $\beta = 1$ or in the limit $\alpha \to \infty$. The limit $\alpha \to \infty$ is trivially mapped on the single species TASEP model. For $\beta = 1$ a phase transition is found although no spontaneous symmetry breaking takes place. According to mean field SSB is expected only at much lower exit rates β.

To demonstrate that the non-symmetric state found in the mean field approximation for small β survives fluctuations, numerical simulations of the dynamics have been carried out and the current difference $J_+ - J_-$ has been measured as a function of time for small β, where the mean field approximation predicts a broken symmetry phase [37]. A typical time evolution of the current difference for a system of size $N = 80$ is given in Figure 4. The figure suggests that the system flips between two macroscopic states: one with a

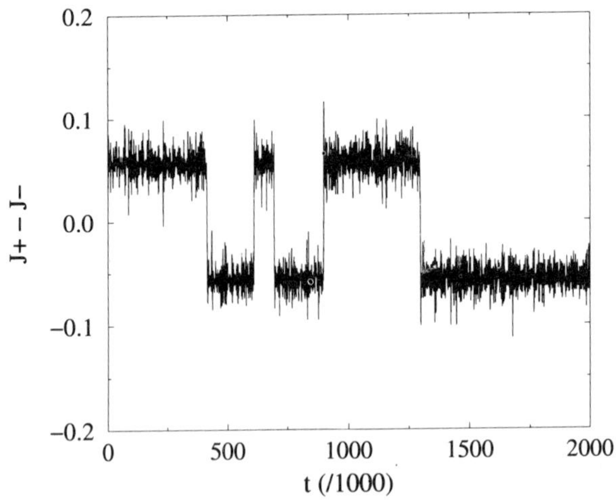

Figure 4. *The time evolution of the current difference in the broken symmetry phase for* $(\alpha = q = 1, \beta = 0.15, N = 80)$. *Time is measured in units of Monte Carlo sweeps. Flips between the two symmetry related states are clearly seen.*

positive net current and the other with negative net current. In the first case the system is predominantly loaded with positive charges moving to the right while in the second case it is loaded with negative charges moving to the left. This time course is characterised by a time-scale $\tau(N)$ which measures the average time between flips. Clearly, when averaged over time, the current difference vanishes, yielding a symmetric state. This is to be expected since we are dealing with a finite system, and one certainly does not expect SSB to take place in a finite system. The question is how does the system behave in the thermodynamic limit $N \to \infty$, and particularly how does $\tau(N)$ grow for large N. Numerical simulations suggest that τ grows exponentially with N. This means that the probability of a flip is negligibly small in a large system, and thus SSB takes place.

In order to gain some insight into the flipping process we consider the limit of very small β [40, 42]. In this limit, particles leave the system at a very small rate, and the system is filled with either positive charges moving to the right or negative charges moving to the left. Starting with a positively charged system, one would like to understand the mechanism by which a system of finite length flips into a negatively charged one. The evolution in the small β limit may be described as follows: with rate β a positive charge leaves the system at the right end. The vacancy created at this end may either move to the left with velocity 1 or may be filled with a negatively charged particle which in turn moves to the left with velocity q. When the negative charge reaches the other end of the system it is delayed for a while, but eventually leaves the system in time of order $1/\beta$. During this time, other negative charges may arrive at the left end forming a small blockage of negative charges. A typical configuration is given in Figure 5. It is composed

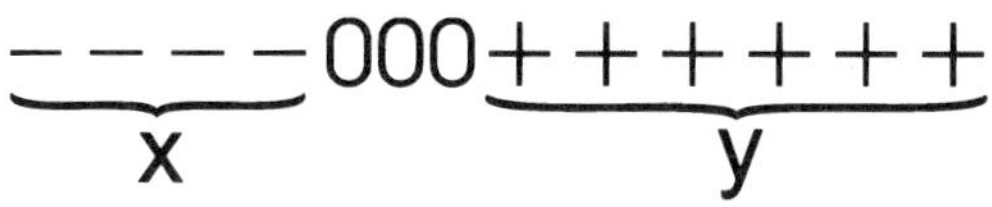

Figure 5. *A typical microscopic configuration of the 'bridge' for small β in a broken symmetry state dominated by + charges. Usually x is a small number of $O(1)$ and y is of $O(N)$ or vise versa.*

of a segment of x negative charges at the left, another segment of y positive charges at the right and in between a segment of $N-x-y$ vacancies. This configuration is denoted by (x, y). In the small β limit other configurations, for example those in which vacancies are present inside the charged segments, do not play a role in the global dynamics and may be neglected.

The dynamics restricted to (x, y) configurations is rather simple. It may be viewed as the dynamics of a single particle diffusing on a square lattice performing the following elementary moves:

$$\begin{aligned} (x,y) &\xrightarrow{b} (x+1, y-1) \\ (x,y) &\xrightarrow{a} (x, y-1) \\ (x,y) &\xrightarrow{b} (x-1, y+1) \\ (x,y) &\xrightarrow{a} (x-1, y), \end{aligned} \tag{31}$$

where

$$a = \frac{1}{2(1+\alpha)}, \qquad b = \frac{\alpha}{2(1+\alpha)}, \tag{32}$$

are the rates of the various moves. Here the first two moves correspond to a positive charge leaving the system at the right end and being replaced by a negative charge (vacancy), respectively. Similarly the last two moves correspond to a negative charge leaving the system at the left end and replaced by a positive charge (vacancy), respectively. A schematic representation of this process is given in Figure 6.

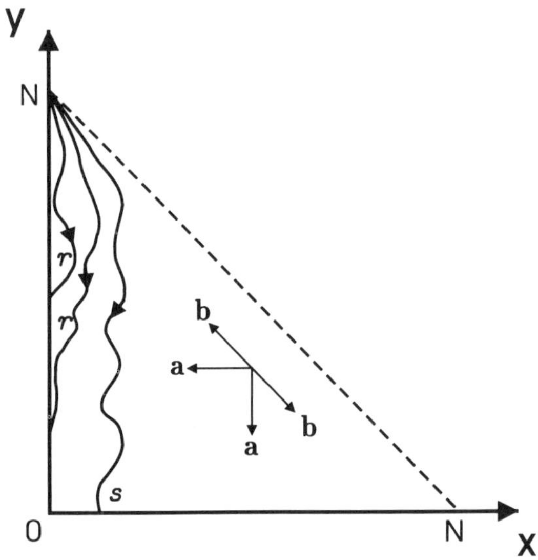

Figure 6. *A representation of the dynamics of the 'bridge' in the limit of small β as a biased diffusion process on a square lattice. The elementary moves and their rates are indicated. Starting with a positively charged system, namely at $(N, 0)$, possible trajectories are given. Typical walks, r, end on the $x = 0$ axis. Walks s, which end on the $y = 0$ axis, are rare and they correspond to a flip.*

The biased diffusion process takes place as long as the particle stays within the triangle $(x \geq 0\ ,\ y \geq 0\ ,\ x + y \leq N)$. When it reaches the boundary of the triangle, for example $x = 0$ the negative charge blockage at the left end disappears and on a very short time scale (as compared with $1/\beta$) the system is filled with positive charges from the left end, moving to $(0, N)$. The evolution of a positively charged system is thus represented by a random walk starting at $(0, N)$ with elementary steps defined by (31). Due to the bias of these elementary steps, a typical walk for large N ends on the $x = 0$ axis. Once it reaches this axis it moves back to $(0, N)$ and the process starts again. This process repeats itself until the diffusing particle performs a walk which starts at $(0, N)$ and ends on the $y = 0$ axis without touching the $x = 0$ axis while diffusing. When this happens the blockage of positive charges at the right end is removed, the system is rapidly filled with negative charges moving to the other end of the triangle $(N, 0)$. This corresponds to a flip. The probability of such a walk taking place has been calculated, yielding the following flipping time [40]:

$$\tau(N) = \frac{C}{\beta} N^{3/2} e^{\kappa N}, \tag{33}$$

where C is a constant and

$$\kappa = \ln\Big[\frac{2}{a}\Big(1 - a - \sqrt{(1-a)(1-2a)}\,\Big)\Big]. \tag{34}$$

The exponential flipping time is a direct result of the fact that the random walk corresponding to the evolution of the model is biased. It takes an exponentially long time to reach a distance of order N against a bias.

A very interesting question is related to the behaviour of nonequilibrium systems with spontaneous symmetry breaking when an *external symmetry breaking field* is introduced. In thermal equilibrium a symmetry breaking field makes the phase that is disfavoured by the field metastable or even (when the field is large) unstable. For example, a positive magnetic field applied to an ordered ferromagnetic Ising system, removes the degeneracy between the two magnetic states. Only the state with positive net magnetisation remains stable. The two magnetic states coexist only at zero field. This is a direct consequence of the Gibbs phase rule.

It is known that in nonequilibrium systems, this is not necessarily the case [43, 44]. Namely when a symmetry breaking field is applied, the state disfavoured by the field may stay as a stable thermodynamic state. This is in violation of the Gibbs phase rule, which does not hold in nonequilibrium.

The 'bridge' model described in this section provides a clear example for this behaviour [40]. To demonstrate this point we introduce a symmetry breaking field by imposing boundary conditions which favour, say, the positively charged state, thus explicitly breaking the symmetry. More specifically, we consider an exclusion model where, instead of having boundary rates α, β for both types of particles, we take α, β_+ for the positive charges and α, β_- for the negative charges. The two exit rates are taken to be of the form $\beta_\pm = \beta(1 \mp H)$, where $0 < H < 1$ is the symmetry breaking field, favouring the positively charged state. The analysis presented above in the limit $\beta \to 0$ may be repeated for non-vanishing field H and the stability of the two phases may be analysed. Here again the dynamics is reduced to a diffusion process of the type (31) but with modifies rates (see Figure 7). Clearly the positively charged state is stable since it is favoured by the

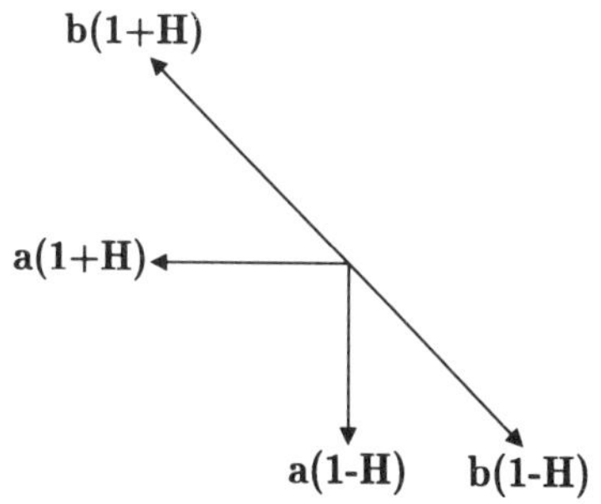

Figure 7. *The elementary moves of the biased diffusion process corresponding to the small β limit when a symmetry breaking field H is present.*

field. The question is whether the negatively charged state is stable when the field H is non-vanishing. To examine this problem, we start with a negatively charged system

$(N, 0)$ and consider a random walk defined by the diffusion process. It is easy to see that as long as $H < a/(1-a) = 1/(1+2\alpha)$ the walk is biased in the negative y direction, yielding a flipping time exponential in the system size. Thus the negatively charged state is *stable* even when it is unfavoured by the symmetry breaking field.

6 Phase separation in one dimension

A phenomenon closely related to spontaneous symmetry breaking is that of phase separation. In $1d$ equilibrium systems with short range interactions, phase separation does not take place and therefore no liquid-gas like transition is expected. The density of particles in such a system is thus macroscopically homogeneous.

Recent studies have shown that driven systems may exhibit phase separation in $1d$ even when the system is governed by local dynamics [39, 45, 46, 47]. Several models have been introduced to demonstrate this behaviour. In these models more than one type of particles is needed for phase separation to take place. In the following we consider in some detail one of these models, and analyse the mechanism leading to nonequilibrium phase separation [39].

The model is defined on a $1d$ lattice of length N with periodic boundary conditions. Each site is occupied by either an A, B, or C particle. The evolution is governed by random sequential dynamics defined as follows: at each time step two neighbouring sites are chosen randomly and the particles of these sites are exchanged according to the following rates

$$AB \underset{1}{\overset{q}{\rightleftarrows}} BA, \qquad BC \underset{1}{\overset{q}{\rightleftarrows}} CB, \qquad CA \underset{1}{\overset{q}{\rightleftarrows}} CA. \tag{35}$$

The rates are cyclic in A, B and C and conserve the number of particles of each type N_A, N_B and N_C, respectively.

For $q = 1$ the particles undergo symmetric diffusion and the system is disordered. This is expected since this is an equilibrium steady state. However for $q \neq 1$ the particle exchange rates are biased. We will show that in this case the system evolves into a phase separated state in the thermodynamic limit.

To be specific we take $q < 1$, although the analysis may trivially be extended for any $q \neq 1$. In this case the bias drives, say, an A particle to move to the left inside a B domain, and to the right inside a C domain. Therefore, starting with an arbitrary initial configuration, the system reaches (after a relatively short transient time) a state of the type $\ldots AABBCCAAAB \ldots$ in which A, B and C domains lie to the right of C, A and B domains, respectively. Due to the bias q, the domain walls $\ldots AB \ldots$, $\ldots BC \ldots$, and $\ldots CA \ldots$, are stable, and configurations of this type are long-lived. In fact, the domains in these configurations diffuse into each other and coarsen on a time scale of the order of q^{-l}, where l is a typical domain size in the system. This leads to the growth of the typical domain size as $(\ln t)/|\ln q|$. Eventually the system phase separates into three domains of the different species of the form $A \ldots AB \ldots BC \ldots C$. A finite system does not stay in such a state indefinitely. For example, the A domain breaks up into smaller domains in a time of order $q^{-\min\{N_B, N_C\}}$. In the thermodynamic limit, however, when the density of each type of particle is non vanishing, the time scale for the break up of extensive domains diverges and we expect the system to phase separate. Generically the system supports particle currents in the steady state. This can be seen by considering, say, the A domain in the phase separated state. The rates at which an A particle traverses a B (C) domain to the right (left) is of the order of q^{N_B} (q^{N_C}). The net current is then of the order of $q^{N_B} - q^{N_C}$, vanishing exponentially with N. This simple argument suggests that for the special case $N_A = N_B = N_C$ the current is zero for any system size.

The special case of equal densities $N_A = N_B = N_C$ provides very interesting insight into the mechanism leading to phase separation. We thus consider it in some detail. Examining the dynamics for these densities, one finds that it obeys *detailed balance* with respect to some distribution function. Thus in this case the model is in fact in thermal equilibrium. It turns out however that although the dynamics of the model is *local*, the effective Hamiltonian corresponding to the steady state distribution has *long range interactions*, and may thus lead to phase separation. This particular mechanism is specific to equal densities. However the dynamical argument for phase separation given above is more general, and is valid for unequal densities as well.

In order to specify the distribution function for equal densities, we define a local occupation variable $\{X_i\} = \{A_i, B_i, C_i\}$, where A_i, B_i and C_i are equal to one if site i is occupied by particle A, B or C respectively and zero otherwise. The probability of finding the system in a configuration $\{X_i\}$ is given by

$$W_N(\{X_i\}) = Z_N^{-1} q^{\mathcal{H}(\{X_i\})}. \tag{36}$$

where $\mathcal{H}$ is the Hamiltonian (setting $q = e^{-\beta}$)

$$\mathcal{H}(\{X_i\}) = \sum_{i=1}^{N} \sum_{k=1}^{N-1} \left(1 - \frac{k}{N}\right)(C_i B_{i+k} + A_i C_{i+k} + B_i A_{i+k}) - (N/3)^2, \tag{37}$$

and the partition sum is given by $Z_N = \sum q^{\mathcal{H}(\{X_i\})}$. The value of the site index $(i + k)$ in (37) is taken modulo N. In this Hamiltonian the interaction between particles is long range, growing linearly with the distance between the particles.

In order to verify that the dynamics (35) obeys detailed balance with respect to the distribution function (36,37) it is useful to note that the energy of a given configuration may be evaluated in an alternate way. Consider the fully phase separated state

$$A \ldots AB \ldots BC \ldots C \tag{38}$$

The energy of this configuration is $E = 0$, and, together with its translationally related configurations, they constitute the N-fold degenerate ground state of the system. We now note that nearest neighbour (nn) exchanges $AB \to BA, BC \to CB$ and $CA \to AC$ cost one unit of energy each, while the reverse exchanges result in an energy gain of one unit. The energy of an arbitrary configuration may thus be evaluated by starting with the ground state and performing nn exchanges until the configuration is reached, keeping track of the energy changes at each step of the way. This procedure for obtaining the energy is self consistent only when the densities of the three species are equal. To examine self consistency of this procedure consider, for example, the ground state (38), and move the leftmost particle A to the right by a series of nn exchanges until it reaches the right end of the system. Due to translational invariance, the resulting configuration should have the same energy as (38), namely $E = 0$. On the other hand the energy of the resulting configuration is $E = N_B - N_C$ since any exchange with a B particle yields a cost of one unit while an exchange with a C particle yields a gain of one unit of energy. Therefore for self consistency the two densities N_B and N_C have to be equal, and similarly, they have to be equal to N_A.

The Hamiltonian (37) may be used to calculate steady state averages corresponding to the dynamics (35). We start by an outline of the calculation of the free energy. Consider a ground state of the system (38). The low lying excitations around this ground state are obtained by exchanging nn pairs of particles around each of the three domain walls. Let us first examine excitations which are localised around one of the walls, say, AB. An excitation can be formed by one or more B particles moving into the A domain (equivalently A particles moving into the B domain). A moving B particle may be considered as a walker. The energy of the system increases linearly with the distance traveled by the walker inside the A domain. An excitation of energy m at the AB boundary is formed by j walkers passing a total distance of m. Hence, the total number of states of energy m at the AB boundary is equal to the number of ways $P(m)$ of partitioning an integer m into a sum of (positive) integers. This and related functions have been extensively studied in the mathematical literature over many years. Although no explicit general formula for $P(m)$ is available, its asymptotic form for large m is known [48]

$$P(m) \simeq \frac{1}{4m\sqrt{3}} \exp\left(\pi(2/3)^{1/2}\, m^{1/2}\right). \tag{39}$$

Also, a well known result attributed to Euler yields the generating function

$$Y = \sum_{m=0}^{\infty} q^m P(m) = \frac{1}{(q)_\infty}, \tag{40}$$

where

$$(q)_\infty = \lim_{n\to\infty} (1-q)(1-q^2)\ldots(1-q^n). \tag{41}$$

This result may be extended to obtain the partition sum Z_N of the full model. In the limit of large N the three domain walls basically do not interact. It has been shown that excitations around the different domain boundaries contribute additively to the energy spectrum [39]. As a result in the thermodynamic limit the partition sum takes the form

$$Z_N = N/[(q)_\infty]^3, \tag{42}$$

where the multiplicative factor N results from the N-fold degeneracy of the ground state and the cubic power is related to the three independent excitation spectra associated with the three domain walls.

It is of interest to note that the partition sum is linear and not exponential in N, as is usually expected, meaning that the free energy is not extensive. This is a result of the long-range interaction in the Hamiltonian and the fact that the energy excitations are localised near the domain boundaries.

Whether or not a system has long-range order in the steady state can be found by studying the decay of two-point density correlation functions. For example the probability of finding an A particle at site i and a B particle at site j is,

$$\langle A_i B_j \rangle = \frac{1}{Z_N} \sum_{\{X_k\}} A_i B_j \; q^{\mathcal{H}(\{X_k\})}, \tag{43}$$

where the summation is over all configurations $\{X_k\}$ in which $N_A = N_B = N_C$. Due to symmetry many of the correlation functions will be the same, for example $\langle A_i A_j \rangle = \langle B_i B_j \rangle = \langle C_i C_j \rangle$. A sufficient condition for the existence of phase separation is

$$\lim_{r\to\infty} \lim_{N\to\infty} (\langle A_1 A_r \rangle - \langle A_1 \rangle \langle A_r \rangle) > 0. \tag{44}$$

Since $\langle A_i \rangle = 1/3$ we wish to show that $\lim_{r\to\infty} \lim_{N\to\infty} \langle A_1 A_r \rangle > 1/9$. In fact it can be shown [39] that for any given r and for sufficiently large N,

$$\langle A_1 A_r \rangle = \frac{1}{3} - \mathcal{O}(r/N). \tag{45}$$

This result not only demonstrates that there is phase separation, but also that each of the domains is pure. Namely the probability of finding a particle a large distance inside a domain of particles of another type is vanishingly small in the thermodynamic limit.

Numerical simulations of the model for the case of unequal densities, where such analysis cannot be carried out, strongly indicate that phase separation takes place as long as none of the three densities vanish. They also indicate that the coarsening process which accompanies phase separation is rather slow, with the characteristic length diverging like $\ln t$ at long times.

7 Summary

In these lecture notes some collective phenomena which occur in one dimensional driven systems have been reviewed. These systems have been extensively studied in recent years by introducing simple models and analysing their steady state properties. Some of these models have been demonstrated to exhibit a rich variety of phenomena which are unexpected in equilibrium one dimensional systems.

Simple asymmetric exclusion processes in open systems were shown to exhibit both first order and continuous phase transitions. Other systems which have in the past been demonstrated to exhibit phase transitions in $1d$ are directed percolation [49, 50] and contact processes [51]. These system, however, possess one or more *absorbing states*. Once the system evolves into one of these states the dynamics is such that the system is unable to exit. Under these conditions, the existence of a phase transition between a trapped and an untrapped states is rather natural. Usually, once the dynamics in these models is generalised to allow for an exit from the absorbing state no phase transition takes place. The phase transitions occurring in the asymmetric exclusion processes discussed in this paper are rather different, as the dynamics in these models does not possess absorbing states.

Mechanisms which lead to spontaneous symmetry breaking and phase separation in one dimensional nonequilibrium systems have been discussed. A common crucial feature of these models is that the dynamics conserves, at least to some degree, the order parameter. In the 'bridge' model, the densities of the two types of particles are conserved in the bulk although they are not conserved at the two ends of the system. In the ABC model, on the other hand, the three densities are fully conserved. When non-conserving processes are introduced into these models spontaneous symmetry breaking and phase separation do not take place. It would be very interesting to consider the possibility of spontaneous symmetry breaking in one dimension when the dynamics does not conserve the order parameter. A related problem has been considered in the context of error correcting computation algorithms. An example of a one dimensional array of coupled probabilistic cellular automata has been constructed and shown to yield breaking of ergodicity, as would a model with spontaneous symmetry breaking [52]. This approach suggests that indeed spontaneous symmetry breaking in $1d$ may exist even when the dynamics is not conserving. However the example given is rather complicated and not well understood.

In spite of the progress made in recent years in the understanding of nonequilibrium collective phenomena, many basic questions remain open, even for the restricted and relatively simple class of systems which evolve into a steady state. For example a classification of continuous nonequilibrium transitions into universality classes, like the one which exists for equilibrium transitions, is not available. Also the dynamical process of the approach to steady state is far from being understood in many cases. The approach outlined in these notes, which involves constructing simple dynamical models and analysing the resulting collective behaviour may prove to be helpful in developing better understanding of some of these complex questions.

Acknowledgments

I thank Martin Evans and Yariv Kafri for many useful comments and for critical reading of the manuscript.

References

[1] Schmittmann B and Zia R K P, 1995, *Phase Transitions and Critical Phenomena* **17**, editors C Domb and J Lebowitz (Academic Press, London).

[2] Krug J and Spohn H, 1991, in *Solids Far From Equilibrium* editor C Godrèche (Cambridge University Press, Cambridge).

[3] Halpin-Healy T and Zhang Y C, 1995, *Physics Reports* **254** 215.

[4] Krug J, 1997, *Advances in Physics* **46** 139.

[5] Proceedings of workshop on *Traffic and Granular Flow* 1995 editors D E Wolf, M Schreckenberg and A Bachem (World Scientific, Singapore).

[6] Nagel K and Schreckenberg M, 1992, *J Physique* **I2** 2221.

[7] Schreckenberg M, Schadschneider A, Nagel K and Ito N, 1995, *Phys Rev* **E 51** 2339.

[8] Alberts B, Bray D, Raff M, Roberts K and Watson J D, 1994, *Molecular Biology of the Cell*, 3rd Edition (Garland Publishing, New York).

[9] Duke T A J, 1990 *J Chem Phys* **93** 9052; Duke T A J and Viovy J L, 1992, *Phys Rev Lett* **68** 542.

[10] Geller S, 1977, *Solid Electrolytes, Topics in Applied Physics*, **21** (Springer, Heidelberg).

[11] Salamon M B, 1979, *Physics of Superionic Conductors, Topics in Current Physics*, **15** (Springer, Heidelberg).

[12] Wilson K G and Kogut J, 1974, *Physics Reports* **12C** 75.

[13] Fisher M E, 1998, *Rev Mod Phys* **70** 653.

[14] Landau L D and Lifshitz E M, 1980, *Statistical Physics I* (Pergamon Press, New York).

[15] Mermin N D and Wagner H, 1966, *Phys Rev Lett* **17** 1133.

[16] de Gennes P G and Prost J, 1993, *The Physics of Liquid Crystals* 2nd edition (Clarendon Press, Oxford).

[17] Wheeler J, 1974, *J Chem Phys* **61** 4474.

[18] Harris A B, 1974, *J Phys C* **7** 1671.

[19] Imry Y and Ma S K, 1975, *Phys Rev Lett* **35** 1399.

[20] Katz S, Lebowitz J L, Spohn H, 1983, *Phys Rev* **B28** 1655; 1984, *J Stat Phys* **34** 497.

[21] MacDonald J T, Gibbs J H and Pipkin A C, 1968, *Biopolymers* **6** 1.

[22] MacDonald J T and Gibbs J H, 1969, *Biopolymers* **7** 707.

[23] Krug J, 1991, *Phys Rev Lett* **67** 1882.

[24] Derrida B, Domany E and Mukamel D, 1992, *J Stat Phys* **69** 667.

[25] Derrida B, Evans M R, Hakim V and Pasquier V, 1993, *J Phys A: Math Gen* **26** 1493; 1993, in *Cellular Automata and Cooperative Systems* editors N Boccara et al, (Kluwer Academic Publishers) 121.

[26] Schütz G M and Domany E, 1993, *J Stat Phys* **72** 277.

[27] Kolomeisky A B, Schütz G M, Kolomeisky E B and Straley J P, 1998, *J Phys A: Math Gen* **31** 6911.

[28] Hinrichsen H, 1996, *J Phys A: Math Gen* **29** 3659.

[29] Evans M R, Rajewsky N and Speer E R, 1999, *J Stat Phys* **95** 45.

[30] de Gier J and Nienhuis B, 1999, *Phys Rev E* **59** 4899

[31] Sandow S, 1994, *Phys Rev E* **50** 2660.

[32] Sasamoto T, 1999, *J Phys A: Math Gen* **32** 7109.

[33] Blythe R A, Evans M R, Colaiori F and Essler F H L, 1999, cond-mat/9910242.

[34] Derrida B, Janowsky S A, Lebowitz L and Speer E R, 1993, *Europhys Lett* **22** 651; Mallick K, 1996, *J Phys A Math Gen* **29** 5375.

[35] Evans M R, 1996, *Europhys Lett* **36** 13.

[36] Arndt P F, Heinzel T and Rittenberg V, 1998, *J Phys A: Math Gen* **31** 833.

[37] Evans M R, Foster D P, Godrèche C, and Mukamel D, 1995, *Phys Rev Lett* **74** 208; 1995, *J Stat Phys* **80** 69.

[38] Derrida B, Evans M R and Mallick K, 1995, *J Stat Phys* **79** 833.

[39] Evans M R, Kafri Y, Koduvely H M and Mukamel D, 1998, *Phys Rev Lett* **80** 425; 1998, *Phys Rev E* **58** 2764.

[40] Godrèche C, Luck J-M, Evans M R, Mukamel D, Sandow S and Speer E R, 1995, *J Phys A: Math Gen* **28** 6039.

[41] Arndt P F, Heinzel T and Rittenberg V, 1998, *J Stat Phys* **90** 783.

[42] Arndt P F and Heinzel T, 1998, *J Stat Phys* **92** 837.

[43] Toom A L, 1980, in *Multicomponent Random Systems* editor R L Dobrushin, in Advanced in Probability, **6** (Dekker, New York) 549.

[44] Bennett C H and Grinstein G, 1985, *Phys Rev Lett* **55** 657.

[45] Lahiri R and Ramaswamy S, 1997, *Phys Rev Lett* **79** 1150.

[46] Lahiri R, Barma M and Ramaswamy S, 2000, *Phys Rev E* **61** 1648

[47] Arndt P F, Heinzel T and Rittenberg V, 1998, *J Phys A: Math Gen* **31** L45, 1999; *J Stat Phys* **97** 1

[48] Andrews G E, 1976, *The Theory of Partitions*, Encyclopedia of Mathematics and its Applications (Addison Wesley, MA), **2** 1-4.

[49] Domany E and Kinzel W, 1984 *Phys Rev Lett* **53** 311.

[50] Kinzel W, 1985, *Z Phys B* **58** 229.

[51] Durrett R and Liggett T M, 1981, *Ann Probab* **18** 186; Jensen I and Dickman R, 1993 *J Phys A: Math Gen* **13** L423.

[52] Gacs P, 1986, *J Comput Sys Sci* **32** 15.

Supercooled liquids and glasses

Walter Kob

Johannes-Gutenberg University, Mainz, Germany

1 Introduction

Glasses are materials that are much more common in our daily life than one might naively expect. Apart from the obvious (inorganic) glasses, such as wine glasses, bottles and windows, we also have the organic (often polymeric) glasses, such as most plastic materials (bags, coatings, *etc.*). In the last few years metallic glasses have also come into daily use in many applications, including, appropriately for us here at St. Andrews, the heads of golf clubs. In view of the widespread use of these materials it might be a bit surprising to learn that glasses are not very well understood from a microscopic point of view and that even today very basic questions such as "What is the difference between a liquid and a glass?" cannot be answered in a satisfactory way. In the present lecture notes we will discuss some of the typical properties of supercooled liquids and glasses, and theoretical approaches that have been used to describe them. Since, unfortunately, it is not possible to review here all the experiments on glasses and the theoretical models to explain them we will discuss only some of the most basic issues and refer the reader who wants to learn more about this subject to other review articles and textbooks [1].

In the following section we will review some of the basic phenomena that are found in supercooled liquids and glasses. Subsequently we will discuss the theoretical approaches to describing the dynamics of these systems, notably the so-called mode-coupling theory of the glass transition. This will be followed by the presentation of results of computer simulations to check to what extent this theory is reliable. These results are concerned with the *equilibrium* dynamics. If the temperature of the supercooled liquid is decreased below a certain value, the system is no longer able to equilibrate on the time scale of the experiments, *i.e.* it undergoes a glass transition. Despite the low temperatures the system still shows a very interesting dynamics, the nature of which is today still quite unclear. Therefore we will present in the final part of these lecture notes a brief discussion of this dynamics and its implication for the (potential) connection of structural glasses with spin glasses.

In what follows, we use for simplicity the language of atomic rather than colloidal systems (see Frenkel, this volume, for a discussion of the analogies). Note that in the context of hard-sphere colloids, which have no temperature dependence, the analogue of supercooling is to raise the colloid volume fraction (or osmotic pressure) to above the one at which crystallisation would occur in equilibrium. With non hard-core (e.g. DLVO) interactions present, the colloidal glass transition may be driven by varying temperature as well as density. However, such interactions are often explicitly temperature dependent (they include entropic as well as enthalpic terms); by sticking to the atomic case, we avoid that complication. The other main difference between colloids and atoms is in the nature of the dynamics (diffusive rather than ballistic): we address this below.

2 Supercooled liquids and the glass transition

In this section we will discuss some of the properties of supercooled liquids and some of the phenomena of the glass transition.

If a liquid is cooled from high temperatures to below its melting point T_m one expects it to crystallise at T_m. However, since the crystallisation process takes some time (critical nuclei have to be formed and have to grow) it is possible to supercool most liquids, *i.e.* they remain liquid-like even below T_m. Some liquids can be kept in this metastable state for a long time and thus it becomes possible to investigate their properties experimentally. For reasons that will become clear below, such liquids are called good glass-formers. It is found that with decreasing temperature the viscosity η of these systems increases by many orders of magnitude. In order to discuss this strong temperature dependence it is useful to define the so-called glass transition temperature T_g by requiring that at T_g the viscosity is 10^{12} Pa s, which corresponds *roughly* to a relaxation time of 100 seconds. (Reminder: water at room temperature has a viscosity around 10^{-3} Pa s). In Figure 1 we show the temperature dependence of $\log \eta$ for a variety of glass-formers as a function of T/T_g. From that plot we see that the viscosity does indeed increase dramatically when temperature is decreased. Furthermore we recognise that this temperature dependence depends on the material in that there are substances in which $\eta(T)$ is very close to an Arrhenius law, *i.e.* are almost straight lines, and other substances in which a pronounced bend in $\eta(T)$ is found. In order to distinguish these different temperature dependencies Angell coined the terms 'strong' and 'fragile' glass-formers for the former and latter, respectively [3]. (Note: this meaning of the word 'fragile' has no obvious connection with one introduced more recently in the context of granular media; see Cates, this volume.)

The strong temperature dependence which is found in $\eta(T)$ is not a unique feature of the viscosity. If other transport quantities, such as the diffusion constant or relaxation times are measured, it is found that they show a similar temperature dependence to the viscosity. On the other hand if thermodynamic quantities, such as the specific heat, or structural quantities, such as density or the structure factor, are measured, they show only a relatively mild dependency in the same temperature interval, *i.e.* they vary by between 10% and a factor of 2–3.

Equipped with these experimental facts one can now ask the main question of glass physics: what is the reason for the dramatic slowing down of the dynamics of supercooled liquids without an apparent singular behaviour of the static quantities? Although this question seems to be a very simple one it has not been possible up to now to find a completely satisfactory answer to it. One obvious response is to postulate the existence of a second-order phase transition at a temperature below T_g. Then the slowing down could

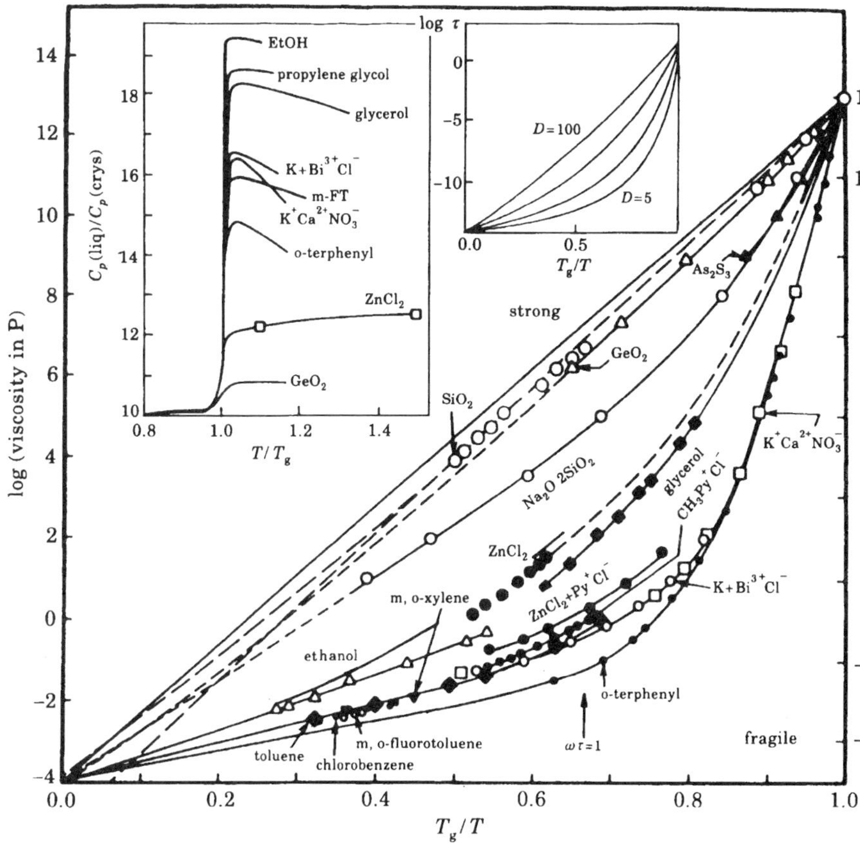

Figure 1. *Main figure: viscosity of different glass-formers as a function of T_g/T, where T_g is the glass transition temperature. Left inset: temperature dependence of the specific heat, normalised to its value for the crystal, for different glass-formers. From reference [2], with permission.*

be explained as the usual critical slowing down observed at the critical point. Although such an explanation is, from a theoretical point of view, very appealing it suffers one big drawback, namely that so far it has not been possible to identify an order parameter which characterises this phase transition or a characteristic length scale which diverges. Thus despite the nice theoretical concept, the phase transition idea is not able to provide a satisfactory explanation for the slowing down of the dynamics.

Things look much better for a different theoretical approach, the so-called mode-coupling theory (MCT) of the glass transition, which we will discuss in more detail below. This theory is indeed able to make qualitative and quantitative prediction for the time and temperature dependence of various quantities and experiments and computer simulations have shown that many of these predictions are true [4]. However, before we discuss the predictions of MCT we return to the temperature dependence of the viscosity or the relaxation times. From Figure 1 it is clear that for every material there will be a temperature at which the relaxation time of the system will exceed by far any experimental time scale. This means that it will not be possible to probe the *equilibrium* behaviour of the system below this temperature. If the system is cooled continously with a given cooling rate from a high temperature to low temperatures there will exist a temperature T'_g at which the typical relaxation time of the system is comparable to the inverse of the cooling rate. Hence, in the vicinity of this temperature the system will fall out of equilibrium and become a glass. (Note that the other glass transition temperature that we have introduced above, T_g, is basically the value of T'_g if one assumes that the relaxation time is on the order of 100 s.) This glass transition is accompanied by the freezing of those degrees of freedom which lead to a relaxation of the system, such as the motion of the particles beyond the nearest neighbour distance. Since below T'_g these degrees of freedom are no longer able to take up energy, the specific heat shows a drop at T'_g, as can be seen in the left inset of Figure 1. Note that empirically it is found that the fragile glass-formers show a large drop in the specific heat whereas the strong glass-formers show only a small one. Note, however, that this correlation is just an empirical one (and it does not hold strictly) and apart from hand-waving arguments it is not understood from a theoretical point of view. The same is also true for the distinction between strong and fragile glass-formers. So far it is not clear what the essential features in a Hamiltonian are that make the system strong or fragile, *i.e.* whether they include the range of the interaction, the coordination number, *etc.*

At the beginning of this section we mentioned that the dynamics of glass-forming liquids becomes slow when they are cooled below the melting temperature. However, it is not a necessary condition for a slow dynamics that the temperature is below T_m. For example, silica has a melting temperature around 2000K and a glass transition temperature around 1450K [5]. From Figure 1 it becomes obvious that at $T = T_m = T_g/0.725$ the viscosity is already on the order of 10^7Pa s! Thus it is clear that slow dynamics has nothing to do with the system being supercooled, or in other words: for the glass transition the melting temperature is a completely irrelevant quantity. Despite this fact we will in the following continue to talk about 'supercooled' liquids, following the normal (imprecise) usage of this term.

We now turn our attention to the MCT, the theory we have briefly mentioned earlier. Here we will give only a very sketchy idea of this theory and refer the reader who wants to learn more about it to the various review articles on MCT [4, 6]. In the MCT the quantities of interest are the correlation functions between the density fluctuations of the particles. If we denote by $\mathbf{r}_j(t)$ the position of the particle j at time t the density fluctuations are given by [7]

$$\delta\rho(\mathbf{q},t) = \sum_{j=1}^{N} \exp(i\mathbf{q}\cdot\mathbf{r}_j(t)) \,, \tag{1}$$

where $\mathbf{q}$ is the wave-vector. From this observable one can calculate the so-called intermediate scattering function $F(q,t)$ which is given by

$$F(q,t) = \frac{1}{N}\langle\delta\rho(-\mathbf{q},t)\delta\rho(\mathbf{q},0)\rangle \,. \tag{2}$$

Here the angular brackets stand for the thermodynamic average. The relevance of the function $F(q,t)$ is given by the fact that it can be directly measured in neutron and light scattering experiments. (See also Pine, this volume.) From a theoretical point of view this correlation function is important since many theoretical descriptions of (non-supercooled) liquids are based on it, or its time and space Fourier transforms [7].

Using the Mori-Zwanzig projection operator formalism [7] it is now possible to derive *exact* equations of motion for the $F(q,t)$. This is done by assuming that $F(q,t)$ are the only 'slow' quantities of interest. Because of the continuity equation, which relates the time derivative of the density of the particles to the currents, the currents also become slow variables. Using these assumptions one arrives at the equations:

$$\ddot{F}(q,t) + \Omega^2(q)F(q,t) + \int_0^t d\tau M(q,\tau)\dot{F}(t-\tau) = 0. \tag{3}$$

Here $\Omega^2(q)$ is given by $q^2 k_B T/mS(q)$, where m is the mass of the particles and $S(q)$ is the static structure factor, *i.e.* $S(q) = F(q,0)$. The function $M(q,\tau)$ is called the memory function and formally exact expressions exist for it. However, because of their complexity, these formal expressions are basically useless for a real calculation and thus in MCT one approximates $M(q,\tau)$ by a quadratic form of the density correlators. In particular it is found that $M(q,t)$ is given by

$$M(q,t) = \frac{1}{2(2\pi)^3}\int d\mathbf{k}\, V^2(q,k,|\mathbf{q}-\mathbf{k}|)F(k,t)F(|\mathbf{q}-\mathbf{k}|,t) \tag{4}$$

where the vertex V^2 is given by

$$V^2(q,k,|\mathbf{q}-\mathbf{k}|) = \frac{n}{q^2}\left(\hat{\mathbf{q}}\cdot[\mathbf{k}c(k) + (\mathbf{q}-\mathbf{k})c(|\mathbf{q}-\mathbf{k}|)]\right)^2 \tag{5}$$

and the so-called direct correlation function $c(k)$ can be expressed via the structure factor by $(S(k)-1)/nS(k)$, where n is the particle density. Thus we see that within MCT the static structure factor determines the vertex V^2, which in turn determines the memory function for the time dependent correlation function. Or in other words: the statics determine the dynamics.

Note that similar equations of motion as the one for $F(q,t)$ exist for the incoherent intermediate scattering function, $F_s(q,t)$. This correlation function is given by

$$F_s(q,t) = \frac{1}{N}\left\langle \sum_j^N \exp\left[i\mathbf{q}\cdot\{\mathbf{r}_j(t) - \mathbf{r}_j(0)\}\right]\right\rangle , \tag{6}$$

i.e. it is just the self (or diagonal) part of $F(q,t)$. Also this time correlation function is important since it can be measured in scattering experiments.

Instead of making at this point a detailed discussion of the properties of the solutions of these MCT equations, we will postpone this discussion to Section 4 where we will make a detailed comparison of the prediction of MCT with the results of computer simulations. The only thing that we mention at this point is that it has been shown that *at long times* there are two types of solution of the MCT equations. The first one is the solution $\lim_{t\to\infty} F(q,t) = 0$. This solution is the only one at high temperature and it corresponds to the physical situation that the system is ergodic, *i.e.* all time correlation functions decay to zero. (Note that temperature enters though the temperature dependence of the static structure factor $S(q)$.) The second solution has the property that $\lim_{t\to\infty} F(q,t) > 0$

and it occurs only below a critical temperature T_c. Since, in this case, the correlation functions do not decay to zero even at long times, the system is no longer ergodic, *i.e.* it is a glass. Thus within the MCT the system undergoes a glass transition at T_c. MCT now makes an asymptotic expansion of the dynamics around this critical point, *i.e.* it treats the quantity $\epsilon = (T_c - T)/T_c$ as a small parameter. Hence all the predictions of the theory are, strictly speaking, only valid very close to T_c and it is difficult to say *a priori* how far away from T_c they are still useful. However, our experience of analysing data has shown that the theory can be used for values ϵ as large as 0.5 or so [4], thus with respect to this the situation seems to be much better than the case of critical phenomena.

3 On computer simulations

In the last few decades computer simulations have been shown to be a very powerful tool to gain insight into the behaviour of statistical mechanic systems and thus can be considered to be a very useful addition to experiments and analytical calculations. Present days computer codes for such simulations are usually quite complex and thus we are not going to discuss the various tricks used in such simulations but refer the reader to some textbooks [8] and the lectures of Kremer (this volume).

Simulations of supercooled liquids and glasses pose special problems for computer simulators since at low temperatures the relaxation times are large, see the previous section, and thus the simulations have to be done for many (microscopically small!) time steps. Fortunately it is usually not necessary to use very large system sizes, a few hundred to a few thousand particles are adequate for most cases, and thus all the computer resources are spent to simulate the system over a time span which is as large as possible. Therefore present day state of the art calculations extend over 10–100 million time steps which corresponds to between several months and several years of CPU time on a fast processor. Note that despite this effort the length of such a run corresponds to only about 10^{-7} seconds, since each time step is on the order of 10^{-15} seconds. However, it should be noted that the time window of these simulations, *i.e.* 7–8 decades, exceeds the one of most experimental techniques, such as neutron or light scattering. A more extensive discussion of advantages and disadvantages of computer simulations of supercooled liquids and glasses and references to the original literature can be found in [9].

We now discuss some of the details of the simulations whose results will be discussed in the next few sections. As mentioned in the previous paragraph the main issue of computer simulations of supercooled liquids is to investigate the system at a temperature which is as close as possible to the glass transition temperature, *i.e.* in that temperature range where the relaxation times of the system are large. Therefore it is advisable to use a system that can be simulated as efficiently as possible. Hence many investigations have been done for so-called 'simple liquids', *i.e.* systems in which the interaction between the particles is isotropic and short ranged. One possible example of such a system is a one-component Lennard-Jones liquid. The main drawback of this system is that it is prone to crystallisation, *i.e.* something which in this business has to be avoided at all costs. Therefore it has become quite popular to study *binary* liquids, since the additional complexity of the system is sufficient to prevent crystallisation, at least on the time scale accessible to today's computer simulations.

The system we study is hence a binary Lennard-Jones liquid and in the following we will call the two species of particles 'type A' and 'type B' particles. The interaction between two particles of type α and β, $\alpha, \beta \in \{\mathrm{A}, \mathrm{B}\}$, is thus given by:

$$V_{\alpha\beta} = 4\epsilon_{\alpha\beta}\left[\left(\frac{\sigma_{\alpha\beta}}{r}\right)^{12} - \left(\frac{\sigma_{\alpha\beta}}{r}\right)^{6}\right].$$

The values of the parameters $\epsilon_{\alpha\beta}$ and $\sigma_{\alpha\beta}$ are given by $\epsilon_{AA} = 1.0$, $\sigma_{AA} = 1.0$, $\epsilon_{AB} = 1.5$, $\sigma_{AB} = 0.8$, $\epsilon_{BB} = 0.5$, and $\sigma_{BB} = 0.88$. This potential is truncated and shifted at a distance $\sigma_{\alpha\beta}$. In the following we will use σ_{AA} and ϵ_{AA} as the units of length and energy, respectively (setting the Boltzmann constant $k_B = 1$). Time will be measured in units of $\sqrt{m\sigma_{\mathrm{AA}}^2/48\epsilon_{\mathrm{AA}}}$, where m is the mass of the particles.

In the following we will study two types of dynamics for this system: a Newtonian dynamics (ND) and a stochastic dynamics (SD). The reason for investigating the ND is that it is a realistic dynamics for an atomic liquid. Thus it is possible to study at low temperatures the interaction of the phonons with the relaxation dynamics of the system. On the other hand the SD is a good model for a colloidal suspension in which the particles are constantly hit by the (much smaller) particles of the bath. In such systems the phonons are strongly damped and thus such a dynamics is one way to 'turn off' the phonons. Hence, by comparing the results of the two types of dynamics it becomes possible to find out which part of the dynamics is universal, *i.e.* does not depend on the microscopic dynamics, and which part is non-universal.

In both types of simulations the number of A and B particles were 800 and 200, respectively. The volume of the simulation box was kept constant at a value of $(9.4)^3$, which corresponds to a particle density of around 1.2. The temperatures used were 5.0, 4.0, 3.0, 2.0, 1.0, 0.8, 0.6, 0.55, 0.5, 0.475, 0.466, 0.456, and 0.446. At each temperature the system was thoroughly equilibrated for a time span which significantly exceeded the typical relaxation times of the system at this temperature. At the lowest temperatures this took up to 40 million time steps. As we will see, the relaxation times for the SD are, at low temperatures, significantly longer than the ones for the ND. Therefore we used in all cases the ND to equilibrate the sample and used the SD only for the production runs. For the ND we used at low temperatures a time step of 0.02, whereas for the SD a smaller one, 0.008, was needed in order to avoid systematic errors in the equilibrium quantities. In order to improve the statistics of the results we averaged over eight independent runs.

4 The equilibrium relaxation dynamics

In this section we will discuss the relaxation dynamics of the system *in equilibrium*. The main emphasis will be to find out to what extent this relaxation dynamics depends on the microscopic dynamics and which aspects of it can be understood within the framework of the mode-coupling theory.

Before we study the *dynamical* properties of the system it is useful to have a look at its *static* properties. In Section 2 we have mentioned that in the supercooled regime thermodynamic quantities and structural quantities show only a weak temperature dependence. That this is the case for the present system as well is demonstrated in Figure 2, where we show the static structure factor for the A particles, $S_{\mathrm{AA}}(q)$, for different temperatures

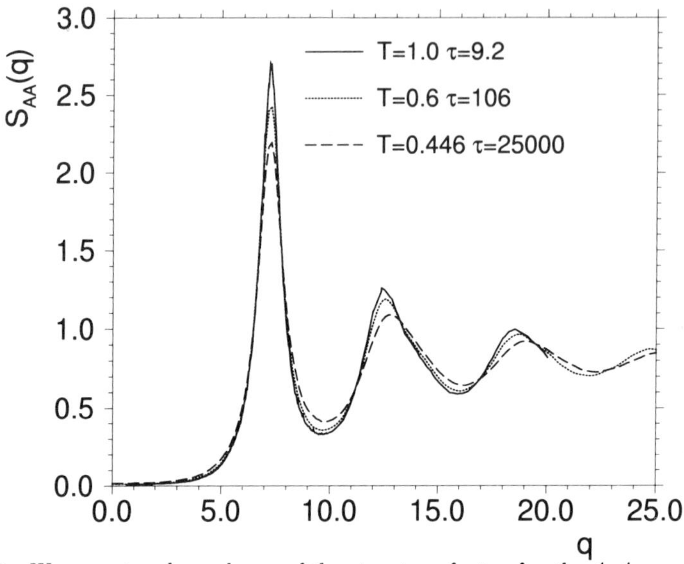

Figure 2. *Wave vector dependence of the structure factor for the A-A correlation for different temperatures. Also included are the relaxation times for the intermediate scattering function for the A-particles.*

T. From this figure we recognise that the T-dependence of $S_{\mathrm{AA}}(q)$ is quite mild, in that the main effect of a decreasing temperature is that the various peaks become more pronounced and narrower. A similar weak T-dependence is also found for the pressure and the total energy of the system [10]. In order to demonstrate that, in the temperature range shown, the *dynamics* of the systems changes strongly we have also included in the figure the typical relaxation times, defined more precisely below, at the different temperatures. From these numbers we see that in this temperature range the relaxation dynamics slows down by about two and a half orders of magnitude, a huge amount compared with the weak temperature dependence of the static structural quantities.

One of the simplest possibilities for studying the dynamics of a liquid is to investigate the time dependence of the mean-squared-displacement (MSD) which is defined by

$$\langle r^2(t)\rangle = \frac{1}{N_\alpha}\sum_{i=1}^{N_\alpha}\langle|\mathbf{r}_i(t) - \mathbf{r}_i(0)|^2\rangle \tag{7}$$

Note that here the sum over the particles of type α is not really needed since in principle all particles of the same kind are statistically equivalent. However, in order to improve the statistics for the MSD it is advisable to make the additional average over the particles of the same kind.

In Figure 3 we show the time dependence of the MSD for the A particles at the different temperatures. Let us start our discussion for the high temperatures, curves to the left. For very short times the particle flies ballistically, since on this time scale it does not even realise that it is part of a many body system. Thus its position is given by $\mathbf{r}_i(t) = \mathbf{r}_i(0) + \mathbf{v}_i(0)t$, where $\mathbf{v}_i(0)$ is its initial velocity. Thus the MSD is proportional to t^2, which is the time dependence seen at short times (see figure).

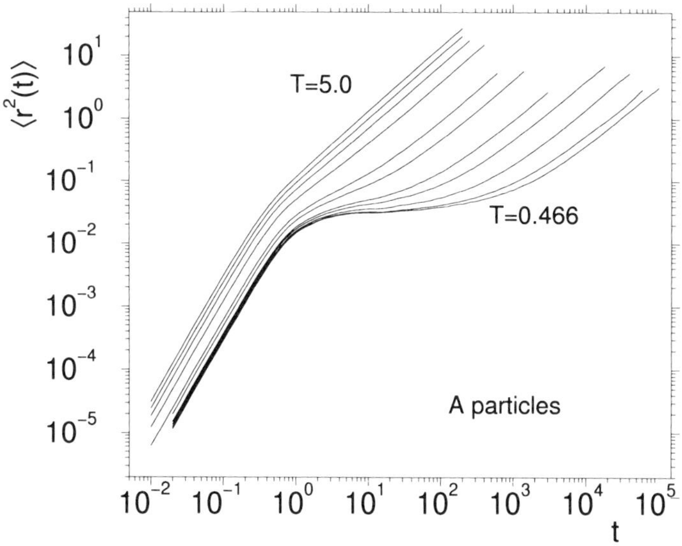

Figure 3. *Time dependence of the mean squared displacement for the A particles for different temperatures.*

After the ballistic flight the particle collides with its nearest neighbours and thus its motion becomes diffusive, *i.e.* $\langle r^2(t)\rangle = 6Dt$, where D is the diffusion constant. This diffusive behaviour is readily seen in the curves at long times. The two time regimes just discussed are also found in the MSD for low temperatures. In addition to them we see from the figure that a third regime is present, in that the ballistic and diffusive regime are separated by a time window in which the MSD shows a plateau. This means that in this time regime the particle does not significantly increase its distance from the point it was at time zero. The physical picture behind this behaviour is the so-called 'cage effect', *i.e.* the fact that on this time scale the particle is trapped by its surrounding neighbours. Only at long times is the particle is able to escape this cage and to become diffusive again. Note that the particles forming the cage are of course also caged since they are surrounded by *their* neighbours. Hence it becomes clear that in order to obtain a correct description of the dynamics of the particles inside the cage and the breaking up of this cage, it is necessary to make a *self-consistent* ansatz for the motion of the particle and its cage and MCT is one way to do this.

Since the intermediate scattering function $F(q,t)$ and its self part $F_s(q,t)$ are of experimental relevance and are also the main focus of MCT, it is of course interesting to investigate their time and temperature dependence. In Figure 4 we show the time dependence of $F_s(q,t)$ for different temperatures. The wave-vector q is 7.25, the location of the maximum in the static structure factor for the A-A correlation. (For other wave-vectors the correlation functions look qualitatively similar [11].) Also in this figure we find the different time regimes that we have discussed in the context of the MSD. At very short times the correlator shows a quadratic time dependence, which corresponds to the ballistic motion in the MSD. At high temperatures we see that, after this time regime, $F_s(q,t)$ decays rapidly to zero, and it is found that this decay is described well by an

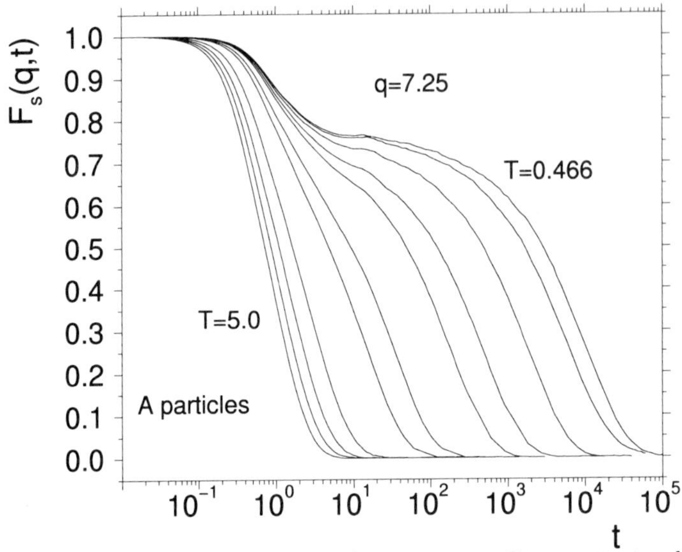

Figure 4. *Time dependence of the incoherent intermediate scattering function of the A particles for all temperatures studied.*

exponential. This behaviour is typical for a liquid at *high* temperatures and is not specific to the present system. At low temperatures the quadratic time dependence is also found at short times. However, in contrast to the high temperature case we find at intermediate times a plateau, the origin of which is again the cage effect that we have discussed before. Only at very long times does the correlation function decay to zero. This ultimate decay is *not* given by an exponential, but by a so-called Kohlrausch-Williams-Watts (KWW) law (also called stretched exponential), *i.e.* by $A\exp(-(t/\tau)^\beta)$, where the amplitude A, the time scale τ and the Kohlrausch-exponent β depends on the wave-vector.

For the following discussion a bit of nomenclature is useful: the time range in which the correlation function is *close* to the aforementioned plateau is called the β-relaxation regime. The time window in which the correlator falls *below* the plateau is called the α-relaxation. Note that the *late* β-relaxation coincides with the *early* α-relaxation regime.

MCT predicts that in the vicinity of the critical temperature T_c the so-called time-temperature superposition principle holds in the α-relaxation regime. This means that a time correlation function $\phi(t)$ can be written as follows:

$$\phi(t,T) = \Phi(t/\tau(T)), \tag{8}$$

where Φ is a master function which depends on ϕ, and $\tau(T)$ is the α-relaxation time at temperature T, which also depends on ϕ. In order to check the validity of this prediction, we define the α-relaxation time as that time at which the correlator has fallen to $1/e$ of its initial value. If Equation 8 does indeed hold, a plot of the correlation function versus the rescaled time $t/\tau(T)$ should give, *in the α-relaxation regime*, a master curve.

That this is indeed the case is shown in Figure 5, where we show the same data as in Figure 4, but this time versus $t/\tau(T)$. We clearly see that the curves at low temperatures

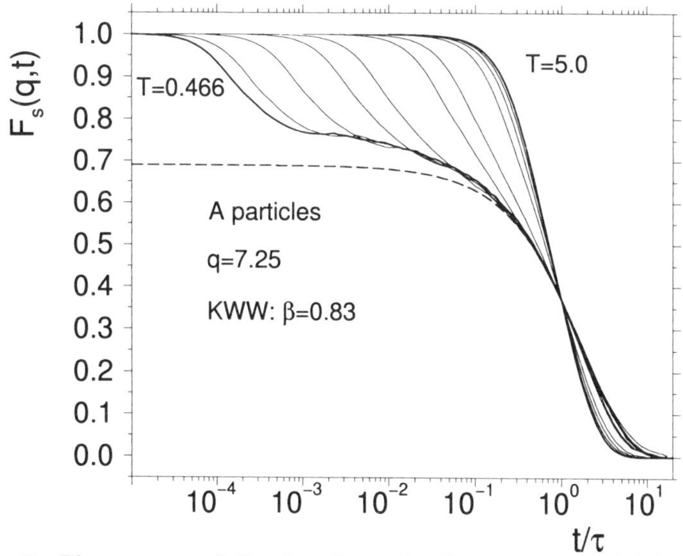

Figure 5. *The same correlation function as in Figure 4 versus* $t/\tau(T)$, *where* $\tau(T)$ *is the* α*-relaxation time of the system. The dashed curve is a fit with the KWW function to the master curve in the* α*-relaxation regime.*

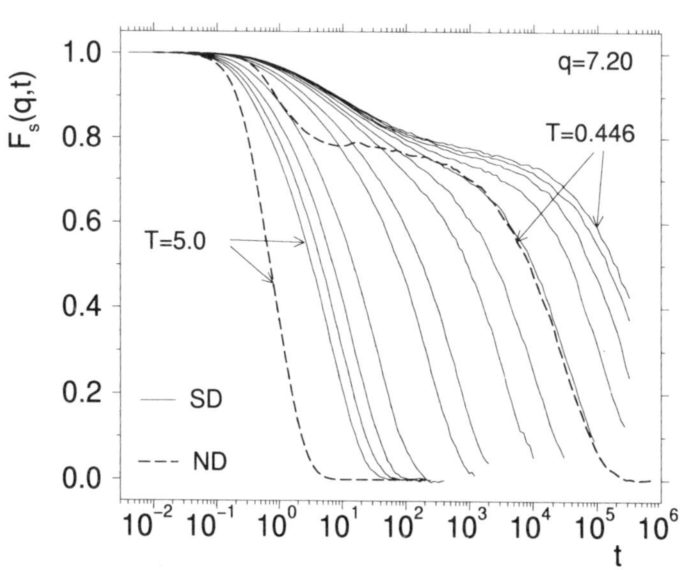

Figure 6. *The incoherent intermediate scattering function for the stochastic dynamics for all temperatures investigated (solid lines). Bold dashed line:* $F_s(q,t)$ *for the Newtonian dynamics at* $T = 5.0$ *and* $T = 0.446$.

fall nicely onto a master curve. In addition MCT predicts that the shape of this master curve can be fitted well by the aforementioned KWW law, and a fit with this functional form is included in the figure as well, showing that this law does indeed fit our master curve very well.

The results discussed so far are all for the ND, *i.e.* the dynamics in which the microscopic motion of the particles is not damped. In order to see how the relaxation dynamics changes if we have a strong damping, we show in Figure 6 the self intermediate scattering function for the stochastic dynamics (solid lines). The wave-vector is the same as the one in Figure 4. We see that the time and temperature dependence of the correlator is qualitatively the same as in the case of the ND. However, a closer inspection shows important differences between the two types of dynamics and in order to see them better we have also included Figure 6 two curves for the ND (dashed lines). First of all the time scale for the α-relaxation is significantly larger for the SD than for the ND. Whereas at high temperature the SD relaxes slower by a factor of about seven, this factor increases to a value around 30 at the lowest temperature, and then stays constant [12]. Note however, that apart from this change of time scale the α-relaxation is the same, in that the shape of the curves as well as the height of the plateau in Figure 6 is the same for ND and SD. This is exactly what is expected within MCT in that the theory predicts that at temperatures around T_c the temperature dependence of the dynamics is independent of the microscopic dynamics, *apart from a system-universal constant factor.* By this it is meant that the time scale of the α-relaxation will depend on the microscopic dynamics, but that this dependence is of a very special form, in that the α-relaxation time of *all* observables scales by the *same, temperature-independent* factor.

Although the relaxation of the curves away from the plateau is independent of the microscopic dynamics, the approach of the curves *to* the plateau depends on it. In particular we see from Figure 6 that for the ND this approach is very abrupt whereas it is very gentle for the SD. The reason for this difference is that in the SD the phonon-like motion of the particles is strongly damped and thus the particles explore their cage in a much gentler way than they do in the ND.

In order to investigate this part of the β-relaxation dynamics in more detail we show in Figure 7 the SD curves from Figure 6 and the ND curve at low temperature versus $t/\tau(T)$. From this figure we see that the curves for the two different kinds of dynamics do indeed fall on top of each other in the α-relaxation regime but that they show the mentioned differences in the early β-relaxation regime.

MCT predicts that *in the β-relaxation regime* the shape of the master curve is not arbitrary, but is given by the so-called β-correlator, a function which is the solution of a certain integral equation [4, 6]. This integral equation, and hence its solution, depend on one parameter λ, the so-called 'exponent parameter'. The value of λ can be calculated from the structure factor and for the present system has the value 0.708 [13]. Using this value of λ it is possible to solve the aforementioned integral equation, and thus to calculate the β-correlator. In Figure 7 we have included (bold dotted line) the best fit with this β-correlator and we recognise that this functional form gives a very good description of the correlators in the vicinity of the plateau. In particular we see that in the case of the SD the fit is also good in the early β-regime, thus showing that the damping of the motion leads to a much better agreement with the theory. The reason for this is that if no damping is present, the dynamics at short times, which is governed by phonon-like motion,

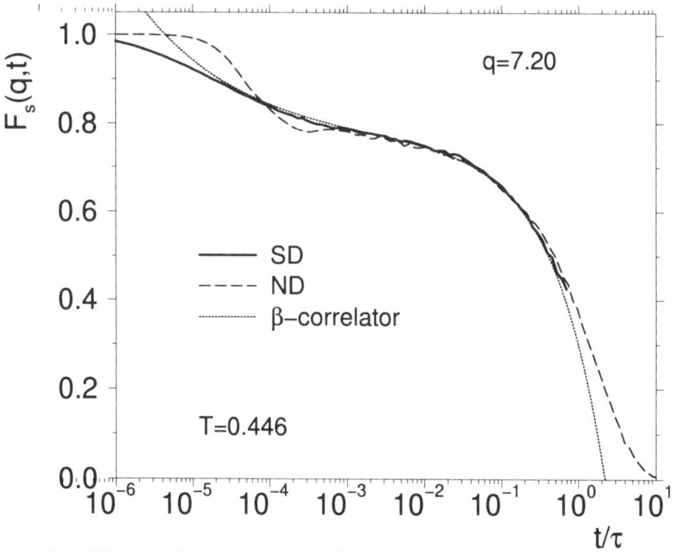

Figure 7. *The incoherent intermediate scattering function for the ND and SD, dashed and solid lines, respectively, versus the scaled time $t/\tau(T)$. Here $T = 0.446$. The dotted curve is a fit with the β-correlator from MCT.*

strongly interferes with the relaxation in the early β-relaxation regime and thus leads to the observed discrepancy between the β-correlator and the curve from the ND. However, if one takes account of this phonon-like dynamics in the theory, a good agreement between the theory and the ND curves is found also [14]. Thus we can conclude that MCT is able to give correct description of the β-relaxation dynamics on a quantitative as well as qualitative level.

We now turn our attention to the temperature dependence of the diffusion constant and the relaxation times. MCT predicts that in the vicinity of T_c these quantities should show a power-law dependence, *i.e.*

$$D(T) \propto \tau^{-1}(T) \propto (T - t_c)^{-\gamma} \tag{9}$$

where the exponent γ can be calculated from the exponent parameter λ and is found in our system to be 2.34 [13].

In Figure 8 we show the temperature dependence of the diffusion constant and the α-relaxation time τ for the A particles for the ND and SD cases. In order to check for the presence of the power law given by Equation 9 we plot these quantities versus $T - T_c$, where the critical temperature T_c was used as a fit parameter. (We mention in passing that in principle it is possible to calculate the value of T_c within MCT. However, it has been found that the theoretical value, $T_c = 0.92$, is very far from the one determined from the correlation functions ($T_c = 0.435$) [13]; this discrepancy is not a particularity of the present system but reflects the fact that MCT seems to have difficulty in estimating this quantity with high accuracy.) Returning to Figure 8 we see that in the supercooled regime the data can be fitted very well by such a power law. In particular we find that the exponent γ of the power law for the relaxation time is independent of the microscopic

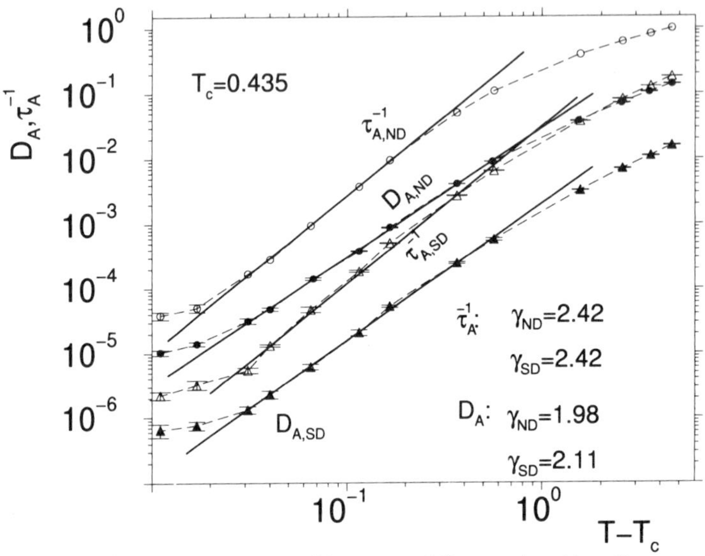

Figure 8. *Diffusion constant and inverse of the α-relaxation time versus $T - T_c$. Open and filled symbols correspond to the Newtonian and stochastic dynamics, respectively. The bold straight lines are fits to the data with power laws.*

dynamics, see the values for γ in the figure, and the same is true also for the exponents for the diffusion constants. However, in contrast to the prediction of the theory (see Equation 8), the exponent for the relaxation time is *not* the same as the ones for the diffusion constant. It is likely that the reason for this lies in the fact that the system is dynamically heterogeneous [15, 16], *i.e.* it has regions in which the dynamics of the particles is significantly faster than in other regions. Since it is not possible to take into account such dynamical differences within MCT due to the mean-field like nature of the theory, the prediction of MCT for the temperature dependence of the product $D(T)\tau(T)$ is, *for the present system*, not correct.

So far we have only tested the applicability of MCT on a *qualitative* level. These types of checks, and many more, can be done for all systems for which the dynamics has been studied in a temperature range in which the time scale of the dynamics changes considerably and in Reference [4] many of these tests are discussed. For simple liquids also *quantitative* tests are possible if the static structure factor is known with sufficiently high accuracy (*e.g.* 1% accuracy for wave-vectors between $0.1q_0 \leq q \leq 3q_0$, where q_0 is the location of the maximum in $S(q)$). For this one has to solve the wave-vector dependent mode-coupling Equations 3–5, using the static structure factor as input. This has been done for hard sphere systems and the theoretical results compare nicely with the ones from experiments on colloidal particles [4, 17]. Similar calculations have also been done for soft sphere systems [18] and water [19]. Here we will discuss the results for the present Lennard-Jones mixture. One quantity which is relatively simple to calculate is the value of the so-called critical non-ergodicity parameter, which is the height of the plateau in some time correlation function *at T_c*. Note that for the case that the correlation function

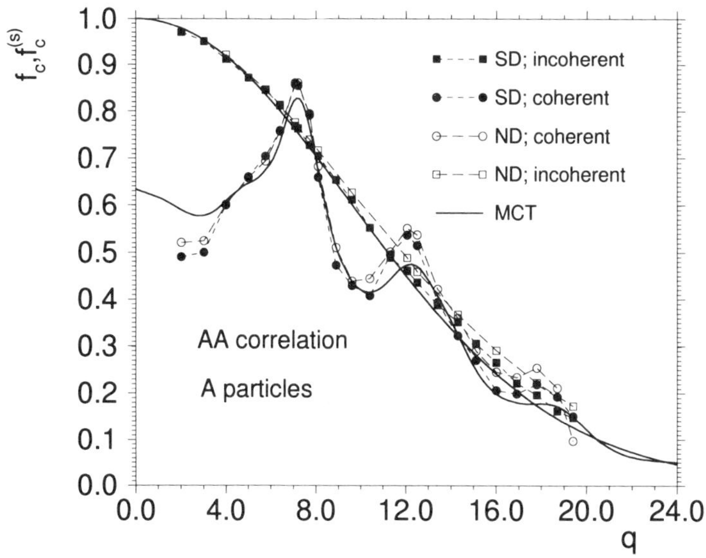

Figure 9. *Wave-vector dependence of the non-ergodicity parameter for the coherent and incoherent intermediate scattering function (circles and squares, respectively). The open and closed symbols correspond to the Newtonian and stochastic dynamics, respectively.*

is the intermediate scattering function this parameter will depend on the wave-vector as well as on the type of particle.

In Figure 9 we show the q dependence of the non-ergodicity parameter for the case of $F_s(q,t)$ for the A particles as well as for $F(q,t)$ for the A-A correlation (ND, open symbols). We see that the coherent part shows an oscillatory behaviour which is in phase with the structure factor. The reason for this is that the structure of the liquid is very stiff on the length scale of the interparticle distances, thus leading to a high plateau in the time correlation function, *i.e.* a large non-ergodicity parameter.

In order to check whether the value of the non-ergodicity parameters depend on the microscopic dynamics we have also included in Figure 9 the data for the SD. We see that the curves for the SD are very close to the ones for the ND and thus conclude that the height of the plateau is independent of the microscopic dynamics, in agreement with the prediction of MCT. Also included in the figure is the theoretical prediction from MCT for the non-ergodicity parameter (solid lines) [13]. We see that these theoretical curves fall nicely onto the data points from the simulations thus demonstrating that the theory is indeed also able to make correct quantitative predictions. It should be appreciated that no free fit parameter of any kind was used to calculate the theoretical curves. We also mention that a similar good agreement between simulation and theory is obtained for the non-ergodicity parameters of the intermediate scattering function for the A-B and B-B correlation as well as for the B particles.

As a further test of a quantitative prediction of the theory we will discuss finally some results of the dynamics in the β-relaxation regime. MCT predicts that as long the time correlation function is close to the plateau it can be written as follows:

$$\phi_l(t) = f_l^c + h_l G(t)\,, \tag{10}$$

where l is just an index labelling the correlator, f_l^c is the non-ergodicity parameter discussed above, h_l is the so-called critical amplitude, and $G(t)$ is a system-universal function, *i.e.* it is independent of l. The physical content of this equation is that *in the β-regime* all time correlation functions have the same time dependence, namely the one given by the function $G(t)$. Therefore Equation 10 is also sometimes called the 'factorisation property'. In order to check the validity of this prediction we have used for the functions $\phi_l(t)$ the distinct parts of the van Hove correlation functions, $G_d^{\alpha\beta}(r,t)$. (Note that therefore the space variable r plays the role of the index l in Equation 10.) These space-time correlations are defined by

$$G_d^{\alpha\alpha}(r,t) = n g_{\alpha\alpha}(r,t) = \frac{N_A+N_B}{N_\alpha(N_\alpha-1)}\left\langle \sum_{i=1}^{N_\alpha}\sum_{j=1}^{N_\alpha} \delta\left(r - |\boldsymbol{r}_i(0)-\boldsymbol{r}_j(t)|\right)\right\rangle \tag{11}$$

for $\alpha \in \{A, B\}$, and

$$G_d^{AB}(r,t) = n g_{AB}(r,t) = \frac{N_A+N_B}{N_A N_B}\left\langle \sum_{i=1}^{N_A}\sum_{j=1}^{N_B} \delta\left(r - |\boldsymbol{r}_i(0)-\boldsymbol{r}_j(t)|\right)\right\rangle\,, \tag{12}$$

where n is the particle density of the system. Note that for $t=0$ these functions are just the usual (partial) radial distribution functions and hence $G_d^{\alpha\beta}(r,t)$ can be considered as a generalisation of the latter to the time domain. In Reference [10] we have shown that for this set of correlation functions the factorisation property is indeed fulfilled, *i.e.* that in the β-relaxation regime the correlators have the form given by Equation 10. Using this equation for the cases $t=t'$ and $t=t''$, and identifying $\phi_l(t) = G_d^{\alpha\beta}(r,t)$, where now the label l runs through r, it follows immediately that the following equation holds for all values of r:

$$\frac{G_d^{\alpha\beta}(r,t') - G_d^{\alpha\beta}(r,t'')}{G_d^{\alpha\beta}(r',t') - G_d^{\alpha\beta}(r',t'')} = \frac{H^{\alpha\beta}(r)}{H^{\alpha\beta}(r')}\,, \tag{13}$$

where $H^{\alpha\beta}(r)$ is the critical amplitude for the function $G_d^{\alpha\beta}(r,t)$, r' is arbitrary, and t' and t'' are arbitrary times in the β-regime. Since the factorisation property holds it becomes possible to determine from the simulation the r dependence of the ratio $H^{\alpha\beta}(r)/H^{\alpha\beta}(r')$. In Figure 10 we show an upper and lower bound for this function (for the case of the A-A correlation), as it was determined from the simulation and we see that this is a non-trivial function of r. Also included in the figure is the theoretical value for this ratio and we see that this curve reproduces well the one from the simulation. (Again in this case, no free fit parameter exists.) Thus this is more evidence that MCT is not only able to make correct qualitative predictions but also quantitative ones.

Many more tests have been done in order to find out to what extent MCT is able to predict the dynamics of this system at low temperatures. The outcome of these tests is that the theory is indeed able to give a good description of this dynamics. Since a similar conclusion has been reached for the case of hard spheres, where the theoretical predictions have been compared with experiments on colloidal systems, we thus can conclude that MCT is able to describe the dynamics of simple liquids on a qualitative as well as

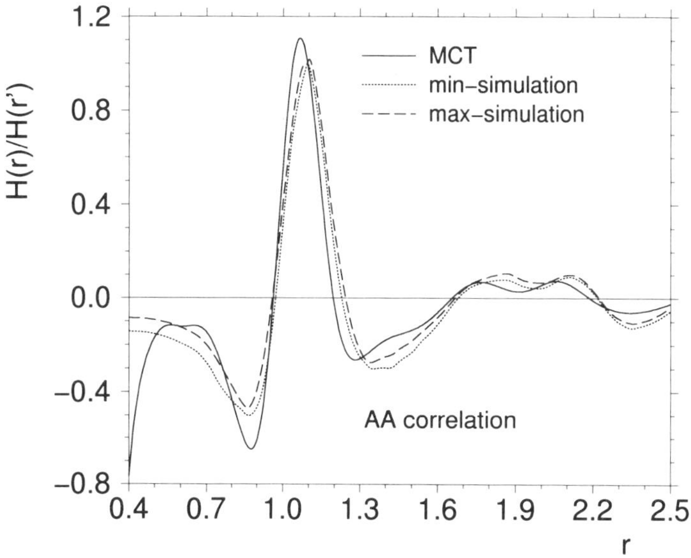

Figure 10. *r−dependence of the critical amplitude for $G_d^{\mathrm{AA}}(r,t)$ for the A-A correlation as determined from the simulation and the mode-coupling theory.*

quantitative level. To what extent this is the case also for more complex systems, such as molecules with odd shapes or systems with long range interactions, is currently still a matter of investigation. The results for water [20], a triangular shaped molecule, and silica [21], a system with long range interactions, however, look promising.

5 Out-of-equilibrium dynamics

The results discussed in the previous section are concerned with the dynamics of the supercooled liquid *in equilibrium.* We have seen that with decreasing temperature this dynamics slows down and hence it is clear that there will exist a finite temperature below which the system cannot be equilibrated any more within the time scale of the experiment or the computer simulation. Hence the system will fall out of equilibrium, *i.e.* undergo a glass transition. As this is a purely kinetic phenomenon, the temperature at which this happens is not intrinsic to the system such as, *e.g.* its melting point, and thus can be changed by choosing a different experimental time scale. For the sake of convenience we will call this temperature the glass temperature T_g, despite the fact that we have defined this term differently in Section 2. Since below T_g the system is no longer able to relax one might expect that the motion of the particles essentially stops, apart from their vibration inside the cage, *i.e.* that relaxation no longer takes place. In order to check whether this expectation is borne out, we investigate in this section the dynamics of a system after a quench below T_g. As we will see, even below T_g relaxation takes place but its nature is very different from that in equilibrium (above T_g). In particular it is found that the properties of the system, such as its structure or relaxation times, change with time. Therefore it is customary to say that the system is aging.

Experiments on aging materials have been done for many years, mainly on polymeric systems, but since they often show very pronounced aging effects such as the material becoming more brittle with time [22], their theoretical description has been considered only on a phenomenological level. Only in recent times strong efforts have been undertaken in order to understand this situation within a well-defined theoretical framework [23, 24] (see Bouchaud, this volume). However, from a theoretical point of view these aging systems are still understood in much less detail than is the case for the (supercooled) equilibrium system and very often only predictions of a very general nature can be made.

In the following we will discuss some results of simulations which have been done in order to investigate the dynamics of a simple glass-former which has been quenched below T_g. The system of interest is the same binary Lennard-Jones mixture whose equilibrium properties we have discussed in the previous section. There we have seen that for this system the relaxation times close to the MCT temperature T_c start to become comparable with the longest runs of present days simulations. Thus from a practical point of view the glass transition (on the computer!) takes place around T_c (=0.435). To investigate the dynamics of the system below T_c we equilibrated it at a temperature $T_i > T_c$ and then quenched it at time zero to a final temperature $T_f \leq T_c$. This quench was done by coupling the system every 50 time steps to a stochastic heat bath, *i.e.* all the velocities of the particles are substituted with ones drawn from a Maxwell-Boltzmann distribution corresponding to a temperature T_f. Note that after such a substitution the *kinetic* energy of the system corresponds to an equilibrium system at temperature T_f but that the *potential* energy does not. Hence, after the substitution, the dynamics of the system will pump potential energy into the kinetic energy and thus the system will (very!) slowly relax. However, if T_f is sufficiently small this relaxation will not be observed on the time scale of the simulation, *i.e.* the system will always be observed to age.

When one investigates the properties of an aging system it is useful to distinguish between two types of observables: the so-called 'one-time quantities' and the 'two-time quantities'. The former term refers to observables which *in equilibrium* are constants, such as the density (in a constant pressure experiment), the total energy of the system, or the structure (as measured, *e.g.*, by the structure factor). In the out-of-equilibrium situation the values of such observables depend on the time since the quench, and hence they depend on *one time*. Two-time quantities are time correlation functions which in equilibrium depend on a time difference, such as the mean squared displacement or the intermediate scattering function. Since in nonequilibrium the time elapsed since the quench has also to be taken into account, such quantities will depend on *two* times in the aging system.

In agreement with theoretical predictions it has been found, see [25, 26, 27], that most one 'one-time quantities' depend only weakly on time. Examples investigated were the total energy of the system, the radial distribution function or the pressure. (In passing we mention, however, that certain one-time quantities can show a sufficiently strong time dependence that they can be used to characterise the aging system very well. Examples of such observables are discussed in Reference [28]). In contrast to this the two-time quantities showed a very strong time dependence (see also Reference [29] for similar results for a soft sphere system). A typical example of a correlator that has such such a strong time dependence is $C_k(t_w + \tau, t_w)$, the generalisation of the incoherent intermediate scattering function, see Equation 6, to the out-of-equilibrium situation. Thus

$C_k(t_w+\tau,t_w)$ is given by

$$C_k(t_w+\tau,t_w)=\frac{1}{N}\left\langle\sum_j^N \exp\left[i\mathbf{k}\cdot\{\mathbf{r}_j(t_w+\tau)-\mathbf{r}_j(t_w)\}\right]\right\rangle , \tag{14}$$

where t_w is the time between the quench and the start of the measurement and hence is also called the 'waiting time'. Thus the meaning of this time correlation function is that a density fluctuation which is present at a time t_w after the quench is correlated with a density fluctuation at a time τ later.

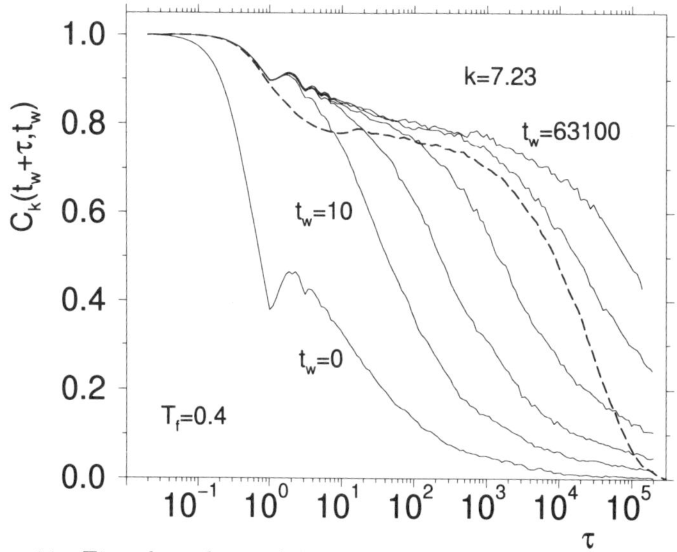

Figure 11. *Time dependence of the generalisation of the incoherent intermediate scattering function to the out-of-equilibrium situation for waiting times* t_w = *0, 10, 100, 1000, 10000, and 63100.* T_f *=0.4. The dashed line is the equilibrium correlation function at* T *=0.446.*

In Figure 11 we show the τ dependence of C_k for different waiting times and the A particles. The wave-vector is $k = 7.23$, *i.e.* the location of the maximum in the static structure factor for the A-A correlation. For small values of t_w the curves rapidly decay to zero. With increasing t_w the curves show at intermediate times a plateau and go to zero only at long times. We see that if t_w is not too small the approach of the curves *to* the plateau is independent of t_w, whereas the time at which they start to fall below the plateau depends on the waiting time. In Reference [26] we have shown that the time at which the curves leave the master curve is approximately proportional to t_w^α, with $\alpha = 0.9$. Thus we find that C_k does indeed show a strong waiting time dependence, as is theoretically expected for a two-time quantity. (Note that the oscillation at $t = 1$, and multiples of it, originate in the coupling of the system to the external heat bath and thus are not really an intrinsic feature of the aging system).

In view of the fact that we are at a very low temperature it might be a bit surprising to see that all the curves approach zero at long times. That is, from the relaxation behaviour at *equilibrium* one would expect that within the time span shown the curves should just fall on the plateau and then stay in its vicinity (see Figure 4). However, one should recall what is happening during the quench: at time zero the configuration of the particles corresponds to one which is typical for the high temperature T_i. Due to the quench the system now tries to equilibrate and to do this it has to move to a part of configuration space which is typical for configurations at T_f. It is this motion of the system in configuration space which leads to the rapid decay of the correlation function. If the waiting time since the quench is large, the system is able to find configurations which are already closer to the ones typical for T_f and thus the driving force for further exploration decreases. Hence the (out of equilibrium) relaxation becomes slower and slower and thus it takes the correlation functions more and more time to decay to zero.

Also included in the figure is the *equilibrium* curve at $T = 0.446$ (bold dashed line). Although the shape of this curve is *qualitatively* similar to the aging curves for long waiting times, a closer inspection shows that there are important differences. For example the approach of the curves to the plateau is much more rapid in the equilibrium case than in the nonequilibrium case. Also at long times significant differences are found. In Figure 5 we have shown that at long times the equilibrium curve can be fitted well with a KWW law. This is not the case for the out of equilibrium case where it is found that the correlators show a power-law dependence on time with an exponent which decreases with decreasing wave-vector [27], but, which is independent of the waiting time.

The results discussed so far are for a quench to $T_f = 0.4$, *i.e.* a temperature which is quite close to the critical temperature of MCT. If the final temperature is significantly lower, the relaxation behaviour can be quite different from the one with higher T_f. This is shown in Figure 12 where we show the same correlation function as in Figure 11, but this time for $T_f = 0.1$. From this figure we see that, for long waiting times, the correlators at *short* times look qualitatively similar to the ones for $T_f = 0.4$. The main difference is that the plateau is higher, which is reasonable since its height is, even in the out-of-equilibrium situation, related to the size of the cage that each particle feels, and it can be expected that this size is proportional to $1 - T_f$.

For long times τ the curves for high and low values of T_f are different also on a qualitative level, in that the ones for $T_f = 0.1$ show a *second* plateau. In the inset of Figure 12 we show the curves for $t_w = 10^3$ for the *individual* runs. We now recognise that most of these curves show at a time between $10^2 - 10^4$ time units one or more sharp drops which are then followed by a regime in which the curves are almost constant. It is this constant part which gives rise to the second plateau in the average curve shown in the main figure whereas the sharp drops average out to a much less sharp decrease in the mean curve.

In order to investigate the microscopic reason for the sharp drops and the subsequent plateaus we have compared the configurations just before the drop with the ones just after the drop [27]. We have found that the fast relaxation is due to the fact that around 10% of the particles (*i.e.* about 100 of them) undergo a sudden, quite co-operative motion in which the particles move by around 0.2–0.5 of their diameter. Despite the smallness of this motion, its cooperative nature leads to the observed fast drop in the correlation function. The likely reason for the occurrence of this cooperative motion is that, due to the quench, the system has built up an internal stress and it seems that the most efficient way to release this stress is to rearrange the particles in a cooperative way. Thus the situation is similar to an earthquake where stress is released in a similar way.

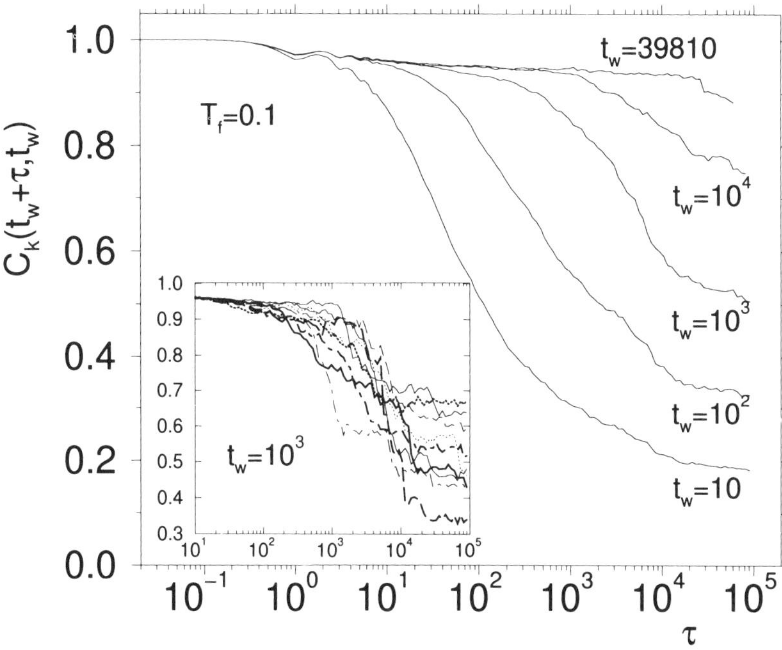

Figure 12. *Time dependence of $C_k(t_w + \tau, t_w)$ for different waiting times. $T_f = 0.1$.*

Finally we discuss some very interesting results concerning the connection between the time correlation functions and the response of the system to an external perturbation. In equilibrium this connection is given by the fluctuation dissipation theorem (FDT) which says the following: consider an observable A and the associated normalised time auto-correlation function

$$C(t) = \frac{\langle A(t)A(0)\rangle}{\langle A(0)A(0)\rangle}.$$

If the system is perturbed with a field conjugate to the observable A the response function $R(t)$ is given by

$$R(t) = -\frac{1}{k_B T}\frac{dC}{dt},$$

where T is the temperature of the system. Thus in equilibrium the FDT relates the time derivative of the correlation function with the response and the factor is the inverse temperature.

In the derivation of the FDT it is required that the system is time translation invariant, an assumption which is clearly not fulfilled in the out of equilibrium situation. Hence the FDT does not hold any more and it has been proposed that the FDT should be generalised as follows [23]: since the correlator depends on two times, also the response will depend on two times. Thus we have, assuming $t' \geq t$,

$$R(t', t) = \frac{1}{k_B T} X(t', t) \frac{\partial C(t', t)}{\partial t}, \tag{15}$$

where the function $X(t, t')$ is defined by this equation. In the context of mean-field spin glass models it has been shown that in the limit $t_w, \tau \to \infty$, $X(t_w + \tau, t_w)$ is a function of the correlation function C only, *i.e.*

$$X(t_w + \tau, t_w) = x(C(t_w + \tau, t_w)), \tag{16}$$

where the function x is now a function of *one* variable only. (Here t_w is again the time since quench.) Within mean-field it is expected that the function $x(C)$ is equal to -1.0 if C is larger than the plateau value, *i.e.* that for these short times the FDT holds. For times such that C has fallen below the plateau it is expected that x is larger than -1.0, *i.e.* the FDT is 'violated'. (The quotes reflect the fact that of course the FDT is not violated, since it is not supposed to hold.) The reason for the interest in the function $X(t', t)$ is twofold: firstly, we see from Equation 15 that $-k_B T / X$ is something like an effective temperature. Thus, if the time and temperature dependence of X is known, it might become possible to use thermodynamics concepts also for the out-of-equilibrium system. Secondly, in the context of spin glasses it has been found that the dependence of $x(C)$ can be used to classify various types of spin glasses (see, *e.g.*, reference [30] for a nice discussion on this). Thus by measuring $x(C)$ for a structural glass, it might become possible to connect the properties of a *structural* glass, such as the present Lennard-Jones system, with a *spin* glass.

Since the correlation function of interest is the generalisation of the incoherent intermediate scattering function, *i.e.* the correlation of a density fluctuation, we need a method to measure the response function to such fluctuations. Theoretically one could apply an external field with wave-vector $\mathbf{q}$ which couples to the position of one particle and see how this perturbation affects the density distribution. However, this approach would lead to very poor statistics and thus a more efficient method has to be used, the details of which are described in [27, 31]. That procedure allows one to measure the integrated response $M(t_w + \tau, t_w)$ with reasonable accuracy, where $M(t_w + \tau, t_w)$ is given by:

$$M(t_w + \tau, t_w) = \int_{t_w}^{t_w + \tau} R(t_w + \tau, t) dt \; . \tag{17}$$

Using Equations 15 and 16 one can rewrite this as

$$M(t, t') = M(C) = \frac{1}{k_B T} \int_C^1 x(c) dc \; . \tag{18}$$

From this equation it becomes clear that a parametric plot of the integrated response versus the correlator will give us the information about the integrant $x(c)$ and hence the factor $X(t_w + \tau, t_w)$.

In Figure 13 we show such a parametric plot for different waiting times and from it we can recognise the following things. For short times, *i.e.* those points at which $C(t_w + \tau, t_w)$ is large, the data points are compatible with a straight line with slope -1, *i.e.* $x(c)$ is -1 and the FDT holds. Thus for these short times the system does not really realise that it is not in equilibrium, since the fast degrees of freedom, such as the vibrations, are still able to follow the dynamics of the system. This is not the case for those processes that relax on longer time scales. In the figure we see that for times that correspond to the aging regime, *i.e.* where the correlator C has fallen below the plateau, the data no longer follow the FDT line, but are well below it. We find that, within the accuracy of our data, this

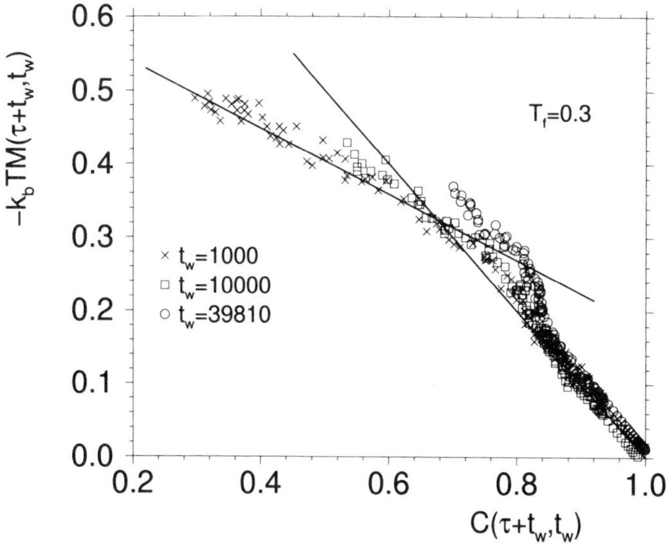

Figure 13. *Parametric plot of the integrated response $M_k(t_w+\tau,\tau)$ versus the correlation function $C_k(t_w+\tau,\tau)$ for different waiting times and $T_f=0.3$. The straight lines have slopes around -1 and -0.62.*

part of the data is compatible with a straight line with slope $-m$, with $m<1$. Such a functional form has been found in mean-field spin glasses (with one-step replica symmetry breaking) and thus we have now evidence that our structural glass is compatible with this type of spin glass. At the moment such a connection is, of course, only a tenuous one. Furthermore one might wonder whether it is really justified to draw from a comparison of *nonequilibrium* properties between systems any conclusion about *equilibrium* properties. Surprisingly, for spin glasses this conclusion has been shown to be correct [32], and thus it is not completely crazy to assume that a similar connection can be made for structural glasses as well. Therefore, at the moment there are strong efforts to test these links since they would allow us to gain a much more unified picture of disordered systems. If such connections are shown to be present we would have learned that there is no fundamental difference between disordered systems in which the disorder is *quenched*, such as in spin glasses, and systems in which it is *self-generated*, such as in structural glasses.

A different important effort to extend our understanding of aging systems is work along the lines of that which has been so successful for the equilibrium dynamics. For temperatures above T_c we have seen that mode-coupling theory is able to describe the dynamics of supercooled liquids not only qualitatively, but also quantitatively. Thus it is natural to try to extend this approach also to the out-of-equilibrium case since only then can certain non-universal features be discussed on a quantitative basis. Although in such calculations one is faced with formidable technical problems, some progress has recently been made [33] and thus it can be hoped that in the not too far future we will have also a quantitative theory for aging systems.

Acknowledgments

I thank H. C. Andersen, J.-L. Barrat, K. Binder, T. Gleim, and M. Nauroth for the fruitful collaborations that led to some of the results discussed in this review, and L. Cugliandolo, W. Götze, J. Kurchan, A. Latz, G. Parisi, and F. Sciortino for many insightful discussions on various aspects of this work. Part of this work was supported by the SFB 261 of the DFG and the Pole Scientifique de Modélisation Numérique at ENS-Lyon.

References

[1] Zallen R, 1983, *The Physics of Amorphous Materials* (Wiley, New York);
Jäckle J, 1986, *Rep Prog Phys* **49**, 171;
Uhlmann D R and Kreidl N J (Eds.), 1986, Vol 3 of *Glass Science and Technology: Viscosity and Relaxation* (Academic Press, Orlando);
Zarzycki J (Ed.), 1991, Vol 9 of *Materials Science and Technology*, (VHC, Weinheim);
Feltz A, 1993, *Amorphous Inorganic Materials and Glasses* (VCH, Weinheim);
Debenedetti P G, 1997, *Metastable Liquids* (Princeton University Press, Princeton)

[2] Angell C A, Poole P H, and Shao J, 1994, *Nuovo Cimento D* **16**, 993

[3] Angell C A, 1985, in *Relaxation in Complex Systems*, edited by Ngai K L and Wright G B (US Dept Commerce, Springfield)

[4] Götze W, 1999, *J Phys Condens Matter* **10A**, 1

[5] Mazurin O V, Streltsina M V, and Shvaiko-Shvaikovskaya T P, 1983, *Handbook of Glass Data, Part A: Silica Glass and Binary Silicate Glasses* (Elsevier, Amsterdam)

[6] Götze W, 1991, in *Liquids, Freezing and the Glass Transition* (page 287), edited by J P Hansen, D Levesque and J Zinn-Justin, Les Houches Session LI, 1989, (North-Holland, Amsterdam);
Götze W and Sjögren L, 1992, *Rep Prog Phys* **55**, 241;
Kob W, 1997, *Experimental and Theoretical Approaches to Supercooled Liquids: Advances and Novel Applications* (page 28) edited by J Fourkas, D Kivelson, U Mohanty, and K Nelson (ACS Books, Washington)

[7] Hansen J-P and McDonald I R, 1986, *Theory of Simple Liquids* (Academic, London)

[8] Allen M P and Tildesley D J, 1990, *Computer Simulation of Liquids* (Oxford University Press, New York);
Frenkel D and Smit B, 1996, *Understanding Molecular Simulation - From Algorithms to Applications* (Academic Press, San Diego);
Binder K and Ciccotti G (Eds.), 1996, *Monte Carlo and Molecular Dynamics of Condensed Matter Systems* (Italian Physical Society, Bolognia);
Binder K and Heermann D W, 1997, *Monte Carlo Simulation in Statistical Physics* (Springer, Berlin, 1997)

[9] Angell C A, Clarke J H R, and Woodcock L V, 1981, *Adv Chem Phys* **48**, 397;
Barrat J-L and Klein M L, 1991, *Ann Rev Phys Chem* **42**, 23;
Kob W, 1995, in Vol III of *Annual Reviews of Computational Physics*, edited by D Stauffer (World Scientific, Singapore);
Kob W, 1999 *J Phys Condens Matter* **11**, R85

[10] Kob W and Andersen H C, 1995a, *Phys Rev E* **51**, 4626

[11] Kob W and Andersen H C, 1994, *Phys Rev Lett* **73**, 1376;
Kob W and Andersen H C, 1995b, *Phys Rev E* **52**, 4134

[12] Gleim T, Kob W, and Binder K, 1998, *Phys Rev Lett* **81**, 4404

[13] Nauroth M and Kob W, 1997, *Phys Rev E* **55**, 657

[14] Nauroth M, Kob W, and Götze W, unpublished;
Nauroth M, 1999, PhD Thesis, TU Munich

[15] Schmidt-Rohr K and Spiess H W, 1991, *Phys Rev Lett* **66**, 3020;
Böhmer R, Hinze G, Diezemann G, Geil B, and Sillescu H, 1995, *Europhys Lett* **36**, 55;
Heuer A, Wilhelm M, Zimmermann H, and Spiess H W, 1995, *Phys Rev Lett* **95**, 2851;
Cicerone M T, Blackburn F R, and Ediger M D, 1995, *J Chem Phys* **102**, 471;
Cicerone M T, and Ediger M D, 1996, *J Chem Phys* **104**, 7210;
Sillescu H, 1999, *Non-Cryst Solids,* **243**, 81, and references therein

[16] Kob W, Donati C, Plimpton S J, Glotzer S C, and Poole P H, 1997, *Phys Rev Lett* **79**, 2827
Donati C, Douglas J F, Kob W, Plimpton S J, Poole P H and Glotzer S C, 1998, *Phys Rev Lett* **80**, 2338
Donati C, Glotzer S C, Poole P H, Kob W, and Plimpton S J, 1999 *Phys Rev E* **60**, 3107

[17] van Megen W and Pusey P N, 1991 *Phys Rev A* **43**, 5429;
Bartsch E, Antonietti E, Schupp W, and Sillescu H, 1992, *J Chem Phys* **97**, 3950;
van Megen W and Underwood S M, 1994, *Phys Rev E* **49**, 4206;
Götze W, and Sjögren L, 1991, *Phys Rev A* **43**, 5442

[18] Barrat J-L and Latz A, 1990, *J Phys Condens Matter* **2**, 4289

[19] Fabbian L, Latz A, Schilling R, Sciortino F, Tartaglia P, and Theis C, 1999, preprint cond-mat/9905091, Phys Rev E (in press)

[20] Gallo P, Sciortino F, Tartaglia P, and Chen S-H, 1996, *Phys Rev Lett* **76**, 2730;
Chen S-H, Gallo P, Sciortino F, and Tartaglia P, 1997, *Phys Rev E* **56** 4231;
Sciortino F, Fabbian L, Chen S-H, and Tartaglia P, 1997, *Phys Rev E* **56** 5397;
Fabbian L, Latz A, Schilling R, Sciortino F, Tartaglia P, and Theis C, 1999, *Phys Rev E* (in press)

[21] Horbach J, Kob W, and Binder K, 1998, *Phil Mag B* **77**, 297;
Kob W, Horbach J, and Binder K, 1999, in AIP Conference Proceedings, Vol 469, p441, *Slow Dynamics in Complex Systems, Tohwa University 1998* (AIP, Woodbury);
Horbach J and Kob W, 1999, *Phys Rev B* **60**, 3169

[22] Struik L C E, 1978, *Physical Aging in Amorphous Polymers and Other Materials* (Elsevier, Amsterdam);
McKenna, G B, 1989, in *Comprehensive Polymer Science, Vol 12, p311, Polymer Properties,* edited by C Booth and C Price (Pergamon, Oxford) and references therein

[23] Cugliandolo L and Kurchan J, 1993, *Phys Rev Lett* **71**, 173;
Cugliandolo L F and Kurchan J, 1994, *J Phys A* **27**, 5749

[24] Bouchaud J-P, Cugliandolo L F, Kurchan J, and Mézard M, 1996 *Physica A* **226**, 243;
Monthus C and Bouchaud, J-P, 1996 *J Phys A Math Gen* **29**, 3847

[25] Andrejew E and Baschnagel J, 1996, *Physica A* **233**, 117

[26] Kob W and Barrat J-L, 1997, *Phys Rev Lett* **78**, 4581

[27] Kob W and Barrat J-L, 1999a, *Physica A* **263**, 234;
Barrat J-L and Kob W, 1999, *J Phys Condens Matter* **11** A247;
Kob W and Barrat J-L, 1999b, *Eur Phys J* **B13** 319

[28] Kob W, Sciortino F, and Tartaglia P, 2000, *Europhys Lett* in press

[29] Parisi G, 1997a, *Phys Rev Lett* **79**, 3660;
Parisi G, 1997b, *J Phys Math Gen* **30**, L765;
Parisi G, 1997c, *J Phys Math Gen* **30**, 8523

[30] Parisi G, Ricci-Tersenghi F, and Ruiz-Lorenzo J J, 1999, *Eur Phys J B* **11**, 317

[31] Barrat J-L and Kob W, 1999, *Europhys Lett* **46**, 637

[32] Franz S, Mézard M, Parisi G, and Peliti L, 1998, *Phys Rev Lett* **81**, 1758

[33] Latz A, 2000, to be published in *J Phys Condens Matter*; preprint cond-mat/9911025

Aging in glassy systems: new experiments, simple models, and open questions

J-P Bouchaud

CEA Saclay, France

1 Introduction

A great variety of systems exhibit slow 'glassy' dynamics [1]: glasses of all kinds, but also spin-glasses and dipolar glasses [2, 3]. Another interesting class of systems is that of pinned 'defects' such as Bloch walls, vortices in superconductors, charge density waves, dislocations, *etc.* interacting with randomly placed impurities [4]. Yet another class of system, where surprisingly slow dynamics can occur, is that of soft glassy materials such as foams or dense emulsions [5]. Of particular interest is the so called *aging* phenomenon observed in the response function of these glassy systems. This response is either to a magnetic field in the case of disordered magnets, to an electric field in the case of dipolar glasses, or to an applied stress in the case of, *e.g.* glassy polymers or dense emulsions. The basic phenomenon is that the response is *waiting-time* dependent, *i.e.* depends on the time t_w one has waited in the low temperature (glass) phase before applying the perturbation. Qualitatively speaking, these systems *stiffen* with age: the longer one waits, the smaller the response to an external drive, as if the system settled in deeper and deeper energy valleys as time elapses.

More precisely, consider a magnetic system which is cooled, in zero field, to below the glass temperature T_g, understood as the temperature below which the relaxation time becomes larger than the experimental time scales: this temperature may or may not correspond to a true phase transition (see Kob, this volume). If the system is then left at T_1 (less than T_g) during a time t_w before applying an external field (or stress), then the time dependent magnetisation saturates after a time comparable to t_w itself. In other words, the time dependent magnetisation (or strain) takes the following form:

$$M(t_w + t, t_w) \simeq M_\infty(t) + M_{\rm ag}\left(\frac{t}{t_w}\right), \tag{1}$$

where $M_{\rm ag}$ is the slow, aging part and $M_\infty(t)$ is a 'fast' part. (Strictly speaking, $M_\infty(t)$ should be written $M_\infty(t/\tau_0)$, where τ_0 is a microscopic time scale). Analogously, if a polymer glass is left at T_1 during a time t_w before applying an external stress, the subsequent strains develop on a time scale given by t_w itself. Note that the decomposition of the dynamics into a fast part and a slow part often also holds *above* (but close to) T_g, where one speaks of β-relaxation and α-relaxation, respectively. The time scale $\tau(T)$ for the α-relaxation is, however, finite and waiting-time independent above T_g. This time becomes effectively infinite for $T < T_g$, and is therefore replaced by the age of the system t_w. In other words, Equation 1 is the continuation in the glass phase of the precursor two-step relaxation commonly observed above T_g [6] (see also Kob, this volume).

Aging in correlation functions

One can also, at least numerically, observe aging in the *correlation functions*. For example, the dynamic structure factor of a Lennard-Jones system, defined as:

$$S_q(t_w + t, t_w) = \langle \exp\{i\mathbf{q} \cdot [\mathbf{r}(t + t_w) - \mathbf{r}(t_w)]\}\rangle, \tag{2}$$

where $\mathbf{r}(t)$ is the position of a tagged particle at time t, exhibits interesting aging properties [7]. Note that the corresponding experiments are much harder to realise, since the measurement of the structure factor typically takes several minutes or so. In order not to mix together different waiting times, one should redo the experiments several times, heating back the system above T_g, and cooling down again, until the number of runs is sufficient to obtain a good averaging for a *fixed* t_w. This point will be further discussed below—see Equation 6.

In equilibrium, the correlation and the response function are related through the *fluctuation-dissipation theorem* (see Kob, this volume). Interestingly, this theorem does not hold in general for the aging part of the correlation and response. A generalisation of this theorem has been provided in [8, 9] in the context of some mean-field spin-glass models, where the true temperature of the system is replaced by an effective temperature, higher than that of the thermal bath (and possibly time dependent). Quite a lot of effort has been devoted to measure this effective temperature in glassy systems, both numerically [10], and experimentally [11].

2 Different types of aging

2.1 Aging in the a.c. susceptibility

Aging can also be seen in measurements of the a.c. susceptibility, defined by the response to an oscillating field at frequency ω. (Its analogue, for mechanical perturbations, is the compliance $J^*(\omega) = 1/G^*(\omega)$ where $G^*(\omega)$ is the frequency-dependent modulus; see McLeish, this volume). These measurements have the advantage that the perturbing field can then be extremely small. On the other hand, since one must wait for at least one period before taking a measurement, the time sector available in these experiments is confined to $\omega t_w{>}1$ (corresponding, in the language of the time dependent magnetisation discussed above, to the short time region $t < t_w$). The a.c. susceptibility typically takes the following form:

$$\chi''(\omega, t_w) = \chi''_\infty(\omega) + f(t_w)\chi_{\text{ag}}(\omega) \tag{3}$$

where $\chi_\infty(\omega)$ is the stationary contribution, obtained after infinite waiting time. Depending on the system, the aging part behaves quite differently. For example, in spin glasses (SG), both the functions f and $\chi_{\rm ag}$ behave similarly, as a *power-law* [3]:

$$f(t_w)\chi_{\rm ag}(\omega)\big|_{\rm sg} \simeq A(\omega t_w)^{x-1} \qquad x \sim 0.7\text{–}0.9. \tag{4}$$

This behaviour is the counterpart, in frequency space, of the t/t_w scaling reported above. A spin glass is a system of magnetic spins on a lattice in which an individual spin experiences both ferromagnetic and antiferromagnetic interactions with its neighbours, leading to frustration. A dipolar glass contains instead local electric dipoles at random sites which interact with each other and with the polarisable surroundings. Some dipolar glasses (DG), close to a ferroelectric transition [12, 13], reveal a very different behaviour, since in this case one has:

$$f(t_w)\chi_{\rm ag}(\omega)\big|_{\rm dg} \simeq At_w^{x-1} \quad \text{or} \quad -B\log t_w, \tag{5}$$

i.e. an aging part which is nearly frequency independent. Glycerol, on the other hand, shows an intermediate behaviour [14]: the aging part is frequency dependent, but the frequency dependence is weaker than the waiting time dependence.

2.2 Role of the thermal history

If a system ages, then by definition it is out of equilibrium. Therefore, one might worry that the properties that one measures at T_1 actually strongly depend on the whole thermal history of the system. Here again, different systems behave very differently. A naïve argument, based on the idea that thermal activation over high energy barriers plays a central role in aging, would suggest that the 'age' of a system cooled very slowly before reaching T_1 should be much larger than the 'age' of the same system, but cooled more rapidly. In other words, cooling the system more slowly allows energy barriers to be surmounted more efficiently at higher temperatures, and thus brings the system closer to equilibrium at T_1. This is precisely what happens for DG systems [12]. One can actually use non-uniform cooling protocols, where the cooling rate is either slow or fast only in the vicinity of T_g, and then fixed to a constant value for the last few Kelvins—see Figure 1. In the case of DG, one sees very clearly that the cooling rate when crossing T_g is the crucial quantity which determines how well the system is able to equilibrate [12]. When the system is cold, then the dynamics is essentially frozen on experimental time scales, and therefore the precise value of the cooling rate there is of minor importance.

Surprisingly, the situation is completely reversed for SG. In these systems, the value of the cooling rate in the vicinity of T_g is completely irrelevant, and the observed a.c. susceptibility is to a large degree independent of the cooling rate, except for the very last few Kelvins [15]. This is illustrated in Figure 1. The naïve picture of a system crossing higher and higher barriers to reach an optimal configuration therefore certainly needs to be reconsidered for these systems.

2.3 Rejuvenation and memory in temperature cycling

An even more striking effect has been observed first in spin-glasses [16, 17], and more recently in a variety of other systems. Upon cooling, say from T_1 to $T_2 < T_1$, the system

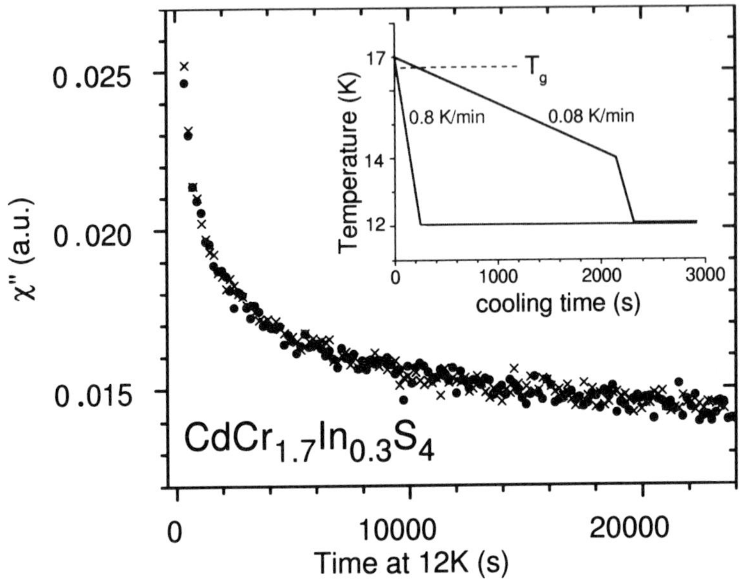

Figure 1. *Effect of the cooling rate (shown in the inset) on χ'' (circles and crosses), for a spin glass, showing that the value of the cooling rate in the vicinity of T_g is completely irrelevant. Only the cooling rate during the very last Kelvins actually affect the contribution of the aging part of the susceptibility.*

rejuvenates and returns to a zero-age configuration even after a long stop at the higher temperature T_1. This is tantamount to saying that the thermal history is irrelevant, as we pointed out above: stopping at a higher temperature is more or less equivalent to cooling the system more slowly. The interesting point, however, is that the system at the same time remembers perfectly its past history: when heating back to T_1, the value of the a.c. susceptibility is seen to match precisely the one it had reached after the first passage at this temperature, as if the stay at T_2 had not affected the system at all: see Figure 2. The paradox is that the system *did* significantly evolve at T_2, since a significant decrease of $\chi''(\omega, t_w)$ is also observed at T_2. The effect would be trivial if no evolution was observed at T_2: in this case, one would say that the system is completely frozen at the lowest temperature, and then all observables should indeed recover their previous value when the system is heated back. The puzzle comes from the coexistence of *perfect memory* on the one hand, and *rejuvenation* on the other. (This rejuvenation has often been identified with some kind of 'chaotic' evolution of the spin-glass order with temperature. As we shall discuss below, this might be misleading and we prefer calling this effect 'rejuvenation' rather than 'chaos'.)

Similar effects have now been seen in different systems: in different spin glasses [15], some dipolar glasses [12], PMMA [19], and very recently disordered ferromagnets [20, 21]. This last case is interesting because the system is a so-called reentrant spin-glass: the system is ferromagnetic at high temperatures, and then becomes a spin-glass at lower temperatures. One can thus compare in detail the aging effects in both phases. The 're-juvenation and memory' turns out to be very similar in both phases, except that memory

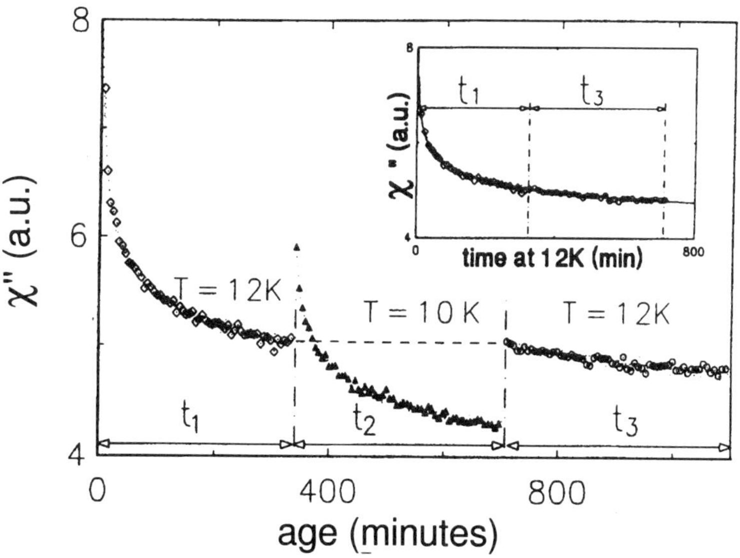

Figure 2. *Effect of a temperature cycle for $T_1 = 12K$ to $T_2 = 10K$ ($< T_1$) and back to T_1. One sees that after a long stop at T_1, aging is restarted when the system is cooled to T_2, as if it never had stopped at T_1 (the central part of the curve would be the same after a direct quench from $T > T_g$ to T_2.) At the same time, a perfect memory of the value of χ'' reached at the end of the stay at T_1 is kept somewhere in the system. The inset shows that after 'cutting' the central part, one recovers precisely the usual aging curve at T_1. (From [16]).*

is only partial in the ferromagnetic phase: only when the system is left at T_2 for a rather short time does one keep the memory. In spin glasses, as soon as $T_1 - T_2$ is greater than a few degrees, the memory is kept intact, at least over experimentally accessible time scales.

2.4 Deviations from t/t_w scaling

We have mentioned above that in many systems such as spin-glasses or polymer glasses, the aging part of the response function scales approximately as t/t_w, meaning that the effective relaxation time of a system below the glass transition is set by its *age*, t_w. This scaling is actually only approximate. It appears that a better rescaling of the experimental data is achieved by plotting the data as a function of the difference $(t+t_w)^{1-\mu} - t_w^{1-\mu}$, with $\mu < 1$ [3]. For $\mu = 0$, t_w drops out, and this describes a usual time-translation invariant response function, while the limit $\mu \to 1$ corresponds to a t/t_w scaling. It is easy to see that the effective relaxation time grows as t_w^μ, *i.e.* more slowly than the age t_w itself for $\mu < 1$ ('sub-aging'). Note that for dimensional reasons, t_w^μ should in fact read $t_w^\mu \tau_0^{1-\mu}$, where τ_0 is a microscopic time scale. Therefore, deviations from the simple t/t_w scaling would mean that the microscopic time scale is still relevant to the aging dynamics, even for asymptotically long times. Such a behaviour is predicted in some mean-field models of spin-glasses, and also in simpler 'trap' models, to be discussed below.

However, one should be very careful in interpreting any empirical value of μ less than one, because several artefacts can induce such an effective sub-aging behaviour. A first possible artefact is due to the external field applied to the system to measure its response. Experimentally, one finds that the higher field (or the external stress), the smaller the value of μ: an external field destroys the aging effect [2, 22]. The extrapolation of μ to zero field is somewhat ambiguous, especially because of a second possible artefact, which is a bad separation between the 'fast' relaxation part in Equation 1 and the slow aging part. This comes from the fact that the so-called 'fast' part $M_\infty(t)$ is not that fast. In spin glasses, it decays as $(\tau_0/t)^\alpha$, where α is a very small exponent, of the order of 0.05. Hence, this leads to a rather substantial 'tail' even in the experimental region where $t = 10^{12}\tau_0$, which pollutes the aging part and leads to an effective value of $\mu < 1$. However, even after carefully removing this fast part, the value of μ in spin-glasses still appears to be stuck slightly below the value 1 [3]. A third reason for this to be so is *finite size effects*. One expects that for a system of finite size, aging will be interrupted after a finite time, when the system has fully explored its phase space. Therefore, after a possibly very long 'ergodic' time, the response of the system has to revert to being time-translation invariant, corresponding to $\mu = 0$. An idea, advocated in [23] and recently reconsidered by Orbach [24], is that a sample made of small grains of different sizes will lead to an effective value of $\mu < 1$ because the ergodic times of some of the grains enter the experimental time window.

As a last possible artefact, let us mention the case of dynamic light scattering experiments where the signal is recorded *while the system is aging*. In this case, instead of measuring $S_q(t_w + t, t_w)$, one actually measures:

$$\widetilde{S}_q(t_w + t, t_w) = \frac{1}{\mathcal{T}} \int_0^{\mathcal{T}} dt' S_q(t_w + t' + t, t_w + t') \tag{6}$$

where $\mathcal{T}$ is the integration time needed to obtain a reliable signal. It is easy to see that even if $S_q(t_w + t, t_w)$ is a function of t/t_w, the averaging over t' will lead to an effective value of μ smaller than one, tending to one only if $\mathcal{T} \ll t_w$. It is possible that this mechanism can explain the value of $\mu \sim 0.5$ determined in [25] for a colloidal glass, where $\mathcal{T} \sim t_w$.

Before ending this section, let us finally add an extra comment concerning the case $\mu > 1$ ('super-aging'). In some simple coarsening models of aging, discussed below, it can be argued that the relevant scaling variable should be $\log(t + t_w)/\log(t_w)$ rather than t/t_w. This corresponds to an effective value of μ greater than one. Other mechanisms leading to $\mu > 1$ can be found, see [26] and below.

3 Simple models of aging

3.1 Domain growth in pure systems

A simple case where aging effects do appear is phase ordering in pure systems (see Bray, this volume, and [27]). Take for example an Ising ferromagnet suddenly quenched from high temperatures to a temperature below the Curie temperature. The system then wants to order, and has to choose between the up phase and the down phase. Obviously, this takes some time, and the dynamics proceeds via *domain growth*. After a time t_w after the quench, the typical distance between domain walls is $\xi(t_w)$, which grows as a power-law of time, *i.e.* relatively quickly, even for small temperatures. A given spin, far from the domain walls, will thus have to wait for a time t such that $\xi(t_w + t) \sim 2\xi(t_w)$ to flip between the up and down phases and decorrelate. For a power-law growth, this means that the effective relaxation time is of the order of t_w itself, corresponding to $\mu = 1$. This behaviour is confirmed by several exactly soluble models of coarsening, such as the Ising model on a chain, or the 'spherical' model, where one can compute explicitly the correlation function to find [27]:

$$C(t_w + t, t_w) = C_\infty(t) + C_{\text{ag}}\left(\frac{t}{t_w}\right). \tag{7}$$

One can also compute several other quantities, such as the effect of an external magnetic field h on the aging properties. One then sees that when $\xi(t_w)^{-1}$ is smaller than h, the driving force due to the curvature of the domain walls is superseded by the driving force due to the external field [28]. In this situation, the favoured phase quickly invades the whole system and aging is stopped. This leads, for very small fields and moderate time scales, to an effective value of the exponent $\mu < 1$, as discussed in Section 2.4. The main quantity of interest to compare with experiments, however, is the response function. The result is that the aging part of the response function vanishes like $\xi(t_w)^{-1}$ as t_w becomes large. In terms of the a.c. susceptibility, one finds that [28, 29]:

$$\chi''(\omega, t_w) = \chi''_\infty(\omega) + \xi(t_w)^{-1}\chi''_{\text{ag}}(\omega t_w). \tag{8}$$

Intuitively, this result means that the aging part of the susceptibility only comes from the domain walls, while the spins in the bulk of domains contribute to the stationary part $\chi''_\infty(\omega)$. Since the density of spins belonging to domain walls decreases as $\xi^{d-1}/\xi^d = \xi^{-1}$, the aging contribution decreases with time as the density of walls.

Domain growth in pure systems is driven by surface tension and does not require thermal activation; the aging effects in these systems are therefore hard to detect experimentally, since the typical size of the domains soon reaches its maximum value (set either by the size of the system or by magneto-static or other considerations). Similarly, one does not expect the cooling rate to have a major influence on the coarsening of the system.

3.2 Domain growth in random systems

More interesting is the situation in disordered ferromagnets (for example in the presence of quenched random fields or random bonds). In this case, the impurities act as pinning sites for the domain walls. The problem of elastic objects (such as domain walls, but also vortices in superconductors, dislocations, *etc.*) pinned by random impurities has been the subject of intense work in the past decade, both from a static point of view (where the typical equilibrium conformation of such objects is investigated) and to understand their dynamics (relaxation to equilibrium, response to an external driving force, creep and depinning transition)—for reviews, see [30, 4, 31, 32]. Actually, these systems constitute 'baby' spin glasses: frustration is present because of the competition between pinning, which tends to distort the structure, and elasticity which tends to reduce the deformations. The main result of the theory is the appearance of a typical pinning energy $E_p(\ell)$

associated to the linear size ℓ of the piece of domain wall that attempts to reconform. This energy scale *grows as a power* of ℓ:

$$E_p(\ell) \sim \Upsilon(T)\ell^\theta, \tag{9}$$

where Υ is a (temperature dependent) energy scale, and θ an exponent which depends on the dimensionality of the structure (1D for dislocations, 2D for domain walls, *etc.*) and on the correlations of the pinning field. Using a very naïve Arrhenius law for thermal activation, this means that the typical time associated with reconformation events that occur on a scale ℓ take a time:

$$t(\ell) \sim \tau_0 \exp\left[\frac{\Upsilon(T)\ell^\theta}{T}\right]. \tag{10}$$

Equivalently, the length scale over which the system can equilibrate after a time t_w only grows *logarithmically*:

$$\ell(t_w) \sim \left[\frac{T}{\Upsilon(T)} \log\left(\frac{t_w}{\tau_0}\right)\right]^{1/\theta} \tag{11}$$

In particular, the typical size of the growing domains in expected to grow logarithmically with time, that is, *extremely slowly*. This very slow growth means that the density of domain walls only decays slowly with time, and therefore that the aging contribution to the susceptibility is still significant even after macroscopic times. (Actually, for domain growth, the exponent of the logarithm is probably larger than $1/\theta$, but this does not matter for the present qualitative discussion. In particular, one expects the aging part of the correlation function tc be a function of $\log(t_w+t)/\log t_w$.)

Although very simple, the exponential relation between length scales and time scales given by Equation 10 has far-reaching consequences: the dynamics becomes, in a loose sense, hierarchical. This is illustrated in Figure 3. The object evolves between metastable configurations which differ by flips of regions of size ℓ_1 on a time $t(\ell_1)$ that, because of the exponential dependence in Equation 10, is much shorter than the time needed to flip a region of size $\ell_2 > \ell_1$. Therefore, the dynamics of the short wavelengths happens on a time scale such that long wavelengths are effectively frozen. As we shall explain below, this feature is, in our eyes, a major ingredient in understanding the coexistence of rejuvenation and memory. Another important consequence is the fact that domain growth becomes a very intermittent process: once an event on the scale of the domain size ξ has happened, the details of the conformation on scales $\ell < \xi$ start evolving between nearby metastable states, while the overall pattern formed by the domains on scale ξ hardly change.

Equation (10) also allows one to define a very important quantity which we call, by analogy with the glass temperature T_g, the 'glass length' ℓ_g, through $\Upsilon(T)\ell_g^\theta = \mathcal{A}T$, introduced in this context in [33, 34, 32]. The factor $\mathcal{A}$ is rather arbitrary; the choice $\mathcal{A} = 35$ corresponds to a time of 1000 seconds if $\tau_0 = 10^{-12}$ seconds. In analogy with the glass temperature T_g, one sees that length scales larger than ℓ_g cannot be equilibrated on reasonable time scales, while length scales smaller than ℓ_g are fully equilibrated. Qualitatively speaking, the equilibrated modes contribute to the stationary part of the correlation and/or response function, while the glassy modes $\ell > \ell_g$ contribute to the aging part. Therefore, the strong hierarchy of time scales induced by the exponential activation law allows equilibrated modes and aging modes to coexist.

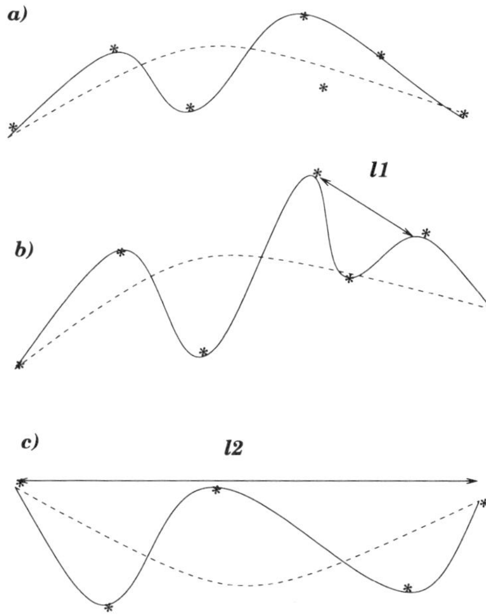

Figure 3. *Schematic evolution of a pinned object on well separated time scales. On scale $t(\ell_1)$, the object reconforms by flipping a small portion of size ℓ_1 from one favourable configuration to another ($a \to b$). On a much longer time scale $t(\ell_2) \gg t(\ell_1)$, the conformation on scale ℓ_2 (dotted lines) has evolved ($b \to c$). The dynamics of the short wavelengths happens on a time scale such that long wavelengths are effectively frozen.*

Finally, it is easy to understand that the logarithmic growth law, Equation 11, leads to a strong *cooling-rate* dependence of the typical size of the domains [13]: since the growth law is essentially that of pure systems as long as $\xi \ll \ell_g(T)$, a longer time spent at higher temperatures (where ℓ_g is large) obviously allows the domains to grow larger before getting pinned at lower temperatures.

3.3 Diffusion in a random potential: the Sinai model

It is useful to consider a toy model for the dynamics of a pinned domain wall by considering the motion of a *point particle* in a random potential. This can be thought of as a reduction of the problem to the dynamics of the centre of mass of the pinned object. From the study of these pinned objects, it is known that reconformations on scale ℓ typically change the position of the centre of mass X by an amount $\propto \ell^\zeta$, where ζ is a certain exponent, analogous to the exponent θ defined above. Since the energy changes by an amount $\Upsilon\ell^\theta$, the statistics of the random potential $V(X)$ acting on the centre of mass X must be such that:

$$\left\langle [V(X) - V(X')]^2 \right\rangle \propto \Upsilon^2 |X - X'|^{2\theta/\zeta} \qquad \text{for} \quad |X - X'| \ll L^\zeta, \tag{12}$$

where L is the total transverse size of the object. For larger distances, $\langle [V(X) - V(X')]^2 \rangle$ saturates to a finite value, since the impurities encountered by the pinned object become uncorrelated.

An interesting example is provided by a *line* in a plane (*i.e.* a domain wall in two dimensions) in the presence of impurities. In this case, the exponents ζ and θ are exactly known: $\zeta = 2/3$, $\theta = 1/3$, leading to $\langle [V(X) - V(X')]^2 \rangle \propto \Upsilon^2 |X - X'|$. If one assumes that the statistics of $V(X)$ is Gaussian, then the potential $V(X)$ is a random walk, and the model under consideration is precisely the well known Sinai model, for which a large number of analytical results are known (for reviews, see [35, 36, 37]).

One can also define in this model a 'glass scale' X_g such that, at temperature T_1:

$$\Upsilon(T_1)\sqrt{X_g} = \mathcal{A}T_1 \qquad \text{or} \qquad X_g = \left(\frac{\mathcal{A}T_1}{\Upsilon(T_1)}\right)^2, \tag{13}$$

where we have taken into account a possible temperature dependence of Υ. For $X \ll X_g$, the diffusion motion is quasi-free, and the dynamics is 'fast' (diffusive). For times corresponding to distances larger than X_g, the energy barriers strongly impede the motion, and lead to a slow logarithmic sub-diffusion:

$$t \sim \tau_0 \exp\left(\frac{V(X)}{T_1}\right) \qquad \longrightarrow \qquad X(t) \sim X_g \log^2\left(\frac{t}{\tau_0}\right). \tag{14}$$

More precisely, all the particles initially launched at $X = 0$ will, after time t, be located in the deepest energy well available at time t, which is at an approximate distance $\log^2 t$ from the initial point. The relative distance between these particles, however, does not grow with time. The deepest trap available is so much more favourable than the others that the relative distance between all particles remains typically of order X_g: this is the Golosov phenomenon (see [38, 36, 39]). Within the deepest well, on the other hand, the probability is roughly uniform, since the energy landcape is shallow in comparison with T_1. One can actually argue, using the beautiful results of [37], that the (intra-well) response of the particle to a small oscillating external field should behave as:

$$\chi_{\rm ag}(\omega, t_w) \sim \frac{\log^3 t_w}{\omega t_w} \tag{15}$$

where t_w is the time since the quench from very high temperatures. This shows that within this simplified model, the response is not exactly a function of ωt_w, but that there are logarithmic corrections. This means that the scaling variable is again, over a limited range of t_w, effectively of the form ωt_w^μ with $\mu < 1$. In Figure 4, we show the results of a numerical simulation performed with Yoshino [40], where $\chi(\omega, t_w)$ is computed for the Sinai model at various frequencies, as a function of t_w. We have shown the slope $1/t_w$ for comparison, and rescaled the different curves using the value $\mu = 0.9$ in the inset.

Now, the very interesting property of the Sinai model is the fact that the potential $V(X)$ is strictly self-affine. This means that the statistics of the potential at small scales is identical, up to a scaling factor, to the statistics of the potential at larger scales (see Figure 5). Therefore, when the temperature is lowered from T_1 to $T_2 < T_1$ after a time t_{w1}, the particle has to restart its search for the most favourable well, much as it had done when the temperature first reached T_1 from high temperatures. Since the probability distribution within the well is uniform at the moment of the second temperature change, it indeed corresponds, effectively, to a high temperature quench. The point however is that the particles cannot leave the deep well they had reached at T_1 before a time exceedingly

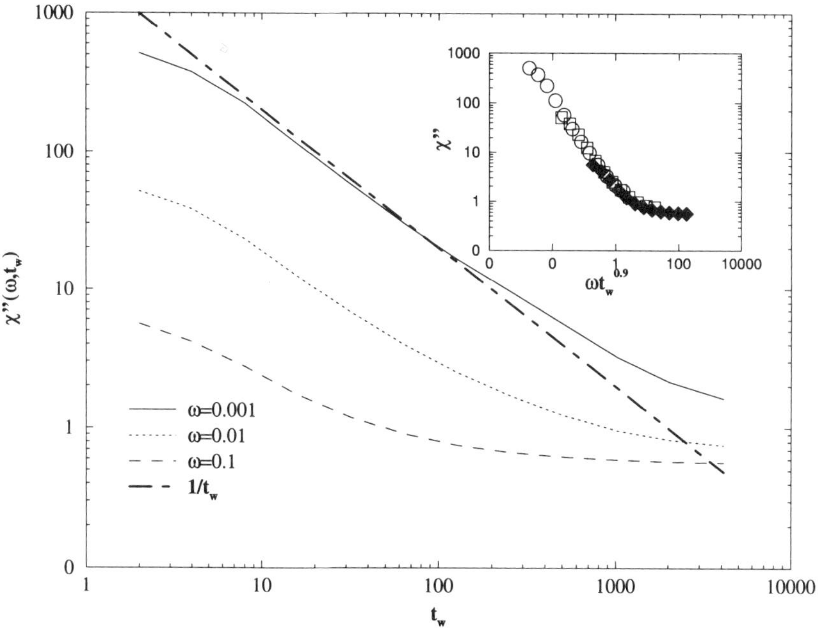

Figure 4. *Dissipative part of the response function to a small oscillating field in the Sinai model, as a function of waiting time, for different frequencies. The slope $1/t_w$ is shown for comparison. In the inset, we show that these curves can be rescaled using an effective value of $\mu = 0.9$.*

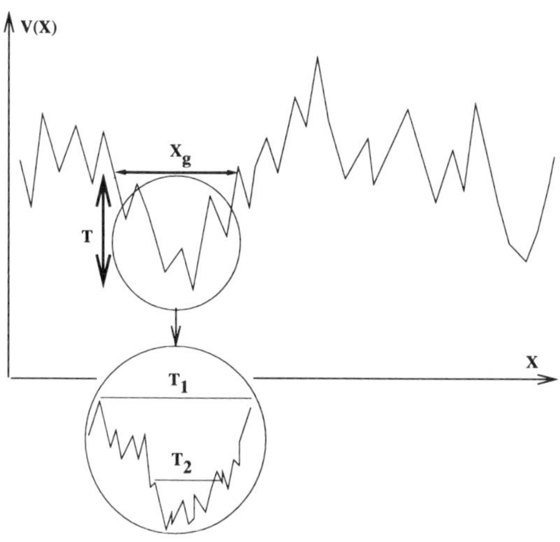

Figure 5. *The potential $V(X)$ in the Sinai model is self-affine. A zoom of a small portion of it, of size $X_g(T_1)$, reveals the same statistical features at a smaller scale. Correspondingly, the dynamics at a smaller temperature T_2 will exhibit properties after a quench from T_1 very similar to the initial dynamics after a quench from high temperature. On the other hand, the dynamics at T_2 mostly takes place within the valley reached at T_1.*

long compared to t_{w1}. This time t^*_{w2} is the time needed to overcome the depth of the well reached at t_{w1}, and is given by:

$$t^*_{w2} = \tau_0 \left(\frac{t_{w1}}{\tau_0}\right)^\beta \qquad \text{with} \qquad \beta = \frac{T_1 \Upsilon(T_2)}{T_2 \Upsilon(T_1)}. \tag{16}$$

Consider the case where the pinning energy vanishes above a certain temperature T_c, as $\Upsilon(T) \sim (T_c - T)^\omega$, with a certain exponent ω. (This is the case, for example, for domain walls near the Curie temperature). Then taking $\omega = 1$, $T_1 = 0.9T_c$ and $T_2 = 0.8T_c$, one finds that $\beta = 2.25$. Therefore, if $\tau_0 = 10^{-12}$ second, and $t_{w1} = 1$ second, one finds that t^*_{w2} is astronomically long, equal to 10^{15} seconds! In other words, the particle is completely trapped, even when the temperature only changes by 10%.

Therefore, this model provided a tantalising scenario for the 'rejuvenation and memory' effect: as the temperature is cooled down, new details of the potential appear, in a self-similar manner, and the aging dynamics over the barriers starts afresh. On the other hand, the particle remains effectively forever in the well that it had reached at T_1. Therefore, when the system is heated back to T_1, perfect memory is recovered. This scenario is very similar to the one advocated for spin glasses on the basis of a hierarchical landscape, inspired by Parisi's mean field solution [41, 33]. We believe that the Sinai model offers a precise basis for such a picture. Note that there is no 'chaos' involved in this model: rejuvenation occurs because previously equilibrated modes are thrown out of equilibrium, but in the very same energy landscape.

3.4 Diffusion in a random potential: the trap model

The above model assumes that the random potential is Gaussian, with a correlation function compatible with direct scaling arguments, which lead to $V(X) \sim X^{\theta/\zeta}$. A slightly different picture emerges from the 'replica variational theory' which predicts a highly *non-Gaussian* effective pinning potential, acting on the centre of mass of the pinned object. As detailed in [34, 42], the potential is a succession of local parabolas (corresponding to locally favourable configurations, or 'traps') matching at singular points (see also [43]). The full 'replica symmetry breaking' scheme needed to reproduce the correct scaling of the potential (*i.e.* $V(X) \sim X^{\theta/\zeta}$) means that the potential is actually a hierarchy of parabolas within parabolas, *etc.* This hierarchy directly corresponds to the existence of different length scales ℓ, each one characterised by an energy scale $E_\ell \sim \Upsilon \ell^\theta$. An important difference with the above Sinai model is the statistics of the depth of the different valleys of the pinning potential: the prediction of the replica approach is that the deepest valleys obey the so-called Gumbel extreme value statistics:

$$\text{for} \quad E \gg E_\ell: \qquad P_\ell(E) \sim \frac{1}{E_\ell} \exp\left(-\frac{E}{E_\ell}\right). \tag{17}$$

Assuming that the time needed to hop out of a trap of depth E is $\tau = \tau_0 \exp(E/T)$, one finds that the above exponential distribution of trap depths induces a *power-law* distribution of trapping times:

$$\text{for} \quad \tau \gg \tau_0: \qquad P_\ell(\tau) \sim \frac{\tau_0^{x_\ell}}{\tau^{1+x_\ell}}, \qquad x_\ell = \frac{T}{E_\ell} \sim \left(\frac{\ell_g}{\ell}\right)^\theta. \tag{18}$$

The 'trap model', which consists of a random walk between independent traps such that their release time is given by Equation 18, has been investigated in detail in [44, 33], and more recently, in the context of the rheology of soft glassy materials, in [5]. The interesting result is that the dynamics of this simple model is time-translation invariant as long as $x_\ell > 1$ (high temperature phase, corresponding to small length scales), but becomes *aging* when the average trapping time diverges, *i.e.* when $x_\ell < 1$ (large length scales). In this case, the dynamics becomes extremely intermittent, since the system spends most of its time in the deepest available trap.

One can compute, within this simple model, the a.c. susceptibility (or frequency dependent elastic compliance) to find [44, 33, 5]:

$$\begin{aligned} \chi''_\ell(\omega, t_w) &= A(x_\ell)(\omega\tau_0)^{x_\ell - 1} \qquad (x_\ell > 1) \\ \chi''_\ell(\omega, t_w) &= A(x_\ell)(\omega t_w)^{x_\ell - 1} \qquad (x_\ell < 1). \end{aligned} \tag{19}$$

(Note that the result quoted above for the Sinai model formally corresponds to $x_\ell = 0$, a result that could have been anticipated [35], up to logarithmic corrections.)

The case where all length scales evolve in parallel therefore leads to a total susceptibility given by $\chi''(\omega, t_w) = \chi''_\infty(\omega) + \chi''_{\rm ag}(\omega t_w)$ with:

$$\begin{aligned} \chi''_\infty(\omega) &= \textstyle\sum_{\ell<\ell_g} \mathcal{G}(\ell) A(x_\ell)(\omega\tau_0)^{x_\ell - 1} \\ \chi''_{\rm ag}(\omega t_w) &= \textstyle\sum_{\ell>\ell_g} \mathcal{G}(\ell) A(x_\ell)(\omega t_w)^{x_\ell - 1} \end{aligned} \tag{20}$$

where $\mathcal{G}$ is a 'form factor' counting the number of available modes at scale ℓ. A very interesting consequence of Equation 20 is that in the low frequency, long waiting time limit (more precisely when $\omega\tau_0 \ll 1$ and $\omega t_w \gg 1$), the sum over ℓ is dominated by the region $\ell \sim \ell_g$, for which $x_\ell \sim 1$. For a fairly general function $\mathcal{G}$, one expects both the stationary and aging part of χ'' to behave asymptotically as $1/\log\omega$. This can be translated, using the fluctuation dissipation theorem, into a noise spectrum $S(\omega) \propto 1/\omega\log\omega$, *i.e.* a so-called $1/f$ noise, ubiquitous in glassy systems. This mechanism suggests that $1/f$ noise should generically exhibit a slowly evolving component. Following the same argument, one also expects that the aging contribution to χ'' decays as a small power of t_w, eventually reaching a $1/\log(\omega t_w)$ behaviour for $\log(\omega t_w) \gg 1$. As mentioned above (see Equation 4), data on spin glasses typically give $1 - x_\ell \sim 0.1 - 0.3$ for ωt_w in the range 1–10000.

The strictly hierarchical nature of the landscape also leads, by construction, to the 'rejuvenation and memory' effect [41, 33]: when the temperature is reduced from T_1 to T_2, the 'glass length scale' moves down from ℓ_{g1} to ℓ_{g2}. Modes corresponding to $\ell_{g2} < \ell < \ell_{g1}$, which were equilibrated at T_1, become aging at T_2 (hence the 'rejuvenation'), while the modes such that $\ell > \ell_{g1}$, which were aging at T_1, become effectively frozen at T_2 (hence the 'memory'). This is illustrated in Figure 6.

Let us finally discuss how values of $\mu \neq 1$ can arise within the trap model. If one assumes that the traps visited during the evolution of the system are all different, as implicitly done above, then $\mu = 1$. This is not true if, for example, the geometry of the trap connectivity is of low dimension, for example if the traps are arranged on a one-dimensional array. In this case, one can show [45] that $\mu = 1/(1 + x)$ for $x < 1$. One can also look at simpler models, when a particle makes directed hops on a line where traps have a deterministic lifetime $\tau(n)$ which grows with the distance n to the origin. The

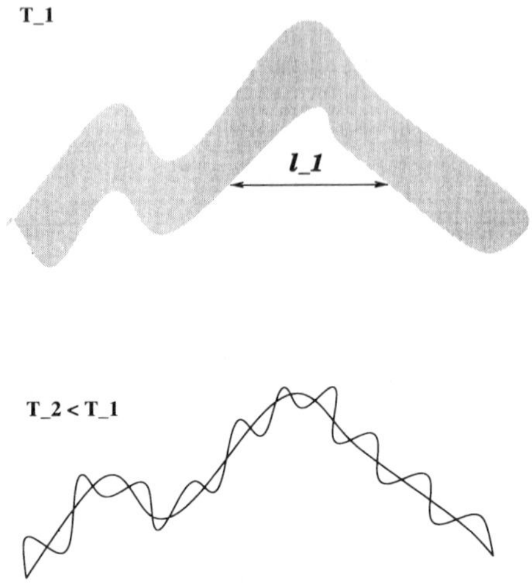

Figure 6. *Scenario for rejuvenation and memory for a pinned object: at temperature T_1, the length scale ℓ_1 is slowly evolving, while the smaller length scales are 'fast', and thus lead to a 'blurred shaped' object if one looks at it on the time scale $t(\ell_1)$. Upon cooling down to T_2, the smaller length scales start 'condensing': the object is pinned on smaller length scales.*

position $N(t_w)$ of the particle at time t_w is therefore given by:

$$t_w = \sum_{n=1}^{N} \tau(n), \tag{21}$$

since we have assumed that the walk is directed towards $n > 0$. The subsequent evolution of the particle, for a time $t \ll \tau(N)$, will be a function of $t/\tau(N)$. Taking $\tau(n) \propto n^\beta$ immediately leads to $\mu = \beta/(\beta+1) < 1$. If $\tau(n)$ grows faster than exponentially with n, then of finds that the effective value of μ is larger than one, because of the presence of logarithmic corrections which lead to super-aging.

4 Back to experiments

4.1 Disordered Ferromagnets

Equipped with the above theoretical ideas, one can return to the experimental results presented in Section 2, and see how far one can go in their interpretation. From the discussion above, one expects that aging in disordered ferromagnets can be understood in terms of the superposition of slow domain growth, and domain walls reconformations in the pinning field created by the disorder. Correspondingly, the aging part of the susceptibility is expected to behave as:

$$\chi''_{\text{ag}}(\omega, t_w) = \frac{1}{\xi(t_w)} \left[\chi''_{w,\infty}(\omega) + \chi''_{w,\text{ag}}(\omega, t_w) \right], \tag{22}$$

where $\xi(t_w)$ is the (slowly growing) domain size, $\chi''_{w,\infty}$ the stationary mobility of the domain walls at frequency ω, and $\chi''_{w,\text{ag}}$ the aging contribution coming from the reconformation modes, expected to scale as ωt_w. This expression accounts well for the observations made in random-ferromagnet-like systems, such as the ones studied in [13, 21], in particular:

- The aging part of the response is quite sensitive to the cooling rate, and decreases when cooling is slower and/or when the waiting time increases. This dependence is logarithmic, and directly reflects the behaviour of $\xi(t_w)$.

- For small frequencies, $\chi''_{w,\mathrm{ag}}$ dominates and one observes an approximate ωt_w scaling of the aging part of χ, up to log-corrections coming from the $1/\xi$ factor. On the other hand, for larger frequencies, the reconformation contribution becomes negligible and the ωt_w scaling breaks down completely, as observed in [13]. The fact that aging in glycerol bears some resemblance with aging in disordered ferromagnets suggests that some kind of domain growth might also be present in structural glasses. The precise nature of this domain growth is however at this stage not very clear, but is certainly a very interesting subject to explore further.

- One observes rejuvenation and memory, induced by slow reconformation of the walls. However, memory is recovered only if the time spent at the lower temperature T_2 is short enough, such that the overall position of the domain walls has not changed significantly. In the other limit, *i.e.* when the walls can move substantially, impurities interacting with the walls are completely renewed, and memory is lost [21].

4.2 Spin-glasses

The interpretation of the aging experiments in spin-glasses is not as transparent; this is directly related to the fact that the correct theoretical picture for spin-glasses in physical dimensions (as opposed to mean-field models, which are effectively infinite-dimensional) is still very much controversial. The simplest description is the droplet theory, where one essentially assumes that a spin-glass is some kind of 'disguised ferromagnet', in the sense that there are only two stable phases for the system, that one can (by convention) call 'up' and 'down'. The dynamics of the system after a quench can then be again thought of in terms of domain growth in a disordered system, with the difference that the 'pattern' that is progressively invading the system is itself random. Correspondingly, the energy of a 'domain' grows with its size ℓ as ℓ^θ, where θ is smaller than the value $d-1$ which holds in a pure ferromagnet. This is the scenario proposed (in a dynamical context) by Fisher and Huse [46], and further investigated by Koper and Hilhorst [47]. However, this scenario immediately stumbles on a first difficulty, in that it would predict a very strong cooling rate dependence of the susceptibility which, as shown in Figure 1, is definitely not observed.

A way out of this contradiction is to argue that even if *at any given temperature*, there are only two stable phases in competition (*i.e.* one pattern and its spin reversed), the favoured pattern changes chaotically with temperature. More precisely, if the temperature changes by ΔT, then the precise relative arrangement of the spins is preserved for lengths scales less than a 'chaos' length $\ell_{\Delta T} \propto \Delta T^{-y}$, and completely destroyed at larger length scales [48]. (Here, y is a new exponent related to θ.) If this is the case, then all the aging achieved at higher temperatures is useless in bringing the system closer to equilibrium at

T_1, where domain growth has to restart from scratch. This also explains how the system rejuvenates upon a small temperature change.

There are two problems with this interpretation. Firstly, chaos with temperature has neither been convincingly established theoretically, nor numerically. In particular, there seems to be no 'chaos' with temperature in the mean-field Sherrington-Kirkpatrick model [49]. Secondly, if the evolution at T_2 consists in growing new domains of the T_2 phase everywhere in space, it is difficult to imagine how this does not destroy completely the correlations built at temperature T_1. A way to do this would be to say that the new T_2-phase only nucleates around particular nucleation sites and grows very slowly, in such a way that a substantial region of space is still filled by the T_1 phase. However, this would mean that it is impossible to describe the dynamics of the system in terms of two phases only, as assumed in the droplet model. At any temperature, the configuration would necessarily be a mixture of all the phases corresponding to the previous temperatures encountered during the thermal history of the system [15].

The 'droplet' theory is radically different from the picture emerging from the Parisi solution of the mean-field Sherrington-Kirkpatrick spin-glass model [50]. There, one can show that the number of 'phases' in which the system can organise is very large. More precisely, there are configurations of nearly equal energy which differ by the flip of a finite fraction of the total number of spins. A consistent picture for how this scenario applies for finite dimensional spin-glasses is however still missing. A recent interesting suggestion made by Houdayer and Martin is that these different phases differ by the reversal of large non-compact, sponge like objects [51]. These objects have a linear dimension equal to the size of the system, but are not space-filling. Rather, their boundary separates an 'interior' from an 'exterior' which form a bicontinuous structure. If this is the case, then by definition this boundary is a kind of domain wall with zero surface tension. This is crucial in the sense that these 'domain walls' can hop from one metastable configuration to another (much as in a disordered ferromagnet) but with *no overall tendency to coarsen.*

In other words, the ordered phase of spin-glasses is, in a sense, full of *permanent* domain wall-like defects (which are probably very similar to the 'active' droplets of Fisher and Huse). These domain walls can only be precisely defined in reference to the true ground state of the system; their physical reality should however be thought of as particularly 'mobile' regions of spins; the precise position of these 'walls' define the possible metastable states of the sample.

After quenching the system to low temperatures, the system coarsens to get rid of the excess intensive energy, as has been beautifully demonstrated numerically in [52] (see also [53, 54]). However, the state left behind is still a 'soup' of walls with zero surface tension. The density of these walls is large and does not decay with time, and therefore provide the major contribution to the aging signal. The rejuvenation and memory effect can be understood in terms of the progressive quenching of smaller and smaller length scales of those walls as the temperature decreases, as we argued above for disordered ferromagnets. This interpretation is actually motivated by the fact that the temperature cycling experiments in the ferromagnetic phase and in the spin-glass phase of a single 'reentrant' spin-glass sample reveal very similar features [21]. The only difference is that memory is perfect in spin-glasses, *as if* no coarsening was present.

Let us insist once more on the fact that the above mechanism for rejuvenation is very different from the 'chaos' hypothesis. Within the trap model, aging is the phenomenon which occurs when the Boltzmann weight, initially uniformly scattered among many microstates, has to 'condense' into a small fraction of them. It is the dynamic counterpart of the entropy crisis transition that takes place in the Random Energy Model [55, 42].

Obviously, the above discussion is very speculative and a deeper understanding of aging in spin-glasses is very much needed. We however believe that the idea that the ordered phase of a spin-glass contains a large number of pinned, zero tension walls, which reconform in their disordered landscape, is a useful picture.

5 Conclusion

In these lectures, we have tried to review the most striking experimental results on aging in a variety of disordered systems, which reveal similar features but also important differences. We have argued that a generic model that reproduces many of these features is that of *pinned defects* in a disordered environment. The fact that energy barriers grow with the size of the reconforming regions immediately leads to a strong hierarchy of time scales. In particular, long wavelength aging modes and short wavelength equilibrated modes coexist in the system and offer a simple mechanism to explain the rejuvenation/memory effect. These properties can be discussed within simplified models where the dynamics of a whole pinned object is reduced to that of its centre of mass in a disordered potential. The main difference between random ferromagnets and spin-glasses seems to lie in the fact that while domains slowly grow in the former case (and therefore progressively reduce the density of domain walls), the fraction of 'domain walls' in spin-glasses appears to remain constant in time.

We have not attempted to discuss here the dynamical mean-field models, which have been much studied recently (for a review, see [1]). These models are also able to reproduce many of the interesting features of aging, including the rejuvenation/memory effect [56]. Furthermore, these models allow one to make precise predictions on the possible violations of the fluctuation-dissipation theorem [8, 9], and the relation between this violation and the non-trivial overlap distribution function which appears in the static solution of these models [57]. Finally, the exact dynamical equations of these models at high temperature are very similar to those of the Mode-Coupling Theory for fragile glasses (Kob, this volume). Therefore, one can study, by analogy, the extension of the Mode-Coupling equations to the glass phase, and obtain, within this framework, interesting results on the aging properties of fragile glasses [58]. However, the relevance of these mean-field theories to finite dimensional systems is not obvious, especially at low temperatures. This is because these models actually describe the diffusion of a single particle in a very high dimensional disordered potential [59, 60]. In high dimensions, the particle is never trapped in the bottom of a valley: there are always directions to escape. (This is true at least for discontinuous spin-glasses. The geometrical interpretation for 'continuous' spin-glasses is less clear [1]). The aging dynamics is dominated by the fact that the average number of unstable directions is decreasing with time, not by the fact that typical energy barriers are growing with time [61]. The inclusion of true activated effects in these mean-field (or Mode-Coupling) equations is still very much an open problem.

Acknowledgments

The ideas presented here owe a lot to many colleagues and friends with whom I have discussed these matters over the years, in particular F. Alberici, L. Cugliandolo, D. Dean, V. Dupuis, P. Doussineau, J. Hammann, J. Kurchan, A. Levelut, M. Mézard, P. Nordblad, E. Vincent and H. Yoshino. Interesting discussions with J. L. Barrat, A. Bray, M. Cates, D. S. Fisher, J. Houdayer, W. Kob, Ph. Maass, E. Marinari, O. Martin, R. Orbach, E. Pitard and P. Sollich must also be acknowledged. Finally, I want to thank Mike Cates for inviting me to give these lectures in St Andrews, and for a most enjoyable and fruitful collaboration over the years.

References

[1] J-P Bouchaud, L Cugliandolo, J Kurchan and M Mézard, in 'Spin-glasses and Random Fields', edited by A P Young (World Scientific) 1998.

[2] L C E Struick, *Physical Aging in Amorphous Polymers and Other Materials* (Elsevier, Houston) 1978.

[3] E Vincent, J Hammann, M Ocio, J-P Bouchaud and L F Cugliandolo, in *Complex Behaviour of Glassy Systems*, Springer Verlag Lecture Notes in Physics **492**, edited by M Rubi, 1997, pp184–219.

[4] T Giamarchi and P Le Doussal, in 'Spin-glasses and Random Fields', edited by A P Young, (World Scientific) 1998.

[5] P Sollich, F Lequeux, P Hebraud and M Cates, *Phys Rev Lett* **70** 2020 (1997);
P Sollich, *Phys Rev E* **58** 738 (1998);
S Fielding, P Sollich and M Cates, *J. Rheol*, in press, cond-mat/9907101.

[6] W Götze, in *Liquids, Freezing and the Glass Transition*, edited by J P Hansen, D Levesque and J Zinn-Justin, Les Houches 1989 (North Holland);
W Götze and L Sjögren, *Rep Prog Phys* **55** 241 (1992).

[7] W Kob, J L Barrat, *Fluctuations, response and aging dynamics in simple glass forming liquids out of equilibrium*, cond-mat/9905248;
W Kob, this volume.

[8] L Cugliandolo and J Kurchan, *Phys Rev Lett* **71** (1993) 173, *J Phys A* **27** (1994) 5749;
S Franz, M Mézard, *Europhys Lett* **26** (1994) 209, *Physica* **A209** (1994) 1;
H Horner, H Sommers and A Crisanti, *Z Phys* **B92** 257 (1993).

[9] L Cugliandolo, J Kurchana and L Peliti, *Phys Rev* **E 55** 3898 (1997).

[10] E Marinari, G Parisi, F Ricci-Tersenghi and J J Ruiz-Lorenzo, *J Phys A* **31** 2611 (1998).

[11] T Grigera, N Israeloff, cond-mat/9904351.

[12] F Alberici, P Doussineau and A Levelut, *J Phys I (France)* **7** 329 (1997); *Europhys Lett* **39** 329 (1997).

[13] F Alberici, J-P Bouchaud, L Cugliandolo, J Doussineau and A Levelut, *Phys Rev Lett* **81** 4987 (1998).

[14] R L Leheny and S Nagel, *Phys Rev* **B 57** 5154 (1998).

[15] K Jonason, E Vincent, J Hammann, J-P Bouchaud and P Nordblad, *Phys Rev Lett* **31** 3243 (1998); K Jonason, P Nordblad, E Vincent, J Hammann and J-P Bouchaud: to appear in *Europ Phys Jour B*.

[16] F Lefloch, J Hammann, M Ocio and E Vincent, *Europhys Lett* **18** 647 (1992).

[17] J O Andersson, J Mattsson and P Nordblad, *Phys Rev B* **48** 13977 (1993).

[18] P Doussineau, T de Lacerda-Arôso and A Levelut, *Europhys Lett* **46** 401 (1999).

[19] L Bellon, S Ciliberto and C Laroche, e-prints cond-mat/9905160 and 9906162.

[20] K Jonason, J Mattsson and P Nordblad, *Phys Rev Lett* **77** 2562 (1996); K Jonason and P Nordblad *Europ Phys Jour B* **10** 23 (1999).

[21] E Vincent, V Dupuis, M Alba, J Hammann and J-P Bouchaud, *Comparing aging in the spin-glass and ferromagnetic phases of a disordered ferromagnet*, cond-mat/9908030.

[22] E Vincent, J-P Bouchaud, D S Dean and J Hammann, *Phys Rev* **B 52** (1995) 1050.

[23] J-P Bouchaud, E Vincent and J Hammann, *J Phys I*, **4** 139 (1994).

[24] Y Joh, R Orbach,G Wood, J Hammann and E Vincent, cond-mat/0002040.

[25] W van Megen et al., *Phys Rev* **E 58** 6073 (1998).

[26] H Takayama, H Yoshino and T Komori, cond-mat/9909228 (08078).

[27] A J Bray *Adv Phys* **43** 357 (1994).

[28] L F Cugliandolo and D S Dean, *J Phys* A **28** 4213 (1995).

[29] L Berthier, J L Barrat and J Kurchan, *Eur Phys J B* **11** 635 (1999).

[30] D S Fisher, *Phys Rep* **301** 113 (1998).

[31] H Leshorn, T Nattermann, S Stepanow and L H Tang, *Annalen der Physik,* **6** 1 (1997).

[32] H Yoshino, in preparation; see also H Yoshino, *Phys Rev Lett* **81** 1483 (1998).

[33] J-P Bouchaud and D S Dean, *J Physique I (France)* **5** 265 (1995).

[34] L Balents, J-P Bouchaud and M Mézard, *J Physique* **I 6** 1007 (1996).

[35] J-P Bouchaud, A Comtet, A Georges and P Le Doussal, *Annals of Physics* **201** 285 (1990).

[36] J-P Bouchaud and A Georges, *Phys Rep* **195** 127 (1990).

[37] P Le Doussal, C Monthus and D S Fisher, *Phys Rev E* **59** 4795 (1999).

[38] A Golosov *Comm Math Phys* **92** 491 (1984).

[39] A Compte and J-P Bouchaud, *J Phys* **A 31** 6113 (1998).

[40] H Yoshino, J-P Bouchaud, in preparation.

[41] Ph Refregier, E Vincent, J Hammann and M Ocio, *J Phys France* **48** 1533 (1987); E Vincent, J-P Bouchaud, J Hammann and F Lefloch, *Phil Mag B* **71** 489 (1995).

[42] J-P Bouchaud and M Mézard, *J Phys A* **30** 7997 (1997).

[43] M Feigel'man and L Ioffe, *Z Phys B* **51** 237 (1983).

[44] J-P Bouchaud, *J Phys I (France)* **2**, 1705 (1992); C Monthus and J-P Bouchaud, *J Phys A* **29** 3847(1996).

[45] B Rinn, Ph Maass and J-P Bouchaud, cond-mat/0001161.

[46] D S Fisher and D A Huse, *Phys Rev B* **38** 373 and 386 (1988).

[47] G J M Koper and H J Hilhorst, *J Phys France* **49** 429 (1988).

[48] A J Bray and M A Moore, *Phys Rev Lett* **58** (1987) 57.

[49] A Billoire and E Marinari, e-print cond-mat/9910352.

[50] M Mézard, G Parisi and M A Virasoro, *Spin Glasses and Beyond*, World Scientific (1987).

[51] J Houdayer and O C Martin, *A geometrical picture for finite dimensional spin-glasses*, e-print cond-mat/9909203.

[52] T Komori, H Takayama and H Yoshino, cond-mat/9908078.

[53] H Rieger, *Ann Rev of Comp Phys* II, edited by D Stauffer (World Scientific, Singapore, 1995);
E Marinari, G Parisi, F Ritort and J J Ruiz-Lorenzo, *Phys Rev Lett* **76** 843 (1996);
E Marinari, G Parisi, F Ricci-Tersenghi and J J Ruiz-Lorenzo, cond-mat/9910232.

[54] Y G Joh, R Orbach, J J Wood, J Hammann and E Vincent, *Phys Rev Lett* **82** 438 (1999).

[55] B Derrida, *Phys Rev Lett* **45** 79 (1980), *Phys Rev* B **24** 2613 (1981).

[56] L Cugliandolo and J Kurchan *Phys Rev B* **60** 922 (1999).

[57] S Franz, M Mézard, G Parisi and L Peliti, *Phys Rev Lett* **81** 1758 (1998).

[58] J-P Bouchaud, L Cugliandolo, J Kurchan and M Mézard, *Physica* **A226** 243 (1996).

[59] J Kurchan and L Laloux, *J Phys A* **29** 1929 (1996).

[60] J-P Bouchaud and M Mézard, *Prog Th Phys (Supp)* **126** 181 (1997).

[61] A Cavagna, *Fragile versus Strong liquids: a saddles ruled scenario*, cond-mat/9910244.

Phase separation and aggregation in colloidal suspensions

Henk Lekkerkerker

Utrecht University, The Netherlands

1 Introduction: the legacy of Perrin and Onsager

Colloidal particles have a size in the range of 1nm to 1000nm and are encountered in many systems both in nature as well as in technology. Their size is such that the thermal motion is sufficiently vigorous to counteract gravitational forces. A good measure to characterise this balance is the gravitational length. This gravitational length χ_g is defined as the distance over which a colloidal particle can be displaced in the gravitational field at the expense of a mean thermal energy equal to Boltzmann's constant k_B times the temperature T. This leads to

$$\chi_g = \frac{k_B T}{m^* g}. \tag{1}$$

Here m^* is the mass of the colloidal particle, corrected for buoyancy effects. In the earth's gravitation field the gravitational length varies from 1μm to 1mm for particles in the colloidal size range. In space experiments under microgravity conditions the gravitational length can reach values of many metres.

With simple yet brilliant experiments, Perrin (1939) studied the competition between thermal motion and gravity in the period 1907–1913. From quantitative measurements of the gravitational distribution of well-characterised monodisperse colloidal particles, Perrin was able to determine Boltzmann's constant k_B. For this work he received in 1926 the Nobel prize for Physics. The interpretation of the experiments of Perrin is based on the idea that colloids can be considered as super-atoms which obey the same thermostatistics as atoms in atomic systems. The statistical mechanical foundation for this idea was provided by Onsager (1933) in the thirties. Onsager showed that in the statistical mechanical laws the interaction between atoms must be replaced by the potential of mean force between the colloidal particles. The potential of mean force $W(h)$ between two colloidal particles can be written as (see Frenkel, this volume)

$$W(h) = \Omega(h) - \Omega(\infty) \tag{2}$$

where $\Omega(h)$ is the Grand Potential for the system when the distance of the surfaces of the two colloidal particles is h, and $\Omega(\infty)$ is the Grand Potential at infinite separation. This potential of mean force can be manipulated by changing the nature of the colloidal particles or the nature of the solvent in an almost continuous way from steeply repulsive to deeply attractive.

This variability of the potential leads to new and fascinating phase behaviour, which is frequently enriched (or some may say corrupted) by long-lived metastable and non-equilibrium states such as gels and glasses (Kob, this volume).

In addition to the variability of the potential of mean force, colloidal suspensions offer unique advantages for the study of phase behaviour which are related to the fact that these systems are intrinsically soft and slow. A typical modulus for a colloidal system consisting of colloidal particles of radius a can be written as k_BT/a^3 which leads to values of 1–100 Pa. This should be compared with a typical modulus of $10^{10} - 10^{11}$Pa for atomic systems. A typical time scale for a colloidal system is the time a^2/D it takes for a particle to diffuse over its own radius. This time is of the order of 10^{-3}s–1s which should be compared to characteristic times of the order 10^{-12}s for atomic systems. In addition to being soft and slow, which leads to considerable experimental advantages, colloidal systems also offer the attraction of being visible with a variety of optical microscopy techniques.

The extension of the conceptual framework pioneered by Perrin and Onsager coupled with the possibilities to prepare well defined colloidal particles has led to exciting developments in the area of phase transitions in colloidal systems in the last 25 years. Some of these developments will be briefly discussed here. For further details of some of these, and a discussion of related topics, see Frenkel, this volume.

2 Hard sphere crystallisation

One of the most intriguing predictions of statistical mechanics which up till today gives rise to fundamental discussions is the fluid-crystal transition in assemblies of hard spheres. This possibility was first speculated upon by Kirkwood (1939) and Alder and Wainwright (1957) obtained definite indications for this transition in their pioneering computer simulations. The volume fractions ϕ_f, ϕ_c between which fluid and crystal co-exist were subsequently determined by Hoover and Ree (1968) with computer simulations. They found $\phi_f = 0.494$ and $\phi_c = 0.545$.

As, in real life, hard-sphere atoms do not exist, the experimental verification of the hard-sphere freezing transition was to be made in a different kind of physical system. Colloidal suspensions are well suited to such experiments as the repulsive interactions can be manipulated from gently soft to steeply repulsive. The hard-sphere freezing transition does indeed occur in suspensions of sterically stabilised colloids as was first demonstrated by De Kruif, Rouw, Jansen and Vrij (1985) and by Pusey and Van Megen (1986). In addition to the fluid-crystal transition these experiments also indicate the existence of a glass transition at a volume fraction of about 0.58. Above this glass transition homogeneous nucleation of crystals was no longer observed. In recent years experiments under micro-gravity conditions have led to new and unexpected results concerning the crystallisation kinetics, crystal structures and the glass transition (Zhu *et al.* 1997). These developments are discussed in the lectures of Chaikin (this volume).

3 Colloids with attractive interactions

In principle the inclusion of an attraction additional to a repulsive core interaction leads to the possibility of a system undergoing a phase transition between a gas and a liquid phase. This is commonly observed in atomic systems. The condensation of a gas was already understood by Van der Waals (1873), who was the first to put forward a simple theoretical model which allowed for the quantitative prediction of the gas-liquid equilibrium.

The generic phase diagram of an atomic fluid is sketched in Figure 1. The region in the phase diagram in which the liquid exists as a stable phase is bounded by the critical temperature T_c above and the triple temperature T_t below. (It is conventional to call the disordered phase above the critical temperature a fluid, which then below the critical temperature splits into a gas region and a liquid region). Whereas in atomic systems the gas-liquid transition is universally observed, the equivalent transition in colloidal suspensions has only been established in recent years. The reason is that the long-ranged attractive interaction between particles required to induce a gas-liquid transition is not commonly encountered in colloidal systems.

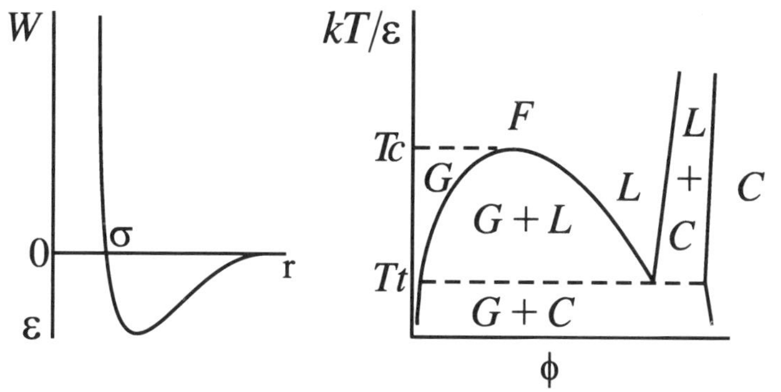

Figure 1. *Potential W(r) and the generic phase diagram for an atomic system. The region in which gas (G) and liquid (L) phases coexist is bounded by the critical temperature T_c and the triple temperature T_t. (F and C denote the fluid and crystal phase respectively.)*

In suspensions of charge stabilised particles the superposition of the repulsive electric double layer interaction and the attractive London–Van der Waals interaction leads to the celebrated DLVO potential (Derjaguin and Landau 1941, Verwey and Overbeek 1948). A characteristic of this potential, which is sketched in Figure 2, is a primary minimum at very short inter-particle distances, a repulsive barrier at intermediate distances and a secondary minimum at larger distances. Upon increasing the salt concentration (typically above 10^{-2}M) or lowering the surface potential the repulsive barrier becomes very low or disappears altogether. This gives rise to the aggregation of the colloidal particles in the primary minimum. As early as 1917 Smoluchowski (1917) gave an admirable theory for the kinetics of this type of aggregation. This long-established field received a new and strong impetus in the 1980s when it was found that the clusters formed by the irreversible aggregation exhibit fractal structures (Weitz and Huang 1984; see also Pine, this volume). Also the relation between aggregation and gelation was clarified by these studies. At

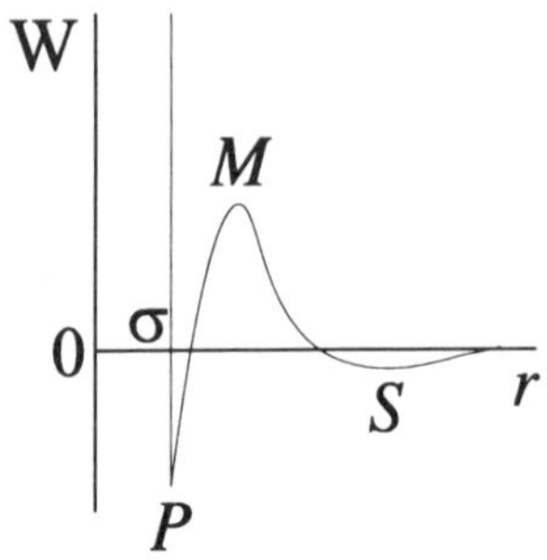

Figure 2. *The DLVO interaction potential, where P is the primary minimum, M the repulsive barrier, and S the secondary minimum.*

the same time computer simulations by Ziff (1984) indicated that colloidal aggregation phenomena lead to fractal structures locally while showing spinodal-like modulation of the fractal structures at long wavelength. Later on it was found experimentally that fractal aggregation in dense colloidal solutions leads to spinodal type dynamics (Carpineti and Giglio 1992). Rouw, Woutersen, Ackerson and De Kruif (1989) surmise that spinodal decomposition (see Bray, this volume) is a general process for any aggregating system be it reversible or irreversible.

For large colloidal particles the secondary minimum may become sufficiently deep to allow in principle for phase separation while the repulsive barrier which separates the primary and secondary minima remains sufficiently high to prevent irreversible aggregation of the dispersion in the primary minimum. Nevertheless Kotera, Furusawa and Kudo (1970) who studied suspensions of large latex particles which fulfil these conditions, observe reversible aggregation rather than phase separation. However, recently Hachisu (1998) observed a gas–liquid and a liquid–solid transition in a charged stabilised suspension of gold colloids with a diameter of 120nm. Apparently the strong London—Van der Waals interaction between gold particles is sufficient to induce the long-range attractive interaction required for a gas–liquid transition.

In dispersions of sterically stabilised colloidal particles attractive interactions can be induced by changing the quality of the solvent. Silica particles coated with stearyl alcohol suspended in benzene provide an example of such a system. Here the interaction between the particles can be tuned by means of the temperature. At sufficiently high temperatures the particles show hard-sphere-like behaviour. Lowering the temperature gives rise to an attraction between the particles which is rather short-ranged due to the small length of the stabilising layer compared to the size of the core. Verduin and Dhont (1995) observed that the gas-liquid phase separation was pre-empted by gelation. Here again the short-range character of the attraction prevents the observation of the colloidal equivalent of a gas-liquid transition.

4 Colloid–polymer mixtures

An interesting way to induce a tunable attractive interaction between colloidal particles is provided by the addition of non-adsorbing polymer. This leads to a so-called 'depletion interaction' (Asakura and Oosawa 1954; Vrij 1976). When two colloidal particles approach each other to a distance smaller than the size of a polymer (roughly twice the radius of gyration of the polymer) the polymers are almost totally excluded from the region between the colloids. The colloidal particles are then forced together by the unbalance in the osmotic pressure exerted by the polymers. The depletion interaction allows one to tune both the range and the depth of the attractive interaction. The range of the attractive interaction is determined by the size of the polymer and the depth of the potential is proportional to the number density of the polymers (which in turn determines the osmotic pressure). Modelling the polymers as penetrable hard spheres with diameter σ_p, and the colloids as hard spheres with diameter σ_c, the depletion potential can be written as

$$W(h) = -\frac{3}{2}k_BT\phi_p\frac{\sigma_c}{\sigma_p}\left(1 - \frac{h}{\sigma_p}\right)^2 \qquad (3)$$

Here σ_p is the volume fraction of the polymer coils and h is the distance between the surfaces of the two colloidal spheres. An attraction of between 2 and 3 k_BT will lead to phase separation. Equation 3 then implies that for a size ratio $\sigma_c/\sigma_p = 10$ the depletion interaction will cause phase separation when the volume fraction of the polymer coils is about 20%.

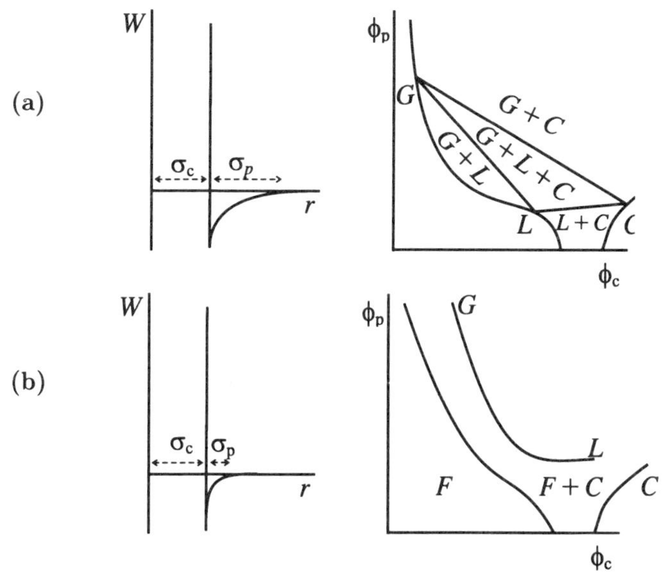

Figure 3. *The interaction potential and phase diagram for colloid+polymer mixtures in the case of (a) a long-range attraction ($\sigma_p/\sigma_c > 0.3$) and (b) a short-range attraction ($\sigma_p/\sigma_c \le 0.3$). (The letters C, F, G, L indicate phases as in Figure 1.)*

Calculations of the phase diagram of colloid–polymer mixtures taking into account the depletion interaction were first performed by Gast, Hall and Russel (1983), and later on extended by Lekkerkerker *et al.* (1992). These calculations show that the range of the attraction determines the global features of the phase diagram, as depicted in Figure 3. When the range of attraction is less than about 30% of the hard core of the colloidal particles the gas-liquid transition becomes metastable and only a widening of the fluid-crystal transition is observed. However, when the range of attraction is larger than 30% of the hard core of the colloidal particles the gas-liquid transition becomes stable. These theoretical predictions have indeed been confirmed experimentally (Ilett, Orrock, Poon and Pusey 1995).

The kinetics of phase transitions in colloid-polymer mixtures has also been studied extensively. In a system of small colloidal particles with a polymer of the same size, Verhaegh, van Duijneveldt, Dhont and Lekkerkerker (1996) studied the fluid-fluid phase separation kinetics and morphology with small angle light scattering and light microscopy. Typical spinodal decomposition patterns were observed. In systems with large colloidal particles and a relatively small polymer, where the colloidal gas liquid transition is metastable, colloidal crystallisation only occurs in a narrow region just above the fluid-crystal binodal (Smits *et al.* 1990). A deeper quench into the two phase region gives rise to transient gelation followed by the formation of an amorphous precipitate (Pusey, Pirie and Poon 1993). The role of the metastable colloidal gas-liquid binodal has been speculated upon by Poon, Pirie and Pusey (1995). Recently Poon *et al.* (1999) studied the kinetics of phase-separation in a colloid-polymer mixture evolving to gas-liquid-crystal coexistence. Several distinct regimes were found which were interpreted by appealing to the underlying free energy landscape of the system.

5 Colloid–colloid mixtures

The depletion interaction mechanism is not limited to colloid-polymer mixtures. Using the same arguments as for colloid-polymer mixtures, mixtures of colloidal particles of different size and/or shape are expected to exhibit similar phase behaviour. In the case of binary mixtures of large and small colloidal hard spheres the depletion potential is to lowest order in the density given by

$$W(h) = -\frac{3}{2}k_BT\phi_s\frac{\sigma_l}{\sigma_s}\left(1-\frac{h}{\sigma_s}\right)^2 . \tag{4}$$

Here ϕ_s is the volume fraction of the small colloidal spheres, σ_l (σ_s) the diameter of the large (small) colloidal spheres and h the distance between the surfaces of the two large spheres. Again it is estimated that an attraction between 2 and 3 k_BT will lead to phase separation which for a size ratio $\sigma_l/\sigma_s = 10$ implies a volume fraction of about 20% of the smaller species. Indeed in mixtures of large and small spheres phase transitions have been observed experimentally (Van Duijneveldt, Heinen and Lekkerkerker 1993, Kaplan, Rouke, Yodh and Pine 1994, Imhof and Dhont 1995, Dinsmore, Yodh and Pine 1995). Again interesting non-equilibrium phenomena in the phase separation behaviour have been observed. For example Imhof and Dhont (1995) observed two types of glassy phases. In one of them only the large particles are immobilised and in the other both types of spheres are immobile.

An even more interesting colloidal mixture from a depletion interaction point of view is that of colloidal spheres and colloidal rods. This was already recognised by Asakura and Oosawa (1958) who mentioned rod-like macromolecules as highly efficient depletion agents. Auvray (1981) derived an expression for the depletion interaction potential due to thin rods in the Derjaguin approximation. More recently Mao, Cates and Lekkerkerker (1997) calculated the depletion interaction of rods of length L and diameter D up to third order in the density of the rods. In the limit that the length L of the rod-like particles is much smaller than the diameter σ of the colloidal spheres, the depletion potential to the lowest order in density is given by

$$W(h) = -\frac{1}{2} k_B T \phi \frac{L\sigma}{D^2} \left(1 - \frac{h}{L}\right)^2 \tag{5}$$

where ϕ is now the volume fraction of the rods and σ the diameter of the spheres. From Equation 5 it follows that for the case $L/\sigma = 0.2$ and $L/D = 20$ a minimum of the depletion interaction of $-2.5k_BT$ is obtained when the volume fraction ϕ of rods is only about 0.4%. Clearly, rod-like particles are efficient depletion agents; very low concentrations of rods are predicted to lead to phase separation. Recently phase separation in mixed suspensions of rods and spheres was observed by Koenderink *et al.* (1999). Colloidal crystals of silica spheres were observed upon addition of a small amount of boehmite rods. Using fluorescent silica spheres of about 740nm diameter, the crystallisation process was followed at particle level in real time and real space by confocal microscopy (Vliegenthart, Van Blaaderen and Lekkerkerker 1999). The phenomena observed in this mixed colloidal system, suggest a two-stage crystallisation process. Rapid aggregation of the colloidal spheres is followed by slow rearrangement into crystals (Figure 4).

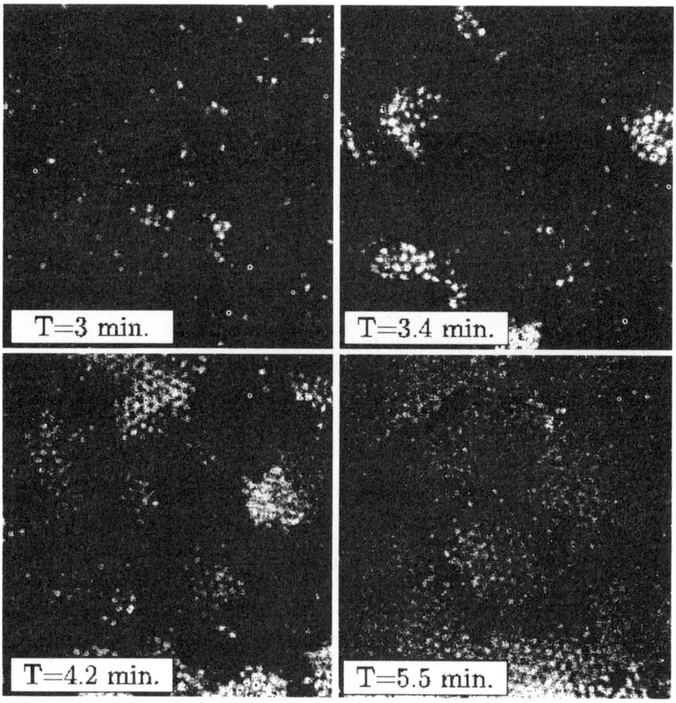

Figure 4. *Confocal microscopy images of the crystallisation process in a mixture of 2.5%(v/v) fluorescent spheres and 0.5%(v/v) rods. The size of the images is 50×50μm.*

6 Concluding remarks

The effect of attractive interactions on the phase behaviour of colloidal suspensions is rich and varied. In addition to the traditional pathways of phase separation, nucleation and growth and spinodal decomposition, one also frequently observes aggregation and gelation. As we have seen, the interplay of these phenomena sometimes leads to macroscopic phase separation but often non-equilibrium and metastable states arise. These phenomena are interesting in their own right and at the same time also provide analogies with phase separation in systems of biological macromolecules such as protein crystallisation (Rosenbaum, Zamora and Zukoski 1996; Muschol and Rosenberger 1997; Ten Wolde and Frenkel 1997; Haas and Drenth 1999).

Acknowledgment

I am grateful to Felix van der Kooij and Gerrit Vliegenthart for their participation in the preparation of this contribution.

References

Alder B J and Wainwright T E, 1957, *J Chem Phys* **27** 1208.
Asakura S and Oosawa F, 1954, *J Chem Phys* **22** 1255.
Asakura S and Oosawa F, 1958, *J.Pol Sci* **33** 183.
Auvray L, 1981, *J Physique* **42** 79.
Carpineti M and Giglio M, 1992 *Phys Rev Lett* **68** 3327.
de Kruif C G, Dhont J K G and Mos H J, 1985, *J Phys (Paris)* **46** 295.
Derjaguin B V and Landau L, 1941, *Acta Physicochim URSS* **14** 633.
Dinsmore A D and Yodh A G, 1995, *Phys Rev E* **52** 4045.
Gast A P, Hall C K and Russel W B, 1983, *J Coll Int Sci* **96** 251.
Haas C and Drenth J, 1999, *J Cryst Growth* **196** 388.
Hachisu S, 1998, *Croatica Chem Act* **71** 975.
Hoover W G and Ree F M, 1968, *J Chem Phys* **49** 3609.
Ilett S M, Orrock A, Poon W C K and Pusey P N, 1995, *Phys Rev E* **51** 1344.
Imhof A and Dhont J K G, 1995, *Phys Rev Lett* **75** 1662.
Kaplan P D, Rouke J L, Yodh A G and Pine D J, 1994, *Phys Rev Lett* **72** 582.
Kirkwood J G, 1939, *J Chem Phys* **7** 911.
Koenderink G H, Vliegenthart G A, Kluijtmans S G J M, van Blaaderen A, Philipse A P and Lekkerkerker H N W, 1999, *Langmuir* **15** 4693.
Kotera A, Furusawa K and Kudo K, 1970, *Kolloid Z u Z Polymere* **240** 837.
Lekkerkerker H N W, Poon W C K, Pusey P N, Stroobants A and Warren P B, 1992, *Europhys Lett* **20** 559.
Mao Y, Cates M E and Lekkerkerker H N W, 1997, *J Chem Phys* **106** 3721.
Muschol M and Rosenberger F, 1997, J*Chem Phys* **107** 1953.
Onsager L, 1933, *Chem Rev* **13** 73.
Perrin J, 1939, *Les Atomes* (Presses Universitaires de France, Paris)
Poon W C K, Pirie A D and Pusey P N, 1995, *Faraday Disc* **101** 65.
Pusey P N, Pirie A D and Poon W C K, 1993, *Physica A* **201** 322
Pusey P N and van Megen W, 1986, *Nature* **320** 340.

Rosenbaum D Zamora P C and Zukoski C F, 1996, *Phys Rev Lett* **76** 150.
Rouw P W Outersen A T J M, Ackerson B J and Dekruif C G, 1989, *Physica A* **156** 876.
Smits C, Vanduijneveldt J S, Dhont J K G, Lekkerkerker H N W and Briels W J, 1990, *Phase transitions* **21** 157.
ten Wolde P R and Frenkel D, 1997, *Science* **277** 1975.
Van der Waals J D, 1873, *On the continuity of the gaseous and liquid states, studies in statistical mechanics* 1988 edn., edited by J S Rowlinson, (North-Holland, Amsterdam)
van Duijneveldt J S, Heinen A W and Lekkerkerker H N W, 1993, *Europhys Lett* **2** 369.
Verduin H and Dhont J K G, 1995, *J Coll Int Sci* **172** 425.
Verhaegh N A M, vanDuijneveldt J S, Dhont J K G and Lekkerkerker H N W, 1996, *Physica A* **230** 409.
Verwey E J W and Overbeek J T G, 1948, *Theory of the stability of colloids* (Elsevier, Amsterdam)
Vliegenthart G A, van Blaaderen A and Lekkerkerker H N W, 1999, *Faraday Disc* **112** 173.
Von Smoluchovski M, 1917, *Zeits f Physik Chemie* **92** 129.
Vrij A, 1976, *Pure Appl Chem* **48** 471.
Weitz D A and Huang J S, 1984, in *Kinetics of aggregation and gelation*, editors Family F and Landau D P; (Elsevier, Amsterdam).
Zhu J, M Li, R Rogers, W Meyer, R.H Ottewill, STS-73 Space Shuttle Crew, W B Russel and P M Chaikin, 1997, *Nature* **387** 883.
Ziff R M, 1984, in *Kinetics of aggregation and gelation*, editors Family F and Landau D P; (Elsevier, Amsterdam).

Thermodynamics and hydrodynamics of hard spheres: the role of gravity

Paul Chaikin

Princeton University, USA

1 Introduction

Hard spheres are the 'fruitfly' of condensed matter physics. Whenever we want to know the physical properties of a system, or understand interesting phenomena, the first question we tend to ask is: "What would a set of hard spheres do?" They are the next simplest system we can imagine after an ideal gas, and already they are extremely interesting. They exhibit a number of unusual phenomena including phase transitions, complex motion, structure, and correlations (Pusey 1987). The lessons we learn from studying hard spheres are often found to generalise to more complicated systems, and they give us a basic intuition and scale and often the hard sphere solutions are sufficient in themselves to explain the sought-after phenomena.

The particular problems we want to look at in this set of lectures involve thermodynamics and entropy of packings (including metastable ones) on the one hand, and the role of hydrodynamic interactions (particularly in nonequilibrium dynamics, such as motion under gravity) on the other. Our basic understanding of viscous drag and diffusion of particles in a fluid comes from the exact solutions for hard spheres by Stokes (1851) and Einstein (1905). The results:

$$f_{\text{drag}} = 6\pi\eta a v \quad (1)$$

$$D = k_B T/(6\pi\eta a) \quad (2)$$

tell us the scaling with viscosity η and radius a, and qualitatively hold for very different situations; polymers, disks, rods, bubbles, *etc.*, which at first glance are not at all like hard spheres. (The reason that hard spheres contain the essence of the problem for hydrodynamics of suspensions is that the Laplacian at large distances cares only about the largest length scale.) Our basic thermodynamic understanding of the first order transition from the liquid to crystalline solid is also based on the remarkable studies which have been done on hard spheres (Alder and Wainwright 1957, Hoover and Ree 1968, Pusey and van Megen 1986); see also Frenkel and Lekkerkerker, this volume. The simple property that hard spheres do not interpenetrate is sufficient to cause thermodynamic crystallisation as particle density is increased. It also gives rise to metastable, glassy packings.

A good deal of the work reported below is related to our NASA project to study the crystallisation of hard sphere samples in microgravity. As part of this study we had to investigate alternative ways to counteract the effects of gravity. This led us to perform a series of experiments on the particle dynamics and other effects of sedimentation. They involve both the roles of hard spheres in hydrodynamic interactions and thermodynamics. Thus although there is an enormous amount of work that has been done on hard spheres, our selection of topics has been coloured by the problems which led up to, and include, a number of experiments performed on space shuttle flights.

1.1 Hard spheres

The contribution of condensed matter physics to science lies in the many body problem (Chaikin and Lubensky 1995). Given a number of interacting particles, what can they do? What states will they form? The typical microscopic two-particle interaction, that we have studied since our most elementary education, consists of a long range attractive interaction and a short range repulsion. Most of the effort is spent on the various forms of the attraction: for neutral atoms and molecules there is the Van der Waals ($1/r^6$) interaction; for metals there is an (almost) isotropic metallic bond (similar in many aspects to a covalent bond); for oppositely charged ions there is the Coulomb attraction. Less instructional time is usually paid to the microscopic repulsive interaction. It is related to the energy cost in pushing two electronic wavefunctions together. There is the direct Coulomb repulsion between the like charges and there is the kinetic energy that must be paid in order to put the electrons in a smaller 'box' as we push them together. (The uncertainty principle demands a higher momentum, and hence energy, as the size of the wavefunction is decreased. The exclusion principle can further decrease the size of the wavefunction and increase the energy, as like-spin electronic wavefunctions cannot overlap.) The equations for the repulsive terms are not readily obtained, and the tendency in most books and in calculations is to parameterise the repulsion by a simple analytic form that increases, at short distances, much faster than whatever attractive term is being used. Hence the often-used $6-12$ (Lennard Jones) potential has a $1/r^{12}$ repulsion to overcome the $1/r^6$ Van der Waals attraction at short range.

In atomic and molecular systems the attractive interaction is not very interesting. Its main role is to set the density and it can give rise to a transition between a less dense and more dense fluid phase, the gas-liquid transition. (However, in colloidal systems the attractive interaction can play a more important role in the structure and dynamics. See Lekkerkerker, this volume.) The structure and correlations of most fluids (and molecules), the structure of solids, and the liquid-to-crystal transition are primarily results of the repulsive interaction. The simplest form for the short range repulsion is the hard sphere potential:

$$\begin{aligned} U(r) &= \infty \qquad r > a\,, \\ U(r) &= 0 \qquad r < a\,. \end{aligned} \tag{3}$$

This form has some peculiarities. Since the potential at less than a diameter is infinite the particles don't interpenetrate; there is 'excluded volume'. When the particles don't touch the potential is zero. Since the only energies are zero or infinity there is no characteristic energy with which to compare k_BT. We cannot raise the temperature T above some energy scale and expect a phase transition. The phase diagram is therefore athermal, the only control parameter is the particle density n, or the volume fraction $\phi = nv_0$ where $v_0 = (4/3)\pi a^3$. Temperature, as the only energy scale, sets all others. The pressure and elastic constants (bulk and shear moduli) are measures of the energy density and thus scale as nk_BT times a function of the volume fraction (*e.g.* $P = nk_BTf(\phi)$). The internal energy is zero, the Helmholtz free energy is $F = -TS$ and the maximisation of the entropy S is what determines the state of the system, independent of T.

Now we want to gain some intuition about the entropy. Entropy is associated with the number of ways we can arrange the particles in a system given certain constraints. Part of the entropy is simply related to the volume that is available for the particle motion and this is related to how well spheres at a certain volume fraction fill space. (See Frenkel, this volume, for a related discussion.)

1.2 Packings

Here we focus on random packings; these include metastable alternatives to the crystalline state of hard spheres at high density. The packing of spheres has been an important problem since the dawn of civilization[1] (Bernal 1965, Zallen 1983). In its earliest form it concerned how much grain (assumed spherical) would fill a certain container. If grain is sold by the bucket and you could figure a way to less densely pack it than your neighbour, you would make a lot of money. Modern packing problems are also driven by commerce: identifying the best sphere packing in an n-dimensional space leads to more efficient data transmission, storage and error correcting codes (Conway and Sloane 1992). Since we are mostly concerned here with colloids and physics we restrict our attention to three or fewer dimensions. A typical set of hard spheres is shown in Figure 1. The figure is from a lecture by Bernal (1964) who performed extensive and clever experiments to find the density, configurations and correlations of lattice packings but particularly of random packing of hard spheres. He was interested in the structure of glass and liquidlike states of elemental and molecular systems. Toward the centre of Figure 1 we see the stacking of ball bearings in the familiar, 'cannon-ball' or 'fruit stand' arrangement that physicists since the time of Kepler (and everyone else since the discovery of fruit) has associated with the densest packing of spheres. The lattice shown is FCC, face centered cubic, formed by stacking two-dimensional hexagonal planes. This packing fills space to a volume fraction of $\pi/3 \times \sqrt{2} = 0.7404...$ The mathematical proof that FCC is the densest packing of spheres was given in the late 1990's, about three hundred years after the conjecture of Kepler who was in fact interested in how to pack cannon-balls on warships. For the previous half century there was merely a limit set by the local packing fraction of four spheres in a tetrahedron, $\phi = 0.7796$, proved to be an upper limit by Roger and generalisable as an upper limit to n dimensions (Conway and Sloane 1992).

[1] Bernal quotes a comment on random packing which Saint Luke attributes to Jesus, "Give, and it shall be given unto you; good measure, pressed down, and shaken together, running over, shall men give into your bosom. For with the same measure that ye mete withal it shall be measured to you again." Note that Nagel (this volume) refers to the need to repeatedly tap granular material in order to have it arrive at its densest packing.

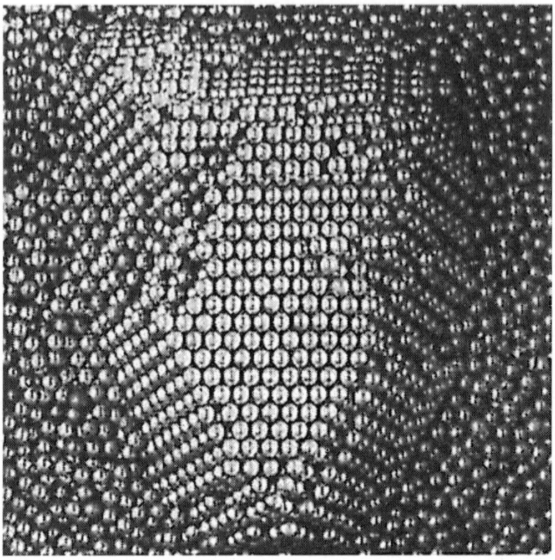

Figure 1. *Lattice and random packing of ball bearings (from Bernal, 1964). The crystalline structure in the centre of the figure is* FCC *with a volume fraction of* $\phi = 0.7404$. *The random packing is similar to random close packing* RCP *with a packing fraction of* $\phi \approx 0.63$.

The disordered ball bearings on the sides of Figure 1 fill space much less densely ($\phi \simeq 0.63$) than the ordered FCC packing ($\phi = 0.74$) (Berryman 1983; Torquato 1999). The particle configurations in the disordered packings are much as we would imagine for a glass or liquid structure and in fact that the random packing of hard spheres is among the best representations of the structure of glasses and liquids (Andersen *et al.*, 1976).

How densely can the disordered ball bearings be packed? This is the old question of filling containers with grain (or with spherical sand, for that matter). The problem itself is not well defined, but the answer is. Try some algorithms. Put a random set of points in three dimensional space. Let them expand as uniform sized spheres and as they touch allow them to move so that the radii can continue to grow. When they can no longer grow they fill space to $\phi \simeq 0.63$. A similar computation takes spheres with no interaction when not touching and harmonic forces on interpenetration. As the system is compressed from a dilute gas at infinitesimal temperature, the highest ϕ at which the pressure is zero is $\phi \simeq 0.63$ (Mason *et al.* 1997). A different algorithm requires that all particles are completely constrained—each sphere has at least four touching spheres not all in the same hemisphere, so that it cannot move. In terms of some known order parameters, the order is seen to rise above $\phi \simeq 0.63$ (Torquato 1999). Another is to simulate a gas of spheres in thermal motion within the metastable amorphous state (most simulations don't run long enough to observe the equilibrium crystal state) and calculate the pressure. As the volume fraction is increased the pressure appears to diverge at $\phi_c = 0.63$. Experimentally, you can take ball bearings as in Figure 1 and shake them up in a large box, or put marbles in a sack or pour sand into a cylinder and tap it many times (Nagel 1999). The well defined answer to all these problems is $\phi = 0.63$, which is the density of 'random close packing', RCP.

However, not all algorithms give this result. Dropping spheres onto a disordered surface and allowing them to find the lowest position by rolling produces a disordered arrangement with a lower volume fraction, $\phi \simeq 0.60$. Another example—start from a single sphere and use the local rule that the next sphere added is always placed so that the density is maximised. After four spheres you have a tetrahedron. But after six spheres you have a structure that is incompatible with a close packed lattice. Continuing to add spheres with the local rule that you place the particle to maximise the density leads to a structure with $\phi \simeq 0.60$. Note that in two dimensions the local rule leads to a hexagonal lattice which is the maximum packing configuration $\phi = \pi/\sqrt{12} = 0.9069$. In two dimensions the local and global packing maxima are consistent with one another. *In three dimensions the local close packing rule is frustrated.* A similar result would obtain if we tried packing pennies on the surface of a spherical balloon. It would not be possible to have six pennies touch the centre one and the hexagonal lattice is frustrated (see introduction in Steinhardt *et al.*, 1983). The geometry and curvature of space plays an important role. The frustration of local dense packing rules, which favour tetrahedral and icosahedral configurations, is presumably one of the reasons we have glasses in three dimensions and one of the reasons that people look for icosahedral local order in glasses.

The reason the random close packing problem is ill-defined mathematically has to do with defining 'random'. Periodic ordering is easy to define: a long range density-density correlation function and/or delta functions in its Fourier transform would suffice. But trying to define 'random' to mean 'not ordered' doesn't quite work. If we take chunks of FCC crystal, randomly orient them, and fill the remaining spaces with the $\phi \simeq 0.63$ configuration we will have a system which is not ordered but which can be made with any volume fraction between 0.63 and 0.7404.

Historically there have been many interesting experiments aimed at measuring some of the properties of the random close packed state. Among the most famous are those of the botanist Stephan Hales (Hales 1727). He was mostly interested in the forces that plants could produce. In one set of experiments he filled a container with English peas, filled the remaining volume with water and then drained the water and measured its volume. This gave him the volume taken up by the peas and thus the volume fraction at random close packing. He then did something extremely clever. He again filled the canister with water, covered it with a tight lid and cooked the peas. The peas absorbed the water, swelled, and became faceted when they encountered the neighbouring peas. He could then remove the peas, count the facets and find the local structure and number of near neighbours (see Zallen 1983).

This is such an elegant experiment that it had to be repeated. Unfortunately, when we tried (Variano 1999), we failed. Most of the peas would not swell sufficiently to facet. As it turns out several other researchers since Hales have tried to repeat his experiment without success. We decided to try a different foodstuff. Israeli couscous did the trick. After swelling we added some ink which went preferentially to the narrow interstices. The separated couscous are shown in Figure 2. It is now easy to count facets, to see the pentagonal/icosahedral shapes and to find that the statistics are very similar to those of ball bearings in sacks and hard sphere RCP simulations.

Some properties of the RCP state are: $\phi \simeq 0.6366 \pm .0002$ (Berryman 1983). Average number of contacts per sphere $\simeq 6$ (Frost *et al.* 1993; Lubachevsky *et al.* 1991). Voronoi polyhedra have $\simeq 14$ faces, predominantly pentagonal (Zallen 1983). The relevance of the

Figure 2. *Israeli couscous taken from an ink-decorated swollen random close packing to illustrate the local structure and configurations of near neighbours in liquid- and glass-like structures. Note the occurrence of pentagonal facets which suggest some local icosahedral structure in amorphous close packed hard sphere systems. (Variano 1999)*

packing fractions to our study of the crystal-fluid transition comes from both exact and approximate treatments of the free volume, entropy and pressure of a system of classical particles. For an ideal gas the translational degeneracy per particle is proportional to the volume of the system, V. The entropy is the logarithm of the degeneracy, $S = k_B N \ln V$ and the pressure is $P = T\partial S/\partial V = (N/V)k_BT = nk_BT$. Van der Waals suggested that for a finite volume per particle, v_0 the effective or 'free' volume obeys $V - Nv_0 = V(1-\phi/\phi_c)$ where ϕ_c is the volume fraction at close packing (*i.e.* where there is no free volume, and the particles have no allowed motions). The entropy and pressure then vary as $S \propto N \ln V(1-\phi/\phi_c)$, $P \propto nT/(\phi - \phi_c)$. While it is now known that these results are exact only in one dimension, the asymptotic form is correct in any number of dimensions. For example, near random close packing $P \simeq 3nk_BT/(1-\phi/0.63)$, near crystal close packing $P \simeq 3nk_BT/(1-\phi/0.74)$.

We can now intuitively understand why hard spheres have an entropically driven liquid-to-crystal transition as volume fraction is increased. Suppose we take a hard sphere sample in the RCP state in a filled box. The volume fraction is $\phi = \phi_c \simeq 0.63$. Even if we turned on thermal motion, since all of the spheres are touching, there is no motion and no free volume and the entropy is zero. On the other hand if we took the same number of particles, and placed them uniformly in the same box in a periodic FCC structure, there would be space between the particles and free volume and entropy. This is clear from the fact that FCC with touching spheres fills space to $\phi = 0.74$, but we have only added enough particles to fill the box to $\phi = 0.63$. Thus at $\phi = 0.63$ the entropy of the ordered crystal is clearly higher than the entropy of the disordered state ($S = 0$) and the crystalline state wins. The crossover from entropy favouring periodic over random will happen at some lower value of ϕ than 0.63. What we have neglected in this argument is the configurational entropy—the different ways of arranging the spheres in the disordered structure. This favours the random (liquid) phase at low ϕ.

The thermodynamics of the hard sphere transition is one of the most studied problems in statistical mechanics and Monte Carlo and molecular dynamics simulations (Alder and Wainwright 1957, Hoover and Ree 1968); see Frenkel, this volume. The phase diagram that results from these studies is shown in Figure 3. For concentrations less than $\phi = 0.494$ there is liquid, from $0.494 < \phi < 0.545$ there is co-existence of liquid and crystals and for $0.545 \leq \phi \leq 0.7404$ there is crystal. However, the simulations tend to slow down when $\phi \gtrsim 0.58$. Starting with a disordered initial configuration, nucleation often does not take place and it has been suggested that there is a glass phase above this volume fraction. Experimentally, among the first indications that colloidal systems follow this phase diagram is the work by Pusey and van Megen who studied the PMMA system as model hard spheres (Pusey and van Megen 1986).

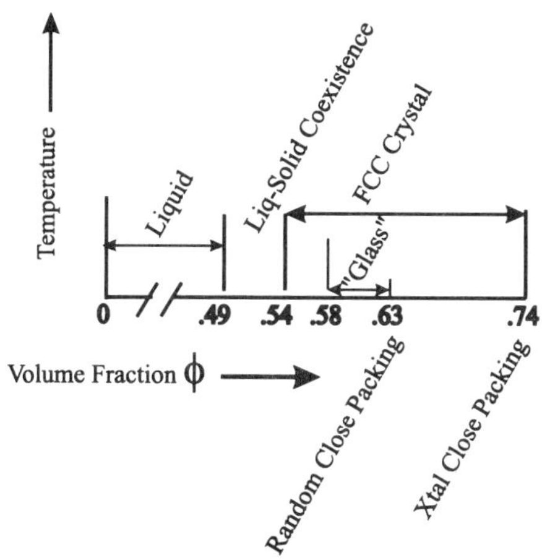

Figure 3. *The equilibrium phase diagram of hard spheres at any finite temperature from computer simulations and colloidal experiments. The 'glass' phase represents a possible metastable state found in early simulations and experiments (Hoover and Ree 1968; Pusey and van Megen 1986).*

1.3 Sedimentation equilibrium

The arrangement of the (macroscopic) spheres in Figure 1 has very little to do with thermodynamics or controlling the volume fraction and everything to do with gravity. As discussed by Frenkel (this volume) we can gauge the relative effects of gravitational and thermal energies by the gravitational height

$$h = k_B T/mg, \tag{4}$$

the distance to which thermal motion can raise a particle from the floor of a container, or the exponential length in a simple isothermal atmosphere with density $n = n_0 e^{-z/h}$. If this height is smaller than the length scale of interest, gravity is important. For the ball bearings in Figure 1, $h \simeq 10^{-20}$m, much smaller than the size of the particles $\simeq 10^{-2}$m. Temperature plays no role for the ball bearings. For air molecules $h \simeq 10^3$m so for metre-scale experiments in the lab everything is thermal and gravity plays no role. For colloids

$h = k_BT/(\Delta\rho(4\pi a^3 g/3)$ where $\Delta\rho$ is the difference in solvent and colloid density, and a is the particle radius. Thus $h \simeq 2\mu$m for 1μm polystyrene spheres ('polyballs') in water, and is relevant on the scale of most experiments. The particles sediment, and there is a variation in concentration. To do macroscopic, fixed density, fixed ϕ experiments we have to get rid of gravity. However, we can also use the distribution caused by gravity to test our understanding of the liquid-solid transition.

When the colloidal particles are allowed to settle to their equilibrium distribution in a gravitational field, the density has a height profile $n(z)$ determined from the interparticle interactions, temperature and gravity. Take an imaginary boundary at z_1. A simple force analysis tells us that at each value of z the total weight (per unit area) of the particles at larger z, $W(z_1) = \int_{z_1}^{\infty} n(z)mgdz$ must be supported by the pressure of the particles below the boundary: $P(z_1) = W(z_1)$. The local pressure should only be a function of the density. If we measure the density as a function of height then we have the density at z_1, and from the integral above z_1, the pressure. We can therefore experimentally determine $P(z_1) = P(n(z_1)) = P(n)$, which is the equation of state. Our experiment was done with 6μm polystyrene spheres in water with sufficient electrolyte that the Debye layer (see Frenkel, this volume) was 10nm, much less than the particle radius, and the effective interaction was therefore essentially hard sphere-like. The density profile was measured with X-ray absorption in a more sophisticated apparatus, but similar to what is used for dental X-rays. Since we are dealing with a colloid and forces between the particles, the pressure is actually an osmotic pressure (see Frenkel, this volume).

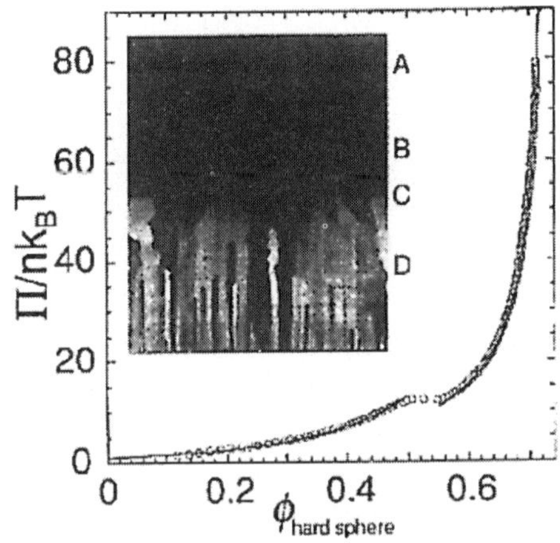

Figure 4. *The equation of state measured by sedimentation equilibrium (Rutgers* et al. *1996) of polystyrene spheres compared with simulations and approximate formulae (Carnahan and Starling 1969, Hall 1972). The break in the curve corresponds to the liquid-to-crystal transition. The inset is a picture of the sample with the liquid region visible above a sharp boundary with the crystalline region.*

The results are shown in Figure 4 (Rutgers *et al.* 1996). The points are the experiment and the solid lines the Carnahan-Starling and Hall equations, which represent two very good approximations to the hard sphere equation of state found in computer simulations. The discontinuity at 0.49–0.54 corresponds to the liquid-solid transition. Above this height

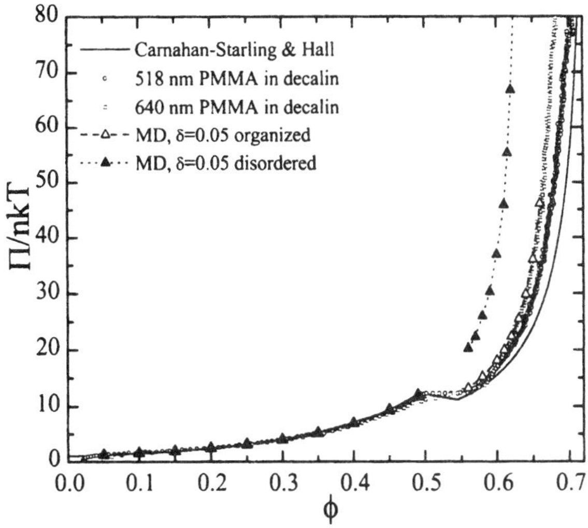

Figure 5. *Measured equation of state for PMMA colloids in decalin. Circles are experimental data from (Phan* et al. *1996) and triangles are from simulations for polydispersity of 0.05 (Phan* et al. *1998).*

(lower density) the sample is cloudy: a liquid. Below this height (higher density) the sample is opalescent and exhibits Bragg scattering: it is crystalline. The osmotic pressure is a direct measure of the interparticle forces, and this experiment essentially proves that particles interacting with a hard-sphere potential have the transition predicted by theory and simulations. Our experiments are quite straightforward, the problem is that it takes about a year for the sample to reach equilibrium and another year to check that it has reached equilibrium.

We have also studied the sedimentation equilibria of the PMMA samples that were used in Pusey and van Megen's original work on hard sphere colloids (Pusey and van Megen 1986, Pusey 1987) and in our experiments in microgravity. The results are shown in Figure 5 (Phan *et al.* 1996). Here we see that this system looks less ideal, especially at high volume fractions. Deviations from the ideal hard sphere equation of state can arise in many ways: softness in the repulsive interaction tends to lower the pressure relative to the ideal, while a small attractive component tends to aggregate the particles into clusters that can inhibit the close packed crystal structure and reduce the free volume increasing the pressure.

1.4 Polydispersity

One question that often arises in the study of hard sphere colloids is the effect of *polydispersity* on any or all of the physical properties. In order to investigate the effects of polydispersity we decided to perform molecular dynamic simulations which can easily yield the equation of state (Phan *et al.* 1998). The pressure of the hard sphere system is simply proportional to the radial distribution function at contact (Russel *et al.* 1989).

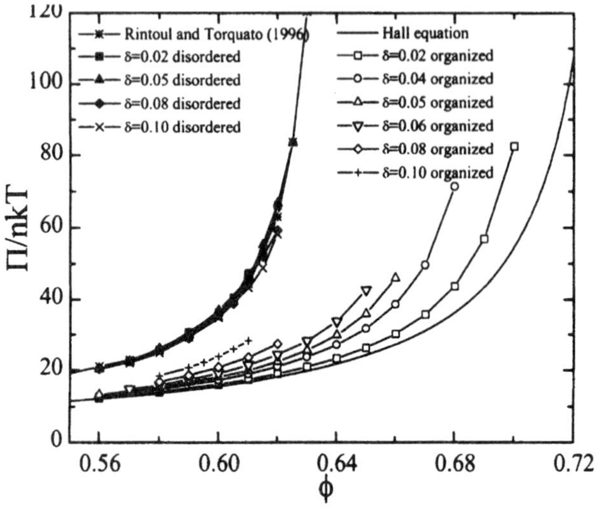

Figure 6. *Equation of state for different polydispersities from molecular dynamics simulations (Phan* et al. *1998). The disordered, liquid branch is almost independent of polydispersity, while the crystalline branch apparently diverges at a value which depends on polydispersity.*

In Figure 6 we show the equation of state for various degrees of polydispersity obtained with a simulation involving 500 particles. In order to quickly obtain the properties of the crystal we choose an initial configuration with the particles arranged on the FCC lattice. For the monodisperse system the system remains crystalline for $\phi > 0.54$ but melts to a disordered (liquid) state for $\phi < 0.54$. However for the polydisperse systems, putting the particles on lattice sites is not sufficient to arrive at a crystalline state even at high volume fraction. The configuration evolves to a disordered state. To maintain a crystal the initial configuration must be 'organised' by arranging the larger spheres to be surrounded by the smaller spheres—a spatial correlation with alternating shells of contrasting size. Thus the equations of state shown in Figure 6 are in the disordered state for a 'disorganised' initial arrangement (whether placed randomly or on a lattice) and crystalline only when they are both 'organised' and initially placed on a lattice.

What is clear from this figure is the strong dependence of the osmotic pressure on polydispersity for the crystalline phase and the near-independence for the disordered phase. If we plot the reciprocal of the osmotic pressure against volume fraction we find an asymptotic linear behaviour extrapolating to zero at finite ϕ (Phan et al 1998). This can be associated with the divergence in $\Pi \propto 1/(1 - \phi/\phi_c)$ as one approaches close packing. The close packing volume fraction ϕ_c as a function of polydispersity for the disordered and crystalline states is shown in Figure 7. Strikingly, the disordered close packed (RCP) volume fraction is very robust against polydispersity. On the other hand the crystal close packed phase has a maximum packing fraction which decreases rapidly with polydispersity.

We can understand the main effects of small amounts of polydispersity on the maximum packing fractions by some simple arguments. Basically, if we require the lattice to

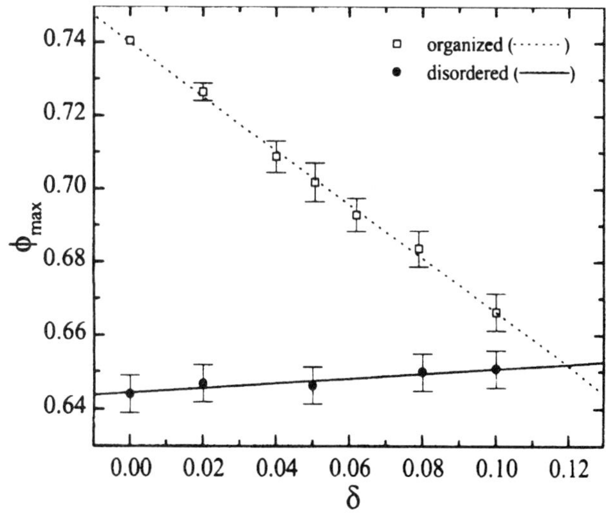

Figure 7. *Maximum packing fraction as a function of polydispersity for organised (crystalline) and disordered (liquid) arrangements of hard spheres (Phan* et al. *1998).*

remain then the sizes of the unit cells should all be the same. But the unit cells surround the particles. The biggest particles swell the lattice and the small ones rattle around in their expanded cells. On the other hand, for the disordered system the large particles can roll until they touch other particles and fill the space left from taking an average sphere and decreasing its size to make a small one.

Let's start with the crystal. Suppose we have a deviation in radius δ, so that $r = a \pm \delta$ for a bimodal distribution, and that the unit cells remain undistorted, as we might expect for a crystal with long range order. The volumes of the spheres are given by the two values $v_s = (4/3)\pi(a \pm \delta)^3 = v_{s0}(1 \pm 3(\delta/a) + 3(\delta/a)^2 \pm (\delta/a)^3)$, but the volume of the large cells only takes the positive signs, $v_c = v_{c0}(1 + 3(\delta/a) + 3(\delta/a)^2 + (\delta/a)^3)$. If we average over sphere volumes, keeping only lowest order terms we have $\langle v_s \rangle = v_{s0}(1 + 3(\delta/a)^2)$ and the total volume of spheres in the system is $N\langle v_s \rangle$. On the other hand, half of the cells increase their volume by $\langle v_c \rangle = v_{c0}(1+3(\delta/a)+3(\delta/a)^2)$ but the other half have no change. The total change in the volume of the crystal to lowest order is $\delta V_c = (N/2)v_{c0}3\delta/a$ and the total volume of the crystal is $V_c = Nv_{c0}(1 + (3/2)\delta/a)$. The volume fraction is then $\phi_{c\delta} \simeq N\langle v_s \rangle / N\langle v_c \rangle \simeq (v_{s0}/v_{c0})(1+(3/2)(\delta/a)) \approx \phi_{c0}(1-(3/2)\delta/a)$, where $\phi_{c0} = \langle v_s \rangle / \langle v_c \rangle$, the volume fraction of the original unit cell. Thus the maximum ϕ of the crystal state decreases like $(3/2)\delta/a$ which slightly overestimates the dependence found in Figure 7 from the simulations.

In order for this model to be correct the cells with the large spheres must be able to 'hold' the structure, that is, the cells with the large spheres must percolate (Zallen 1983, Ziman 1979). In three dimensions percolation on an FCC lattice occurs at an occupancy fraction $p_c \simeq 0.16$, so that half of the cells (the ones with large spheres) can easily support a structure where the small spheres are not touching their neighbours and are free to move in their cells. It might even be possible for the small spheres to freely diffuse throughout the system since the low threshold p_c allows simultaneous percolation of both large and small cells. (This contrasts with the two-dimensional case where any percolation path connects opposite faces of a bulk sample and prevents any other paths from doing the same.)

In the simulations we noticed that the Lindemann criteria for some of the polydisperse samples was violated. According to the Lindemann criterion a solid melts when the RMS displacement of a particle exceeds $\simeq 0.2$ of the distance to its nearest neighbour. In fact we found an RMS displacement in the solid phase that diverged with time—the particles were diffusing. We then calculated separately the displacements for the larger-than-average and the smaller-than-average spheres. The small particles were diffusing while the large particles obeyed Lindemann and had a maximum displacement of less than $0.2a$.

2 Sedimentation and fluidisation

Sedimentation problems have a long history, and persist to the present day. The method of *particle fluidisation* is a common technique in chemical reactors, and is used to prevent sedimentation by imposing an upward flow of solvent (or, in granular systems, air) to maintain artificial buoyancy. In many cases the particles and suspending fluid have a large density mismatch and a high Reynolds number. Both of these effects contribute to highly chaotic and turbulent flow. For use in our study of colloidal crystalisation we were interested in precisely the opposite limit: very slow steady flow at low Reynolds number, and with particles similar in density to the surrounding fluid (see Table 1). Originally we wanted to counteract gravity with an upward flow, but it soon became evident that there was exciting physics in the fluidisation problem on its own. The particles interact via the hydrodynamic interaction—a velocity-dependent force. The essence of this study became a matter of finding a system where the physics is governed almost entirely by the hydrodynamic interaction, and we wanted to study the consequences in the most basic way. The particular question we are interested in here is whether this non-potential interaction can produce non-trivial interparticle positional correlations which will act back on the flow field. Note also that a sedimenting or fluidised bed is a 'driven system' (see Mukamel, this volume): there is a constant input of energy through gravity or the upward pumping of fluid and the steady state is not in Boltzmann equilibrium.

Stokes velocity	$v_{\rm st} = \frac{g\Delta m}{6\pi\eta a}$		$\sim 1.11 \times 10^{-5} \left(\frac{a}{1\mu\rm m}\right)^2$ cm/sec
Diffusion Constant	$D_0 = \frac{k_B T}{6\pi\eta a}$		$\sim 2.2 \times 10^{-9} \left(\frac{a}{1\mu\rm m}\right)^{-1}$ cm^2/sec
Reynolds number	${\rm Re} = \frac{vl\rho}{\eta}$	$\sim \frac{v_{\rm st} a\rho}{\eta}$	$\sim 1.11 \times 10^{-7} \left(\frac{a}{1\mu\rm m}\right)^3$
Peclet number	${\rm Pe} = \frac{vl}{D}$	$\sim \frac{v_{\rm st} a}{D_0}$	$\sim \sim 0.5 \left(\frac{a}{1\mu\rm m}\right)^4$
Gravitational height	$h = \frac{k_B T}{g\Delta m}$	$= \frac{a}{Pe}$	$\sim 2 \times 10^{-4} \left(\frac{a}{1\mu\rm m}\right)^{-3}$ cm

Table 1. *Some possibly relevant numbers for polystyrene particles in water.*

2.1 Physics of the sedimentation velocity

Let's try to gain some intuition for the sedimentation problem by working our way up from one to two to many particles (Happel and Brenner 1986). A single particle settling in a fluid far from any boundary falls at the Stokes velocity, a balance between gravity and Stokes drag, $v_{st} = g\Delta m/6\pi\eta a$, with Δm the effective mass. Note that it is not possible to stably suspend a single particle. If the upward fluid velocity is slightly greater than the Stokes velocity, the particle is driven up and out of the system; if the fluid velocity is slightly less, the particle falls to the bottom. A particle of radius a, moving through an infinite fluid at velocity $\mathbf{u}_0$, sets up a velocity field which behaves as (Oseen 1927):

$$\mathbf{u} = \frac{3a}{4r}\left[\mathbf{u}_0 + \frac{(\mathbf{u}_0\cdot\mathbf{r})\mathbf{r}}{r^2} + O\left(\frac{a}{r}\right)^2\right] \tag{5}$$

where r is the distance measured from the particle centre. The field is very long range, decaying at large distances like $1/r$.

The case of two particles is already interesting and nonintuitive, but once we understand what they do, we can figure out some of the more complex flows at finite density. The Stokes equation, which holds at zero Reynolds number,

$$\nabla P = \eta\nabla^2\mathbf{v}, \tag{6}$$

is linear. For two particles under gravity, we can simply consider each particle settling in the flow field of the other. Two particles therefore fall faster than their Stokes velocity. For particles separated vertically by one diameter the sedimentation velocity is $1.357v_{st}$. In general the drag force in the vertical direction is reduced by a factor of $(1-(3/4)a/d)$ (to leading order in a/d) where d is the horizontal separation of the spheres (Happel and Brenner 1986). If the particles are not on top of each other then the flow field from one exerts a torque on the other and they rotate. The downward velocity is faster on the side facing the opposite particle so the rotation is in the sense shown in Figure 8.

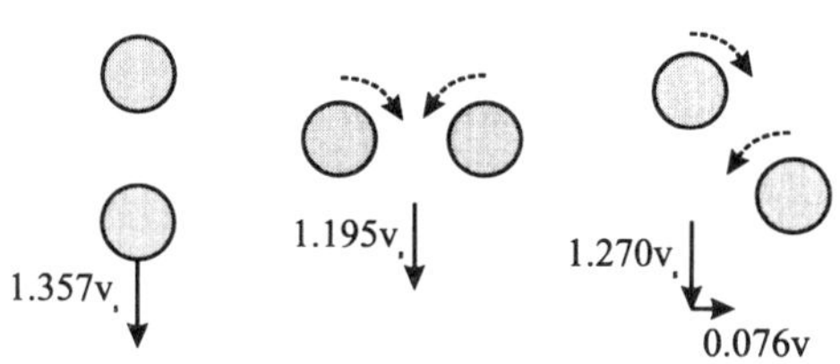

Figure 8. *Sedimentation of pairs of spherical particles in different configurations. The sedimentation velocity as a multiple of the Stokes velocity is indicated.*

Suppose the line of centres of the two particles makes an angle θ to the horizontal. For the vertical configuration the particles fall at an enhanced velocity, but for general θ they drift sideways (Figure 8). What is unusual from the point of view of Newtonian physics is that the line-of-centre forces are in the same direction (rather than opposing) so that the particles neither attract nor repel. This is an interesting consequence of the reversibility of the Stokes equations: if there is a solution for a particular flow then there is a solution when all the velocities and forces are reversed. It is like running a film backwards and turning the projector upside down (to reverse gravity) (Herbolzheimer 1999). If the particles attracted, the reverse film would have them repel. The only situation which doesn't break this symmetry is if they remain at fixed separation.

Three particles is already too complicated for complete analysis in all but a few special cases (where the three particles oscillate relative to one another as they fall). In the more usual case two particles are above a third particle. They fall faster than the lone particle and catch up to it, perform a complex dance, and two of the particles (not necessarily the original two) take off and leave a third behind. Recent computer calculations by Janosi *et al.* (1997) show that the motion is chaotic, in the sense that the final positions of the particles have exponential sensitivity to small displacements of the initial positions.

The problem we are really after, for understanding sedimentation and fluidisation, is to know what happens with a finite density of particles. The naive approach is to just calculate what happens when each particle falls in the flow field of all the others. This doesn't work for two reasons. The main one is that the sum or integral diverges, $\int_a^\infty (1/r)dr \to \infty$. The second is that, with a finite density, we have to include the back-flow which results from conservation of mass: particles move down, fluid must move up. There is a non-trivial but exact result from Batchelor (1972) for the case of a dilute random distribution of settling particles (whose position can change with the flow). The problem is especially complicated by the fact that setting the density or volume fraction is not sufficient to define the outcome: the interparticle correlations and the stress conditions are fundamentally important. Table 2 shows the limiting behaviour expected for sedimentation of a cubic lattice; that of a random distribution of particles held fixed in their relative positions; and Batchelor's calculation of random particles free to move. The easiest case to understand is for the flow through a lattice (Zick and Homsy 1982, Sangani and Acrivos 1982). This can be found by solving the flow through a unit cell, and the result has been used widely to validate new numerical hydrodynamic computational methods. The dependence of the sedimentation velocity v_{sed} on $\phi^{1/3}$, where ϕ is the volume fraction, comes from the fact that there is a well defined separation, the lattice spacing, which varies as $a\phi^{-1/3}$. The $\phi^{1/2}$ dependence in the case of random fixed particles is less obvious, but the result arises from the fact that momentum can be redistributed by whatever constraint forces hold the particles in their relative positions. This results in 'screening' of the hydrodynamic forces (Saffman 1973).

Random, free to move	$1 - 6.55\phi$
Random, fixed positions	$1 - 2.12\phi^{1/2}$
Periodic FCC lattice	$1 - 1.76\phi^{1/3}$

Table 2. $v_{\text{sed}}(\phi)/v_{\text{st}}$ *to lowest order in* ϕ.

Batchelor's calculation involves a choice of reference frame which eliminates the diverging integral by including the backflow, and has three pieces to the linear term. The backflow contributes -5.5ϕ to $v_{\text{sed}}/v_{\text{st}}$, which allows for the volume of each particle plus the average volume of fluid dragged downward by it. The other terms relate to the hydrodynamic interactions and the flow field from each particle: 0.5ϕ from the average $\nabla^2 u$ term in a gradient expansion (the linear term averages to zero), and -1.55ϕ from the probability of a near-neighbour encounter as opposed to the average separation. The

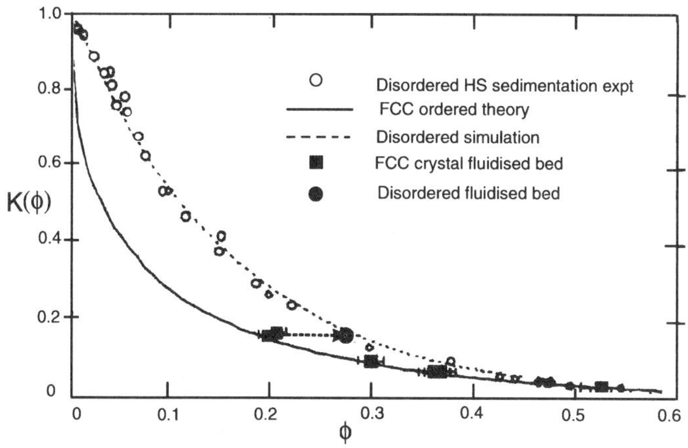

Figure 9. *Reduced sedimentation velocity as a function of volume fraction ϕ for 'random' particle configurations and for sedimentation of particles in an* FCC *crystal. The* FCC *theory line is from Zick and Homsy (1982); disordered simulation (similar to Batchelor's result at low ϕ) from (Ladd 1990), disordered experimental data from Paulin and Ackerson (1990) and Buscal* et al. *(1982). Dashed arrow from squares to circle is the effect of adding electrolyte to destroy the crystal.*

net result $v_{\text{sed}}(\phi) = v_{\text{st}}(1 - 6.55\phi)$ is a much bigger correction for volume fraction than is found in either the Einstein viscosity ($\eta(\phi) \simeq \eta_0(1-2.5\phi)$) or the self diffusion coefficient, $(D_s(\phi) \approx D_s(0)(1 - 1.65\phi))$. In the latter two cases the particle motions are random and there is no net backflow, which is actually the dominant effect in sedimentation.

The normalised sedimentation rate as a function of ϕ for the FCC periodic case, and for Batchelor's limiting calculation for the random case, are shown in Figure 9 along with experimental data on a number of different colloidal systems. Note that the mean sedimentation rate decreases as volume fraction increases. This is what allows stable fluidised beds to arise, and produces sharp sedimentation fronts. Suppose that we are trying to stabilise a sedimenting colloid at $\phi = 0.22$. We read off the appropriate sedimentation velocity from Figure 9 and set up a fluid velocity to exactly match that value. Suppose we make a mistake and set the fluidisation flow velocity v_{fl} too slow, say to the value for $v_{\text{sed}}(\phi = 0.23)$. Then there is still a net downward velocity of the particles, and since there is a bottom to the container the volume fraction increases as particles settle and occupy a smaller volume. In fact the volume fraction will increase precisely to the value of ϕ which gives the set $v_{\text{sed}}(\phi)$, because at that value there is no net flow of particles and no further change in concentration. If our initial upward fluidisation velocity was too high, the particles flow up, and the density decreases until we again meet the criterion $v_{\text{sed}}(\phi) = v_{\text{fl}}$. So ϕ stably adjusts itself to the fluidisation velocity (in the range $v_{\text{st}} > v_{\text{fl}} > v_{\text{st}}/50$, where the lower limit is set by the flow which can just lift a close packed, porous sediment). We can use this to control the volume fraction continuously in a fluidised bed by just turning the velocity knob on the pump that controls v_{fl}.

Suppose now we have a sample sedimenting with no fluidisation. Consider the density profile at the top of the sedimenting region. If some particles get left behind they find themselves surrounded by too few particles; the local ϕ is lower, the velocity higher and they catch up with the rest and produce a very sharp sedimentation front. A similarly sharp transition occurs for the fluidised bed, where the same description holds in the reference frame of the fluid. Overall, the fluidised bed has a uniform density and a sharp transition to a particle-free region. The width of the transition region usually involves broadening due to diffusion, counteracted by the difference in sedimentation velocity of the concentrated and dilute regions. This gives a length scale $l = D/v_{\rm sed}$. If D is set by thermal motion (rather than hydrodynamic dispersion arising from the chaotic flow of the fluid) and $v_{\rm sed} \simeq v_{\rm st}$ then $l = h$ the gravitational length. Note that, in contrast, the *bottom* of a fluidised bed is completely unstable unless it is supported. If we have a dense region and a less dense region below it, the particles in the less dense region settle faster, move away from the dense region, and broaden the interface. There is also a mechanical instability (the Rayleigh instability) when we have a dense fluid above a less dense fluid. That is why we always need a bottom supporting the fluidised bed.

Now let's return to Figure 9. One question that arises is whether either of these *configurations*, the random arrangement or the crystal arrangement, are stable in their own right; the arguments above suggest only that the average *density* is stable (self-correcting) under sedimentation. Will a periodic array of spheres remain periodic as they sediment? Will a random arrangement of spheres remain random? The answer to both of these questions seems to be *no*. The instability of the periodic case was studied by Crowley (1976) and is illustrative of many aspects of sedimentation. At first he considered a horizontal line of particles with separation d, $x_n = nd$, and calculated what happens if there is a modulation of the spacing $x_n = nd + \delta \cos qnd$. He found that there is an instability with fastest growing wavevector $q = 2\pi/4d$ and growth rate $\gamma = av_{\rm sed}/d^2$. This is illustrated in Figure 10. If two particles move slightly toward each other, they

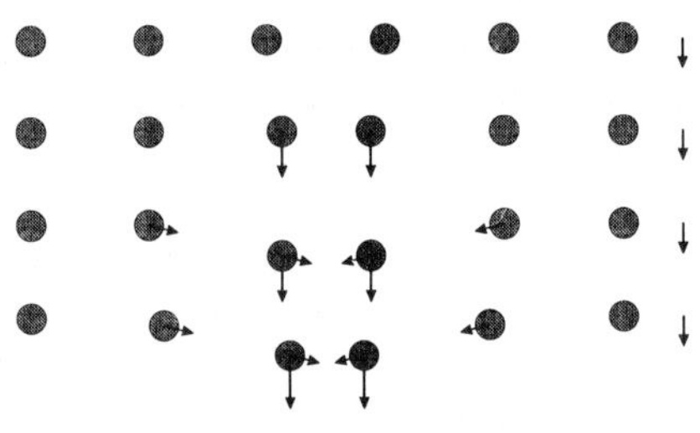

Figure 10. *Crowley's instability of a periodic line of particles in sedimentation. Each line is at a later time. A fluctuation brings two spheres closer together so they sediment faster. Since they have an angle to the horizontal with neighbouring spheres there is a horizontal drift which brings them closer together, as in Figure 8. This accelerates their sedimentation and leads to clumping.*

fall faster than the average. As they advance each forms a non-horizontal pair with the former neighbour in the line. These non-horizontal pairs will drift, and draw the initial two particles closer together which reinforces the initial fluctuation, and creates a denser region which sediments faster. Thus the periodic horizontal line is unstable, and the strongest instability is to the formation of quartets of particles. Crawley also looked at the problem of a horizontal plane of particles in a square arrangement. He found that it has a similar instability, but that the growth rate is maximum at $q = 0$ and decreases rather gradually to values of about $|q| = \pi/2d$. This means that large-scale (low q) density fluctuations will break the horizontal periodic lattice into big clumps. The three-dimensional problem, say a cubic lattice, has not been solved but it is expected that this is also unstable. That is why the sedimentation velocity that was exactly calculated for an FCC lattice (for all volume fractions), has not been experimentally verified until recently (Rutgers 1995). One of the results of our experiments, not yet theoretically understood, is that the uniformly random arrangement is also not stable. Particle correlations build up from the hydrodynamic interactions which screen the flow fields and produce density correlations at large length scales.

In order to create a stable periodic lattice for sedimentation studies we decided to use charged colloidal particles which form a colloidal 'Wigner crystal' at low volume fraction. The particles we chose are in the interesting regime where their interaction energy is sufficient to overcome thermal motion and crystallise, but not sufficient to overcome gravity. In the absence of flow the particles sink to the bottom of the fluidisation cell, but with flow we can control the density, suspend the particles and allow crystallisation. In fact the system allows many interesting tests and controls for both colloidal crystals and fluidised beds. We could use the Bragg scattering from the crystal to accurately measure the volume fraction, which we found to be constant from the top to the bottom of the bed. We could use a single sample, control the concentration with the velocity and measure the dependence of elastic properties on ϕ. We could also measure the sedimentation rate for different volume fractions to compare with the prediction for the FCC periodic structure (whose presence was confirmed by the Bragg scattering). The data points are shown in Figure 9 and there are significant differences from the curve for the disordered spheres. We could not reduce ϕ to zero without melting the crystal so we were limited to $\phi > 0.2$. To make sure that we had not made a mistake somehow, we added electrolyte to the fluid that was injected into the fluidised bed while keeping the same flow rate. We observed the bed collapse to a lower height (a higher density). The result for the melted colloid created by adding the electrolyte is also shown in Figure 9 (by the dashed arrow) and agrees with the random distribution. More recently Ackerson *et al.* (1995, 1996) measured sedimentation for low ϕ for charged spheres that were not in a crystal but still strongly correlated. They found that the leading order correction was indeed $\phi^{1/3}$ indicating that this is the behaviour whenever there is a characteristic distance in the pair distribution function.

So, the sedimentation velocity depends very much on positional correlations between the particles. This leads you to think that if there are fluctuations in the positions of the particles during sedimentation, there would likely be fluctuations in the local sedimentation velocity. These fluctuations are of considerable interest since they determine effective diffusion or hydrodynamic dispersion and mixing in the beds.

2.2 Velocity statistics

There have been several studies of the velocity deviation and particle dispersion in sedimentation (Nicolai and Guazzelli 1995, Nicolai *et al.* 1995b, Xue *et al.* 1992, Ham and Homsy 1988). They tended to show that the velocity deviation $\delta v = \sqrt{\langle (\delta v)^2 \rangle}$ is comparable to the sedimentation velocity itself, and that it is anisotropic with a larger variance in the direction of gravity. Our contribution to this data is shown in Figure 11. The

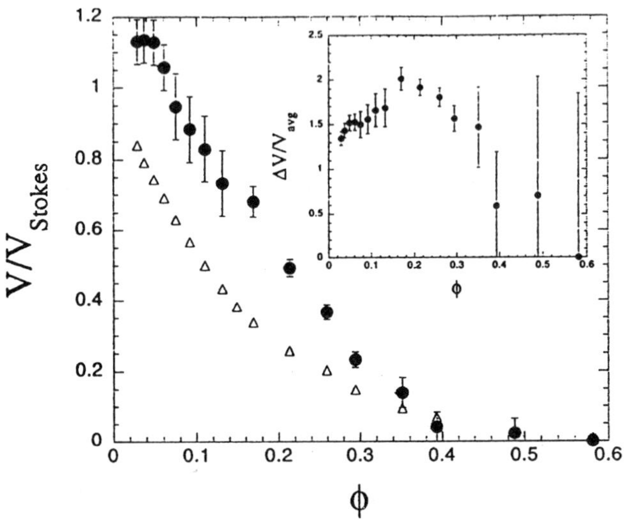

Figure 11. *Sedimentation velocity (triangles), measured by the fluidisation speed, for 15 micron polystyrene spheres as a function of volume fraction. The* RMS *deviation* δv *of velocity as measured by DWS is also shown (circles). The inset shows the normalised* $\delta v(\phi)/v_{\text{sed}}(\phi)$.

measurements were made by DWS (diffusing wave spectroscopy—see Pine, this volume) which allows a measurement of the RMS particle motion over wide-ranging time scales. Over the range of ϕ studied the deviation almost tracks the sedimentation velocity and is the same order of magnitude.

Theoretically, the problem of the velocity deviation has drawn a spectrum of conclusions. The most fundamental limit is $\phi \to 0$ where we might expect to be able to calculate the leading term. One argument (an isolated sphere falls with the Stokes velocity) leads to $\delta v/v_{\text{sed}} \to 0$. A second argument says that the characteristic velocity is v_{sed} so that $\delta v/v_{\text{sed}} \to 1$. The third argument has $\delta v/v_{\text{sed}} \to \infty$ (Caflisch and Luke 1985). This is the most interesting argument, as follows.

Consider a region of length scale r. The number of particles in this region goes as volume times mean particle density: $N \propto nr^3$. Assuming the particles are uncorrelated, *i.e.* random, the deviation in the number of particles in this region goes as the square root of the number, $\delta N = N^{1/2} \propto \sqrt{nr^3}$. Since for sedimentation or fluidisation the particles are not exactly density matched, the region is lighter or heavier than the average and it should sediment faster or slower than the average. Its weight deviation is $\delta W = N^{1/2} g\Delta m$. This weight must be supported by viscous drag (or the region accelerates). The drag force is the viscous stress ($\sigma \propto \delta v/r$) times the area ($\propto r^2$) giving $f_{\text{drag}} \propto \eta r \delta v$. (To within a factor of 6π this is the same result we would have from considering the region as a sphere with Stokes drag.) Setting $f_{\text{drag}} = \delta W$ we have

$$\delta v_r = \sqrt{\langle|\delta v|^2\rangle_r} \propto \frac{N^{1/2} g\Delta m}{\eta r} \propto \frac{g\Delta m}{\eta a}\frac{a}{r}\sqrt{\phi\frac{r^3}{a^3}} \propto v_{\text{sed}}\sqrt{\phi\frac{r}{a}} \tag{7}$$

If this is correct, the fluctuations in the velocity of a region grows with the size of that region. The velocity fluctuation would be dominated by the largest region in the system, which would be the size of the container (Hinch 1987). For an infinite system δv diverges.

Although this appears as a strange result it is an inevitable consequence of having a random particle distribution. For larger regions the mass fluctuations increase (as $\sqrt{N} \propto r^{3/2}$) even as the density fluctuations decrease (as $\sqrt{N}/V \propto \sqrt{N}/N = 1/\sqrt{N} = r^{-3/2}$). But the buoyancy forces are determined by the total buoyant mass so there is no normalisation by the volume.

The prediction of this divergence in velocity deviation was first presented (in a different form) by Caflisch and Luke (1985). Since the result challenges our intuition, questions were raised as to whether the hydrodynamic interaction would be self-screened by the particles: perhaps the particles would become correlated in such a way as to prevent the divergence. Koch and Shaqfeh (1991) considered three particle interactions and proposed such a screening mechanism, suggesting that the screening length would scale inversely with the volume fraction and there would not be a divergence. In contrast, Ladd (1996) used a lattice Boltzmann model to simulate the sedimentation of 40,000 particles and concluded that the hydrodynamics were unscreened and that the velocity fluctuations would indeed diverge with system size.

In order to study this very fundamental problem of whether the hydrodynamic interaction could induce density-density correlations (positional correlations) between the particles, we prepared a series of samples designed to approximate the ideal limits of infinite Peclet number (no contribution from thermal motion), zero Reynolds number (no inertial effects), and low volume fractions (no particle correlations from excluded volume). The particle dynamics were studied by particle image velocimetry (PIV) first with a home-made system and software (Rutgers 1995b) and later with a commercial instrument. In PIV, an image, usually a 2D slice of the sample defined by a plane of illumination or the depth of field of the camera lens, is digitally taken and stored in memory. Images are recorded and stored at fixed time intervals. Each image is then computationally divided into squares which contain about 1 particle per box. A cross correlation of the intensity (density) in one box with that in the same box but in the subsequent image, gives the displacement of the centre of mass of the particles in that box in the time interval. A displacement field or velocity field is thus generated on the grid which defines the boxes. With the velocity field we can then study $\langle \delta v(\mathbf{0},0)\delta v(\mathbf{r},t)\rangle$.

In Figure 12 we show the velocity fields for two different concentrations and the velocity fields with the average sedimentation velocity subtracted (Segre *et al.*. 1997). In the $\phi = 0.03$ sample the velocity fluctuations are evident even without subtraction, the deviation is comparable to the sedimentation velocity. From the $\phi = 0.001$ sample it is clear that the fluctuations decrease as the volume fraction is lowered. From the expanded velocity figures it is also clear that there is a characteristic length scale for the velocity fluctuations, which we call 'swirls', and that the length scale increases as the volume fraction is reduced.

In order to quantitatively analyse the flow fields we calculated a series of velocity-velocity correlation functions. As shown in Figure 13 (insets), the velocity correlations decay over a finite distance in a non-power-law way and with a length scale that increases as the volume fraction decreases. Now the only intrinsic length scale in the system is the particle radius, a, and we expect that any other length scale will scale as a times some function of the volume fraction ϕ. For example, the mean free path is $a\phi^{-1}$. In fact that is the form predicted for the hydrodynamic screening length in the model of Koch and

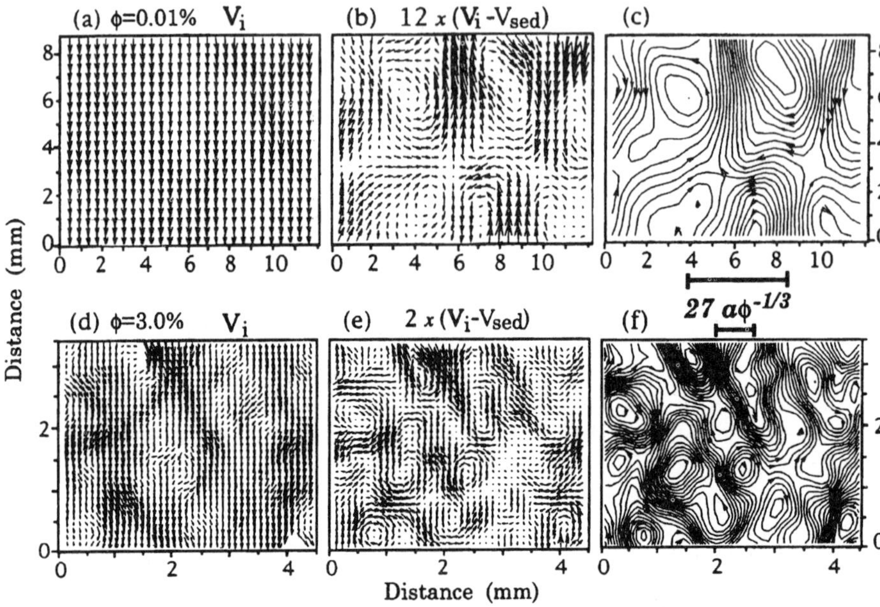

Figure 12. *Velocity fields for 15 micron polystyrene spheres in water for volume fractions of 0.03 (upper) and 0.0001 (lower). The leftmost curves are direct velocity fields from PIV. The centre curves have the average subtracted and are multiplied by different factors. The rightmost curves are the streamlines. Note the difference in fluctuation length scale and fluctuation amplitude for the different concentrations. Larger amplitudes and smaller length scales characterise the more concentrated sample.*

Shaqfeh (1991). However, the scaling that seems to give good data collapse of our data is a correlation length $\xi \simeq 20a\phi^{-1/3}$, related to the mean interparticle distance, but larger by the factor $\simeq 20$. In Figure 13 we see that both horizontal and vertical correlation functions lie on the same curves, with this distance normalisation; $\xi_\perp \simeq 17a\phi^{-1/3}$, $\xi_\parallel \simeq 21a\phi^{-1/3}$. This is not *a priori* a trivial result. There is a mean interparticle separation $(a\phi^{-1/3})$ for a ideal gas as well, but there is no feature in the radial distribution function $g(r)$ or in (its Fourier transform) the structure facture $S(q)$ corresponding to this distance, either in the ideal gas or in our dilute suspensions (Chaikin and Lubensky 1995). This would be a natural length scale only if the particles were on a lattice or correlated apart by a repulsive interaction. Moreover the length scale we have found is twenty times greater.

The velocity correlation functions reflect what is evident in the flow fields shown in Figure 12. The change in sign in $\langle \delta v_z(0)\delta v_z(x)\rangle$ shows the effect of the 'swirls' that we see in the velocity fields or in time-lapse photos, and the size of these swirls increases as particle density decreases. What is unusual however is that the scaling with ϕ implies that the swirls contain the same number of particles ($\simeq 8000$), independent of concentration.

Given that there is a length scale in the system we must rethink the earlier argument about the divergence in the velocity deviation. In fact the velocity variance as a function of

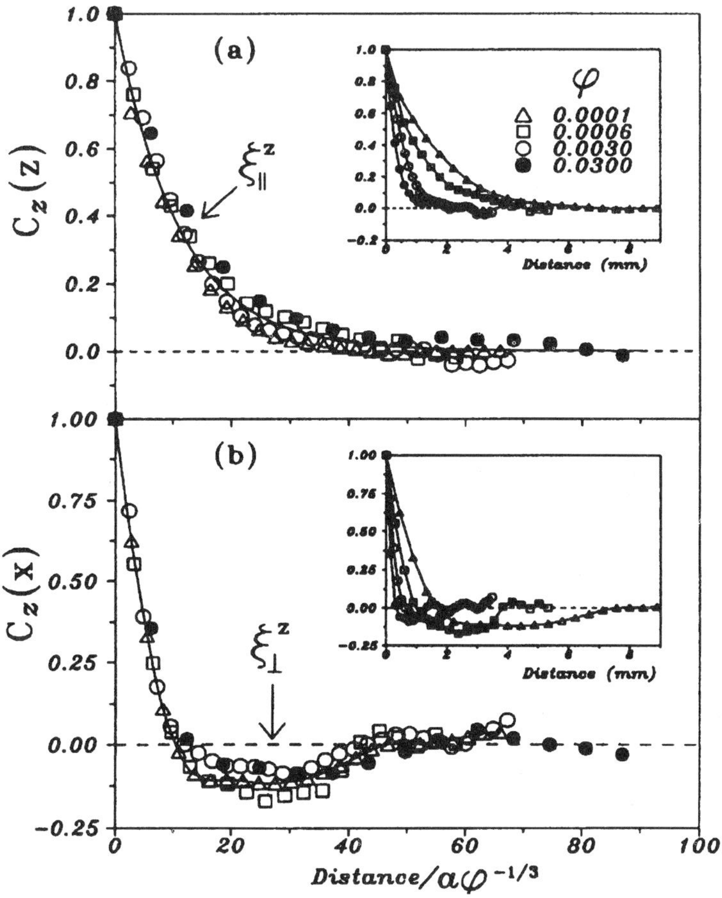

Figure 13. *Velocity-velocity correlation functions for different volume fractions. The upper curves are for the vertical velocity correlated in the vertical direction and the lower curve for the vertical velocities correlated in the horizontal direction. The insets are unscaled while the main figures are scaled by the mean interparticle separation.*

length scale is directly related to the velocity correlation function; the finite characteristic length $\xi = a\phi^{-1/3}$ sets a cutoff on the r dependence in Equation 8, so that at large distances the velocity deviation saturates as

$$\delta v_{\infty} \simeq \delta v_{\xi} \simeq v_{\text{sed}}\phi^{1/3}. \tag{8}$$

Here we find that the deviation goes to zero with infinite dilution but with a power law $\delta v \sim \phi^{1/3}$. We can directly test this result from the velocity fields we have measured by taking the average $\langle (v - v_{av})^2 \rangle$. This is plotted in Figure 14 and we see that the $\phi^{-1/3}$ prediction is obeyed over three orders of magnitude in ϕ. How general are these results? According to our findings the velocity deviation scales with the average sedimentation velocity and with $\phi^{-1/3}$, and there is a length scale for the system ξ. If we take all of our data and all of the data previously published by other groups we can plot $(\delta v / v_{\text{sed}})/\phi^{1/3}$

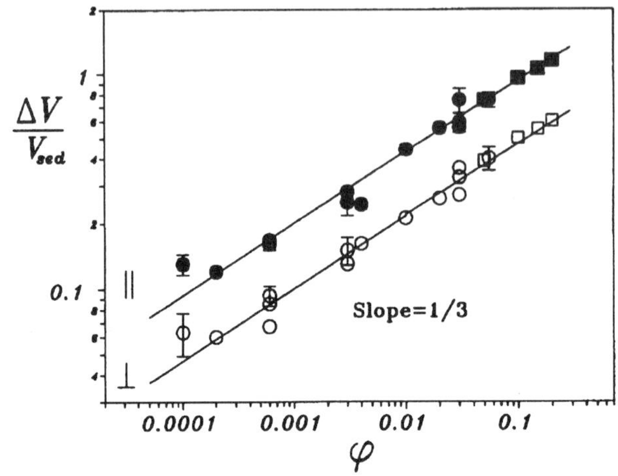

Figure 14. *Velocity deviation as a function of volume fraction.*

against W/ξ where W is the size of the container for the fluidised bed. We have also included the data from the simulation (Ladd 1996). We see that the data fall on a 'universal curve'. The velocity deviation depends on the size of the container only when the size is comparable to or less than the correlation length ξ. At larger sizes there is no scaling with container size, the velocity variance saturates and there is no divergence. There is hydrodynamic screening but not of the form envisioned by Koch and Shaqfeh (1991). As the concentration goes to zero the variance goes to zero as we might have expected from the simplest argument (that an isolated sphere falls with the Stokes velocity), but again not in a trivial way. The hydrodynamic interaction is important even in this limit in its contribution to hydrodynamic dispersion (as distinct from thermal diffusion). At long times the particles undergo trajectories which are diffusive with diffusivity $D \simeq \delta v\, \xi \approx 40 a v_{\mathrm{st}}$; note that this is independent of volume fraction and persists as $\phi \to 0$. However, the diffusivity is caused by hydrodynamic chaos, not thermal motion.

The hydrodynamic screening that we observe is theoretically unexpected. So, the first thing we should look for is whether some of the basic assumptions for our experiment have not been met. We might expect a cutoff length ξ if we are no longer in the Stokes regime, so that inertia becomes important. It is especially suspicious that the product of $\delta v\, \xi$ is constant. This is part of the expression for the Reynolds number $\mathrm{Re} = \rho \delta v \xi / \eta$ where η is the viscosity. As we see in Table 1, the particle Reynolds number is $\approx 10^{-4}$ for our system, the Reynolds number on the size of the swirls is $\approx 10^{-2}$ and the Reynolds number on the scale of our container is $\approx 10^{-1}$, so at first glance we are always at low Re. However, $\mathrm{Re} \sim 1$ is just a guideline and inertial effects may come in at any finite Reynolds number. (Usually turbulence and other high Re effects typically are found at $\mathrm{Re} > 10^3$.) So to test whether Re or Pe ($\mathrm{Pe} = \delta v \xi / D_T$, where $D_T = k_B T / 6\pi\eta L$) are important we (we being Phil Segre) performed experiments on different sized particles (factor of ten variation) and with different viscosity solvents (factor of ten) and found the same scaled results. This is not surprising since the results of other groups, shown in Figure 15, range from 1 μm spheres in water to 'ball bearings in molasses'. Thus it appears that our results

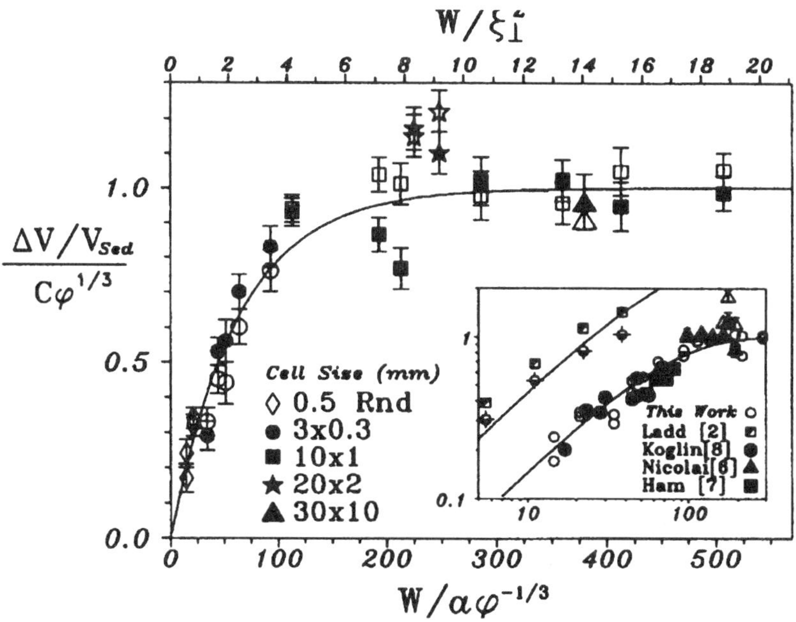

Figure 15. *'Universal plot' of deviation in sedimentation velocity normalised by* $\phi^{1/3}$ *as a function of the width of the sample cell normalised by the correlation length, for a number of different systems ranging from polystyrene spheres to ball bearings. Inset also compares with the simulations of Ladd (1996). (From Segre* et al., *1997.)*

are intrinsic, and representative of the behaviour of high Pe, low Re sedimentation at low volume fractions.

2.3 Origin of hydrodynamic screening

What has to be explained is that some form of hydrodynamic screening cuts off the random density fluctuations at large scales (greater than ξ), leading to interparticle correlation so that the particle distribution is less than random, and produces swirls which contain not a few but many ($\simeq$ 8000) particles per swirl. The problem has not yet been solved, but there are some intriguing possibilities. My own attempt is based on the idea that the swirls are constantly created and destroyed, and the length scale is given by the steady state creation and annihilation rate being set equal. The annihilation process is fairly straightforward to estimate. Within a swirl there is a horizontal density imbalance which is relieved in the 'turnover time' $\tau_{\text{annihilation}} = \xi/\delta v$ by collisions with other swirls or a rearrangement into a vertical density imbalance. We have some evidence of this time by observing the time correlation function $\langle \delta v(0,0)\delta v(0,t)\rangle$ which decays with time constant $\xi/\delta v$ (if we translate along with the average sedimentation velocity); or we can simply observe a video of the sedimenting particles. But how can we estimate the creation time for the swirls? The one simple process we have for envisioning the disruption of a uniform density of sedimenting particles is that of Crowley. If we imagine that the system consists

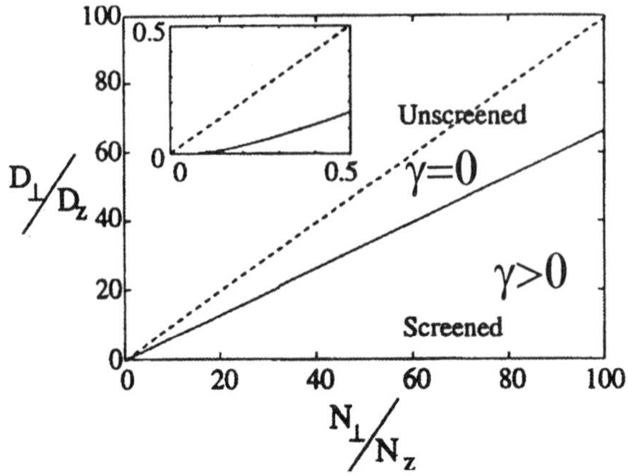

Figure 16. *Nonequilibrium phase diagram for a sedimenting suspension in the model of Levine* et al. *(1998). The coordinates are noise and diffusion anisotropies. Above the solid line there are unscreened spatial particle correlations and the sedimentation velocity variance diverges. Below the line the velocity does not diverge with system size and the particle distribution is random.*

of a lattice of swirls separated by ξ then the rate of growth of the instability away from the lattice scales as $\tau_{creation} \propto \xi^2 v_{\rm sed}/a$. Setting the two rates equal to one another we find $\xi \propto a\phi^{-1/3}$. So the scaling is correct but we would imagine that the coefficient would be of order unity rather than twenty, as observed.

Probably the most interesting explanation of the hydrodynamic screening is found in the phenomenological model of Bruinsma, Levine and Ramaswamy (Levine *et al.* 1998). They suggest that there is a dynamical transition in the system as a function of the anisotropies of the (self-generated) noise, $\mathcal{N}_{||}$ and $\mathcal{N}_{\perp}$, and the diffusion, $D_{||}$ and $D_{\perp}$. In this gravity-driven system there is no requirement for a fluctuation-dissipation theorem which relates $\mathcal{N}$ and D. They set up and solve the convection-diffusion equations for the system and find that there is a phase diagram as illustrated in Figure 16. Above the dashed line there is conventional behaviour and no screening. Below the dashed line there is a correlation length and screening at long distances. The basic idea is that advection tends to kill the horizontal density fluctuations when the noise anisotropy exceeds a threshold. Along with this, the density density correlation function $S(q)$ changes from flat (as for a random system) to one which has a maximum at the scale $1/\xi$ and decays to zero at low wavenumbers.

These results suggest that there is a strong interplay between the hydrodynamic interaction and the interparticle correlations. The sedimenting system of particles does not have a random distribution at all length scales. There is some form of hydrodynamic screening. This brings us back to the 'exact' calculation of Batchelor (1972) for the sedimentation velocity at low volume fractions. The first order correction to the Stokes velocity is no longer straightforward if the particles have long range correlations even as the volume fraction goes to zero.

3 Colloidal crystallisation in microgravity

Having established from sedimentation equilibrium studies that we had a good hard sphere system, and from fluidisation studies that we could not directly compensate gravity by using a fluidised bed, we and NASA decided to prepare a space shuttle experiment. The first thing that happens is that you get an acronym and a badge. Our acronym was CDOT (colloidal disorder-order transition). The main idea of the experiment was to understand the fluid-crystal transition. Two questions that interested us most at that time were the equilibrium crystal structure and the shear modulus. Pusey *et al.* (1989) had reported that the structure of hard sphere crystals on the ground was a random stacking of hexagonal closed-packed layers, rather than the FCC structure, and we wondered whether the samples were influenced by sedimentation. The shear modulus had not been previously measured for the hard sphere system, although it had been calculated by Frenkel and Ladd (1987). But the real reason for the experiment was an intuition that there were other effects in the formation of the solid that were influenced by gravity and sedimentation. We wanted a simple experiment in microgravity to see whether more elaborate experiments would be worth doing. Therefore CDOT was conceived and executed as a small 'glovebox' experiment, a proof of concept and initial trial.

As we shall see, what we found from CDOT is not what we expected. The random stacking structure remained. But we found that the natural growth mechanism, a dendritic growth instability, had been masked by gravity, and the samples which were in the glass phase and did not crystallise on earth, quickly formed crystals in microgravity. This raised our interest sufficiently that we set up and performed a much more elaborate set of experiments, codenamed 'phAse', which provided probably the most detailed data yet on the nucleation and growth of hard sphere colloidal crystals.

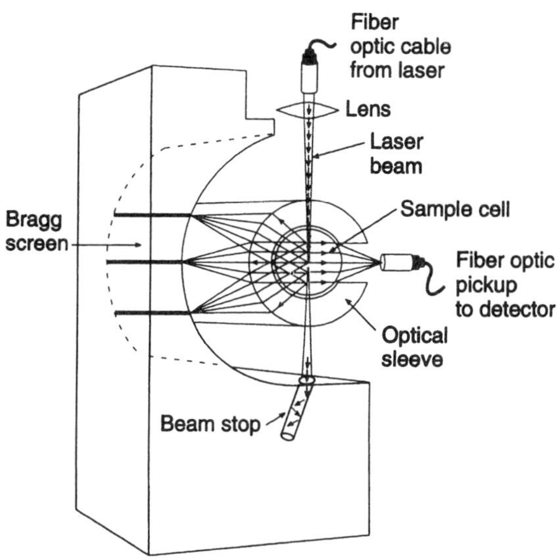

Figure 17. *Schematic of the CDOT glovebox apparatus.*

A schematic of the CDOT apparatus is shown in Figure 17. The sample is contained in a glass cylinder cell which can be translated along its axis and oscillated about its axis. Light from a laser diode is fibre-optically brought to the cell and is focussed in a 1mm beam diametrically through the centre of the cell. Another fibre collects light at 90° and brings it to an avalanche photodiode for use in dynamic light scattering studies (see Pine, this volume). The cell is surrounded by a plastic cylinder which acts as a cylindrical lens and focuses all light emitted from the sample at a given angle to a spot on the translucent screen. A video camera is focussed on the screen and stores or transmits the images during the experiment. The screen collects scattered light over the range 20–160° and is used for Bragg scattering to determine the crystal structure. The samples are hand loaded by the astronauts into the apparatus while it is mounted in the glovebox.

Along with the light scattering and rheology apparatus on the CDOT experiments, there was also a set up where the astronauts photographed all of the samples at various times after mix/melting in microgravity. It was observed that all of the samples in the coexistence region show dendrites and that all of the samples in the 'glass' regime on earth, $\phi = 0.58 - 0.63$, readily crystallised in space. Both of these observations were new and unexpected (Zhu 1997, Cheng 1998). After the first flight we analysed the growth equations originally written by Ackerson and Schatzel (1995,1996) for hard sphere colloids, and found that in the coexistence region where the growth is diffusion limited, a spherical crystal growth mode is unstable against the well-known Mullins-Sekerka instability (Russel *et al.* 1997).

The original $\phi = 0.619$ 'glass' sample which crystallised in space survived reentry. (All coexistence/dendritic samples did not withstand the random $\approx 20g$ RMS shuttle re-entry.) After it had sat in our lab for about six months as a crystal, we decided to remix the 'glass' region sample to see whether it had somehow changed. But we didn't want to destroy the crystal if it would never grow back on earth. So we used the magnetic mixing bar to mix one half of the sample. The crystal mix-melted in this region and then formed a glass-like structure as evidenced by the fact that the stir bar did not settle to the bottom of the sample cell. Over a period of 1 year we observed the sample. The space-grown crystal region slowly grew into the glass region, but there was no nucleation and growth of crystals in the glass region.

The CDOT missions also gave us preliminary Bragg scattering data to determine the crystal structure. We were hoping to find crystallisation of the FCC phase in microgravity, but the first images we obtained from space destroyed any such hope. A typical image of the laser beam scattered from the sample onto a screen is shown in Figure 18.

The image is dominated by large streaks which indicate 'Bragg rods' rather than Bragg spots. Rods result from the Fourier transform of a two dimensional lattice, while spots are from a three dimensional lattice. The random stacking of hexagonal planes produces an incoherent superposition of two dimensional structures and thus Bragg rods. Seemingly the crystallisation is sufficiently rapid that the crystal structure is determined by kinetics rather than by thermodynamics. After the CDOT experiments several theoretical papers showed that the stable phase at all crystalline volume fractions is FCC, but that the transformation of a random stacking sequence to FCC is slow (Pronk and Frenkel 1999, Mau and Huse 1999).

The much more elaborate 'phAse' (physics of hard sphere experiment) experiment was constructed to quantitatively investigate nucleation and growth, structure and rheology of the PMMA hard sphere samples. The most effective piece of the flight apparatus was

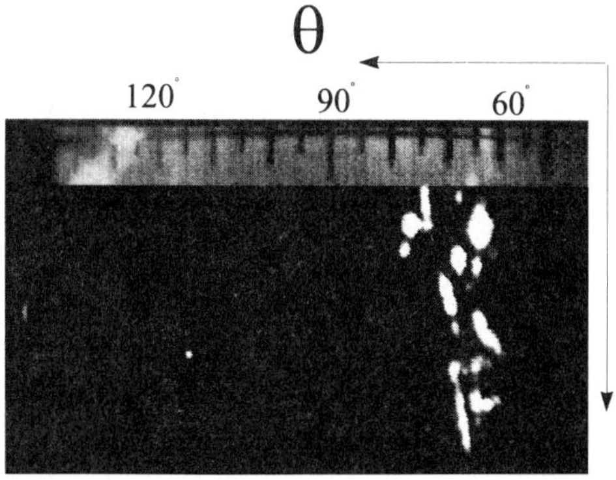

Figure 18. *Image from the Bragg screen on CDOT. The predominant scattering into Bragg rods or streaks rather than Bragg spots indicated the presence of two dimensional structures—the uncorrelated stacking of two dimensional hexagonal planes in the random hexagonal stacking structure.*

the Bragg imaging section. The sample cells were $r = 10$cm glass hemispheres with a 1cm radius by 1cm high cylindrical cavity, filled with the colloidal suspension, placed axially at the sphere centre. An 8mm diameter (Gaussian) laser beam entered axially through the flat region of the cylinder. Any light from the scattering volume which leaves at a particular angle is focused to a spot that is at distance $3r$ from the sphere centre. A fluorescent screen is located at this position and is imaged by a CCD camera. A typical Bragg scattering pattern recorded during flight is shown in Figure 19 (upper panel). When averaged azimuthally and plotted as a function of the scattering angle from the central beam stop we have the powder average (unnormalised for form factor) scattering function which is shown plotted at several times after initial melting in Figure 19 (lower panel). Here we see the liquid-like structure factor $S(q)$ evolving into a crystal form with the development of Bragg peaks.

As the crystals evolve from the metastable liquid, we must determine what fraction of the sample is liquid and solid, and the structures in each phase. The analysis follows that of van Megen (1998). From previous calculations we know the scattering function for different close packed crystal structures and for the concentrated liquid phase (which fits a modified Percus–Yevick form quite well). The contribution of the liquid is determined mostly from the scattering at qr below the solid 'shoulder'. The liquid contribution is subtracted and the result compared with different solid structures. Adjustments in relative liquid/solid fractions are made until acceptable liquid and crystal $S(q)$'s can be matched. Further analysis of the crystalline phase is accomplished by evaluating the main Bragg peak, which corresponds to scattering from the stacked hexagonal planes. The position of the peak, q_m, gives the plane spacing, hence the lattice spacing, density of particles and volume fraction of growing crystallites. The peak width, Δq gives the size of the crystallites, while the integral under the peak tells the fraction of the sample which is crystallised.

The difference between growth in gravity and microgravity shows up both in the raw scattering data (Figure 20) and in the analysed quantities shown in Figure 21. The data are from the same sample and the same apparatus on the shuttle and on the ground.

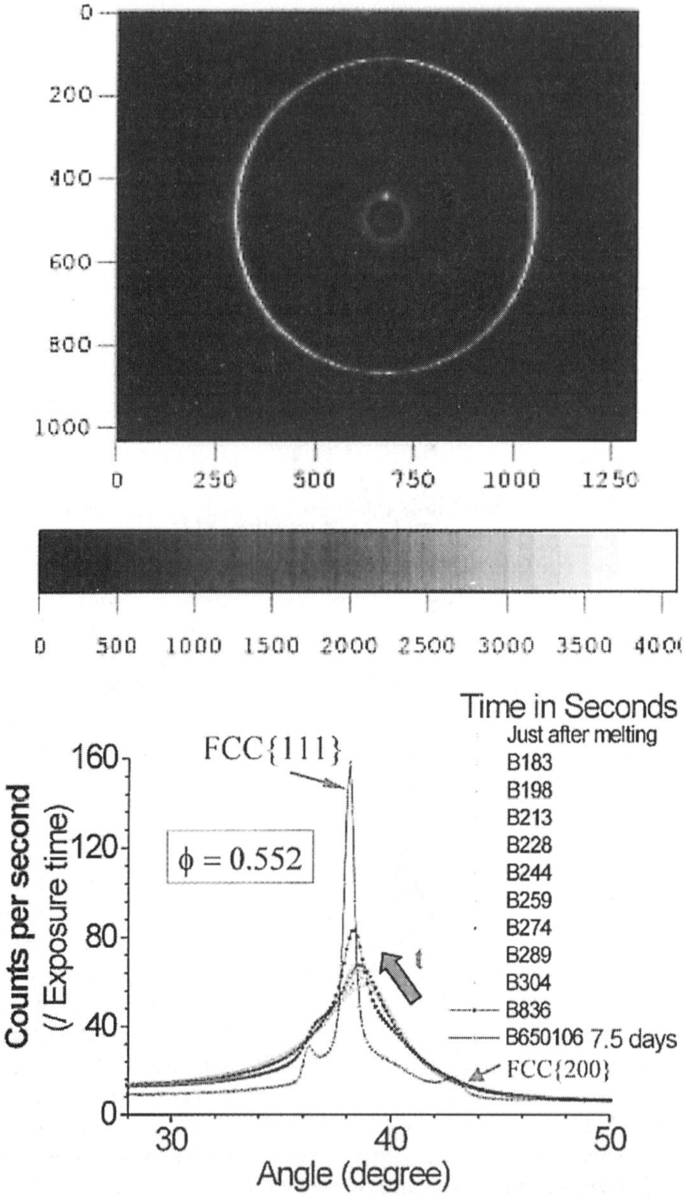

Figure 19. *Upper: Image from the Bragg screen on the phAse experiment. Lower: The aximuthally averaged intensity as a function of scattering angle is plotted for several times after mix/melt. Note how the liquid structure evolves to the crystal and the emergence of the* FCC *200 peak at long times.*

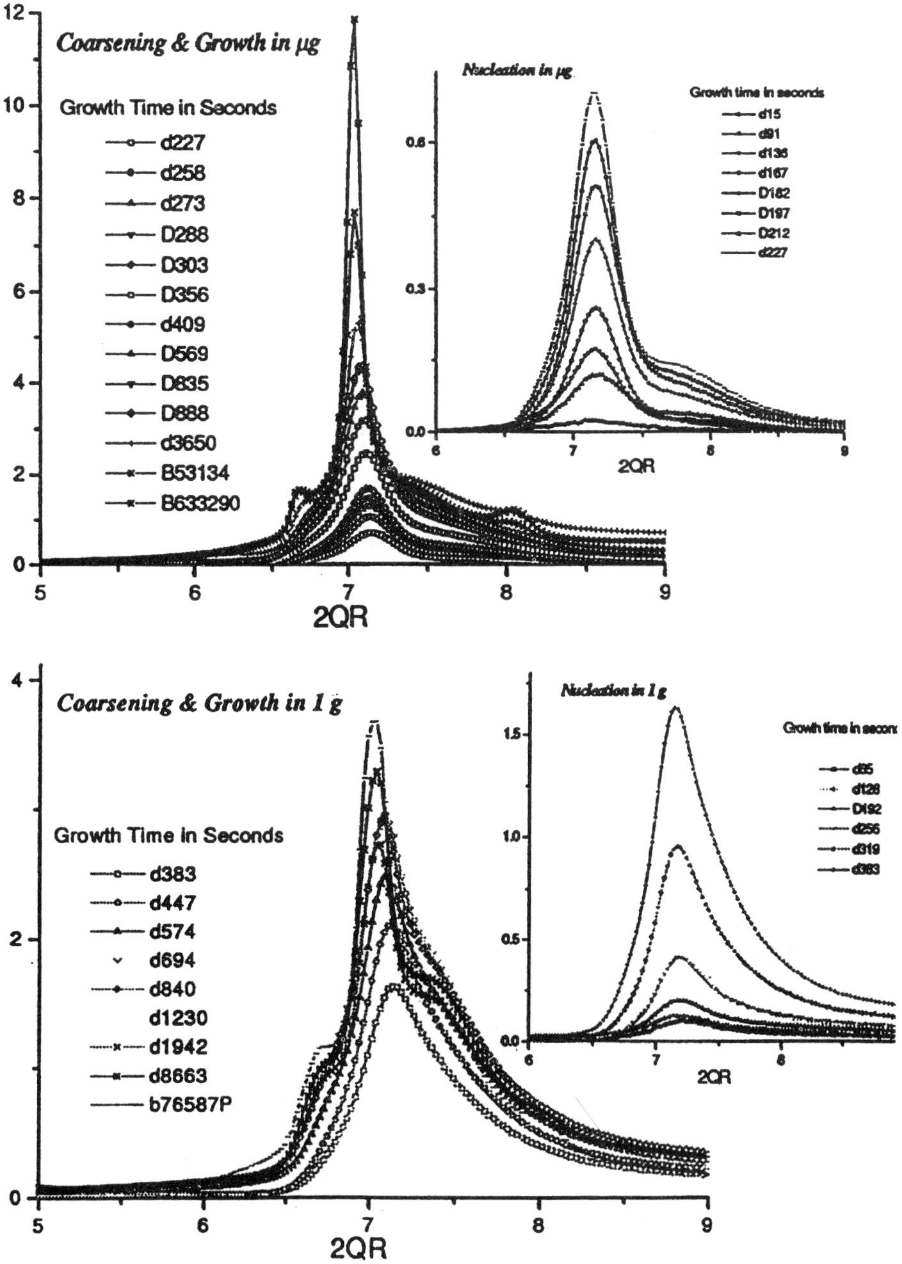

Figure 20. *Scattering functions for the same sample ($\phi = 0.552$) in the Phase apparatus ingravity and microgravity. Note the sharper structure overall and evolution of the* FCC *200 peak in the microgravity data.*

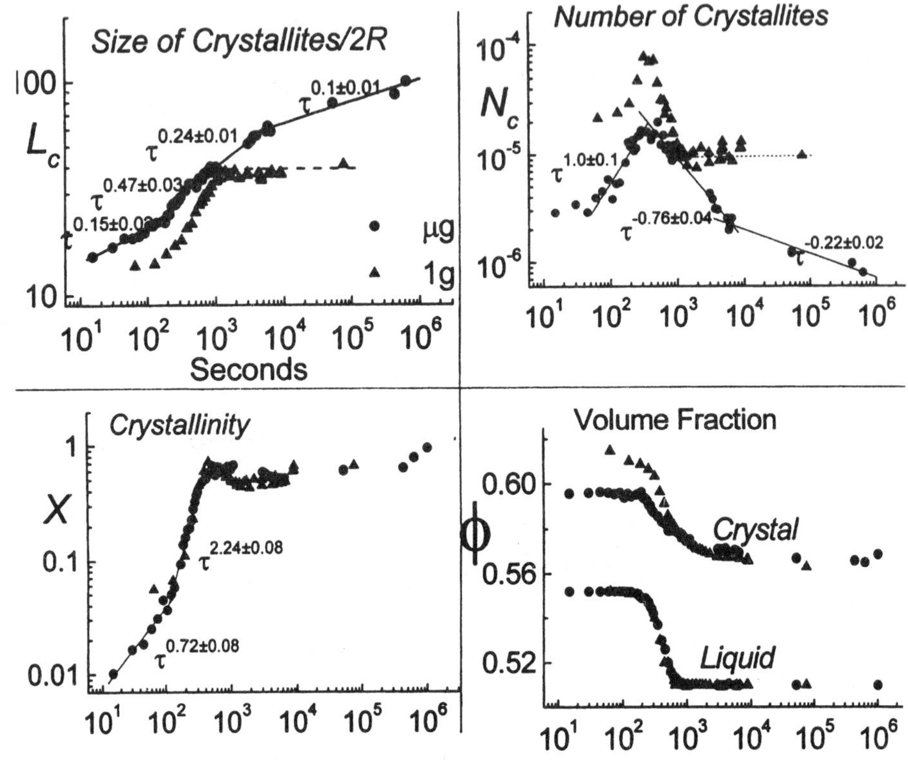

Figure 21. *For samples in the fully crystalline region (this sample has $\phi = 0.552$) there is not much difference between gravity and microgravity save for the faster growth and coarsening of microgravity crystallites.*

Although the fraction of crystal in the sample grows at about the same rate in space and on the ground, many of the other characteristic properties are different. In particular the microgravity crystals, with diffusion fields undisturbed by sedimentation, show the classic diffusion-limited growth behaviour, with the size of the crystallites growing as $L_c \propto t^{1/2}$. At later stages the microgravity crystallites show a slow coarsening to a larger size, while gravitational stresses seem to suppress the further growth of ground-based crystals.

It was also possible to oscillate the sample cells in the 'phAse' apparatus, while using light scattering to measure the local strain field in the centre of the cell. With fixed small amplitude drive and different frequencies, we could detect shear resonances in the sample which are associated with the cylindrical geometry. From these resonances we measured the velocity of shear sound and hence the shear modulus. The results are shown in Figure 22 and agree well with the hard sphere simulations of Frenkel and Ladd (1987).

The shuttle experiments show that the natural growth of the shear-melted hard spheres is kinetically dominated and yields the random hexagonal stacking structure. From Fig-

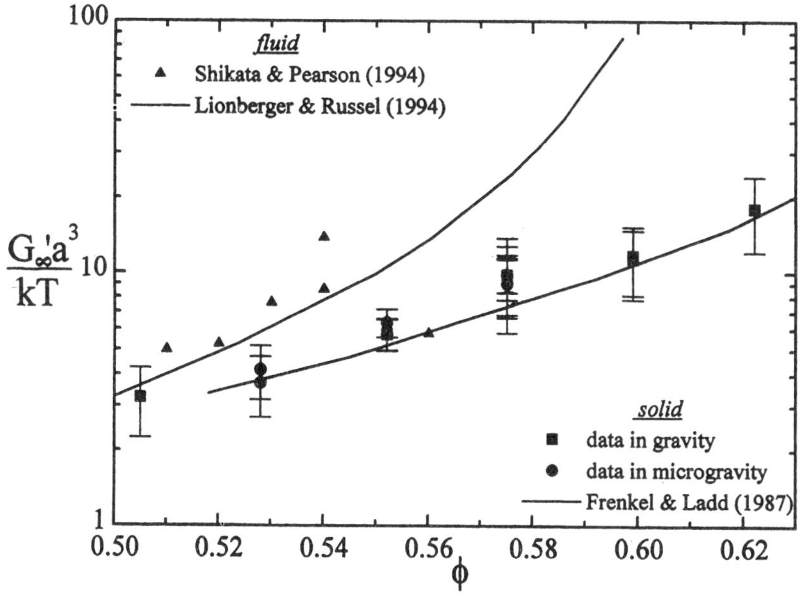

Figure 22. *Shear modulus of hard sphere colloidal crystals which have been grown and measured in microgravity and on the ground.*

ures 19 and 20 we see that structure related to the FCC (200) peak begins to appear after a period of about 2 weeks. This is roughly what we would expect from the arguments of Frenkel (this volume). However, it would also be nice to have a way to control the growth of the crystal rather than to do the effective fast quench from the metastable liquid. To do this, we need another control parameter. Temperature is what is used in more conventional systems, but hard spheres are athermal and temperature is not relevant for the phase transition. However, temperature controls the osmotic pressure and a temperature gradient might therefore be used to produce a density/volume fraction gradient. We could take a sample which is in the liquid phase ($\phi < 0.49$) and the heat one end. This would drive the concentration higher at the unheated end and once it reached $\phi > 0.545$ we would have crystals. The experiment should really be done in microgravity to prevent sedimentation and thermal convection. We tried on earth with a density matched sample and we found that we could grow large controlled crystals with no dendrites (Cheng *et al.* 1999), but the control was not ideal due to thermal convection (both colloid and solvent expand on heating).

The most intriguing result from our shuttle experiments is the crystallisation of the 'glass' samples. What does gravity have to do with the glass transition? We don't yet know. But if we follow along some of the themes of this Summer School we might find the answers in 'jamming'. (See Nagel and Cates, this volume.) Aside from the random dense packing of hard spheres (RCP), if stresses such as shear, or body forces such as gravity are applied, a system can jam up—most particles are held in position by neighbours at a volume fraction below the RCP value of 0.63.

The idea of random loose packing, RLP, is even less well defined than RCP. However it is very important and useful. Letting particles settle or pouring sand into a cylinder one finds a typical packing fraction of around 0.58 (see Nagel, this volume). If you start above $\phi = 0.58$ and shear the sand it dilates to $\phi \simeq 0.58$. If you start below 0.58 and shear sand it densifies to $\phi \simeq 0.58$. (Therefore it is better to build your house on soil of $\phi > 0.58$ if you don't want it to sink in the densifying fluidised soil during an earthquake.) In soil mechanics, a porosity of $\simeq 0.42$ (porosity $= 1 - \phi$) is known as critical porosity for this reason. So, one scenario is that in gravity, at volume fractions above 0.58, the particles are jammed in an RLP state and cannot move. In the absence of gravity there is no jamming and the particles are free to find their equilibrium configuration—to crystallise. This can be tested in the future by measuring diffusion in these samples with and without gravity.

Acknowledgments

The work which is reported here could not have been done without the collaboration of Bill Russel and Eric Herbolzheimer. However, the real work was done by the students and postdocs who worked on these projects over the past several years. In particular acknowledgments are due: Maarten Rutgers, Juizhi Xue, Phil Segre, Jixiang Zhu, and Zhengdong Cheng. The collaboration with NASA owes a great deal to Rick Rogers for the CDOT experiments and to Bill Meyer for all of the microgravity experiments, the crew of shuttle missions STS 73, 83 and 94 and the entire phAse team. Research was supported by NASA and Exxon.

References

Ackerson B J and Schätzel K, 1995 *Phys Rev E* **52**, 6448

Ackerson B J and Schätzel K, 1996, *J Chem Phys* **105**, 9258

Alder B J and Wainwright T E, 1957, *J Chem Phys* **27**, 1208

Andersen H C, Chandler D, Weeks J D, 1976, *Adv. Chem Phys* **34**, 105

Batchelor G K, 1972, *J Fluid Mech* **52**, 245

Bernal J D, 1964, *Proc R Soc London Ser.* **280**, 299

Bernal J D, 1965, 'The geometry and structure of liquids' in *Liquids: Structure, Properties, Solid Interactions*, ed. Hughel. (General Motors Research Laboratories, Warren, MI)

Berryman J G, 1983, *Phys Rev A* **27** 1053

Brenner M P, 1999, *Phys Fluids* **11**, 1

Buscall R, Goodwin J W, Ottewill R H and Tadros T F, 1982, *J Colloid Interface Sci* **85**, 78

Carnahan N F and Starling K E, 1969, *J Chem Phys* **51**, 635

Caflisch R E and Luke J H C, 1985, *Phys Fluids* **28**, 259

Chaikin P M, and Lubensky T C, 1995, *Principles of Condensed Matter Physics* (Cambridge University Press)

Cheng Z, Russel W B, and Chaikin P M, 1999, *Nature* **401**, 893

Conway J H and Sloane N J A, 1992, *Sphere Packings, Lattices and Groups* (Springer-Verlag, New York)

Crowley J M, 1976, *Physics of Fluids* **19**, 1296

Davis R H and Acrivos A, 1985, *Ann Rev Fluid Mech* **17**, 91

Einstein A 1905, *Annalen der Physik* **17**, 549

Frenkel D, and Ladd A J C, 1987, *Phys Rev Lett* **59** 1169

Frost R, Schon J C, and Salamon P, 1993, *Comp Mat Sci* **1**, 343

Guyon E *etal.*, 1987, *Disorder and Mixing*, NATO ASI Summer School, Cargese, (Kluwer Academic, Dordrecht) **153**

Hales S, 1727, *Vegetable Staticks*, reprinted in 'The History of Science Library', ed. Hoskin: Elsevier, 1961

Hall K R, 1972, *J Chem Phys* **57**, 2252

Ham J M and Homsy G M, 1988, *Int J Multiph Flow* **14**, 533

Happel J, and Brenner H, 1986, *Low Reynolds number hydrodynamics*, (Martinus Nijhoff Publishers, Dordrecht)

Herbolzheimer E, 1999, private communication, recalled hearing this argument from A Acrivos

Hinch EJ, 1987, in Guyon E *etal.*, 1987, *Disorder and Mixing*, NATO ASI Summer School, Cargese, (Kluwer Academic, Dordrecht)

Hoover W G and Ree F H, 1968, *J Chem Phys* **49**, 3609-3617

Janosi I M, Tel T, Wolf D E, and Gallas J A C, 1997, *Phys Rev E* **56**, 2858

Koch D L and Shaqfeh E S G, 1991, *J Fluid Mech* **224**, 275

Koglin J E, 1971, Ph.D thesis, Karlsruhe, Germany

Ladd A J C, 1990, *J Chem Phys* **93**, 3484

Ladd A J C, 1996, *Phys Rev Lett* **76**, 1392

Levine A, Ramaswamy S and Bruinsma R, 1998, *Phys Rev Lett* **81**, 5944

Lindemann F A, 1910, *Z Phys* **11**, 609

Lubachevsky B D, Stillinger F H, and Pinson E N, 1991, *J Stat Phys* **64**, 501

Mason TG, Lacasse MD, Grest GS, Levine D, Bibette J, Weitz DA,1997, *Phys Rev E* **56**: 3150-3166

Mau S-C, and Huse D A, 1999, *Phys Rev E* **59**, 4396

Nicolai H and Guazzelli E, 1995, *Phys Fluids* **7**, 3

Nicolai H, Herzhaft B, Hinch E J, Oger L, and Guazzelli E, 1995b, *Phys Fluids* **7**, 12

Oseen C W, 1927, *Hydrodynamik*, Leipzig: Akademische Verlag

Paulin S and Ackerson B J, 1990, *Phys Rev Lett* **64**, 2663

Percus J K, and Yevick G J, 1958, *Phys Rev* **110**, 1

Phan S, Russel W B, Cheng Z, Zhu J, Chaikin P M, Dunsmuir J H, and Ottewill R H, 1996, *Phys Rev E* **54**, 6633

Phan S, Zhu J, Russel W B, and Chaikin P M, 1998, *J Chem Phys* **108**, 9789

Pronk S, and Frenkel D, 1999, *J Chem Phys* **110**, 4589

Pusey P N, and van Megen W, 1986, *Nature* **320**, 340

Pusey P N, van Megen W, Bartlett W P and Ackerson B J, 1989, *Phys Rev Lett*, **63**, 2753-2756

Pusey P N, van Megen W, Underwood SM, Barrtlett P, Ottewill RH, 1990, *J Phys.-Cond Matt* **2**: SA373-SA377, Suppl A DEC

Pusey P N, 1991, in *Liquids, freezing and the glass transition* Chapter 10, Colloidal suspensions (eds Hansen, J P, Levesque D and Zinn-Justin J) 763-942 (Elsevier, Amsterdam)

Pusey P N, 1987, *J Phys France* **48**, 709

Rintoul M and Torquato S, 1996, *J Chem Phys* **105**, 9258

Russel W B, Saville D A and Schowalter W R, 1989, *Colloidal Dispersions* (Cambridge University Press, Cambridge)

Russel W B, Chaikin P M, Zhu J, Meyer W V, Rogers R, 1997, *Langmuir*, **13**, 3871

Rutgers M A, Xue J-Z, Herbolzheimer E, Russel W B and Chaikin P M, 1995, *Phys Rev B* **51**, 4674

Rutgers M A, 1995b, Thesis Princeton University

Rutgers M A, Dunsmuir J H, Xue J Z, Russel W B and Chaikin P M, 1996, *Phys Rev B* **53**, 5043-5046

Sangani A S and Acrivos A, 1982, *Int J Multiphase Flow* **8**, 343

Segre P N, Herbolzheimer E and Chaikin P M, 1997, *Phys Rev Lett*, **79**, 2574
Steinhardt P, Nelson D, and Ronchetti M, 1983, *Phys Rev B*, **28**, 784
Stokes G G, 1851, *Trans Cambridge Phil Soc* **9**, pt II, 8
Stillinger F H, Debenedetti P G and Torquato S, 1999, to be published in *Phys Rev Lett*
Variano E, 1999, Junior Paper, Princeton University Physics Department 1999
Woodcock L V, 1997, *Nature* **384**, 141
Xue J-Z, Herbolzheimer E, Rutgers M A, Russel W B, and Chaikin P M, 1992, *Phys Rev Lett* **69**, 1715
Zallen R 1983 *The Physics of Amorphous Solids*, Wiley, New York
Zick A A and Homsy G M, 1982, *J Fluid Mech* **115**, 13
Ziman J M, 1979, *Models of Disorder*, Cambridge University Press, Cambridge
Zhu J, Li M, Meyer W, Ottewill R H, STS-73 Space Shuttle Crew, Russel W B, Chaikin P M, 1997, *Nature* **387**, 883

Granular materials: static properties as seen through experiments

Sidney Nagel

The University of Chicago, USA

1 Introduction

Granular materials are simple systems—so simple, in fact, that it is shameful that we don't understand them well at all. What could be simpler than a pile of sand at room temperature made up of large, essentially non-deformable and non-cohesive, classical objects? If there were no gravity or confining forces compressing the material into a solid mass, it would simply float apart like a gas in which each particle moves according to Newton's laws and interacts with its neighbours via hard-core repulsion. The element that makes this gas different from other gases is that, because the particles are so large with many internal modes of vibration, when they collide they lose some fraction of their kinetic energy. This dissipation of energy can change the nature of the granular gas dramatically and produce nearly static, long filamentary structures.

A granular material, confined by the force of gravity, can flow and resemble an ordinary liquid. Yet when looked at even cursorily it is clear that the motion of this granular fluid is bizarre. In some cases, the only moving particles are near a surface where, if ordinary hydrodynamics were applicable, the velocity would be zero due to non-slip boundary conditions. Unlike other fluids, a compacted granular material must expand before it can begin to flow.

A static pile at rest, confined by the force of gravity, behaves in many ways like an ordinary solid and can withstand a considerable amount of stress without yielding. Yet a tiny stress in the 'wrong' direction can cause large portions of the pile to give way and start to flow. The material is exquisitely sensitive and continually shifts and rearranges in response to all the minute forces acting upon it. In addition, there is no equation of state for a granular material. This is because its density at a given temperature and pressure is not well-defined. Simply by tapping a container filled with grains, the material will settle to a new density. Thus the density is a consequence of the previous history and handling of the material.

What makes these systems so unusual? At first sight, it seemed that the fact that the particles were large might lead to simplifications in the description of the pile. However, the opposite is the case. Because the particles are large, ordinary temperature plays no role in the description of the medium. In order to be important the thermal energy must be comparable to some energy in the system. For a granular material, there is only one energy with which to compare k_BT and that is the energy mgd required to lift a particle of mass m in gravity g by its own diameter d. This is the energy typically necessary to make any small rearrangement of the positions of the particles in the pile. For a 1mm grain of sand, $k_BT < 10^{-12}mgd$. A room-temperature experiment on a macroscopic granular system actually corresponds to an ultra-low temperature experiment on a molecular system. For all practical purposes $T = 0$ for a medium with grains larger than a few microns.

Without the existence of a useful temperature scale, a number of simplifications that we have grown to rely upon become invalid. A granular system cannot be considered to be in equilibrium. No matter how long one waits, the system is always caught in some metastable state. Any perturbation simply forces the system to go from one metastable state to another. Temperature allows a system to explore phase space but this can happen only if the temperature is sufficiently high so that energy barriers can be traversed. We do not know how to calculate (or even measure) the average value of any property since we do not know over which ensemble we should do the averaging. To what extent does the preparation history of the material affect its 'average' properties? How much do we need to know about its preparation to specify the state of such a system? That is, what aspects of the preparation history are relevant for determining the properties? For the flowing material we must face a different problem since temperature sets a velocity scale for the motion of molecules in an ordinary liquid; the flow velocities may be compared to this thermal motion to tell if the flow is rapid or slow. Since the temperature in a granular fluid is so low, all flows must be considered as rapid and the only velocity scale is the one imposed by the macroscopic flow itself. In another kind of comparison with ordinary fluid dynamics one might ask for any given geometry and flow what is the Reynold's number, $Re = \rho UL/\eta$ (where U is a typical velocity, L is a typical length scale, ρ the fluid density and η its viscosity, which is independent of flow rate). In order for such a categorisation of the flow in a granular material to be useful, one must have a quantity corresponding to the kinematic viscosity. Although several attempts have been made to measure or define such a quantity, these 'granular viscosities' have not been shown to be independent of velocity, frequency (of vibrations) or length scale over which they are measured.

As alluded to above, one characteristic of granular material is that there are mechanisms for the dissipation of energy. In the granular gas mentioned above, some kinetic energy is dissipated upon each collision. The energy of the collision goes to heat up the individual particles or to create sound waves that leave the vicinity of the experiment. There is another form of dissipation which is equally, if not more, important for producing some of the non-intuitive properties of these materials: friction between the particles. When moving, the rubbing of one particle against another, even without inelastic collisions, can dissipate energy. The presence of friction produces other problems as well. For a completely static pile, the presence of friction produces an indeterminacy of how the forces are distributed within the structure. Unlike the conservative forces derived from a potential of interaction, a frictional force between two particles can have a range of values for the same position of the particles. The degree of the frictional force 'mobilisation' depends on the history, that is, on how the particles were placed into position. Just as with a block on an inclined plane, a particle in a bead pack can be stable and not move even though the forces on it vary. This indeterminacy in the frictional force contributes to the difficulty in describing the force propagation within a pile.

As a last introductory comment, I should note that many industries rely on processing and transporting granular materials: the pharmaceutical industry handles powders and pills, the agricultural industry handles foodstuffs, the construction industry handles gravel and sand, and the chemical industry produces many of its products in the form of powders. Nevertheless, the technology for dealing with granular material is poorly developed. Studies have shown that we waste 40% of the capacity of many industrial plants dealing in granular material. This is much worse than is found in plants designed to handle material in the liquid form. It is evident that a better understanding of the science of these materials could have a potential impact on many diverse industrial processes.

In what follows, I will describe some of the experiments that have been performed to elucidate the behaviour of granular materials. In particular, I will discuss some issues that relate to the properties of such materials at rest or subject to only small perturbations. I will not discuss here the behaviour of rapidly flowing material. In the next Section, I will concentrate on investigations of the force distributions in these materials. This includes the visualisation of the so-called force chains. Section 3 will then describe some experiments which show the unusual manner in which sound propagation occurs in these materials. This behaviour is related to the existence of force chains. I will then briefly mention some experiments on how the density of such materials depends on vibration, *i.e.* vibration-induced compaction, and relate the fluctuations of the density to the notion of compactivity defined by Edwards. Finally I will conclude by mentioning in Section 5 some similarities between jamming in granular material and freezing at the glass transition.

2 Force distributions in granular matter

Much of the material in this Section is taken from Reference [1].

Inside a static pile of granular material, each particle is under stress since it helps to support the weight of all the particles above it as well as any extra load placed on the pile's surface. The stresses are, of course, transmitted via the inter-particle contacts. It is this network of contacts, their overall geometry and strength, that determines many of the pile's most salient bulk properties such as its ability to conduct sound [4–6] or shock [7,8] and its ability to bear a load [2,3].

Many of us have developed our intuition about packings based on crystalline solids. However, such intuition is clearly useless for granular systems. In a crystal, the atoms are identical and the load is shared uniformly between them. For granular materials, one finds that the slightest amount of disorder, due to even tiny variations in the grain dimensions or to packing imperfections, can cause large variation in the force transmission. Since the particles interact via essentially hard repulsion even small changes in position will produce large variations in the local network of force contacts. In these materials, it is found that stresses are transmitted along 'force chains' rather than homogeneously. These chains make up a ramified network of particle contacts involving only a portion of all grains [9–11]. It is also important to realise that unlike a regular solid (crystalline or glassy) where the atoms have attractive as well as repulsive interactions, the particles in a *non-cohesive* granular pile only interact via repulsive forces. There is, at least in principle, no attraction between the grains. By working in a humidity-free environment and taking care to keep static electricity to a minimum (not necessarily an easy thing to do) the attractive forces can be kept small. This implies that the inter-particle potential has a discontinuity in one of its derivatives at the distance where they just begin to touch each other. The role that this discontinuity can play in the mechanical properties of the bulk material has as yet not been made fully clear.

One common way to visualise force chains in a material is to use stress-induced birefringence as was done by Dantu [12,13]. In such an experiment a bead pack is placed between a pair of crossed polarisers. If the material used to make the beads is isotropic then no light will be transmitted through this configuration. (The exception is in a three-dimensional packing where multiple reflection from the particle surfaces can cause the polarisation of the light to rotate. However, by back-filling the container with an index-matching fluid, this reflection-induced transmission of light can be minimised [10].) However, when the material is subject to a large stress, then light can be transmitted; the stress induces a birefringence which rotates the polarisation of light propagating in a direction perpendicular to the direction of the applied stress. The amount of rotation depends on the relative orientation of the stress compared to the axes of the linear polarisers as well as the magnitude of the stress. For this reason it is advisable to use two crossed circular polarisers (instead of linear ones) in order to view the entire stress pattern. (One could, alternatively, take multiple exposures of the system with the crossed linear polarisers rotated so that their axes have different orientations from one exposure to the next.) Thus by viewing the bead pack under stress between two crossed circular polarisers one can see, by the strong light transmission, where the stresses are largest and identify the force chains running through the material. A number of experiments on 2D and 3D systems have imaged force chains by exploiting this technique [9-18]. An example of such a picture for a 3-dimensional system made up of spherical glass beads is given in Figure 1, taken from [10].

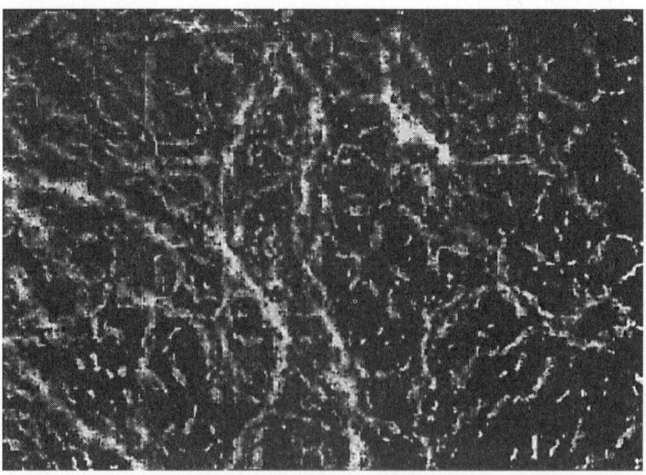

Figure 1. *Visualisation of a 3-d force chain in a granular medium in an index-matched fluid viewed between crossed polarisers. Photograph taken from [10].*

It is very difficult to analyse the data from these stress-induced birefringence experiments in order to obtain the magnitude of the forces at the points of contact. This is particularly true in three dimensional geometries where one must remember that some of the stress may be oriented along the viewing axis, that is along the direction of the light propagation. Those stresses will not contribute to the total light transmission since they do not produce any birefringence for light propagating in that direction.

In order to get more quantitative information about the distribution of forces at the inter-particle contacts another method has been used [1,10]. In this method contact forces are measured using the marks left by carbon paper lining the container walls [1,10,16,17]. (This method so far has been able to obtain the distributions of forces only at the confining boundaries of the container.) Beads are placed into a container and two pistons, at the top and bottom of the container, apply pressure to the bead pack. The beads are then pressed into the carbon paper which leaves a mark, or smudge, on another blank sheet of paper. When the container is disassembled, these sheets are examined and the smudges left by the carbon are digitised on a flatbed scanner. These marks can then be analysed according to their darkness or area. By comparing with a calibration curve for how large or how dark a spot should be for a given force, the magnitude of the force creating each smudge can be determined. As a cross-check of the experiment, the total force applied externally can be compared to the total force obtained by adding up all the forces computed from each individual dot on the upper and lower surfaces. If the technique measures the normal forces only (and not the tangential, or shear, forces) then the computed force at the top and bottom plate should equal each other and both should equal the force applied by the pistons. This is found to be the case within a few percent. One can also worry that during the loading of the system by putting pressure onto the piston, the beads can move, leaving smudges along their path. This would give spurious forces not associated with the final packing configuration. Again such effects can be detected by examining the individual dots to see if they are circular or elliptical in shape. Under careful loading of the apparatus, it is found that the dots are almost perfectly circular indicating that there has been little slip in the process of applying the pressure.

By analysing all the carbon-paper marks left on the container boundaries, a histogram, or distribution $P(F)$, can be made of the number of contacts that have a given force, F. It is convenient to normalise the forces to the average force, $\langle F \rangle$, $f \equiv F/\langle F \rangle$. Remarkably, it has been found that the distribution, $P(f)$ is the same on all the surfaces of the container and is invariant under a number of different conditions. Figure 2 shows the experimental curves for $P(f)$ reported in [1]. The different symbols in the graph indicate the distributions measured at the different walls of the container. (Along the side walls, the average force varies with distance from the top surface. The average force $\langle F \rangle$ was determined at each depth and this value was used to calculate $P(f)$ at each point.) A good fit to the data was obtained with the following form:

$$P(f) = a\left[1 - b\exp\left(-f^2\right)\right]\exp(-cf) \tag{1}$$

with fit parameters $a = 3$, $b = 0.75$, and $c = 1.5$. It was also possible to calculate the lateral force-force pair correlation between forces at different distances, r, along the bottom or top surfaces. This correlation function was flat, showing that little, if any, correlation between forces existed.

In order to understand the behaviour of $P(f)$, Coppersmith *et al.* [10,19] developed the 'q-model' in which the material is modelled as a layered crystal with forces propagating

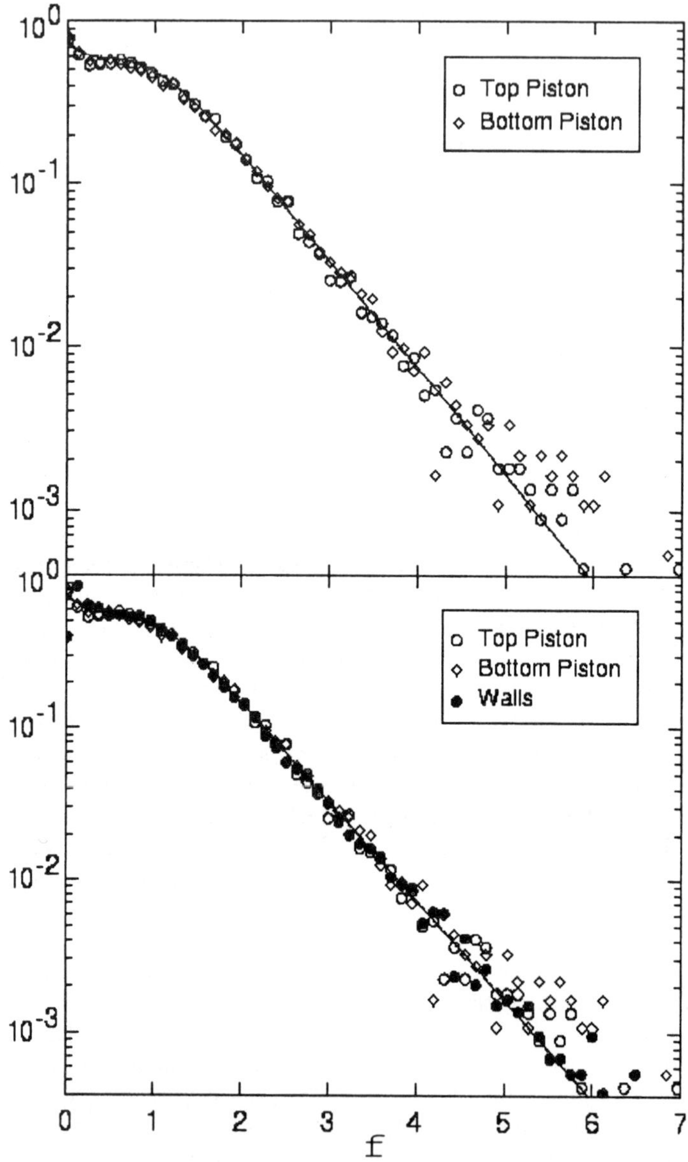

Figure 2. *The distribution of forces, P(f), between particles and the different surfaces of a cylindrical container. The open circles are data for the top piston, the diamonds are the data for the bottom piston, and the solid circles are the data for the cylinder walls. The solid line is a fit using Equation 1. This figure is from [1].*

from an upper layer to the one below it. In order to account for the disorder, each particle distributes its load randomly to the n particles below it. For example, in the case where $n = 2$ a particle rests on top of only two neighbours, the load is distributed with a fraction q transferred to one of the particles and a fraction $(1 - q)$ transferred to the other. The quantity q is chosen randomly from a distribution between 0 and 1. Consistent with Newton's third law, the sum of the vertical forces leaving a particle must equal the total vertical force that particle feels from the ones resting upon it (as well as its own weight which is negligible when that particle is far from the top surface). In some limits this model, $P(f)$ can be solved exactly. Under quite generic conditions (essentially that the distribution from which q is chosen extends all the way to the endpoints of the interval between 0 and 1), it is found that at large forces, $P(f) \propto \exp(-\beta f)$, with β some constant. Thus, compared to a Gaussian distribution, this result implies a much higher probability of finding force values F much larger than the mean value, $\langle F \rangle$.

This result can be understood by a simple analogy with the scattering of particles [10,19]. In the q-model, we can visualise the 'flow' of forces as arrows that come into and then leave each particle site. For example, with $n = 2$ as used above, there will be two forces coming into each site and two forces leaving it. The sum of the forces in the vertical direction coming in equals the sum of those leaving the site (as long as we can neglect the extra weight of the particle at that site). Formally this looks equivalent to a diagram we would draw if we were calculating the kinematics of elastic scattering of particles in a gas. There the conserved scalar quantity is the energy whereas in the q-model it is the vertical force coming into or leaving each site. Just as in a gas in equilibrium, where this scattering leads to a Boltzmann distribution of energy, so too the 'scattering' of forces at each site in the q-model leads to an exponential distribution of forces once there has been sufficient scattering (*i.e.* once one is deep enough in the pile).

This exponential decay of $P(f)$ at large f (that is for $F > \langle F \rangle$) has been observed in the experiments [1,10] as shown in the data of Figure 2 and described by the fitting Equation 1. It has also been seen in the computer simulations of Radjai et al. [20–22], Luding [23] and Thornton [24] as well as in experiments on shear cells [25] and 2D arrays of rods [11]. The behaviour of $P(f)$ at small values of f, by contrast, is not well understood. As can be seen in Figure 2, one sees that $P(f)$ flattens out below $f \approx 1$ and approaches a constant value at low forces.

It is interesting to note the differences between the $P(f)$ distribution in granular matter and those found in other kinds of materials. If the material were a crystal without disorder, then the $P(f)$ distribution would be a sharply peaked function—essentially a delta-function broadened by thermal effects. There would not be any weight in $P(f)$ at forces near zero. In an amorphous material like a glass, one would find a curve for $P(f)$ that might look qualitatively like that in Figure 2 except with one important difference. Although, the $P(f)$ for the glass would have a significant weight near $f = 0$, since there are attractive forces between atoms in the glass, there would not be a discontinuity in $P(f)$ at that point; the function $P(f)$ would continue smoothly to negative forces. In the non-cohesive granular material the discontinuity in $P(f)$ at $f = 0$ is large and it is real, only smoothed out by the degree to which there are cohesive forces.

What is the effect of this discontinuity on the bulk properties of granular matter? It implies that any arbitrarily small perturbation can change the connectivity of the particles. Thus a small force on the surface will, somewhere in the pile, move a particle that was once touching another one, albeit with a tiny force, to a position where it no longer makes contact. Because the original force is indeed tiny one might expect that such rearrangements of the connectivity of the pile would be unimportant. However, there are very many particles which have these tiny forces. Indeed we can show quite generally [26] that for Hertzian contacts, $P(f)$ diverges at low f as $f^{-1/3}$. (Hertzian contact theory describes the interaction between spherical elastic particles under weak deformation, for which the force law between particles is $f(r) \propto (r - r_0)^{3/2}$ where r_0 is the point at which the particles just touch.) Such an argument is based on relating $P(f)\,df = G(r)\,dr$ where $G(r)$ is the radial distribution function of the material. As long as $G(r)$ is non-zero at the point r_0 of first contact, then the quantity dr/df will lead to a divergence in $P(f)$. Other force laws will produce different functional forms at small f.

It is interesting to note [26] that this type of argument can also be applied to an equilibrium liquid where it leads to an essentially exponential decay for $P(f)$ at large f for the case where the liquid has a steep repulsive inter-particle potential $V(r)$. At very small distances, where $g(r)$ is small, one can make the approximation $g(r) \propto \exp(-V(r)/k_BT)$. Here $g(r)$, the pair distribution function, obeys $g(r) = G(r)/4\pi r^2$ This type of argument thus produces a similar result as does the q-model at large forces. It has the advantage that it is an exact result for equilibrium liquids in the asymptotically small $g(r)$ regime.

3 Experiments on sound propagation

In the last Section, we saw that a granular material is inhomogeneous in that force chains extend over many particle diameters. The normal forces between particles are distributed over a wide range, with their number falling off exponentially at large values and possibly diverging at small forces. Sound, the elementary excitation of this unconsolidated (*i.e.* loosely packed) granular material, must propagate on this network of forces. It might be expected that it is particularly sensitive to the nature of the contact forces between particles and to the way that these forces vary within the medium. As I will show in this Section, this expectation is in fact the case and there are many unexpected and striking features of the sound propagation due to the disordered and nonlinear nature of the contacts and the tenuous network that they form.

Depending on the wavelength of the sound wave, different effects are found. At wavelengths, λ, much greater than the length of, or the spacing between, force chains, we expect to be able to treat the medium as homogeneous. At much higher frequency, λ decreases and eventually becomes comparable to the grain diameter, d. In this regime, the sound propagation is more sensitive to the geometry of the contacts. Both regimes have distinctive behaviour.

Low frequency sound will not propagate in the horizontal direction in an unconsolidated granular material [5]. This is basically a very pronounced mirage effect due to the fact that sound propagates at different velocities, c, at different depths in the material. This, in turn, is due to the very non-linear forces between particles. When two spherical particles are held together by the tiniest of external forces, they will touch at just one point and the force constant between them will be negligibly small. When the external pressure is increased the particles distort so that the area of contact increases (to a plane instead of just a point) and the force constant likewise increases. The greater the pressure pushing the particles together, the larger will be the forces and thus also the velocity of sound. Using Hertzian contact theory, one finds that the velocity depends on the pressure [27,28] as $c \propto P^{1/6}$. (Experimental data [28] has shown that the exponent is not precisely 1/6. The argument underlying the mirage effect does not depend on the exact value of the exponent of P as long as it is positive.) In the granular material the pressure will vary with depth. If one is far from any confining walls, the average pressure will increase linearly with depth so that $c \propto h^{1/6}$. (If one is near a wall the pressure will have a more complicated dependence; it will saturate at some value depending on what fraction of the weight of the material is held up by frictional interactions with the wall.) A sound wave, initially propagating in the horizontal direction, will travel at different rates along its wavefront, faster at the bottom than at the top: the bottom will get ahead of the top and the wavefront will be tilted. The wave (moving perpendicularly to the wavefront) has thus gained a component of velocity along the vertical direction. This wave will reach the top surface at which point it is moving completely in the vertical direction. Thus at low frequencies horizontal sound propagation is not allowed in unconsolidated granular material.

In the other extreme, when the wavelength λ is comparable to the grain diameter, even more remarkable behaviour is observed. In this regime one can no longer consider the medium as being homogeneous; the geometry of the particle packing and the force chain network will determine the exact path of the sound. Strong interference effects, as seen for example in the universal conductance fluctuations in sub-micron samples of a conductor at low temperatures, are to be expected. In one study of sound propagation in a box filled with glass beads [5], it was found that the vibrations caused by the sound itself was enough to disturb the placement of the particles. This slight shift in the particle placement, in turn changed the transmission amplitude of the sound measured at the detector giving rise to an enormous amount of noise. With a monochromatic source, run at constant frequency and amplitude, this noise in the measured transmission had a power spectrum: $S(f) \propto f^{-2}$ which persisted over 5 decades in frequency f.

One can see the intricate interference effects if one sweeps the frequency of the source. As one varies the frequency one finds that the amplitude of the sound measured by the detector varies considerably and appears to be very 'noisy'. This is shown in Figure 3. However, one can see that these features in the transmission spectrum are not real noise since the measured response is completely reproducible if the measurement is repeated without disturbing the pile [5,29]. This is shown by the two curves (displaced from each other for clarity) in the upper panel. They are nearly identical in all their features. However, if the pile is disturbed even slightly, then the spectrum changes as shown in the lower panel of Figure 3. Although the overall trend in the data persists, all of the fine structure is now changed.

An even more dramatic effect demonstrating the sensitivity of sound transmission to these fragile force contacts and force networks has to do with the extreme sensitivity of the sound transmission to minute temperature changes in the system [30]. A tiny temperature drift of only 0.03K was observed to cause the measured signal at the detector to increase by a factor of three. To check that this effect was due to a change in temperature (and not, for example, to a change in humidity) a small resistor was placed inside the material. This resistor was made small and rounded so that it was approximately the twice the size of any single bead (d = 5mm) making up the material. The configuration

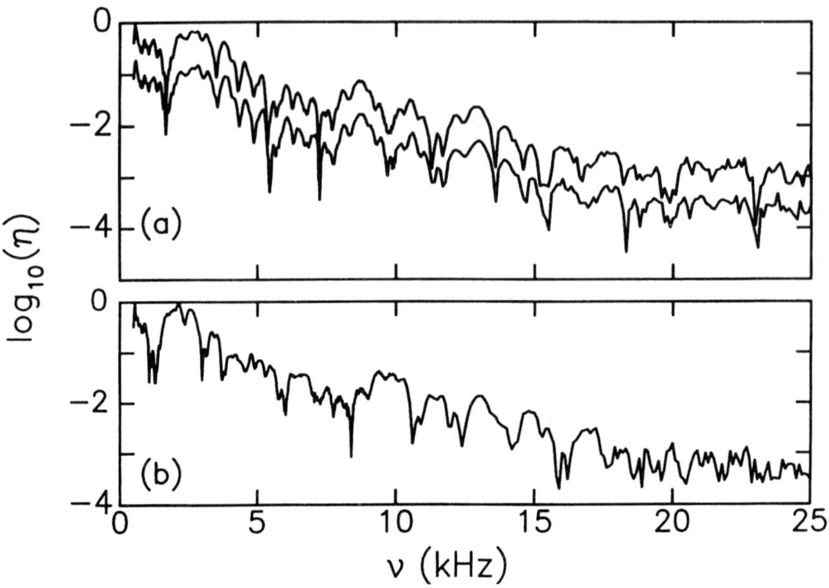

Figure 3. *The frequency response, η, at the detector. (a) The lower curve, displaced for clarity, was measured immediately after the upper one in order to show the reproducibility of the structure. (b) The same measurement as in (a) but with a slightly disturbed arrangement of the beads. Taken from Ref. [5].*

for this experiment is shown schematically in the inset to Figure 4. The heater, marked H in the diagram, was embedded into the pile between the source and the detector but offset to one side. By passing a small current pulse through the resistor lasting only 0.2 seconds, its temperature could be rapidly raised by less than a degree producing in turn a minute thermal expansion of only a few hundred nanometers in the heater. Prior to applying the pulse, a constant signal was measured by detector D_2. The perturbation produced an enormous effect: as shown in the main part of Figure 4, the transmitted sound intensity decreased by approximately 25%. After the rapid response, the signal slowly approached its initial value. To show the reproducibility of the effect, a second pulse was applied approximately 85 seconds after the first. A similar decrease in the signal was observed. The perturbation caused by the heat pulse is both very local and very small: the temperature rise affected only a single particle (out of the 105 or so particles in the region between the source and the detector) and the thermal expansion was almost 5 orders of magnitude smaller than either λ or d. This effect was interpreted [31] as being due to sound propagating predominantly along force chains within the medium. In places where there are no force chains, the effect on the signal of a small thermal expansion of the heater would be small. However, if the heater was placed on, or near, a force chain, its minute thermal expansion could shift the contacts along the chain and produce a large effect.

By embedding within the material an array of heaters, Liu [31] was able to investigate the spatial properties of the low-amplitude vibrations. He found that the disturbance of a single heater could be characterised by a time scale revealing the elapsed time of the

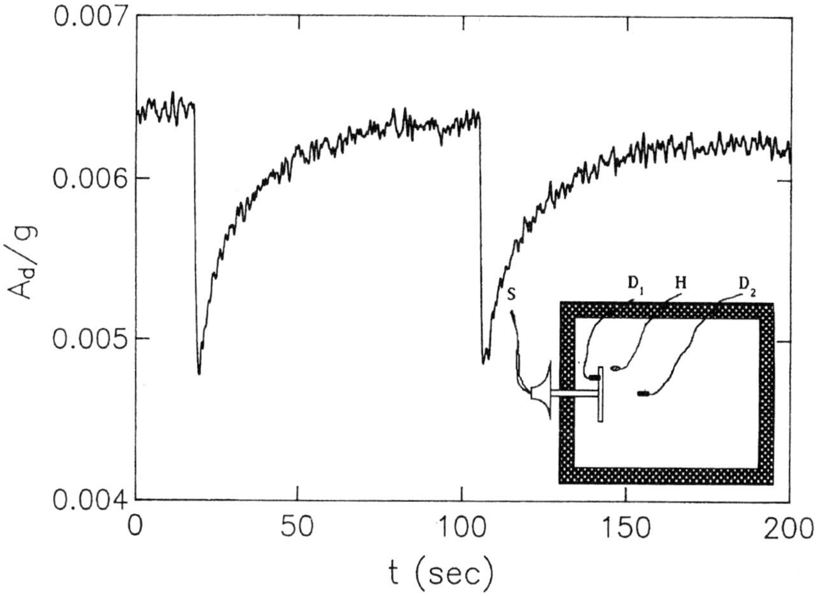

Figure 4. *Inset: a schematic view of the top of the apparatus used to demonstrate the effect of a small temperature change on the transmission of sound. The source, labelled S, was vibrated at fixed amplitude at a frequency of 4kHz. The heater, labelled H, was a distance of 1cm from the driving plate. One detector, labelled D_1, monitors the amplitude and frequency of the source ensuring its stability. The data in the main part of the figure shows the response at the detector, D_2, to two consecutive current pulses at an interval of 85 seconds apart. Taken from [30].*

signal travelling along a path via the heater. To show that the effect was in fact due to inhomogeneous forces within the material, he placed two heaters symmetrically with respect to the source and detector and showed that they produced very different effects. He also showed that the spatial pattern caused by heaters placed at different positions within the medium is very irregular, in that two adjacent heaters can give very different responses. The most natural interpretation of these experiments, as proposed by Liu, is that the sound predominantly travels along the force chains caused by the strong contact force inhomogeneities within the medium.

I now want to address the question of what is the velocity of sound in this medium. (I will consider only the high frequency regime since, as I argued above, at low frequencies there may be a large depth dependence to the velocity.) There are at least two different ways to determine this quantity: a measurement of the time of flight velocity and a measurement of the group velocity [29]. In the time-of-flight method one creates a pulse at the source and measures the time it takes for the first effect to be felt at a detector a given distance away. An experiment of this type yielded [29] the value $c_{\rm tof} \approx 280$m/sec at a depth of 6cm. (Since this value is nearly the velocity of sound in air, the measurement was checked by re-measuring the velocity both in an evacuated box and one filled with

Helium gas where the sound velocity is 970m/sec. The same value for the sound velocity was found in both cases.) The determination of the group velocity, c_g, was done by measuring the phase, ϕ of the detected signal at the detector as a function of frequency, ν. From the slope of $\phi(\nu)$ and the distance L_d between the source and detector one determines [29] $c_g = 2\pi L_d \, d\nu/d\phi \approx 57$m/sec, which is almost five times slower than that measured by the time-of-flight experiment.

The observed discrepancy between c_{tof} and c_g indicate complicated behaviour in the propagation of the sound; the sound must travel a tortuous path between the source and the detector determined by the exact structure of the force chains in the medium. This way of thinking about the sound propagation is different from the assumption that it propagates by diffusing through a random medium. (Such diffusing waves have been studied in a variety of contexts including electron scattering in mesoscopic metals [32] and light propagation in a densely scattering medium [33].) Sound in sand is different from these other cases because force chains span the distance between the source and detector. One might expect that, independent of frequency, all the sound waves must pass through the same set of force bearing contacts. That is indeed what the experiments of Liu [31] indicate is the case.

4 Experiments on granular compaction

One of the characteristic features of a granular material is that it can exist over a wide range of densities between close and loose packing. For spheres, this range is between approximately 0.64 and 0.55 [34]. Thus the density variation in a granular material can be well over ten percent even for spheres; for materials consisting of particles with less uniform shapes, the variation can be much larger. As one vibrates such a material the density will change. If it is initially loosely packed it will become denser. If it is initially tightly packed it becomes less dense. During normal handling, granular materials are subjected to a wide range of vibration intensity so that compaction occurs whether or not one wants a density change. Moreover, in industries which use sand in which to cast metal parts, compaction is used in order to create a dense uniform material in which to pour the molten metal.

In the absence of a useful temperature scale, external vibrations provide the energy that can be used to 'anneal' the system into a configuration closer to its ground state. Compaction occurs at different rates depending on the amplitude of vibrations, the shape of the container, the initial density of the material, and the friction with the walls. Studies by Knight *et al.* [35], have shown that the compaction of granular material due to vibrations occurs logarithmically slowly. In those experiments, a tube filled with monodisperse spheres was subjected to vibrations which consisted of individual 'taps' of a given intensity (acceleration and frequency). Even after 10^5 taps, densification of the material still occurred. Data from this experiment is shown in Figure 5. (In this data the acceleration of the individual taps was normalised by the acceleration of gravity g to $\Gamma = A\omega^2/g$.)

A variety of models have been proposed to account for this extremely slow settling [36,37]. One model that captures many of the features of the experimental results is based on the idea that the rate of increase in the granular volume fraction is exponentially reduced by excluded volume [38,39,40]. The model can be thought of as corresponding to

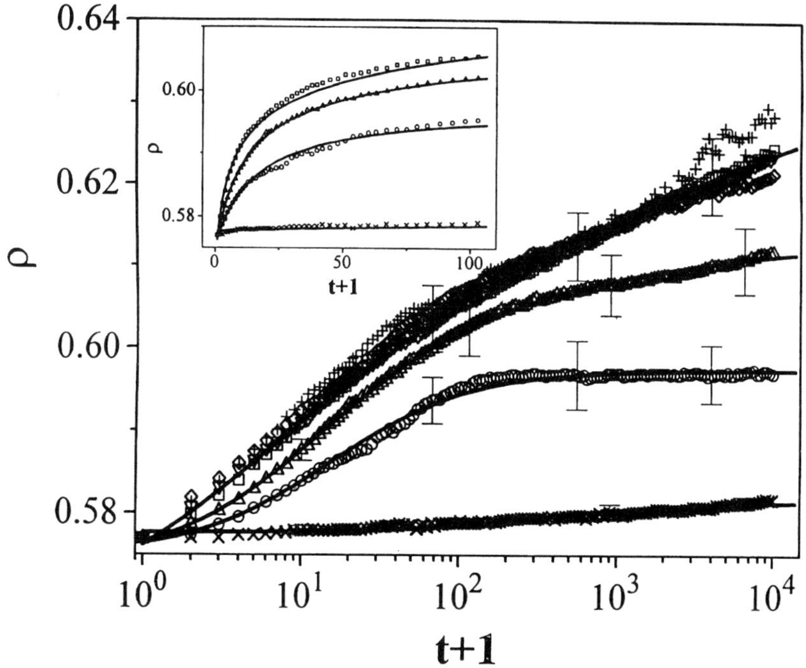

Figure 5. *The packing fraction ρ as a function of the logarithm of the tap number, t, for Γ = 0.75($\times$), 1.0($\circ$), 1.25($\triangle$), 1.5($\square$), 1.75(∇), 2.5($\diamond$) and 3.0(+). Each curve is an average of 5 separate experimental runs; the error bars show the rms variation between runs. The inset shows a subset of the data on a linear scale. Taken from [35].*

the situation where equal-sized cars (which correspond to the particles) are being parked in a parking lot without assigned slots. In the case where there is a high density of cars already parked, the insertion of an extra vehicle (or equivalently the insertion of an extra particle into the bead pack), requires the existence of an unlikely hole large enough to accommodate it. In a completely random situation, the average space depends only on the average density, ρ. The addition of an extra car or particle requires as the density increase an increasingly large number of previously parked cars (or resting particles) to move. This number increases as $\rho_0\rho/(\rho_0 - \rho)$ where ρ_0 is the maximum density (which is unity for a one-dimensional system). If all densification occurs by random movements of the particles, that is random 'parking' and 'unparking' events, then the time it takes to add each extra particle will be approximately exponential in $\rho_0\rho/(\rho_0 - \rho)$. As a result, the approach to the steady state density is logarithmic over a wide range of time. However, a detailed examination of this model [40] shows that at long times a final regime takes over.

As can be seen in Figure 5, the density of the granular material increases as the acceleration of the vibrations is increased. However, at very large accelerations the opposite trend occurs and the density starts to decrease as the acceleration is further increased. The vibration has two effects: it anneals away defects that are already in the material

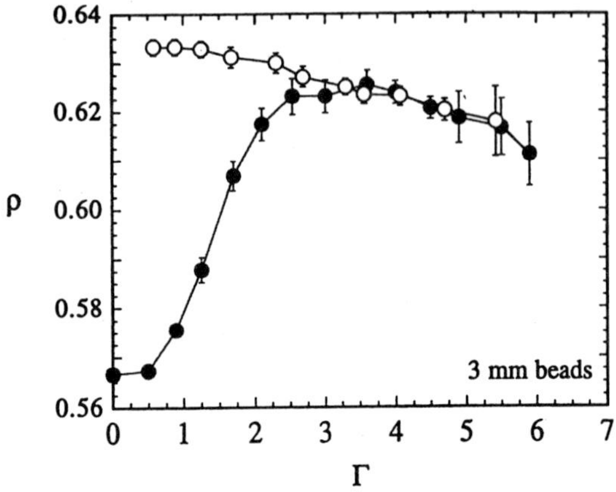

Figure 6. *The density as a function of the vibration history. The beads are prepared in a low density initial state. Γ is then slowly first increased (solid symbols) and then decreased (open symbols). The beads first compact irreversibly. Upon subsequently decreasing Γ the density increases still further. The branch with the higher density (open symbols) is reversible and upon subsequently raising Γ again, the density retraces the previous values obtained on lowering Γ. Taken from [41].*

and it can create new ones. Clearly if the vibrations are very violent, they will disrupt any order already in the pack and make it less dense. The vibrations can profitably be thought of as equivalent to a temperature in an ordinary solid. If the solid exists at low temperature with many defects, then raising the temperature slowly can anneal those defects away and any subsequent lowering of the temperature will produce a low temperature solid with considerably fewer defects than was in the original sample. The same type of phenomenon can be observed in the granular material as seen in Figure 6. When started in a fluffed-up, low-density state, the application of low-amplitude vibrations will slowly increase the density. As the vibration amplitude is increased the density increases and reaches a maximum at an acceleration several times that of gravity. Above this value the density starts to decrease with increasing intensity. Upon the subsequent slow lowering of the vibration intensity the density increases steadily all the way down to the very lowest acceleration values. Thus the material has been effectively annealed and the slow increase of density upon decreasing the acceleration is reminiscent of the behaviour of a liquid supercooled into a glass state where also the density is found to increase slowly as the temperature is lowered.

Experiments on granular materials indicate that even after the system has reached its steady state density, the density will fluctuate about its steady-state value in a manner reminiscent of how a thermal system fluctuates about its thermal equilibrium value [39]. In a thermal system, the fluctuation–dissipation theorem demonstrates that a measurement of fluctuations gives equivalent information as does a measurement of the response of the system to small perturbations.

This analogy between the density fluctuations in the granular material and those in a thermal system has been pursued. After measuring the density fluctuations over an extended interval, the power spectrum (the square of the Fourier transform of the density as a function of time) was computed [39]. This function gives the various frequency components in the relaxation of the system. At very low frequencies, f, the spectrum is flat and at very high frequencies it falls off as $1/f^2$. In between these two regimes there is a more complicated frequency dependence. Similar dependence was found in the simulations of the 'parking model' mentioned above [39,40].

Edwards and coworkers [42] proposed a theory of powders that replaces the ordinary thermodynamic temperature with an effective temperature for a static powder that they call the 'compactivity'. This is defined as being proportional to the derivative of the logarithm of the number of accessible states with respect to the volume (as distinct from the ordinary thermodynamic one in which the derivative is taken with respect to the energy). It is difficult to measure such a quantity since it is apparently impossible to measure the number of accessible states and even more difficult to determine its derivative with respect to volume. However, using the density fluctuations, it is possible to determine the compactivity, χ by making an analogy with ordinary thermodynamics [39]. Recall that the specific heat can be written in two forms:

$$c_v = \frac{dE_0}{dT} = \frac{\langle (E - E_0)^2 \rangle}{k_B T^2}$$

where $\langle (E - E_0)^2 \rangle$ is the fluctuation of the energy around its equilibrium value $E_0(T)$ and k_B is Boltzmann's constant. The analogous equation for the powder is:

$$\frac{dV_0}{d\chi} = \frac{\langle (V - V_0)^2 \rangle}{\chi^2}$$

where $\langle (V - V_0)^2 \rangle$ is the fluctuation of the volume around its steady-state value V_0, χ is compactivity, and the equivalent of Boltzmann's constant is set to unity. We thus have an equation that allows one to measure the compactivity difference between two configurations as a function of the density (or volume) fluctuations integrated along a path between them:

$$\int_{V_1}^{V_2} \frac{dV_0}{\langle (V - V_0)^2 \rangle} = \int_{\chi_1}^{\chi_2} \frac{d\chi}{\chi^2} = \frac{1}{\chi_1} - \frac{1}{\chi_2}.$$

Of course the usefulness of having a measure like temperature or compactivity is predicated upon having a notion of 'equilibrium' such as is embodied in the zeroth law of thermodynamics (which says that two bodies each in equilibrium with a third will be in equilibrium with each other). At present, although we have the ability to measure the compactivity of a granular material, we have no equivalent experimental demonstration that an equivalent of the zeroth law holds. Indeed the data [39] indicate that different regions of material which are in good contact with each other may not have the same variation of compactivity as the state of the system is varied.

5 Jamming and the relation to the glass transition

So far I have concentrated almost entirely on the properties of granular materials. However, there are certain features of these systems which are reminiscent of supercooled liquids. As one example of such a similarity, recall that the data shown in Figure 6 is similar to the annealing behaviour of a glass (see Bouchaud and Kob, this volume). As the temperature is raised close to the glass transition temperature, the glass will anneal into a more dense state. Upon subsequent lowering of the temperature the density will remain high and even increase slightly.

There are a number of other similarities between macroscopic, athermal systems and supercooled liquids and glasses. The ones that I am interested in discussing here are those that deal with how a material loses the ability to flow. In a macroscopic system such as a granular material one calls this process 'jamming'. It occurs because the material develops a yield stress; too small a stress will not be sufficient to force the material to flow. The jamming transition in this case occurs as the external stress is decreased below the yield stress value. An apparently different kind of transition occurs when a liquid is supercooled to the point where it becomes frozen in an amorphous state. As in the granular material, the liquid too develops a yield stress. At this 'glass transition' the viscosity has diverged, the diffusion has vanished, and the material appears in all ways to be solid.

At first glance it appears that there should be no relation between these different ways in which a material can lose the ability to flow. After all, in the case of jamming, the transition to a stuck state is clearly kinetic in origin and has to do with the application of external forces driving the system out of equilibrium. In the case of the supercooled liquid, on the other hand, the transition to the glass appears to be thermodynamic in origin. As the temperature is lowered, the liquid freezes as there is not sufficient thermal energy to allow the material to explore all of phase space. Is there a common conceptual framework with which to address these phenomena? This was the subject of a four month programme at the Institute of Theoretical Physics in Santa Barbara in 1997 which was summarised in Europhysics News [43].

In all out-of-equilibrium systems one question that naturally appears is whether or not there is a useful concept of temperature. In the last section I described one attempt by Edwards and coworkers to define a concept related to temperature for a static powder. Many others have also been proposed. Most notably are those which use the random motion of the grains to define an effective temperature. The problem also appears in a discussion of the glassy state. Since the glass has fallen out of equilibrium, the question arises to what extent the measured temperature can be used to describe the state of the system. It is clear that it is not sufficient and the concept of 'fictive temperature' has been introduced to describe some of the nonequilibrium thermal characteristics of the system [44]. The question has also been addressed of whether there exists for glassy systems (including structural glasses and spin glasses) a fluctuation–dissipation theorem (see Kob, this volume). Such a theorem relies on there being a temperature [45]. As discussed in the last Section, one may ask whether the fluctuations in a driven steady-state system that is far from equilibrium can be used to define a temperature, and whether there is a zeroth law of thermodynamics. There are many different ways of defining an effective 'temperature' for a system out of equilibrium. One must then ask whether or not they are consistent with each other.

One other feature that shows a resemblance between glasses and jammed systems is the appearance of heterogeneities near the transition between the fluid and the stuck state. The entire discussion in the early part of this review was devoted to the description of force chains in granular materials. These are the inhomogeneities that occur naturally as the system becomes jammed and stops moving. Cates, Wittmer, Bouchaud and Claudin

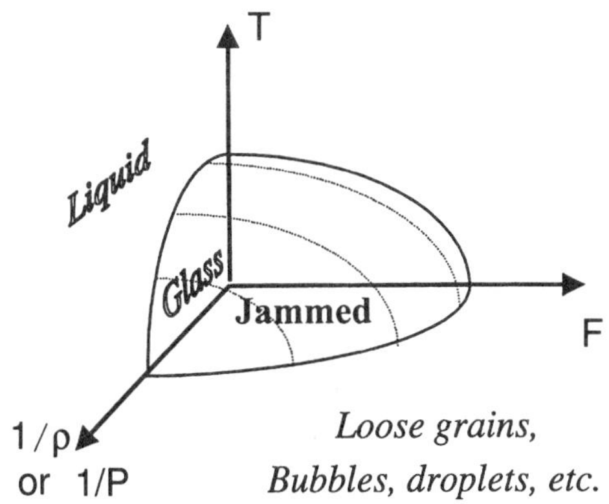

Figure 7. *A schematic phase diagram that attempts to unify the different ways in which the ability of a system to flow can be lost. Taken from [51].*

[46] have contended that jammed systems really belong to a different class of materials than do ordinary solids (see Cates, this volume). They call these systems 'fragile matter' because they are solid only so long as the applied stress is along the same direction as that which created the jam in the first place. If the applied stress varies from this direction then the jam will disappear and the material will flow.

Likewise there are many experiments that indicate supercooled liquids are very heterogeneous as they approach the glass transition. Several recent experiments [47,48] indicate that the time scales for translational diffusion in liquids decouple from those for rotational relaxation and viscosity. Simulations also have indicated that there are dynamic heterogeneities that occur as the glass transition is approached [49,50].

Given these similarities one might ask whether there can be any common underlying physics relating jamming and the glass transition? In what I have talked about so far I have considered jamming in cases where there are no attractive interactions between particles, so that the particles are held together by an externally applied stress. I have also implicitly assumed that the particles are so large that the thermal motion is totally irrelevant. Can one generalise the concept of jamming to systems where there are attractive forces and where thermal motion is relevant? Liu and Nagel [51] have attempted to do so by suggesting a schematic 'jamming phase diagram' that would unify jamming and glass formation. This diagram is sketched in Figure 7. The vertical axis is the temperature. As the temperature is lowered a liquid will become a glass. The axis coming out of the plane is the inverse density or inverse pressure axis. As the density or pressure increases, a system will become so constricted in its phase space that it can no longer move. The horizontal axis is the applied load. (This is the 'incompatible' load in the terminology of Cates *et al.* [46].) For a foam, this could be the shear stress. If this stress were too large the foam would flow. If it is too small, however, the foam will again become stuck or jammed. According to this picture, it is not possible for jamming to occur when the density is too low, the temperature too high, or the external load too great.

The glass transition is normally conceived of as occurring in the temperature/density plane, that is with the load set to zero. Likewise jamming is thought of as occurring in the load/density plane. By combining the three axes we see that these different phenomena may be related. It is important to realise that this diagram is schematic in the sense that the pressure and load axes really just represent two of the different components of the stress tensor. It is also probable that the phase diagram will not be as smooth as drawn. For example the jamming of a material such as corn-starch, which shear thickens, must have re-entrant regions.

This phase diagram suggests that the same types of constraints that lead to a system becoming jammed may also be important for how a liquid becomes frozen into a glass. One consequence of this way of looking at these problems is that it suggests a number of different experiments. It suggests that one should look at how thermal motion (due to Brownian motion, vibrations or ordinary temperature) will alter the ability of a system to become jammed. It also suggests that one might profitably look at how the application of a shear stress changes the glass transition temperature in a supercooled liquid.

The jamming phase diagram and the connection between jamming and the glass transition is tantalising. However, this connection is still not firmly established and much more work needs to be done to see if this way of looking at these phenomena is fruitful and productive. At present we are performing a number of simulations to test whether similar behaviour is observed close to the glass transition as is found near the jamming transition in athermal systems [26]. Likewise, experiments are being planned to test in colloidal systems whether the concept of the jamming phase diagram is useful [52].

Acknowledgments

I am grateful to all of my collaborators on the different topics that I have covered in this review. In particular almost all of the work on granular materials has been done in collaboration with Heinrich Jaeger. I am also indebted to Tom Witten for many discussion concerning the behaviour of these materials and to Susan Coppersmith for discussions of the q-model and the experiments on force distributions. Many students and post docs have worked on the subjects reported here. In particular, I thank Chu-heng Liu, James Knight and Ed Nowak for helping to initiate our work in the areas of compaction and sound propagation. David Schecter and Daniel Mueth did the early measurements on the force distributions. Finally, I am very grateful to Andrea Liu for all of our collaborative work on the subject of jamming. This work was supported by the NSF under NSF DMR-9722646 and the NSF MRSEC grant at the University of Chicago.

References

[1] D M Mueth, H. M Jaeger, and S R Nagel, *Phys Rev E* **57** 3164–3169 (1998).

[2] T Travers, M Ammi, D Bideau, A Gervois, J C Messager, and J P Troadec, *Europhys Lett* **4** 329–332 (1987).

[3] E Guyon, S Roux, A Hansen, D Bideau, J P Troadec, and H Crapo, *Rept Prog Phys* **53** 373–419 (1990).

[4] M Leibig, *Phys Rev E* **49** 1647 (1994).

[5] C-h Liu and S R Nagel, *Phys Rev Lett* **68** 2301–2304 (1992).

[6] S Melin, *Phys Rev E* **49** 2353–2361 (1994).

[7] A V Potapov and C S Campbell, *Phys Rev Lett* **77** 4760–4763 (1996).

[8] R S Sinkovits and S Sen, *Phys Rev Lett* **74** 2686–2689 (1995).

[9] M Ammi, D Bideau, and J P Troadec, *J Phys D: Appl Phys* **20** 424–428 (1987).

[10] C-h Liu, S R Nagel, D A Schecter, S N Coppersmith, S. Majumdar, O Narayan, T A Witten, *Science* **269** 513 (1995).

[11] G W Baxter, in *Powders and Grains 97*, edited by R P Behringer and J T Jenkins (Balkema, Rotterdam, 1997), pp 345–348.

[12] P Dantu, in *Proc 4th Int Conf On Soil Mech and Found Eng* (London, 1957), Vol 1, pp 144–148.

[13] P Dantu, *Ann Ponts Chauss* **IV** 193–202 (1967).

[14] T Wakabayashi, in *Proc. of the 9th Jpn Natl Cong Appl Mech* (Ueno Park, Tokyo, 1959), pp 133–140.

[15] T Travers, M Ammi, D Bideau, A Gervois, J C Messager, and J P Troadec, *J Phys France* **49** 939–948 (1988).

[16] F Delyon, D Dufresne, and Y-E Levy, *Ann Ponts Chauss* **53,54** 22–29 (1990).

[17] D Dufresne, F Delyon, and Y-E Levy, *J Sciences LPC* 209–213 (1994).

[18] D Howell, R P Behringer, and C T Veje, *Chaos* **9** 559–572 (1999).

[19] S N Coppersmith, C Liu, S Majumdar, O Narayan, and T A Witten, *Phys Rev E* **53** 4673–4685 (1996).

[20] F Radjai, M Jean, J J Moreau, and S Roux, *Phys Rev Lett* **77** 274–277 (1996).

[21] F Radjai, D E Wolf, S Roux, and J J Moreau, in *Powders and Grains 1997*, edited by R P Behringer and J T Jenkins (Balkema, Rotterdam, 1997), pp 211–214.

[22] F Radjai, D E Wolf, M Jean, and J J Moreau, *Phys Rev Lett* **90** 61–64 (1998).

[23] S Luding, *Phys Rev E* **55** 4720–4729 (1997).

[24] C Thornton, *KONA* **15** 81–90 (1997).

[25] B Miller, C O'Hern, and R P Behringer, *Phys Rev Lett* **77** 3110–3113 (1996).

[26] C S O'Hern, S A Langer, A J Liu, and S R Nagel (to be published).

[27] L D Landau and E M Lifshitz, *Theory of Elasticity* (Pergamon, NY, 1986) Ch 1.

[28] J Duffy and R D Mindlin, *J Appl Mech (ASME)* **24** 585 (1957).

[29] C-h Liu and S R Nagel, *Phys Rev B* **48** 15646–15650 (1993).

[30] C-h Liu and S R Nagel, *Journal of Physics: Condensed Matter* **6** A433–A436 (1994).

[31] C-h Liu, *Phys Rev B* **50** 782 (1994).

[32] S Washburn and R Webb, *Adv Phys* **35** 375 (1986); P A Lee and A D Stone, *Phys Rev Lett* **55** 1622 (1985); B L Al'tshuler, *Pis'ma Zh Ekspp Teor Fiz* **41** 530 (1985).

[33] D J Pine, D A Weitz, P M Chaikin and E Herbolzheimer, *Phys Rev Lett* **60** 1134 (1988); G Maret and P E Wolf, *Z Phys B* **65** 409 (1987).

[34] G Y Onoda and E G Liniger, *Phys Rev Lett* **64** 2727 (1990).

[35] J B Knight, C G Fandrich, C N Lau, H M Jaeger and S R Nagel, *Phys Rev E* **51** 3957 (1995).

[36] G C Barker and A Mehta, *Phys Rev E* **4** 184 (1993).

[37] D C Hong, S Yue, J K Rudra, M Y Choi, and Y W Kim, *Phys Rev E* **50** 4123 (1994).

[38] E Ben-Naim, J B Knight, E R Nowak, H M Jaeger, and S R Nagel, *Physica D* **123** 380 (1998).

[39] E R Nowak, J B Knight, E Ben-Naim, H M Jaeger, and S R Nagel, *Phys Rev E* **57** 1971 (1998).

[40] A J Kolan, E R Nowak, and A V Tkachenko, *Phys Rev E* **59** 3094 (1999).

[41] E R Nowak, J B Knight, M Povinelli, H M Jaeger, and S R Nagel, *Powder Tech* **94** 79 (1997).

[42] S F Edwards, in *Proceedings of the International School of Physics 'Enrico Fermi'*, Lerici (1988); S F Edwards and R B S Oakeshott, *Physica* **A157** 1080 (1989); A Mehta and S F Edwards, *Physica* **A157** 378 (1989).

[43] S R Nagel, *Europhys News* **29 #2** 58–59 (March/April 1998).

[44] M D Ediger, C A Angell, and S R Nagel, *J Phys Chem* **100** 13200 (1996).

[45] L.F Cugliandolo, J Kurchan and L Peliti, *Phys Rev E* **55** 3898 (1997).

[46] M E Cates, J P Wittmer, J-P Bouchaud and P Claudin, *Phys Rev Lett* **81** 1841 (1998).

[47] F Fujara, B Geil, H Sillescu, G Fleischer, *Z Phys B* **88** 195 (1992).

[48] C-Y Wang and M D Ediger, *J Phys Chem B* **103** 4177 (1996).

[49] R Yamamoto and A Onuki, *Phys Rev E* **58** 3515 (1998).

[50] D N Perera and P Harrowell, *Phys Rev Lett* **81** 120 (1998).

[51] A J Liu and S R Nagel, *Nature* **396** 21 (1998).

[52] D Weitz (private communication).

Stress transmission in jammed and granular matter

Michael Cates

University of Edinburgh, UK

1 Introduction

In what follows, I will review some recent work on the mechanical properties of materials whose structural integrity arises from a jamming process. The basic suggestion is that, because the solidity of such materials arises solely as a result of the applied load, their mechanical properties can be very different from those of simple (elastic) solids. Indeed, in elastic materials, the applied stress modifies but does not fundamentally alter the interparticle forces that are already present. In the jammed case, the network of contacts which comprises the load-bearing 'skeleton' of the material has formed adaptively under whatever stresses caused the system to jam. A linear response to incremental stresses is not guaranteed, and we will see that in some simple models, certain types of incremental loading cannot be supported without irreversible rearrangement of the contact network. I will go on to argue that these distinctive mechanical properties may be shared by dry granular materials (which can also be viewed as jammed in a certain sense). However, I should warn that this view remains controversial, for reasons which should become clearer below. A concise and balanced recent survey of the relevant literature is found in [1].

2 The jamming transition

Jamming is a familiar event. It is what happens when we attempt to force a system of interacting particles (or, in some cases, cars or people) to move faster than conditions will allow. For example, on a busy highway, jamming can occur when everyone is in a hurry to get past everyone else. Drivers cannot adapt to the resulting fluctuations in speed; they have to brake sharply, and everything stops. Paradoxically it is the wish of drivers to go faster that makes them slow down [2]. Something similar happens in a simple kitchen experiment. Take a packet of corn-starch powder, and add *just* enough water to make a fluid paste. At least, it will appear to be a fluid, but if you try to stir it rapidly with a spoon, it will stop moving altogether.

Why does this happen? It is related to the phenomenon of *dilatancy* in dry granular media: for grains to get past one another they require space, which is why flowing sand is less dense than well-settled static sand (Nagel, this volume). But for a colloidal suspension of corn-starch granules, the volume available is fixed; this is because the granules themselves, and the surrounding fluid, are effectively incompressible. Although in slow enough flows the particles are able to get past one another, fast flow becomes almost impossible. Actually, jamming is more than just an inability to flow fast: for while a modest shear stress does make the material flow, a large one is not just ineffective, but directly counterproductive. Large stresses inhibit, rather than assist, the rearrangement of particles, which stop moving almost completely.

It remains an open question whether the kind of jamming seen in colloids can, under some conditions, be viewed as a phase transition, and whether the new state is really solid-like or just very sluggish (and perhaps erratic) in its flow; see, *e.g.* [3]. In fact, not much at all is known about the jamming transition. One way forward is to study driven one-dimensional models (see Mukamel, this volume). In contrast to equilibrium problems, such models can exhibit very rich behaviour, including true phase transitions; so possibly the physics of the jamming process is accessible in one dimension. Indeed, various jamming transitions are found in 1-D traffic models [2]. However, there is no obvious way to introduce *shear flow* in the one-dimensional context (although traffic models with several lanes can approximate it [4]). So, to understand jamming that results from shear stresses, one must resort to either large-scale simulation, or phenomenological models.

Simulations on three-dimensional hard sphere colloids [5] suggest an appealing physical picture of jamming. As the material is sheared, chains of particles in close contact build up along the compressional axis. (In a simple shear flow, $v_x = \dot{\gamma} y$, this is in the $(-1, 1)$ direction, at 45 degrees to the flow; see Pine, this volume.) These roughly linear clusters join up and span the system, creating 'force chains' that allow stresses to be transmitted from one side of the sample to the other (Figure 1a). At this point, the simulation fails for numerical reasons, but one hypothesis is that this now represents a jammed state, which could in principle support a macroscopic load. Note, however, that the simulation [5] is done without Brownian motion. It is possible that a small amount of Brownian motion would give the jammed state a finite lifetime. But equally, it might not do so: the jammed state could instead be a kind of anisotropic glass, in which local Brownian motion is so constrained that bulk flow is completely suppressed (Nagel, this volume). Such a state could, like other glasses, support stresses in static equilibrium; it could also show interesting aging behaviour (see Bouchaud, this volume).

The issue of how glasses form under large shear stresses, remains a completely open avenue to both theory and experiment. It certainly seems that colloidal glasses are a good experimental candidate for such studies since, due to their small elastic modulus, it is relatively easy to apply and maintain the required stresses. As suggested by Chaikin during this Summer School even the force of gravity could be enough to jam a colloid—which might explain why the colloidal glass transition appears be very different in microgravity from on earth (Chaikin, this volume).

(a)

(b)

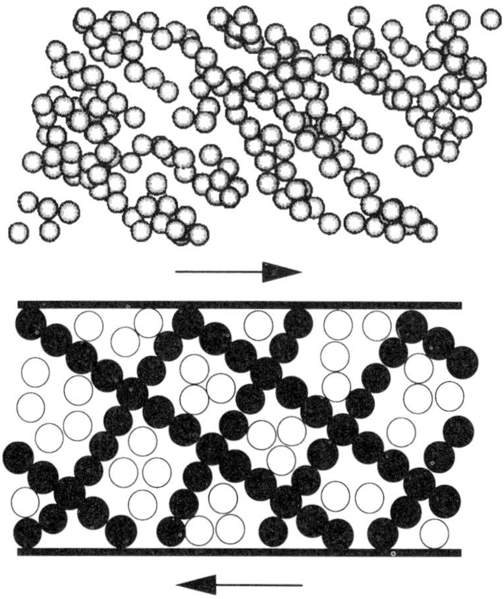

Figure 1. *(a) Simulation (courtesy J. Melrose) of a dense colloidal suspension at volume fraction* $\phi = 0.54$*, strained to* $\gamma = 0.22$*. Shown in the figure are* only *those spheres which have come into close contact (within* 10^{-5} *diameters) with at least one neighbour. The contact network is strongly anisotropic and suggests the onset of 'force chains' running top left to bottom right. (b) Idealised rectangular network of force chains. Black: primary force chains; grey: secondary force chains. (The latter are optional, and absent in the simplest model, Equation 1.) Directors* $\mathbf{n}, \mathbf{m}$ *are shown.*

3 A simple model of jamming in colloids

Consider a concentrated colloidal suspension of hard particles, confined between parallel plates at fixed separation, to which a shear stress is applied. To model the jammed state, we start from a grossly simplified idealisation of a force chain: a linear string of at least three rigid particles in point contact. Crucially, this chain can only support forces *along its own axis*: successive contacts must be co-linear, with the forces along the line of contacts, to prevent torques on particles within the chain. Note that neither friction at the contacts, nor deviations from a spherical shape, can change this 'longitudinal force' rule. Finite particle deformability, however, can allow small transverse forces to arise; this is discussed in detail elsewhere [6]. (It causes the models outlined below to cross over to more conventional, 'elastoplastic' behaviour under large enough loads.) From now on, we consider rigid particles.

As a minimal model of the jammed colloid, we take a parallel array of such force chains, characterised by a unique 'director', which is a unit vector along the chains, $\mathbf{n}$. The chains are surrounded by a sea of 'spectator' particles, and incompressible solvent. This is obviously oversimplified, for we ignore completely any contacts between chains;

the deflections caused by weak interactions with the spectator particles; and the fact that there must be some spread in the orientation of the chains themselves [5]. However, with these assumptions, we obtain a tractable and interesting alternative to conventional elasticity theory. For in static equilibrium, and with no body forces acting (no gravity) the stress tensor σ_{ij} must obey

$$-\sigma_{ij} = P\delta_{ij} + \Lambda\, n_i n_j \tag{1}$$

or equivalently, in the notation of McLeish (this volume) $-\underline{\underline{\boldsymbol{\sigma}}} = P\underline{\underline{\mathbf{I}}} + \Lambda\mathbf{nn}$.

Here P is an isotropic fluid pressure, and Λ (which is positive) is a compressive stress carried by the force chains. The form of the 'deviatoric' stress, $\Lambda\, n_i n_j$, reflects the fact that the force chains can only exert forces in the $\mathbf{n}$ direction. The force density across an arbitrary plane of normal $\mathbf{u}$ is then $\Lambda\,(\mathbf{n}\cdot\mathbf{u})\,\mathbf{n}$, which points along $\mathbf{n}$, as required; the $\mathbf{n}\cdot\mathbf{u}$ factor is proportional to the number of force chains per unit area cutting the plane. (Note that, for a general $\mathbf{u}$, the deviatoric stress gives rise to a *shearing* force.) In static equilibrium, for hard repulsive potentials, the forces along $\mathbf{n}$ can only be compressive, not tensile, which is reflected in our requirement of positive Λ. Were the forces to be tensile instead, the form of the deviatoric stress would closely resemble that of a parallel array of polymeric 'springs' of fixed orientation $\mathbf{n}$ (see McLeish, this volume).

Equation 1 defines a material that is mechanically very unusual. It permits static equilibrium only so long as the applied stress is either isotropic, or is a uniaxial compression along $\mathbf{n}$. So long as this remains true, incremental loads (an increase or decrease in compressive stresses at fixed compression axis) can be accommodated reversibly, by what is, at the particle contact scale, essentially an elastic mechanism. But the material is certainly not an ordinary elastic body, for if instead one tries to compress the sample in some slightly different direction (causing a rotation of the principal stress axes) static equilibrium cannot be maintained without changing the director $\mathbf{n}$. Remember that $\mathbf{n}$ describes the orientation of a set of force chains that pick their ways through a dense sea of spectator particles. Accordingly $\mathbf{n}$ cannot simply rotate in response to the new stress; instead, the existing force chains must be abandoned and new ones created with a slightly different orientation. This entails dissipative, plastic, reorganisation, as the particles start to move but then re-jam in a configuration that can support the new load. The contact network is, in the process, completely rebuilt.

4 Fragile matter

4.1 A tentative definition

Our model of a jammed colloid offers an idealised example of 'fragile matter': it can statically support applied shear stresses (within some range), but only by virtue of a self-organised internal structure, whose mechanical properties have evolved directly in response to the load itself. Its incremental response can be elastic only to *compatible* loads; *incompatible* loads such as those of a different compression axis, even if small, will cause finite plastic reorganisation. The inability to elastically support *some* infinitesimal loads has been proposed [6] as a useful definition of the term 'fragile', which previously had not been given a clear technical meaning in the current context.

Certainly, it seems useful to have distinct meanings for the words 'fragile' and 'soft': the latter usually means low shear modulus (McLeish, this volume) whereas a jammed system need not be soft in this sense at all. Indeed a jammed colloid might in principle support a large load with very little strain, as might a dry granular medium. (If you doubt this, push hard on the spoon in the corn-starch solution, or stand on a bucket filled with sand.) On the other hand, the word 'fragile' already has a somewhat different, and well-established, meaning in the context of glasses (see Kob, this volume). So far, there is no apparent connection (and hopefully not too much confusion!) between these two uses of the word. But if, as suggested by Nagel and Liu [7] jammed materials are themselves glasses in some sense, perhaps there is some deeper link between the two concepts of fragility that remains to be found.

4.2 Does jamming cause fragility?

It has been argued [6] that jamming may lead *generically* to mechanical fragility, at least in systems with overdamped internal dynamics. Any such system is likely to arrest as soon as it can support the external load; since the load is only just supported, one expects the state to be only marginally stable. Any incompatible perturbations then force rearrangement; this will leave the system in a newly jammed but—by the same argument—equally fragile state.

This speculative scenario is related, but not identical, to several others in the literature [8–14], including the emergence of rigidity by successive buckling of force chains [9]; the concept of self-organised criticality (SOC) [10] (see also [15]); and also ideas of mechanical percolation which underly 'hypoplastic' models of granular matter in the recent soil-mechanics literature [11]. Fragility is also connected with recent ideas concerning 'isostatic' packings and marginal mechanics in (frictionless) spheres [12, 13, 14]; for a further discussion of these links, see [6].

4.3 Two types of fragility

Consider again our idealised jammed colloid (Figure 1b). So far we allowed for an external stress field (imposed at the plates) but no body forces, such as gravity. What body forces can it now support *without* plastic rotation of the director?

Various models are possible. One possibility is to assume that Equation 1 continues to apply, with $P(\mathbf{r})$ and $\Lambda(\mathbf{r})$ now varying in space. If so, there is no static solution of Equation 1 for a *general* body force: only for one acting along $\mathbf{n}$. This should not be too shocking; a *general* body force likewise cannot be supported in a simple fluid, where the constitutive equation for hydrostatic equilibrium is just the same as Equation 1, but with $\Lambda = 0$. Of course, the special case of a uniform force density, which corresponds to gravity, can be supported in a fluid, but not a localised force. Equation 1 is the constitutive equation of a medium that can support a localised force, but only so long as it points along $\mathbf{n}$. Here and below, the term 'constitutive equation' means any recipe for finding the stress in a material from its flow history and/or its present structure (McLeish, this volume).

On the other hand, a general body force can be supported, in three dimensions, if there are several different orientations of force chain, possibly forming a connected network [16, 11, 17, 18]. A minimal model for this type of fragility is:

$$-\sigma_{ij} = \Lambda_1 \, n_i n_j + \Lambda_2 \, m_i m_j + \Lambda_3 \, l_i l_j \tag{2}$$

with $\mathbf{n}, \mathbf{m}, \mathbf{l}$ directors along three nonparallel populations of force chains; the Λ's are compressive pressures acting along these. Body forces, if present, then cause $\Lambda_{1,2,3}$ to vary in space, but with no need for reorganisation. Note that Equation 2 is written down as if the three force chain assemblies were in a vacuum, as might be relevant in a dry granular assembly (see below). If there is also a pressure field from a suspending fluid, we need only two directors to support an arbitrary body force.

We can thus distinguish two levels of fragility, according to whether incompatible loads include localised body forces (*bulk* fragility, exemplified by Equation 1), or are limited to forces acting at the boundary (*boundary* fragility, exemplified by Equation 2). It is important to note that the boundary forces always include stresses applied by a rheometer to (say) a jamming colloid. For even though we are used to thinking of such 'bulk stresses' as acting directly on fluid elements, they ultimately originate at the boundary and are quite distinct from 'body forces', in the sense used here.

4.4 Micro-fragility

More generally, we should also distinguish between *macro-fragile* responses involving changes in the *mean* orientation of force chains, as described by the simple models above, and the *micro-fragile* responses of individual contacts under tiny changes in loading. Dry granular matter, for example, certainly seems to be micro-fragile. For example, as shown by careful sound transmission measurements [19], exceedingly small incremental loads can and do cause reconstructions of the contact network, even quite far from the loading point: see Nagel, this volume, for a detailed discussion. For an interesting theoretical model of micro-fragility, see [15].

However, micro-fragility does not guarantee macro-fragility of the type implied by Equation 2 for example; one could imagine that small reorganisations constantly occur, but that the macroscopic properties of the medium are not much affected by these. (All loads would then be macroscopically 'compatible', in the sense that the *mean* properties of the contact network would not have to change to support them, though there might always be some local reconstruction going on.) Nevertheless, the observation of micro-fragility in granular media is strongly suggestive that the ideas of conventional elasticity are not applicable to such materials—at least, not without extremely careful experimental justification.

4.5 Fixed principal axis (FPA) model

Returning to our simple 'boundary-fragile' model, Equation 2, the chosen values of the three directors (two in 2D) clearly should depend on how the system came to be jammed. (We call this the 'construction history' of the jammed state.)

Suppose our colloid, for example, has jammed at time t_0 in response to a stress given by $\underline{\underline{\sigma}}(t) = \lambda(t)\underline{\underline{\sigma}}(t_0)$ where $\lambda(t_0) = 1$ and $\lambda(-\infty) = 0$, with $\lambda(t)$ slowly varying. In other words, a fixed pattern of applied stress has very gradually been increased in amplitude, up to the onset of jamming, and held there. In this case, all tensorial information about the construction history resides in the stress tensor itself. Then, if one director points along the major compression axis (which seems plausible: this is where the strong force chains are likely to point) then by symmetry any others should lie at right angles to it. Applying a similar argument to the intermediate axis leads to the ansatz that all three directors lie along principal stress axes. This ansatz of perpendicular directors gives perhaps the simplest possible fragile constitutive equation in 3D. Note that in this case the effect of a fluid pressure can trivially be absorbed into the Λ's; it doesn't then matter whether there are two directors, or three.

With the ansatz just made, Equation 2 becomes a 'fixed principle axes' (FPA) model [20, 6]: the construction history fixes the orientation of the stress tensor that can be supported by the jammed structure. However the compressions $\Lambda_{1,2,3}$ are free to vary; these now determine the eigenvalues (principal stresses) of the stress tensor. We shall assume that this variation has to lie within a certain range — there is a limit to how anisotropic the stress tensor can become. Both in two-dimensional and in uniaxial problems, a maximum anisotropy of the stress tensor translates into the Coulomb condition [21]

$$\frac{|\Lambda_1 - \Lambda_2|}{\Lambda_1 + \Lambda_2} \leq \sin\phi \qquad (3)$$

where ϕ is called the internal friction angle. (In fact, within simple models of dry granular materials, ϕ can be identified with the 'angle of repose', that is, the slope of a free-standing, poured pile of sand [21].)

Although grossly oversimplified, the FPA ansatz leads to some nontrivial predictions for the jammed state in the colloidal problem, such as a constant ratio of the shear and normal stresses when these are varied within the jammed regime. Such constancy is indeed reported by Laun [3] in "the regime of strong shear thickening"; see [6].

5 Granular matter viewed as a jammed state

We now return from jammed colloids to granular materials. Although the formation of dry granular assemblies under gravity is not traditionally described in terms of jamming, it is a closely related process. Indeed, the filling of silos and the motion of a piston in a cylinder of grains both exhibit jamming and stick-slip phenomena associated with force chains; see [22]. And, just as in a jammed colloid, the mechanical integrity of a sandpile disappears as soon as the load (in this case gravity) is removed. It could be misleading to think of gravity, in such a pile, as a perturbation applied to a pre-existing body having well-defined continuum-mechanical properties; yet this is the basis of most existing calculations of how stresses propagate through the material [23].

Traditionally the continuum mechanics of granular media is described by an 'elasto-plastic' constitutive equation. The simplest of these assumes a linear relation (Hooke's law) between stress and strain, except in regions where the inequality in (3) is saturated. Even simpler is to assume that Equation 3 is saturated everywhere; this is the 'rigid-plastic' constitutive equation [21]. Interestingly, the latter represents a fragile model: any incremental load that increases the ratio on the left hand side of (3) is incompatible.

(a)

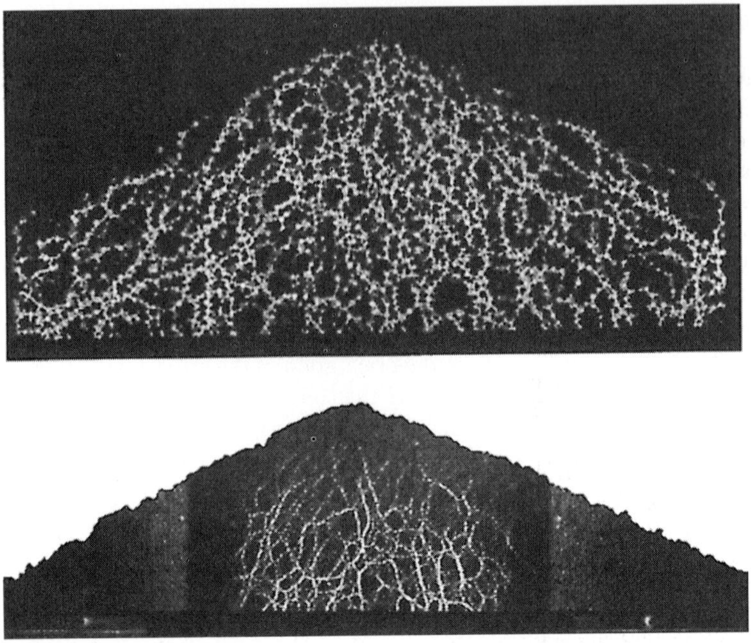

(b)

Figure 2. *(a) Simulation (courtesy J. Wittmer) of a granular pile in two dimensions. The simulation is done on convex elastic particles (whose 'hooked' shape allows macroscopic granular friction, even though they are microscopically smooth). Each particle comprises four small discs glued around the edges of one large one; see [24]. The scale of brightness denotes particle compression (normalised by the mean value at each height). (b) Experiment (courtesy L. Vanel) [25] on a granular packing of photoelastic discs between crossed polars: discs subject to strong compression show up brightly.*

However, it takes no account of the history of the material, nor of the presence of force chains. As discussed by Nagel (this volume) it is clear that force chains are a major factor in how stresses propagate in granular media, both through experiment and simulation [16, 18, 17]; see also Figure 2. So, something beyond the rigid-plastic model is likely to be required.

5.1 Role of construction history

In the context of a sandpile, Equation 2 is interpreted to mean that a fragile granular skeleton of force chains is laid down at the time when particles are first buried at the free surface; so long as subsequent loadings are compatible with this structure, the mean properties of the skeleton will remain intact. If, in addition, the skeleton is rectilinear (perpendicular directors) this forces the principal axes to maintain forever the orientation they had close to the free surface (FPA model). However, we do not insist on this last property and other models, based on an oblique family of directors in Equation 2, have also been developed [6]. It turns out that the FPA model already accounts quite well for the forces measured experimentally beneath conical piles of sand, constructed by pouring cohesionless grains from a point source onto a rough rigid support [20, 26]. A notable feature is the 'dip': the vertical normal stress on the base of such a pile has a local minimum, not a maximum, directly beneath the highest point. Within the FPA model one finds that the primary force chains, lying along the major principal axis of the stress tensor, bisect the free (upper) surface of the pile and the vertical direction. These chains transport the gravitational load outward, like an arch [8] so that the maximum downward pressure lies on a ring, rather than under the apex. One consequence of this modelling approach is that one expects a very strong dependence of the stress propagation on the construction history. For example, if one builds a sandpile, not by pouring grains from a point source, but by sieving a series of concentric discs one on top of another (to make a cone), then the local granular skeleton will be quite different. Within the model of Equation 2, there turns out to be only one choice of parameters that can describe this situation, in which the material properties of the medium should have local isotropy about any vertical axis. (The sieving process imparts no local information about the global conical geometry; in contrast, for a pile poured from a point source, every grain comes to rest on an inclined surface which betrays the overall shape of the growing pile.) This maximally symmetric model does not give a dip, but instead a smooth maximum of the stress beneath the apex of the pile. The absence of a dip is because the local isotropy requires that the weight of a grain is carried toward and away from the central axis in equal measure, rather than preferentially outward as in the FPA case.

Thus, an unambiguous prediction of our work is that, for a sieved pile, the dip should disappear. Recently it was confirmed by experiment that this is indeed what happens [25]. This experiment rules out several alternative explanations of the dip, such as attributing it to a slight deflection of the supporting surface beneath the pile [23].

5.2 Fragile or elastoplastic?

As mentioned previously, our fragile constitutive model (2) shows very different mechanics from conventional forms of elastoplasticity. For example, in 2D, when combined with the equation of stress continuity ($\partial_i \sigma_{ij} = \rho g_j$ for sand under gravity), it gives differential equations for the stress tensor that are everywhere hyperbolic [6]. With a zero-force boundary condition at the upper surface of a pile, this gives a well-posed problem: the forces acting at the base follow uniquely from the body forces by integration.

If *different* forces are now imposed at the base, rearrangement is inevitable, as first recognised by Evesque (private communication 1996). This is boundary-fragile behaviour. The same does not hold within a traditional elastoplastic modelling approach [23] whose equations are elliptic in elastic zones. In such models of a sandpile, the forces acting at the base cannot be found without specifying a displacement field there [6]. To define this displacement, one would normally invoke as a *reference state* the one in which the load (gravity) is removed. For cohesionless poured sand, this state is undefined, just as it is for a jammed colloid which, in the absence of the applied shear stress, is a fluid. For a somewhat different viewpoint, however, see [27].

One route to an elastic reference state is to consider a hypothetical sandpile where each grain becomes firmly 'glued' to its neighbours (or to the base), upon first coming to rest. The resulting medium is surely elastic, and must therefore be governed by elliptic equations. This does not mean that it is a conventional homogeneous elastic continuum (for which the states of zero strain and zero stress coincide). Indeed, a glued pile built under gravity will certainly have non-zero stresses if gravity is later removed [6]. More importantly, for a typical disordered packing of near-rigid, glued grains, there will arise many *tensile contacts* even under a purely compressive external load. Thus the problem of glued and unglued piles might, in practice, be quite dissimilar. This question still requires careful investigation by both computational and laboratory experiments.

On the other hand, the dichotomy between fragile and elastoplastic models may be a false one, at least if we relax our assumption of perfectly rigid grains. This is addressed in [6] where it is shown that, for some fragile models, there is a smooth crossover to more conventional elastoplasticity when grains are deformable, with fragility emerging as the limiting behaviour for rigid particles. However, the elastoplastic models that result from this involve strong anisotropy. Since any elastoplastic model that involves the construction history is also likely to be strongly anisotropic [23], there may be a less sharp distinction between the two approaches than previously thought.

5.3 Outlook

In summary, the task of constructing a continuum description for stress transmission in cohesionless granular media is an unexpectedly tricky one, because (a) grains are typically very rigid; (b) there can be no tensile forces between any grains in the entire system (a nonlinear constraint); (c) the contact network is history dependent. Thus, for granular media (as emphasised also by Nagel, this volume) all states are metastable: there is no ansatz of 'molecular chaos' (or thermal equilibrium) that allows the static properties to be decoupled from the construction history of the material. In other words, for granular matter, even the simplest *static* properties are the direct consequence of nonequilibrium dynamics, metastability, and flow.

For packings of hard cohesionless particles, while it is clear that macro-fragile models like Equation 2 are oversimplified, they may lie closer to the truth than any of the alternatives now available. Not only do these fragile models account for the observed role of construction history in the presence of a stress dip for a poured pile [20] and correctly predict its absence for a sieved one [25]; they also account quantitatively for some recent high-quality measurements on small silos [28].

Obviously, it may be possible to explain these results within more conventional (elastic or elastoplastic) modelling frameworks. However, this presumably requires an explicit treatment of (history dependent) anisotropy. Given that our constitutive equation (2) corresponds to the strong anisotropy limit of an elastoplastic model [6], it is possible that the dichotomy between elastic and fragile approaches will dissolve as both theories are refined, in response to improved laboratory data. Meanwhile, our models provide an interestingly direct link between continuum constitutive equations and the observed microphysics of stress propagation in granular matter. This involves a force chain network that arises adaptively in response to the loadings applied. These models also suggest a generic link with the mechanics of other jammed materials, such as colloids.

Acknowledgments

Much of the work described in this article was done in collaboration with J. Wittmer, J.-P. Bouchaud and P. Claudin. Constructive interactions with E. Clement, S. Edwards, P. Evesque, P.-G. de Gennes, G. Gudehus, J. Jenkins, D. Levine, J. Melrose, S. Nagel, C. Thornton and T. Witten are also acknowledged.

References

[1] E Clement, *Current Opinion in Colloid and Interface Sci* **4** 295 (1999).

[2] For a survey of recent works, see D E Wolf, M Schreckenberg and A Bachem, Eds., *Traffic and Granular Flow* World Scientific, Singapore 1996.

[3] H M Laun, *J Non-Newtonian Fluid Mech* **54** 87 (1994).

[4] See *e.g.* O J O'Loan, M R Evans and M E Cates, *Physica A* **258** 109 (1998).

[5] R S Farr, J R Melrose and R C Ball, *Phys Rev E* **55** 7203 (1997);
R C Ball and J R Melrose, *Adv Colloid Interface Sci* **59** 19 (1995);
J R Melrose and R C Ball, *Europhys Lett* **32** 535 (1995).

[6] M E Cates, J P Wittmer, J-P Bouchaud and P. Claudin, *Phys Rev Lett* **81** 1841 (1998); *Phil Trans Roy Soc Lond Ser A* **356** (1998); *Physica A* **263** 354 (1999).

[7] A J Liu and S R Nagel, *Nature* **396** 21 (1998).

[8] S F Edwards and C C Mounfield, *Physica A* **226** 1, 12, 25 (1996);
S F Edwards, *Physica A* **249** 226 (1998).

[9] S Alexander, *Phys Repts* **296** 65 (1998).

[10] P Bak, *How Nature Works; The Science of Self-Organised Criticality* (Oxford University Press, Oxford 1997).

[11] D Kolymbas *Archive of Appl Mech* **61** 143 (1991).

[12] S F Edwards and D V Grinev, *Phys Rev Lett* **82** 5397 (1999).

[13] C Moukarzel, *Phys Rev Lett* **81** 1634 (1998); J-N Roux, cond-mat/0001246; J-N Roux in *Powders and Grains*, eds. R P Behringer and J Jenkins (Balkema, Rotterdam, 1997).

[14] A V Tkachenko and T A Witten, *Phys Rev E* **60** 687 (1999).

[15] P Claudin and J-P Bouchaud, *Phys Rev Lett* **78** 231 (1997).

[16] P Dantu, *Annales des Ponts et Chaussées* **4** 144 (1967).

[17] C Thornton and G Sun, in *Numerical Methods in Geotechnical Engineering*, I Smith, Ed., Balkema, 143-148 (1994).

[18] F Radjai, D E Wolf, M Jean, and J J Moreau, *Phys Rev Lett* **90** 61 (1998).

[19] C-h Liu and S R Nagel, *Phys Rev B* **48** 15646 (1993).

[20] J P Wittmer, P Claudin, M E Cates and J P Bouchaud, *Nature* **382** 336 (1996).

[21] R M Nedderman, *Statics and Kinematics of Granular Materials*, Cambridge University Press, (1992).

[22] E Kolb, T Mazozi, J Duran and E Clement, *Euro Phys J B* **8** 483 (1999).

[23] S B Savage in *Physics of Dry Granular Media*, H J Herrmann, J P Hovi and S Luding, Eds., NATO ASI Ser E 350 (1998). See also *Proc. 3rd Int. Conf. on Powders and Grains, Durham NC 18-23 May 1997*, R Behringer and J T Jenkins, Eds., Balkema, 1997, especially F Cantelaube and J D Goddard, pp. 185-194.

[24] See *e.g.* M A Scherer, V Buchholtz, T Poschel and I Rehberg, *Phys Rev E* **54** R4560 (1996).

[25] L Vanel, D Howell, D Clark, R P Behringer and E Clement, *Phys Rev E* **60** R5040 (1999).

[26] J Smid and J Novosad, in *Proc 1981 Powtech Conf, Ind Chem Eng Symp* **63** D3V 1-12 (1981);
R Brockbank, J M Huntley and R C Ball, J. Physique II **7** 1521 (1997).

[27] P-G de Gennes, *Rev Mod Phys* **71** S374 (1999);
P Evesque and P-G de Gennes, *Compt Rend de l'Acad des Sci II B* **326** 761 (1998).

[28] L Vanel, P Claudin, J-P Bouchaud, M E Cates, E Clement, J P Wittmer, *Phys Rev Lett* **84** 1439 (2000).

Participants

- Mr Ronojoy Adhikari
  Indian Institute of Science
  Department of Physics
  Bangalore 560 012
  India
  email: rjoy@physics.iisc.ernet.in

- Mr André Aradian
  LPMC
  Collège de France
  11 place Marcelin Bethelot
  75231 Paris Cedex 05
  France
  email: aradian@ext.jussieu.fr

- Dr Peter F Arndt
  Institut für Theoretische Physik
  Universität zu Köln
  Zülpicher Str 77
  D-50937 Köln
  Germany
  email: arndt@thp.uni-koeln.de

- Mr Michael Berni
  Dept of Chemical Engineering
  Imperial College
  Prince Consort Rd
  London SW7 2BY, UK
  email: m.berni@ic.ac.uk

- Dr Jean-Francois Berret
  CC026 GDPC
  Univerité Montpellier II
  F-34095 Montpellier II, cedex 5
  France
  email: berret@gdpc.univ-montp2.fr

- Mr Ludovic Berthier
  ENS
  Lab de Physique
  46, Allee d'Italie
  69007 Lyon
  France
  email: lberthie@enslapp.ens-lyon.fr

- Mr Thomas Bickel
  LDFC
  3 rue de l'Université
  67084 Strasbourg
  France
  email: bickel@ldfc.u-strasbg.fr

- Mr Richard Blackwell
  Dept of Applied Mathematics
  University of Leeds
  Leeds LS2 9JT, UK
  email: amtrjb@amsta.leeds.ac.uk

- Mr Christophe Blanc
  LMCP
  T16, case 115
  4 Place Jussieu
  75252 Paris
  France
  email: blanc@lmcp.jussieu.fr

- Mr Benoit Blottiere
  IRC in Polymer Science and Technology
  University of Leeds
  Leeds LS2 9JT, UK
  email: phybb@phys-irc.leeds.ac.uk

- Dr Jean-Philippe Bouchaud
  Ser de Phys de l'Et. Cond
  CEA
  Ormes der Merisiers
  91191 Gif-Sur-Yvette
  Cedex France
  email: bouchau@amoco.saclay.cea.fr

- Professor Alan Bray
  University of Manchester
  Department of Physics and Astronomy
  Manchester M13 9PL, UK
  email: alan.bray@man.ac.uk

- Dr Alex Bunker
  MPI-P
  Ackermannweg-10
  D-55021 Mainz
  Germany
  email: bunker@mpip-mainz.mpg.de

- Mr Yoram Burak
  School of Physics and Astronomy
  Tel-Aviv University
  Tel-Aviv 69978
  Israel
  email: yorambu@post.tau.ac.il

- Mr João Tavares Cabral
  Dept of Chemical Engineering
  Imperial College
  Prince Consort Rd
  London SW7 2BY, UK
  email: j.cabral@ic.ac.uk

- Mr Vincent Carrier
  Centre de Recherche Paul Pascal
  Av Albert Schweitzer
  F-33600 Pessac
  France
  email: carrier@crpp.u-bordeaux.fr

- Professor Michael Cates
  Dept of Physics and Astronomy
  The University of Edinburgh
  Kings Buildings
  Edinburgh EH9 3JZ, UK
  email: M.E.Cates@ed.ac.uk

- Professor Paul Chaikin
  Princeton University
  Department of Physics
  Princeton, NJ 08544, USA
  email: chaikin@pupgg.princeton.edu

- Dr Michel Cloitre
  Lab Mixte CNRS ELF-Atochem
  95 rue Danton
  92303 Levallois Perret
  France
  email:michel.cloitre@levl.elf-atochem.fr

- Dr John de Bruyn
  Deptartment of Physics
  Memorial Univ of Newfoundland
  St John's, NF
  Canada A1B 3X7
  email: jdebruyn@kelvin.physics.mun.ca

- Ms Karin de Moel
  University of Groningen
  Laboratory of Polymer Chemistry
  Nijenborgh 4
  9747 AG Groningen
  Netherlands
  email: demoel@chem.rug.nl

- Ms Caroline Derec
  Lab de Physica-Chimie Théorique
  ESPCI
  10 rue Vauquelin
  75231 Paris Cedex 05
  France
  email: derec@turner.pct.espci.fr

- Mr Markus Deserno
  MPI-P
  Ackermannweg-10
  D-55021 Mainz
  Germany
  email: deserno@mpip-mainz.mpg.de

- Dr Marjolein Dijkstra
  H H Wills Physics Lab
  University of Bristol
  Tyndall Avenue
  Bristol BS8 1TL
  email: M.Dijkstra@bristol.ac.uk

- Dr Stefan Egelhaaf
  Dept of Physics and Astronomy
  The University of Edinburgh
  Edinburgh EH9 3JZ, UK
  email: S.U.Egelhaaf@ed.ac.uk

- Dr Erika Eiser
  GDPC case 26
  Univerité Montpellier II
  34095 Montpellier II, cedex 5
  France
  email: eiser@gdpc.univ-montp2.fr

- Mr Axel Ekani-Nkodo
  Institut Charles Sadron
  6 rue Boussingault
  67083 Strasbourg Cedex
  France
  email: ekani@ics.u-strasbg.fr

- Dr Martin Evans
  Dept of Physics and Astronomy
  The University of Edinburgh
  Edinburgh EH9 3JZ, UK
  email: M.R.Evans@ed.ac.uk

- Mr Roland Faller
  MPI-P
  Ackermannweg-10
  D-55021 Mainz
  Germany
  email: faller@mpip-mainz.mpg.de

- Mr Oded Farago
  School of Physics and Astronomy
  Tel-Aviv University
  Tel-Aviv 69978
  Israel
  email: farago@orion.tau.ac.il

- Ms Suzanne Fielding
  Dept of Physics and Astronomy
  The University of Edinburgh
  Kings Buildings
  Edinburgh EH9 3JZ, UK
  email: suzanne.fielding@ed.ac.uk

- Dr Jon O Fossum
  NTNU Physics Dept Gloeshaugen
  Sem Saelands vei 9
  N-7034 Trondheim
  Norway
  email: fossumj@phys.ntnu.no

- Professor Daan Frenkel
  FOM, AMOLF
  Kruislaan 407
  NL-1098 SJ Amsterdam
  Netherlands
  email: d.frenkel@amolf.nl

- Ms Linda Galloway
  Dept of Physics and Astronomy
  The University of Edinburgh
  Kings Buildings
  Edinburgh EH9 3JZ, UK
  email: l.a.galloway@ed.ac.uk

- Mr Anthony Gent
  New College
  Oxford OX1 3BN, UK
  email: gent@thphys.ox.ac.uk

- Dr Ryuichi Hasegawa
  Nagoya Univ.
  Research & Education Center 1-4F
  Furou-cho Chikusa-ku
  Nagoya 464-8601
  Japan
  hasegawa@zoom.cse.nagoya-u.ac.jp

- Dr David Head
  Dept of Physics and Astronomy
  The University of Edinburgh
  Kings Buildings
  Edinburgh EH9 3JZ, UK
  email: david@ph.ed.ac.uk

- Mr Dov Hirshfeld
  Dept of Physics
  Bar-Ilan University
  Ramat Gan 52900
  Israel
  email: hirshfld@alon.cc.biu.ac.il

- Dr Larisa Iskakova
  Dept of Mathematical Physics
  Ural State University
  Lenina ava, 51
  Ekaterinburg, 620083
  Russia
  email: Larisa.Iskakova@usu.ru

- Mr Michael Kapnistos
  Univ of Crete and I.E.S.l.-FO.R.T.H
  P.O. Box 1527
  Crete
  Greece
  email: kapnist@rea.edu.physics.uch.gr

- Dr Norbert Kern
  FOM, AMOLF
  Kruislaan 407
  NL-1098 SJ Amsterdam
  Netherlands
  email: kern@amolf.nl

- Professor Alexei Khokhlov
  Moscow State University
  Dept of Physics
  Moscow 117234
  Russia
  email: khokhlov@polly.phys.msu.su

- Dr Walter Kob
  Koma 331
  Institut für Physik
  Universität Mainz
  Staudinger Weg 7
  55099 Mainz
  Germany
  email: kob@moses.physik.uni-mainz.de

- Dr Evelyne Kolb
  LMDH
  Université P. et. M. Curie
  Case 86
  4 Place Jussieu
  75252 Paris
  France
  email: kolb@ccr.jussieu.fr

- Mr Nikolaos Kopsias
  Institute of Physical Chemistry
  NRCPS "Demokrios"
  GR 15310 Aghia Paraskevi Attikis
  Greece
  email: nkopsias@mail.demokritos.gr

- Professor Kurt Kremer
  Max Planck Institut
  für Polymerforschung
  Postfach 3148
  55021 Mainz
  Germany
  email:kremer@th01.mpip-mainz.mpg.de

- Professor Henk Lekkerkerker
  University of Utrecht
  Vant Hoff Lab
  Padualaan 8
  NL-3584 CH Utrecht
  Netherlands
  email: h.n.w.lekkerkerker@chem.ruu.nl

- Dr Alexei Likhtman
  IRC in Polymer Science and Technology
  University of Leeds
  Leeds LS2 9JT, UK
  email: phyall@phys-irc.leeds.ac.uk

- Mr Stefano Lise
  Imperial College
  Dept of Mathematics
  180 Queen's Gate
  London SW7 2BZ, UK
  email: lises@ic.ac.uk

- Dr Alexander Lobkovsky
  Institute for Theoretical Physics
  University of California
  Santa Barbara,
  CA 9310, USA
  email: lobkovsk@itp.ucsb.edu

- Mr Dmitry Lukatsky
  Dept of Materials and Interfaces
  Weizmann Institute of Science
  Il 76100 Rehovot
  Israel
  email: lukatsky@wicc.weizmann.ac.il

- Ms Caroline Mabille
  Centre de Recherche Paul Pascal
  Av Albert Schweitzer
  33600 Pessac
  France
  email: mabille@crpp.u-bordeaux.fr

- Ms Anna V Maidens
  Dept of Physics and Astrophysics
  University of Leeds
  Leeds LS2 9JT, UK
  email: phyam@phys-irc.leeds.ac.uk

- Dr Prabal Kumar Maiti
  MPI-P
  Ackermannweg-10
  D-55128 Mainz
  Germany
  email: maiti@mpip-mainz.mpg.de

- Dr Yong Mao
  Cavendish Laboratory
  Dept of Physics
  University of Cambridge
  Madingley Rd
  Cambridge CB3 OHE, UK
  email: ym101@cus.cam.ac.uk

- Mr Simon Marlow
  IRC in Polymer Science and Technology
  University of Leeds
  Leeds LS6 2JD, UK
  physwm@phys-irc.novell.leeds.ac.uk

- Dr Akihiko Matsuyama
  Dept of Physics and Astronomy
  The University of Edinburgh
  Kings Buildings
  Edinburgh EH9 3JZ, UK
  email: A.Matsuyama@ed.ac.uk

- Professor Tom McLeish
  IRC in Polymer Science & Technology
  University of Leeds
  Leeds LS2 9JT, UK
  email: phy6tcbm@phys-irc.leeds.ac.uk

• Dr François Molino
Groupe de Dynamique des Phases Condensées
Université Montpellier II
Place Eugène Bataillon
F-34095 Montpellier II,
France
email: molino@gdpc.univ-montp2.fr

• Ms Emma Montevecchi
Katholieke Universiteit Leuven
Lab. VSM
Celestijnenlann 200D
B-3001 Heverlee
Belgium
Emma.Montevecchi@fys.kuleuven.ac.be

• Professor David Mukamel
Dept Phys Complex Systems
Weizmann Institute
Il 76100 Rehovot
Israel
email: fnmukaml@wicc.weizmann.ac.il

• Professor Sid Nagel
University of Chicago
James Franck Institute
5640 S Ellis Ave
Chicago, IL 60637, USA
email: sid@rainbow.uchicago.edu

• Dr Taco Nicolai
Laboratory of Physical Chemistry
Faculty of Sciences
Universite du Maine
Av. Olivier Messiaen
F-72085 Le Mans
France
email: Nicolai@aviion.univ-lemans.fr

• Dr Corey Shane O'Hern
Dept of Physics and Astronomy
Univ of Pennsylvania
209 South 33rd St.
Philadelphia,
PA 19104-639, USA
email:ohern@lubensky.physics.upenn.edu

• Mr Florian Ozon
Institut Charles Sadron
6 rue Boussingault
67083 Strasbourg Cedex
France
email: ozon@ics.u-strasbg.fr

• Dr Ignacio Pagonabarraga
Dept of Physics and Astronomy
The University of Edinburgh
Kings Buildings
Edinburgh EH9 3JZ, UK
email: ignacio@ph.ed.ac.uk

• Dr Marian Paluch
MPI-P
Ackermannweg 10,
Postfach 3148, 55021 Mainz,
Germany
email: paluch@mpip-mainz.mpg.de

• Mr Karl Pays
Centre de Recherche Paul Pascal
Av Albert Schweitzer
33600 Pessac
France
email: PAYS@crpp.u-bordeaux.fr

• Dr George Petekidis
Dept of Physics and Astronomy
The University of Edinburgh
Edinburgh EH9 3JZ, UK
email: G.Petekidis@ed.ac.uk

• Professor David Pine
University of California Santa Barbara
Department of Materials Research
Santa Barbara, CA 93106, USA
email: pine@engineering.ucsb.edu

• Mr Olivier Pitois
LMSGC, UMR 193,
LCPC-CNRS
Cité Descartes
2 Alleé Kepler
77420 Champs-sur-Marne
France
email: olivier.pitois@lcpc.fr

• Mr Dariusz Plewczyński
Institute of Physical Chemistry
Polish Academy of Sciences
Kasprzaka 44/52 Street
01-224 Warsaw
Poland
email: darman@ichf.edu.pl

• Dr Wilson Poon
Dept of Physics and Astronomy
The University of Edinburgh
Edinburgh EH9 3JZ, UK
email: W.Poon@ed.ac.uk

- Dr Igor Potemkin
  Physics Department
  Moscow State Univ.
  Moscow
  117234 Russia
  email: igor@polly.phys.msu.su

- Mr S Pronk
  FOM, AMOLF
  Kruislaan 407
  NL-1098 SJ Amsterdam
  Netherlands
  email: massimo@amolf.nl

- Dr Laurence Ramos
  Case Courrier 26
  Univerité Montpellier II
  F-34095 Montpellier II, cedex 5
  France
  email: ramos@gdpc.univ-montp2.fr

- Mr David Reguera
  Dipartament de Fisica Fonamental
  University of Barcelona
  Diagonal 647
  08028 Barcelona
  Spain
  email: davidr@precario.ffn.ub.es

- Mr Frédéric Restagno
  Laboratoire de Physique
  ENS Lyon
  46 Allée d'Italie
  F-69364 Lyon
  Cedex 07 France
  email: frestagn@dpm.univ-lyon1.fr

- Ms Yael Roichman
  Technion
  Haifa 32000
  Israel
  email: nyael@techunix.technion.ac.il

- Mr Andrea Rossi
  SISSA
  Via Beirut 2-4
  34014 Grignano
  Trieste
  Italy
  email: rossi@sissa.it

- Dr Didier Roux
  Centre de Recherche Paul Pascal, CNRS
  Av Schweitzer
  F-33600 Pessac
  France
  email: didier@crpp.u-bordeaux.fr

- Ms Tanja Schilling
  Trierer Strasse 175
  52078 Aachen
  Germany
  email: t.schilling@fz-juelich.de

- Dr Richard Sear
  Dept of Physics
  University of Surrey
  Guildford
  Surrey GU2 5XH, UK
  email: r.sear@surrey.ac.uk

- Ms Kheya Sengupta
  Raman Research Institute
  C V Raman Avenue
  Sada Shivanagar
  Bangalore: 560080
  India
  email: kheya@rri.ernet.in

- Dr Peter Sollich
  Dept of Mathematics
  King's College London
  Strand
  London WC2R 2LS, UK
  email: peter.sollich@kcl.ac.uk

- Mr Martin Steinhauser
  MPI-P
  Ackermannweg-10
  D-55021 Mainz
  Germany
  email: sthauser@mpip-mainz.mpg.de

- Mr Daniel Svensek
  Faculty of Mathematics and Physics
  Jadranska 19
  SI-1000 Ljubljana
  Slovenia
  email: daniel@fiz.uni-lj.si

- Dr Takashi Taniguchi
  Nagoya Univ.
  Dept of Computational Science
  School of Engineering
  Chikusa-ku
  Nagoya 464-8603
  Japan
  email: taniguch@stat.cse.nagoya-u.ac.jp

- Dr Paulo Teixeira
  IRC in Polymer Science and Technology
  University of Leeds
  Leeds LS2 9JT, UK
  email: phypt@phys-irc.leeds.ac.uk

- Dr Fabrice Thallmann
  LEPES-CNRSB
  BP166X
  25, Av. des Martyrs
  38042 Grenoble
  Cedex 9 France
  email: thalmann@lepes.polycnrs-gre.fr

- Dr Emmanuel Trizac
  Lab de Physique Théorique
  Université d'Orsay
  Batiment 210
  91405 Orsay
  France
  email: etrizac@amolf.nl

- Mr Janos Torok
  Technical University of Budapest
  Budafoki, ut 8.
  H-1111 Budapest
  Hungary
  email: torok@planck.phy.bme.hu

- Mr Yoav Tsori
  School of Physics and Astronomy
  Tel-Aviv University
  Tel-Aviv 69978
  Israel
  email: tsori@post.tau.ac.il

- Dr Violeta Uricanu
  ICECHIM
  202 Splaiul Independentei
  77208-Bucharest -6
  Romania
  email: viorica@ibd.dbio.ro

- Mr Loic Vanel
  LMDH
  Université P. et. M. Curie
  Tour 22- Boite 86 - 4,
  4 Place Jussieu
  75252 Paris, France
  email: loic@aomc.jussieu.fr

- Dr René van Roij
  H.H Wills Physics Lab
  University of Bristol
  Tyndall Avenue
  Bristol BS8 1TL, UK
  email: R.VanRoij@bristol.ac.uk

- Mr Joshua Weitz
  Bldg. 54-621,
  MIT
  Cambridge MA 02139, USA
  email: jsweitz@mit.edu

- Ms Carole Jo Whiting
  Polymer IRC
  University of Leeds
  Leeds LS2 9JT, UK
  email: phycw@phys-irc.leeds.ac.uk

- Dr Nigel Wilding
  Dept of Physics and Astronomy
  The University of Edinburgh
  Edinburgh EH9 3JZ, UK
  email: N.B.Wilding@ed.ac.uk

- Ms Anne-Sophie Wunenburger-Krepper
  Centre de Recherche Paul Pascal
  Av Albert Schweitzer
  F-33600 Pessac, France
  email: krepper@crpp.u-bordeaux.fr

- Mr Tatsuya Yamaue
  Nagoya Univ.
  Research & Education Center 1-4F
  Furou-cho Chikusa-ku
  Nagoya 464-8601, Japan
  yamaue@zoom.cse.nagoya-u.ac.jp

- Dr Andrej Zubarev
  Dept of Mathematical Physics
  Ural State University
  Lenina ava, 51
  Ekaterinburg, 620083
  Russia
  email: Andrey.Zubarev@usu.ru

Index

1937